W0112700

DESIGNING ZERO CARBON BUILDINGS

In this significantly revised third edition, *Designing Zero Carbon Buildings* combines embodied and operational emissions into a structured approach for achieving zero emissions by a specific year with certainty.

Simulation and quantitative methods are introduced in parallel with analogue scale models to demonstrate how things work in buildings. Where equations are provided, this is also explained with common analogue objects, pictures, and narratives. A Zero Equation introduced in this book is not only explained as an equation but also as an analogy with a jam jar and spoons, making the book accessible for a range of audiences. Tasks for simple experiments, exercises, discussion questions, and summaries of design principles are provided in closing lines of chapters.

This book introduces new case studies, in addition to an updated case study of the Birmingham Zero Carbon House, applying embodied and operational emissions to assess their status using the Zero Equation. The approach introduced brings about a sense of realism into what true zero emissions mean. Written for students, educators, architects, engineers, modellers, practising designers, sustainability consultants, and others, it is a major positive step towards design thinking that makes achieving zero carbon emissions a reality.

Prof Ljubomir Jankovic, MSc, PhD, CEng, MCIBSE, MASHRAE, FIAP, FIBPSA, has spent the past 35 years as an academic, researcher and practitioner, focusing on how environmental design of buildings can be improved using instrumental performance monitoring, dynamic simulation, advanced computer modelling methods, and utilisation of bio-sourced materials. He holds an MSc from the University of Belgrade and a PhD from the University of Birmingham, both in Mechanical Engineering. He is a Chartered Engineer, a Member of ASHRAE, a Member of CIBSE, a Fellow of the Institution of Analysts and Programmers and a Fellow of the International Building Performance Simulation Association. His professional society activities included roles of President of ASHRAE UK London and South East Chapter, Vice-President of ASHRAE UK Chapter, and an ASHRAE Distinguished Lecturer.

DESIGNING ZERO CARBON BUILDINGS

EMBODIED AND OPERATIONAL EMISSIONS IN ACHIEVING TRUE ZERO

Third Edition

Ljubomir Jankovic

Routledge
Taylor & Francis Group

LONDON AND NEW YORK

Designed cover image: Overall cover design by the author. Individual tiles design by Holly Doron. Author photo courtesy of Ian Scott

Third edition published 2024
by Routledge
4 Park Square, Milton Park, Abingdon, Oxon, OX14 4RN

and by Routledge
605 Third Avenue, New York, NY 10158

Routledge is an imprint of the Taylor & Francis Group, an informa business

© 2024 Ljubomir Jankovic

The right of Ljubomir Jankovic to be identified as author of this work has been asserted in accordance with sections 77 and 78 of the Copyright, Designs and Patents Act 1988.

All rights reserved. No part of this book may be reprinted or reproduced or utilised in any form or by any electronic, mechanical, or other means, now known or hereafter invented, including photocopying and recording, or in any information storage or retrieval system, without permission in writing from the publishers.

Trademark notice: Product or corporate names may be trademarks or registered trademarks, and are used only for identification and explanation without intent to infringe.

First edition published by Routledge 2012
Second edition published by Routledge 2017

British Library Cataloguing-in-Publication Data
A catalogue record for this book is available from the British Library

Library of Congress Cataloging-in-Publication Data
Names: Jankovic, Ljubomir, author.
Title: Designing zero carbon buildings: embodied and operational emissions in achieving true zero / Ljubomir Jankovic.
Other titles: Designing zero carbon buildings using dynamic simulation methods
Description: Third edition. | Abingdon, Oxon; New York, NY: Routledge, 2024. |
Revised third edition of: Designing zero carbon buildings using dynamic simulation methods /
Ljubomir Jankovic. New York: Routledge, 2017. | Includes bibliographical references and index. |
Identifiers: LCCN 2023030241 (print) | LCCN 2023030242 (ebook) | ISBN 9781032378718 (hbk) |
ISBN 9781032378701 (pbk) | ISBN 9781003342342 (ebk) Subjects: LCSH: Sustainable buildings—Design and construction. |
Buildings—Performance—Computer simulation. | Carbon dioxide mitigation.
Classification: LCC TH880 .J36 2024 (print) | LCC TH880 (ebook) | DDC 720/.47—dc23/eng/20230731
LC record available at https://lccn.loc.gov/2023030241
LC ebook record available at https://lccn.loc.gov/2023030242

ISBN: 9781032378718 (hbk)
ISBN: 9781032378701 (pbk)
ISBN: 9781003342342 (ebk)

DOI: 10.4324/9781003342342

Typeset in Gill Sans Std
by codeMantra

Access the supplementary material: www.ljankovic.com

To my family

CONTENTS

ACKNOWLEDGEMENTS

Several people and organisations helped generously in the preparation of this book, and I wish to thank them all, in no particular order. I wish to thank the University of Hertfordshire, and especially Dr Steven Adams, for encouragement and support. Thanks also go to Emission Zero Engineering Architecture Ltd for access to their project material. Special thanks go to Greenpower Education Trust and Fordingbridge plc for their collaboration on the Greenpower Centre; to the Bournville Village Trust (BVT) for financial assistance related to the Rowheath Solar Village project and to Alan Shrimpton of the BVT and to Dr Leslie Jesch, my former PhD supervisor and good friend, for their collaboration on the Rowheath Solar Village; to the Bioeconomy Consultants-NNFCC, to Mrs Cathie Eberlin of Leading Energy, to Ian Pritchett of Greencore Construction Ltd, Hab Oakus LLP and Linford C-Zero Ltd for their collaboration on hemp-lime–related projects; and to Innovate UK, to Ron and Isabel Beattie of Beattie Passive Retrofit Ltd, and to Paul McGrath of Birmingham City Council for collaboration on RetrofitPlus project. Birmingham architect John Christophers helped generously with essential information, photographs, and access for instrumental monitoring of his unique Zero Carbon House on which I have written a case study in one of the chapters. Photographer Martine Hamilton Knight provided high-quality photographs of John's house which stand out in comparison with my own photographs. I am grateful to Powerhouse Company, ASHRAE, Integrated Environmental Solutions Ltd, Snøhetta, Skanska, University of Cambridge Institute for Sustainability Leadership, Architype, and Max Fordham for providing material for case studies of zero carbon projects featured in this book. Special thanks go to Dr Andrew Tindale, David Cocking, and Dr Yi Zhang for proofreading the material on multi-objective optimisation and making useful suggestions. I am grateful to Dr Andy Tindale of DesignBuilder Ltd for supporting this book project. I am also grateful to OneClickLCA for their support. My former student Holly Doron prepared wonderful hand illustrations for the cover, and she and Oliver Mould prepared illustrations for figures which improved the book's graphical appearance. Thanks also go to my numerous other former students who approached their coursework on scale model experiments with enthusiasm when they knew that some of their material may be featured in this book, and whose scale models are acknowledged where they are referred to in the book. My daughters Katarina and Sofia proofread parts of the manuscript and suggested useful style improvements. My wife Leena has provided me with unconditional support and created an environment that enabled me to focus on this book project.

All photographs, images, and illustrations generously provided by others have been acknowledged accordingly, next to where they occur in this book. The screen images of software tools are acknowledged as follows: DesignBuilder, by permission of DesignBuilder Software Limited, www.designbuilder.co.uk; EnergyPlus™ Copyright © 2011, The Regents of the University of California, through Lawrence Berkeley National Laboratory (subject to receipts of any required approvals from the U.S. Dept. of Energy). All rights reserved. Used by permission. EnergyPlus™ was developed with support from the U.S. Department of Energy; IES Virtual Environment, by permission of Integrated Environmental Solutions Ltd, www.iesve.com; JEPlus/JEPlus+EA, by permission of Energy Simulation Solutions Limited, cms.ensims.com; and Microsoft® Excel for Mac is copyright © 2022 by Microsoft.

I have personally created all other material that has not been explicitly acknowledged.

Birmingham, 17 April 2024

DISCLAIMER

The methods and examples in this book are provided as a guide to designers and not as solutions to specific design problems. The Author of this book and the Publishers will not accept any liability expressed or implied arising from the application of the material in this book and the associated supplementary content.

PART I

ZERO CARBON: WHY AND HOW?

CHAPTER 1

INTRODUCTION

1.1 WHAT ARE ZERO CARBON BUILDINGS AND HOW WE DESIGN THEM?

A zero carbon building is a building that achieves nil balance between its carbon dioxide emissions generated from non-renewable energy sources, and its carbon emissions reversed as a result of renewable energy use or other appropriate measures, during its life cycle. In a life cycle of say 60 years, the nil balance may occur near the beginning, near the end, or sometime in between that time period, but it has to occur within the life cycle in order for that building to be a zero carbon building.

The way we design zero carbon buildings (Figure 1) is that we start from the site and climate, then we design building geometry, thermal insulation, solar gain, solar shading, thermal mass, ventilation, and integration of daylight with electrical lighting. By integrating all these aspects and by balancing the need for heating and cooling, we achieve thermal comfort for building

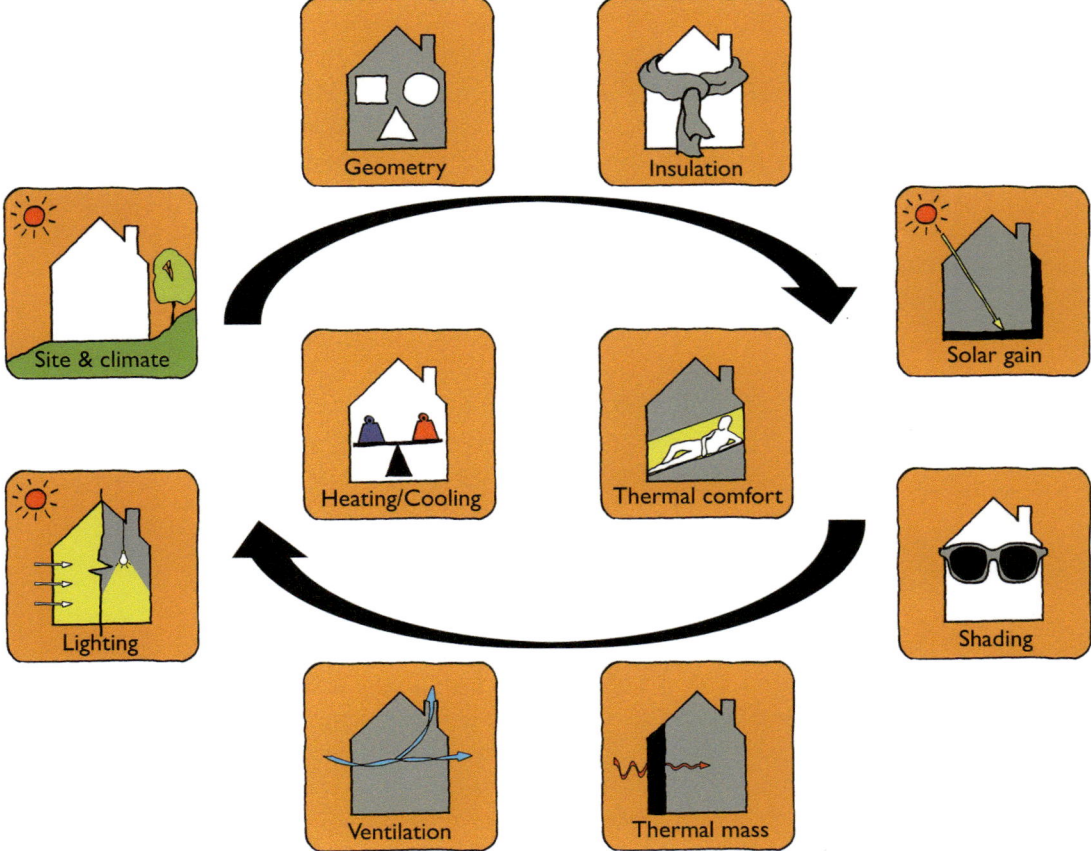

Figure 1   The way we design zero carbon buildings

occupants. We then put all of this into number-crunching simulation and optimisation tools, which enable us to harmonise design parameters. As a result, we obtain renewable energy requirements that balance carbon emissions arising from the combination of design parameters and requirements for heating and cooling (Figure 2). Thus, zero emissions are achieved, and the problem is solved. What more could be there to talk about?

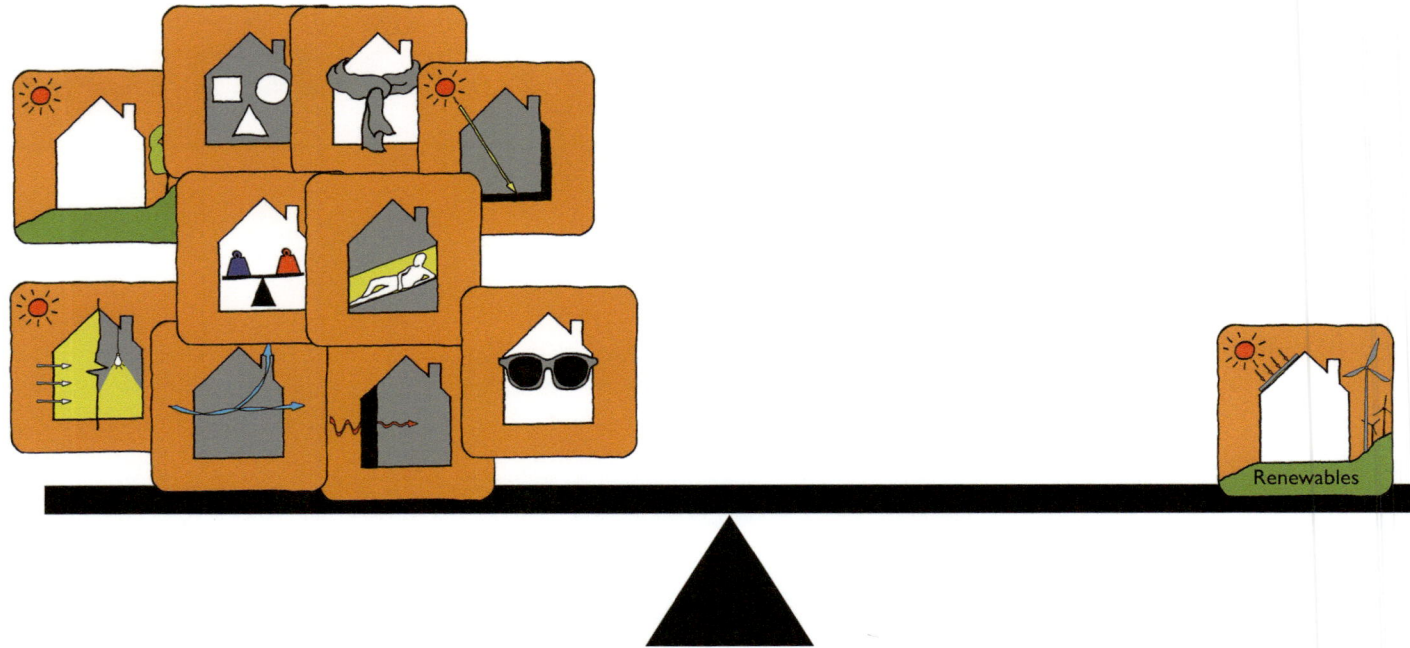

Figure 2   Renewable energy balances operational emissions

Except, there is an elephant in the room. My research shows that not taking into account embodied emissions can delay reaching zero emissions by three to four decades (Jankovic et al., 2021). So, we have a problem if we don't take embodied emissions into account. Renewable energy that we use to balance operational emissions will not be sufficient to balance embodied and operational emissions together (Figure 3a). It is therefore much harder to achieve true zero emissions when accounting for both embodied and operational emissions. We must consider using less carbon-intensive materials, such as bio-sourced materials and/or locally sourced materials, less intensive methods of construction, and adding more renewable energy and optimising designs to make it easier to achieve a balance between embodied and operational emissions on one side and renewable energy on the other side (Figure 3b).

Therefore, this book introduces embodied emissions as the first major part of the overall net zero design method. It then consolidates operational emissions from the second edition (Jankovic, 2017b) as the second major part of the net zero design method, and it combines the two into a method that achieves combined net zero emissions with certainty and by a specified year.

1.2 WHAT ARE EMBODIED EMISSIONS AND WHAT IS THEIR ROLE IN ACHIEVING ZERO CARBON?

Embodied emissions are emissions from making 'stuff'. A newly constructed building will have greater than zero or positive embodied emissions on the first day after it is constructed. This will be a starting point that is usually of quite a high value in terms of tons of CO_2. That starting point represents a significant challenge for zero carbon design, and it needs to be gradually reduced to zero over a number of years during the building life cycle, if the building is to become a zero carbon building.

Embodied emissions are caused by extracting and making building materials, and by transporting them and using them in construction. All these processes require energy and result in positive carbon emissions if the energy comes from fossil fuel sources. As a result, all conventional building materials, such as brick, concrete, glass, and others, have greater than zero embodied

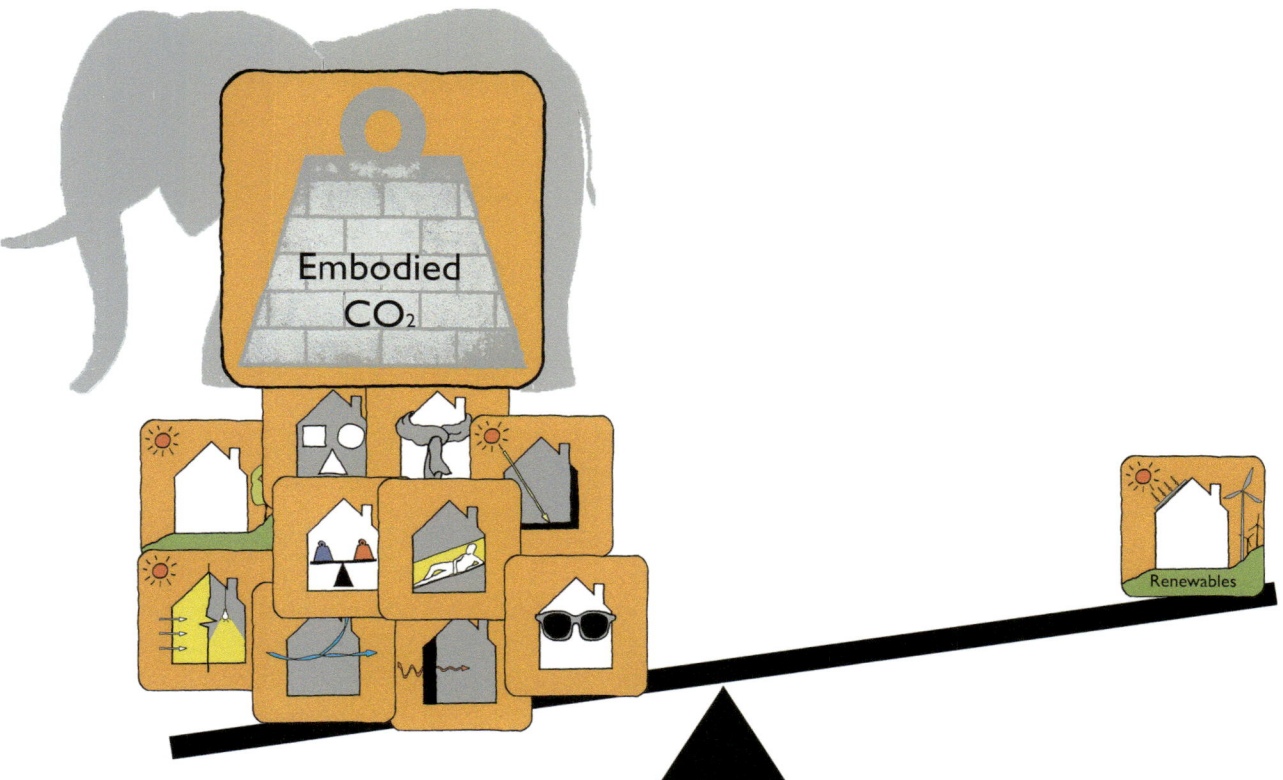

a) Renewable energy that we used to balance operational emissions will not be sufficient to balance combined embodied and operational emissions

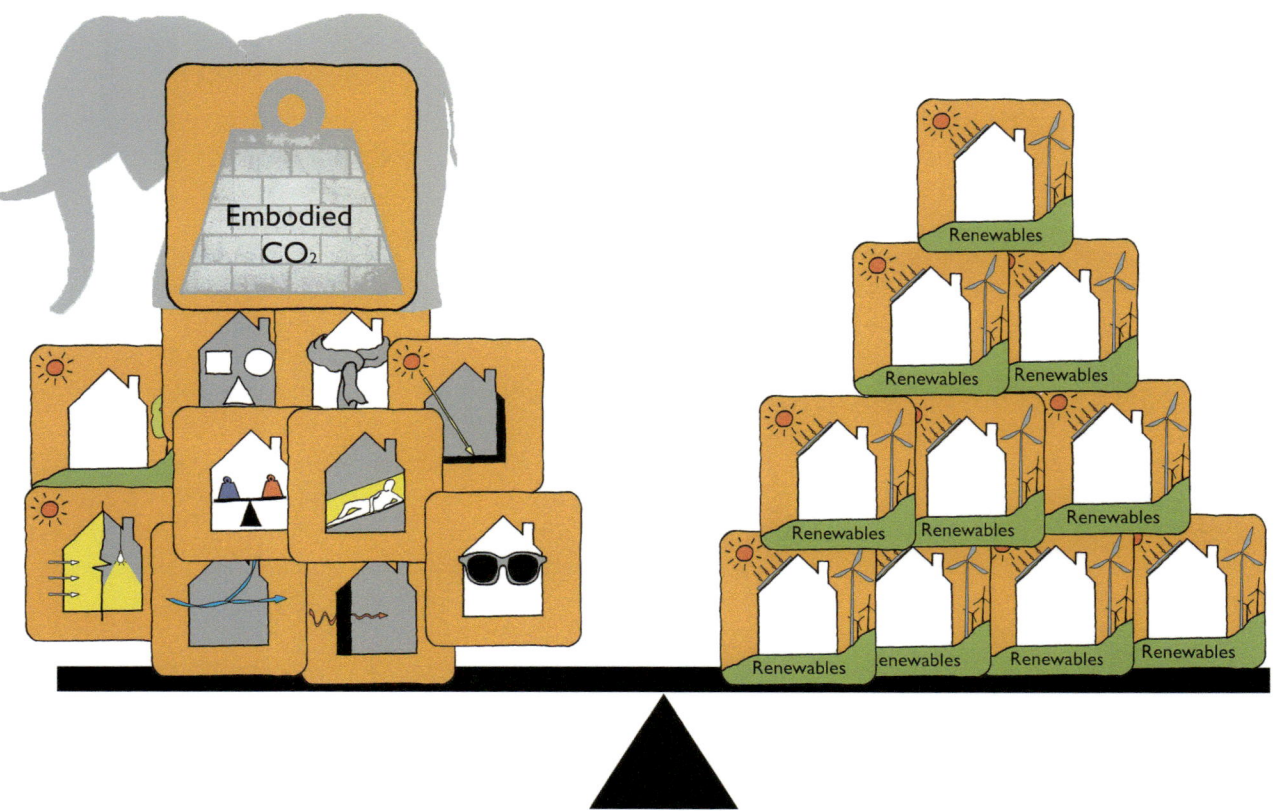

b) Using less carbon intensive materials and adding more renewable energy will be required to achieve true zero

Figure 3   It is much harder to achieve true zero emissions when embodied emissions are considered

emissions. Buildings constructed from conventional materials will be harder to design to be zero carbon, as the starting point for reduction to zero will be higher.

Bio-sourced building materials partially contain negative embodied emissions arising from sequestration of carbon dioxide in the plant from which they originate. These materials include timber, straw, industrial hemp, and others. They also need to be extracted, manufactured, transported and used in construction, causing positive embodied emissions. These positive emissions may be reduced, balanced or completely reversed by the carbon sequestration effect from the bio-source origin, depending on the material and other circumstances. Buildings constructed from bio-sourced materials will be easier to design to be zero carbon, as the starting point for reduction to zero will be lower.

Embodied emissions will be discussed in much detail throughout the book, starting from Part 2.

1.3 WHAT ARE OPERATIONAL EMISSIONS AND WHAT IS THEIR ROLE IN ACHIEVING ZERO CARBON?

Operational emissions are emissions from using 'stuff'. A newly constructed building will have nil operational emissions on the first day it is constructed, before any energy is used for heating or cooling. Thus, operational emissions are emissions arising from using energy in buildings. If energy comes from fossil fuels, operational emissions will be positive, as they create new carbon dioxide. If, however, the building uses renewable energy sources, such as solar photovoltaic energy or wind energy, the resulting emissions from these systems will be negative. This is not only because the use of these systems suppresses the use of fossil fuels, but it also supplies new electricity energy into the electricity grid, thus reversing overall carbon dioxide emissions.

Operational emissions will be discussed in much detail throughout this book, starting from Part 3.

1.4 TOTAL CUMULATIVE EMISSIONS: PUTTING EMBODIED AND OPERATIONAL EMISSIONS TOGETHER – THE ZERO EQUATION

A key concept used throughout this book is what I call the 'Zero Equation'. It combines embodied and operational emissions together to make it easier to understand the design of zero carbon buildings. This and other concepts will be explained diagrammatically using common object analogies and also using formal expressions.

1.4.1 The Zero Equation representation using a jam jar, two spoons, and seesaw

First, let me explain the Zero Equation using a common objects analogy of a jam jar, two spoons, and a seesaw. I am expressing it in this way to help conceptual understanding of what true zero emissions mean.

In this analogy, the jam jar represents a building, and the jam represents a total of embodied and operational emissions. A full jam jar is equivalent to embodied emissions on day one when a new building is constructed or an existing building is retrofitted. The jam in the smaller of the two spoons represents additional (positive) operational emissions arising from using energy in the building. This smaller spoon adds jam (positive operational emissions) to the jar once a year. The jam in the larger spoon represents reversed (negative) emissions arising from using renewable energy systems. These renewable systems, such as solar photovoltaic, not only suppress emissions from fossil fuels but reverse them by supplying new energy to the grid. The larger spoon therefore removes a spoonful of jam (now combined embodied and operational emissions) from the jar once a year (Figure 4).

The question now is: when is the jar going to get empty, in other words, when are the combined embodied and operational emissions going to go down to zero? This depends on the relative sizes of the jam jar, the first spoon that adds jam to it, and the second spoon that removes the jam from it.

Let us now put this on a 'seesaw' to create a balance between the jam in the jar, and a few of the first and second spoon-full portions of jam. We place all positive emissions, embodied and operational on one side of the seesaw, and all negative emissions on the other side of the seesaw (Figure 5).

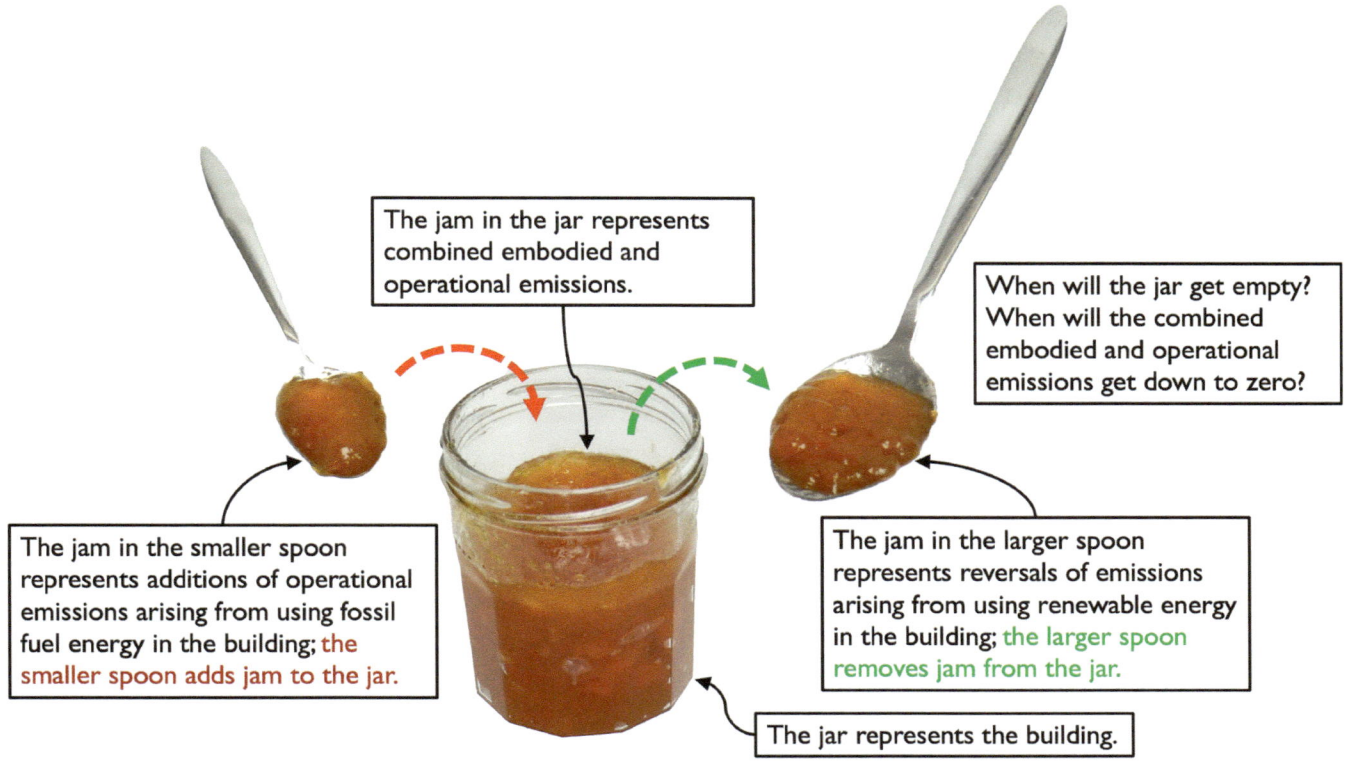

The jam in the jar represents combined embodied and operational emissions.

When will the jar get empty? When will the combined embodied and operational emissions get down to zero?

The jam in the smaller spoon represents additions of operational emissions arising from using fossil fuel energy in the building; the smaller spoon adds jam to the jar.

The jam in the larger spoon represents reversals of emissions arising from using renewable energy in the building; the larger spoon removes jam from the jar.

The jar represents the building.

Figure 4   Jam jar, a small spoon and a large spoon as a representation of building carbon emissions

Figure 5   The Zero Equation: a seesaw with a representation of embodied and additional operational emissions on the left side and the reversed emissions from renewable energy on the right side

A combination of the initial quantity of jam in the jar (embodied emissions), the size of the smaller spoon that adds positive operational emissions to it, and the size of the larger spoon that removes the combined emissions from it results in achieving zero emissions sooner or later, as follows:

• a larger initial quantity of jam in the jar will take a longer time to empty (and thus achieve zero emissions later);
• a smaller initial quantity of jam in the jar will take a shorter time to empty (and thus achieve zero emissions sooner);
• a larger first spoon will add more jam (more operational emissions) to the jar on a regular basis and it will take longer time to empty the jar with the second spoon;
• a smaller first spoon will add less jam (less operational emissions) to the jar on a regular basis and it will take less time to empty the jar with the second spoon;
• a smaller second spoon will remove a smaller amount of jam (smaller amount of reversed emissions) from the jar and would take a longer time to empty the jam jar (and thus achieve zero emissions later), especially if the first spoon that adds to the jar is just a bit smaller than the second spoon;
• a larger second spoon will remove larger amount of jam (larger amount of reversed emissions) and would take a shorter time to empty the jam jar (and thus achieve zero emissions sooner), especially if the first spoon that adds to the jar is much smaller than the second spoon.

This is our Zero Equation, expressed using a jam jar, two spoons, and a seesaw. From this, we can see that the seesaw can be balanced only if the second spoon is larger than the first spoon, and if the total of the contents of the jam jar and the number of the first spoons becomes equal to the contents of the same number of the second spoons. The number of spoons on the left and right of the seesaw required to empty the jam jar will be equivalent to the years that it will take to achieve zero emissions.

1.4.2 The Zero Equation representation using a mathematical formula
Second, let me now explain the Zero Equation using a mathematical formula. I am now expressing it in this way to help quickly evaluate how many years it would take to achieve total zero emissions.

In this mathematical expression, the seesaw from the previous section is replaced by the equals sign. Similarly to the jam jar analogy from the previous section, all embodied and positive operational emissions will go on one side of the equals sign, and all negative operational emissions will go on the other side of the equals sign. Thus, the Zero Equation is as follows:

$$E_{CO_2} + P_{CO_2} \times t = N_{CO_2} \times t \tag{1}$$

where:

E_{CO_2} – Embodied emissions that occur in year $t = 0$ ($kgCO_2$)
P_{CO_2} – Positive operational emissions that occur from year $t = 1$ ($kgCO_2$/year)
N_{CO_2} – Negative operational emissions arising from renewable energy that occur from year $t = 1$ ($kgCO_2$/year)
t – time in years

From the Zero Equation (1) above, we can find out that time in years that will be required to achieve zero emissions will be:

$$t = \frac{E_{CO_2}}{N_{CO_2} - P_{CO_2}} \tag{2}$$

We can conclude from Equation (2) that a pre-requisite for achieving zero emissions in 't' years is that 't' is greater than zero. That is achieved when the absolute value of the negative operational emissions N_{CO_2} is greater than the positive operational emissions P_{CO_2}. Otherwise, the denominator would either be negative or zero, and 't' would, respectively, either be negative or infinity. In either of these two latter cases, zero emissions would never be reached. We can also conclude that the lower the numerator and the larger the denominator in Equation (2), the smaller the 't', and therefore the shorter the time to achieve zero emissions.

We can further modify the Zero Equation (1) as follows:

$$CE_{CO_2} + f \times BVE_{CO_2} + P_{CO_2} \times t = f \times BN_{CO_2} \times t \qquad (3)$$

where:

CE_{CO_2} – Constant embodied emissions, arising from materials, construction process, HVAC system, and maintenance ($kgCO_2$)

BVE_{CO_2} – Base level of variable embodied emissions, arising from changing the size of a renewable energy system, such as a PV system, to target a zero emissions year ($kgCO_2$)

VE_{CO_2} – Variable embodied emissions, arising from changing the size of a renewable energy system, such as a PV system, whilst targeting a zero emissions year ($kgCO_2$)

BN_{CO_2} – Base level of negative operational emissions that occur from year $t = 1$ ($kgCO_2$/year)

f – Base level multiplying factor for renewable energy system, such as a PV system, so that $f \times BN_{CO_2} = N_{CO_2}$ and $f \times BVE_{CO_2} = VE_{CO_2}$. This factor is used to target a zero emissions year by changing the size of a renewable energy system.

The base level of variable embodied emissions will be established by calculating embodied emissions of an initial PV system. The base level of negative operational emissions will be established through computer simulation that achieves negative annual emissions arising from the PV system. That base level may or may not be sufficient for achieving total zero of cumulative emissions by a specified year.

The base level factor will then be used to find out by how much we need to adjust the base level negative emissions in order to achieve total zero emissions by a specified year, whilst also adjusting variable embodied emissions associated with the renewable energy system.

Thus, if all other terms except 'f' are known in Equation (3) after we have set 't' to a certain number of years when we would like our building to achieve total zero, we can find out how much we need to adjust base level emissions by rearranging Equation (3) into another variation of the Zero Equation as follows:

$$f = \frac{CE_{CO_2} + P_{CO_2} \times t}{BN_{CO_2} \times t - BVE_{CO_2}} \qquad (4)$$

The embodied emissions need to include initial embodied emissions as well as embodied emissions from maintenance, such as replacement of HVAC or PV systems in order to obtain accurate calculations.

We will use this approach in a practical example in Chapter 19.

1.5 WHY DO WE NEED ZERO CARBON BUILDINGS?

Having elaborated on the different types of carbon dioxide emissions and their representations, it is useful to take a moment and put this into a wider context: why we need zero carbon buildings?

The need for zero carbon buildings is driven by the climate change, which is caused by greenhouse gasses, including carbon dioxide, methane, nitrous oxide, and others. Carbon dioxide is the most significant contributor, produced by burning fossil fuels. Energy used in buildings and construction contributes to 40% of carbon dioxide emissions worldwide. As the earth receives heat from the sun, greenhouse gasses slow down heat loss into space, making the earth a heat trap.

1.5.1 The Greenhouse effect
The greenhouse effect is explained in simple terms using the diagram in Figure 6. The radiation coming from the sun is characterised with high energy content and short wavelength (1). As the radiation reaches the earth surface, it gets absorbed (2) and re-radiated back into the atmosphere (3). The re-radiated radiation is characterised with lower energy content and longer

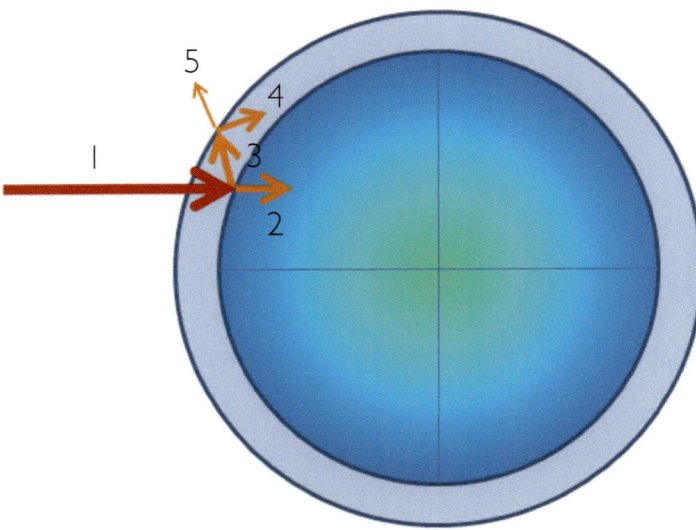

Figure 6   The greenhouse effect

wavelength (orange arrow). As optical transparency of the atmosphere is higher for short wave radiation than it is for long wave radiation, some of the re-radiated radiation is retained within the atmosphere. This is a fortunate fact as it causes the air temperature to be higher than the temperature of the universe and provides suitable conditions for the emergence and maintenance of life.

However, by increasing the amount of greenhouse gasses (such as carbon dioxide, methane, nitrous oxide, and others), the optical transparency of the atmosphere for long wave re-radiation is reduced. Hence more heat is retained within the atmosphere (4), and less heat is dissipated back into space (5), causing the gradual rise of global temperature.

In 2018, the Intergovernmental Panel on Climate Change published a landmark report that explains the risks to the planet and humanity if global warming exceeds 1.5°C temperature rise from the pre-industrial level. Unless there is a sharp decline in greenhouse gas emissions by 2030, the report states that the following decades will see global warming surpassing 1.5°C and *'irreversible loss of the most fragile ecosystems, and crisis after crisis for the most vulnerable people and societies'* (Masson-Delmotte et al., 2018).

How close are we to reaching this tipping point? The Copernicus Climate Change Service reported that the global warming in December 2020 reached 1.18°C and that we are on route to reaching 1.5°C temperature increase by January 2034 (Copernicus, 2022). A subsequent IPCC Sixth Assessment Report (Masson-Delmotte et al., 2021) explained how some changes already reached a tipping point. Thus, a continuing sea level rise is no longer reversible, whilst it could take up to three decades for global temperatures to stabilise under the most optimistic scenario of strong and sustained carbon emissions reduction.

The same report estimates the carbon budgets under different scenarios. A carbon budget refers to the *'maximum amount of cumulative net global anthropogenic CO_2 emissions that would result in limiting global warming to a given level with a given probability, taking into account the effect of other anthropogenic climate forcers'* (Masson-Delmotte et al., 2021, p. 29). According to the carbon budget from this report, generating 900 $GtCO_2$ of emissions would result in keeping global warming to 1.5°C with 17% probability, and generating 300 $GtCO_2$ of emissions would result in keeping global warming to the same temperature with 83% probability. In other words, there is 17% probability that global warming to 1.5°C will not be exceeded if another 900 $GtCO_2$ is generated, or there is 83% probability that global warming to 1.5°C will not be exceeded if another 300 $GtCO_2$ is generated.

To put this into a wider context, Jankovic and Christophers (2022) estimated that it would take 410 $MtCO_2$ of upfront embodied emissions to retrofit all existing UK homes by 2050. This means that the carbon cost of transition within the

UK alone would represent 0.14% of the remaining carbon budget associated with an 83% probability of limiting global warming to 1.5°C.

Clearly, carbon emissions from buildings and construction, being responsible for 40% of the global emissions, remain to be one of the toughest open problems in today's world. Solving this problem requires innovative thinking where nothing should be off the table. This is why we need zero carbon buildings, and this book aims to facilitate the ways toward fulfilling that need.

1.6 THIS BOOK AND COMPLEMENTARY SOURCES

As commitments for CO_2 emission reductions increase through global awareness and legislative pressures, there is an apparent lack of textbooks on the methodology for the zero carbon design of buildings. Dynamic modelling and simulation of buildings are the only means of testing designs before construction, but the material published in this area is reduced to very basic user manuals of corresponding software tools. This book fills the gap as it develops a structured design method underpinned by dynamic simulation.

There are several related titles that provide useful complementary information to this book. *Targeting Zero* by Simon Sturgis (Sturgis, 2017) looks into whole life and embodied carbon strategies for design professionals. However, embodied carbon is considered in isolation from operational carbon and thus an opportunity for combined total zero emissions is missed. *Design Professional's Guide to Zero Net Energy Buildings* by Charles Eley (Eley, 2017) is interview based to outline how building energy use can be minimised. However, it is generally descriptive, not method based. *Heating, Cooling, Lighting: Sustainable Design Strategies Towards Net Zero Architecture* by Norbert M. Lechner and Patricia Andrasik (Lechner & Andrasik, 2022) is now in its 5th edition and has been established over 25 years, dealing with building fabric and systems. Although it uses the term 'Net Zero Architecture', there is a sense that this is used superficially, as there is no specific method to achieve it. *Net Zero Energy Building – Predicted and Unintended Consequences* by Ming Hu (Hu, 2019) brings about considerations of environmental indicators and an environmental multi-regional input-output model. However, it is broadly based, not focused on net zero building design. *Carbon Responsibility and Embodied Emissions* by João F. D. Rodrigues, Tiago M. D. Domingos, and Alexandra P.S. Marques (Rodrigues et al., 2011) has its strengths in awareness raising. However, it uses oversimplified principles, not accounting for embodied emissions but merely referring to these as unintended consequences. *Net Zero Energy Buildings – Case Studies and Lessons Learned* by Linda Reeder (Reeder, 2016) is complementary to this book by introducing in-depth case studies of net zero energy buildings. However, it is not focused on the fundamental principles and the method of how to design such buildings. *CIBSE AM11: Building Performance Modelling* (Awbi et al., 2015) provides guidance on a range of different aspects of modelling, including modelling of energy, thermal environment, ventilation, lighting, plant and renewable systems, and modelling for compliance with regulations. It provides a detailed expert overview of building modelling in these areas and guidance on quality assurance in modelling. Although it contains references to zero carbon buildings, it does not focus on their design. *The Passivhaus Designer's Manual: A Technical Guide to Low and Zero Energy Buildings* by Hopfe and McLeod (Hopfe & McLeod, 2015) gives an in-depth account of principles that underpin a Passivhaus building and describes initiatives for near zero energy and energy plus projects. However, it does not go into details of how to integrate these principles into an overall design. Additionally, Passivhaus buildings are designed using PHPP (Passivhaus Institut, 2022b), a spreadsheet-based calculation tool that uses 12 sets of monthly inputs per year for design calculations. In contrast, this book uses dynamic heat transfer calculations in hourly time steps, hence 8,760 sets of inputs per year, and it provides detailed guidance on how to integrate the principles of building physics into holistic designs of zero carbon buildings.

This book covers a specific angle of the subject that differentiates it from the above titles; however, these other titles cover certain aspects of the broader subject area in more detail. I therefore encourage the reader to consult this other material as background reading.

This is a 'how-to' methodology book focused on dynamic simulation and other quantitative methods for zero carbon design. It is intended to provide detailed guidance to the reader about converting principles of building physics into zero carbon design, combining embodied and operational emissions. The material is provided in a way that enables the reader to repeat the design steps and thus carry out own designs independently.

1.7 RATIONALE FOR THE THIRD EDITION

The first edition of this book, published in 2012, introduced a structured method for designing zero carbon buildings and demonstrated that it was perfectly possible to design new or retrofit existing buildings to zero carbon performance using available technologies. It made a case for change from a perception that we were dealing with an impossible problem to solve to an understanding that zero carbon design would become a mainstream.

The second edition, published in 2017, was expanded from the first edition. It introduced tasks for simple simulation experiments suitable for classroom and independent study work. It also introduced expanded summaries of design principles as the key learning outcomes in the majority of chapters. It included new advanced topics on multi-objective optimisation; reverse modelling; reduction of the simulation performance gap; nature-inspired emergent simulation leading to sketches that become 'alive'; and on an alternative economics for achieving the sustainability paradigm.

Whilst the first and second editions essentially dealt with operational emissions only, the third edition is a step change. It introduces embodied emissions as the first major part of the overall net zero design method. It then consolidates operational emissions from the second edition as the second major part of the net zero design method, and it combines the two into a method that achieves combined net zero emissions with certainty and by a specified year.

The third edition introduces four new case studies, in addition to an updated case study of the Birmingham Zero Carbon House, now applying embodied as well as operational emissions to assess its zero carbon status.

As the number of unsubstantiated claims about net zero buildings has increased in recent time, the fully quantified approach introduced in this third edition will bring about a sense of realism into what true zero emissions mean. This is ultimately a major positive step toward design decision-making that brings carbon emissions from buildings under control.

CHAPTER 2

METHODS AND TOOLS FOR DETERMINING EMBODIED EMISSIONS

2.1 BUILDING LIFE CYCLE AND THE WHOLE-LIFE CARBON

In its professional statement on 'Whole-life carbon assessment for the built environment', the Royal Institution of Chartered Surveyors defines the construction project life cycle (effectively building life cycle) into four stages (RICS, 2017):

1 **Product stage**, consisting of raw material extraction and supply (denoted as and referred to as A1), transport to manufacturing plant (A2), and manufacturing and fabrication (A3)
2 **Construction process**, consisting of transport to site (A4), and construction and installation process (A5)
3 **Use stage**, consisting of use (B1), maintenance (B2), repair (B3), replacement (B4), refurbishment (B5), operational energy use (B6), and operational water use (B7)
4 **End of life**, consisting of deconstruction and demolition (C1), transport to disposal facility (C2), waste processing for reuse, recovery, or recycling (C3), and disposal (C4).

The notion of whole-life carbon includes all of the above, plus:

5 **Carbon benefits and carbon loads beyond the project life cycle**, consisting of reuse, recovery, and recycling (denoted and referred to as D).

The process between the starts of items 1 and 2 above is given the name 'cradle to gate', and the process between the start of item 1 and the end of item 4 is given the name 'cradle to grave'.

The period for the whole-life carbon assessment is set to 60 years for domestic and non-domestic projects, and to 120 years for infrastructure projects (RICS, 2017).

This sets the scene for embodied emissions calculation in construction projects that will be followed in this book.

2.2 LITERATURE, DATA SOURCES, AND UNITS

In the early days of accounting for embodied emissions, an Inventory of Energy and Carbon (ICE) database was introduced by Prof. Geoffrey Hammond and Craig Jones at the University of Bath (Hammond et al., 2011). The most recent database version is named 'Embodied Carbon Footprint Database' (C. Jones & Hammond, 2019) and is available as an open-source Excel spreadsheet. This database will be used to demonstrate manual calculations of embodied emissions in building materials in this book.

In terms of the units of emissions, the database uses $kgCO_2e$, where 'e' stands for 'carbon dioxide equivalent'. This refers to a global warming potential – the release of greenhouse gasses associated with embodied emissions in a specific material. The difference between $kgCO_2$ and $kgCO_2e$ comes from the fact that some processes will release certain amounts of methane or other greenhouse gasses in addition to carbon dioxide during the creation of certain building materials, and methane has 25 times higher global warming potential than carbon dioxide. This is reflected in embodied emissions expressed in $kgCO_2e$, which are generally slightly higher for individual materials than embodied emissions expressed in $kgCO_2$. In some sources, embodied emissions are available in $kgCO_2$ only. As we are always interested in the maximum emissions/the worst-case scenario, we will always use the highest available figure. However, for simplicity, we will express both $kgCO_2$ and $kgCO_2e$ as $kgCO_2$ in this book.

Special materials, such as hemp-lime bio-composite, also known as hempcrete, are not in the ICE database, and their properties are available from Bevan and Woolley (2008). This bio-sourced material comes with negative embodied emissions (Bevan & Woolley, 2008, p. 81) as a result of carbon sequestration in the hemp plant during its growth, and can significantly reduce embodied emissions in newly constructed buildings (Jankovic et al., 2021).

The calculation of embodied emissions in a building also needs to include embodied emissions in building services. A methodology for this is provided by the CIBSE TM65 guide entitled 'Embodied carbon in building services: a calculation methodology' (Harnot et al., 2021). The methodology follows the 'product->construction->in-use->end-of-life' sequence. It provides a consistent approach for manufacturers' data requirements, methodology for embodied carbon calculation in products used in mechanical, electrical, and public health systems, and for reporting of embodied emissions. It is accompanied by supplementary spreadsheets for a manufacturer form and a reporting form, where a material-by-material approach is followed for each product or system.

Embodied emissions in photovoltaic systems are initially calculated on the basis of a fixed value of 2,560 $kgCO_2$ per kW_p (Circular Ecology, 2021). Thus, for instance, a 5 kW_p system will have $5 \times 2,560 = 12,800$ $kgCO_2$. This typical level of embodied emissions is forecast to change in the future (Worboys, 2021). In 2040 for instance, it is predicted that 325 $kgCO_2$ per kW_p will apply, so that a 5 kW_p system will have $5 \times 325 = 1,625$ $kgCO_2$ of embodied emissions. Although this is a broad basis for the calculation in comparison with CIBSE TM65 (Harnot et al., 2021) which uses a material-by-material approach, it demonstrates the scale of embodied emissions in the PV that could make a significant proportion of the overall embodied emissions. This will be supplemented later in the book with data from other sources (Worboys, 2021; Vindian Solar, 2023) to form three different scenarios for embodied emissions in photovoltaic systems.

Embodied emissions in solar thermal systems can be calculated using an approach introduced by Menzies and Roderick (2010). The approach is based on a detailed material inventory for the solar absorber, solar cylinder, and for the pipework, and it uses the ICE database for embodied emission factors for each material. The breakdown of each calculation is shown in detail, and therefore it can be customised for different material inventories corresponding to different products.

In order to account for all embodied emissions, additional calculations need to be carried out for travel of construction workers to the building site (VCA, 2016), material deliveries (TheyWorkForYou, 2013), and operation of construction equipment on site (Heidari & Marr, 2015). These are just some example sources, and the reader is encouraged to find corresponding sources for their specific region.

Emissions from maintenance and from the end-of-life treatment of the materials also need to be taken into account. Periodic replacement of the building services can be taken into account in a straightforward manner by using and if necessary, adjusting the values already calculated for boilers, stoves, solar PV, and solar thermal systems. Emissions from repainting the building internally and/or externally need to be obtained either from existing databases of embodied emissions, or from manufacturers' specifications. If there are no readily obtainable data, there will be uncertainties in the calculations.

Emissions from the end of life can be obtained from the 'Whole-life carbon assessment for the built environment' (RICS, 2017). This includes deconstruction and demolition emissions, transport emissions, and disposal emissions (RICS, 2017, p. 26).

Details of how the calculations are carried out and the uncertainties related to the assumptions made will be explained in Chapter 5.

2.2.1 How negative emissions from carbon dioxide sequestration can be justifiably used

In this section, I investigate how negative emissions from carbon dioxide sequestration associated with timber can be justified and used. This analysis comes in response to the absence of carbon storage values in the initial Inventory of Carbon and Energy (ICE) database (Hammond et al., 2011). Although the more recent ICE database accounts for carbon storage in some timber products (C. Jones & Hammond, 2019), it is important to understand the reasons for its absence in the initial database, for which two reasons are provided by its authors. It is also important to understand the conditions for using carbon storage values, and to be able to calculate these values if they are not given in the database.

First, the database contains information for cradle-to-gate embodied emissions, in other words from extraction of raw material to delivery to a construction site. Whilst timber would qualify for negative emission values in this phase as a result of carbon sequestration during the growth of the green plant from which it originated, there is uncertainty as to what will happen to timber during the operation and at the end of life of the building. If timber is burnt at the end of life or placed into landfill without gas capture, the initially negative embodied emissions would be reversed and carbon dioxide would end up in the atmosphere.

Second, the authors were of the view that assigning negative embodied emissions to timber products would lead to increased use of timber for this reason alone and that it would lead to negative embodied emissions applied inappropriately.

The reasons for not accounting for carbon storage in timber products can be overcome if we take care of a responsible end-of-life disposal that does not reverse the negative embodied emissions and if timber originates from certified sustainable sources (RICS, 2017). This can be done by converting waste timber into usable products and also by ensuring that the same number of trees are planted as those felled for use in construction (RICS, 2017). Moving forward on this basis, we can adjust the database values for timber embodied emissions by the amount of negative embodied emissions per kilogram of timber. These values are calculated using carbon content in carbon dioxide and carbon content in dry timber and adjusted for moisture content in moist timber, as explained below.

Carbon content in carbon dioxide is calculated from the atomic mass of CO_2 divided by the atomic mass of C (British Standards Institution, 2011), expressed as symbols of the corresponding atoms as $(C + 2 \times O)/C$. Replacing the atom symbols with their atomic mass gives $(12 + 2 \times 16)/12 = 3.667$. This means that burning 1 kg of carbon will produce 3.667 kg of CO_2. Absorbing 1 kg of carbon into a dry mass of timber will result in the removal of 3.667 kg CO_2 from the atmosphere.

Carbon content in timber is widely reported to be 50% of the timber mass. However, a detailed analysis of 41 North American species (Lamlom & Savidge, 2003) shows a variation between the species from 46% to 55% of the dry mass of wood in kiln dries samples, where the lower value corresponds to hardwood and the higher value to conifers. I will adopt the figure of 46.5% as a conservative estimate, close to the lower end of the range but just slightly above it, considering that hardwood is more likely to be used as a construction material in buildings. As EN 16449 states that moisture accounts for 12% of timber products (CEN European Committee for Standardization, 2014) the carbon content of 46.5% (or 0.465) in dry timber needs to be further adjusted. Thus, adding 12% of moisture to 100% of the dry mass of timber gives 112% of moist timber mass (or 1.12), of which 46.5% is carbon. The percentage of carbon normalised to 100% of the mass of moist timber is therefore $0.465/1.12 = 0.415$ or 41.5%.

Multiplying the content of C in CO_2 of 3.667 kg CO_2/kg and the percentage of carbon in moist hardwood timber of 41.5% results in approximately 1.52 kgCO_2/kg. This is the amount of CO_2 removed from the atmosphere that contributes to each kilogram of moist hardwood timber. This means that we can reduce values for embodied emissions in timber without carbon storage by this amount with reference to Embodied Carbon Footprint Database values (C. Jones & Hammond, 2019), assuming a responsible end-of-life treatment of timber that does not result in a reversal of CO_2 into the atmosphere and that timber originates from certified sources. In this way, negative emissions from carbon dioxide sequestration of -1.52 kgCO_2/kg can be justifiably used.

Although the recent Embodied Carbon Footprint Database (C. Jones & Hammond, 2019) provides carbon storage values for some timber products, it is useful to be able to calculate these values if they are absent from the database or if they are available from other data sources. To test this approach, we use a database value of 0.493 kgCO$_2$/kg of embodied carbon in 'Timber - Average of all data - No Carbon Storage' and add the calculated carbon storage value of -1.52 kgCO$_2$/kg. The result of -1.03 kgCO$_2$/kg corresponds exactly to the database value of 'Timber - Average of all data - Including Carbon Storage' in the database. The exact correspondence of this result to the database value gives us confidence to calculate carbon storage values for cases where these values are absent from the database, or where the values come from other sources.

In order to account for carbon storage in bio-sourced materials, such as materials originating from growing plants, we need to add negative embodied emissions arising from carbon sequestration in these materials and prescribe the end-of-life treatment for these negative embodied emissions to be sustained.

2.3 DATA SOURCES AND SOFTWARE TOOLS

This section provides a brief overview of efforts in different countries and different approaches to establishing embodied emissions.

Embodied Carbon Footprint Calculators for Construction (Circular Ecology, 2022) references a selection of embodied carbon calculators and assessment tools. IMPACT (BRE Group, 2020), referenced on this site, is enabling technology for software developers that consists of a database and a specification for incorporation into software tools. Two free tools based on IMPACT were available at the time of writing: One Click LCA for Buildings and Infrastructure Projects (One Click LCA Ltd, 2018) and eToolLCD (eTool, 2016).

The United States Green Building Council developed a LEED carbon calculator (Jacques, 2022). Embodied emissions are calculated using Materials and Resources LEED Credit Library (USGBC, 2022), accounted for over the building life cycle. This is one of several enhancements in LEED v4.1 standard for green building design (USGBC, 2019).

A 'GaBi' approach for life cycle modelling of products and systems has been established in Germany over the past 25 years with over 10,000 users (Sphera, 2022). It uses an extensive database to establish life cycle assessment of a range of entities, from consumer products to large construction developments. It reports to be using 'the most accurate LCA databases available', which are regularly updated. Hence, this approach has a potential for accurate and up-to-date information on embodied emissions in building materials.

A 'SimaPro' approach developed in the Netherlands focuses on the product life cycle and is underpinned by the use of extensive databases, including agriculture, packaging, food, industry data, and others. According to its web page, it has been established for over 30 years and used in over 80 countries. It helps to model life cycle of products in a 'systematic and transparent way', whilst dealing with hotspots in supply chains and uncertainty calculations (PRé Sustainability B.V., 2022). Hence, this approach has a complementary potential in relation to the GaBi approach, with respect to product life cycles, including life cycles of building materials.

A 'OneClickLCA' approach developed in Finland appears to be one of the most comprehensive approaches for accounting for embodied emissions (OneClickLCA, 2022b). At the time of writing, it included over 60 data sources from Europe, North America, Asia Pacific, Middle East and South America, covering over 150,000 data points with regular updates. The platform enables whole building life cycle assessment, facilitating sustainable and circular designs and enabling certification by achieving credits in LEED, BREEAM, and numerous other green building certification systems. The platform has direct integration with a number of simulation tools, including IES Virtual Environment and DesignBuilder (OneClickLCA, 2022).

We are now going to use a model developed in the IES Virtual Environment (Figure 7) to illustrate the use of OneClickLCA. The building model is set to a London location and is based on conventional materials, with a heating source set to a ground source heat pump. Thermal properties of the building envelope are shown in Table 1.

TABLE I THERMAL PROPERTIES OF A CONCEPTUAL OFFICE BUILDING USED IN THE ONECLICKLCA ANALYSIS	
Description	**U-value (W/m²K)**
External walls	0.10
Ground floor slab	0.11
Roof	0.11
Vertical glazing	0.81
Atrium glazing	0.83

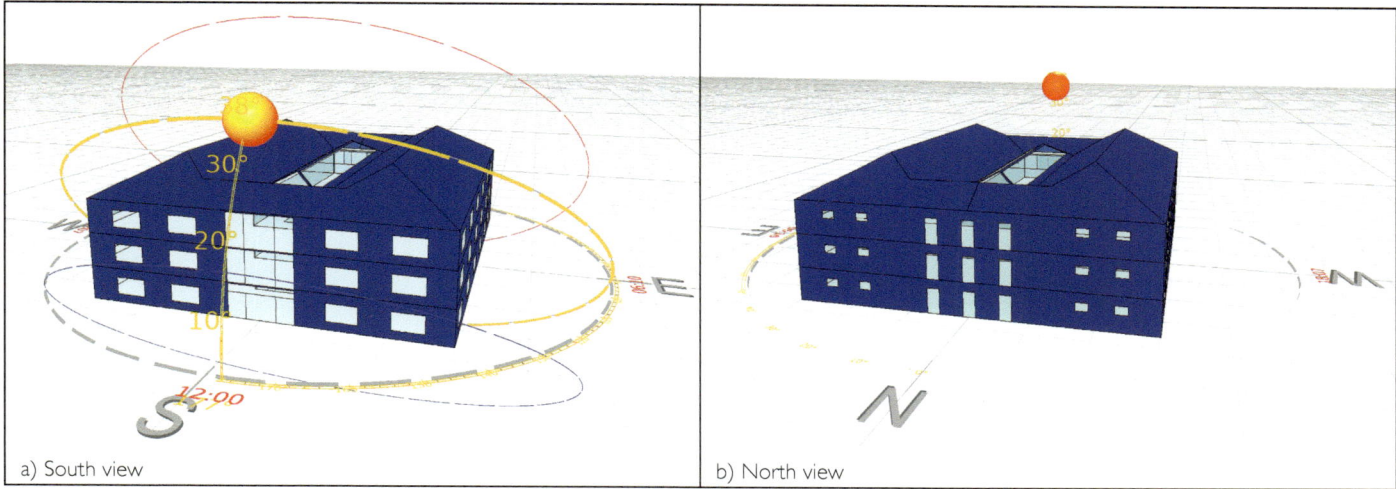

a) South view

b) North view

Figure 7   Conceptual model of an office building in London at noon on 21st March in IES Virtual Environment

Dynamic simulation of this model was carried out on an hourly basis for London – Gatwick location, latitude 51.15N, longitude 0.18W, and altitude of 62 m, using weather data file, 'GBR_London.Gatwick.037760_IWEC.epw' from WMO station number 037760, obtained from Energy Plus Weather data site (DOE, 2023). Operational carbon emissions from electricity use obtained from this simulation were 40,254 $kgCO_2$ without the PV contribution. The PV made negative contribution to operational emission of −13,965 $kgCO_2$, and therefore the total operational emissions were 26,289 $kgCO_2$.

The model was subsequently exported to OneClickLCA, and the results are summarised in Figure 8. As it can be seen from that figure, the amount of OneClickLCA operating carbon emissions in row B1–B7 (3,172,302 $kgCO_2$) is considerably higher than that obtained from the detailed dynamic simulation, even without considering the contribution from renewable energy (40,254 $kgCO_2$). A partial explanation of this difference comes from the difference in time periods between the two sets of results: OneClickLCA was set for analysis of 27 years, which takes the time period to year 2050 assuming the starting point of the year of writing this text in 2023; and the IES Virtual Environment simulation was set to run for a period of one year. Multiplying IES Virtual Environment result by 27 years gives the total operational emissions from the simulation of 1,086,858 $kgCO_2$, still representing operational emissions nearly three times lower than the operational emissions from OneClickLCA.

Whilst an explanation of this difference is beyond the scope of this text, one of the very useful results from OneClickLCA, in my opinion, is the Mass-kilogram Classifications. Thus, Figure 9 gives us where the 'low hanging fruit' is for reducing embodied emissions by using bio-sourced materials instead of conventional materials. As floor slabs, ceilings, roofing decks, beams, and roof ('horizontal structures') account for 84.3% of all upfront carbon, this should be the first port of call for investigating the use of bio-sourced materials as replacement for conventional materials, in order to reduce embodied emissions. Subsequently,

Net Zero Carbon Download Results Summary

Result category		Carbon emissions kg CO$_2$e	Biogenic carbon kg CO$_2$e bio ⑦	Carbon savings from materials reuse kg CO$_2$e	Carbon savings from exported energy kg CO$_2$e	Carbon offsets kg CO$_2$e	Net Carbon kg CO$_2$e	
➕ A1-A5	Upfront carbon	391 324	−91 768	−8 967			290 590	Details
➕ B1-B7	Operating carbon	3 173 916		−1 614	0		3 172 302	Details
➕ C-D	End of life	43 179	0	−78 740			−35 561	Details
	Total	**3 608 419**	**−91 768**	**−89 320**	**0**		**3 427 331**	

Figure 8   Summary of the OneClickLCA analysis

Mass kg - Classifications

● Floor slabs, ceilings, roofing decks, beams and roof - 84.3%
● External walls and facade - 13.1%
● Windows and doors - 1.4%
● Internal walls and non-bearing structures - 1.1%

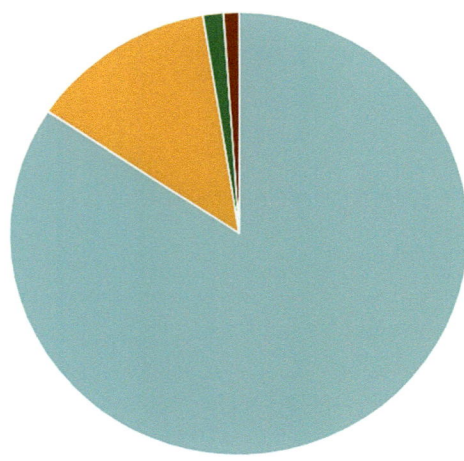

Figure 9   OneClickLCA Mass-kilogram Classifications result

external walls and façade ('external vertical structures'), which account for 13.1% of upfront carbon, should be considered for replacement of conventional materials with bio-sourced materials.

2.4 ACCURACY AND AMBIGUITY OF EMBODIED EMISSIONS DATA

How accurate and unambiguous are embodied emissions databases?

First, in the cases where emissions information is described as 'cradle to gate', where exactly is the cradle and where is the gate? Databases contain general information, which will inevitably differ from the actual locations, and hence individual tracking and labelling of materials to correspond to actual locations of the sources and destinations would improve data accuracy.

Second, and especially in the case of the Inventory of Carbon and Energy (ICE) database (Hammond et al., 2011), carbon emission factors at the time of the development of this database were quite different from the carbon emissions at present, as a result of the gradual grid decarbonisation. Even though the database was updated in 2019 (C. Jones & Hammond, 2019), these values still need to be revisited and updated.

Third, carbon intensity of national grids change with time continuously, as shown in the case of UK by the GridCarbon mobile app (Rogers & Parson, 2022). Unless databases of embodied emissions are linked to live feeds of grid carbon intensities, they will be out of date before they are even published.

Fourth, where a specific material used in construction is not available in a database, the nearest match needs to be found that could be from sources that have been developed with less rigour than the mainstream databases.

Fifth, as shown in the previous section in the example of calculations using OneClickLCA, there are uncertainties related to operational emissions calculations, and these are best dealt with detailed dynamic simulations. There are other uncertainties as a result of applying general rules to specific buildings, relating to transport of materials and end of life emissions.

Therefore, there will be inevitable uncertainties in carbon emissions data and in carbon emission calculations. Hence, emissions data need to be selected with care and with attention to detail in order to minimise these uncertainties. But even with these uncertainties, we will be much closer to true embodied emissions by carrying out detailed calculations, in comparison with not doing these calculations at all.

2.5 TASKS FOR SIMPLE EXERCISES

1. Draw by hand a simple enclosure 10 m wide, 6 m deep, and 3 m high.
2. Calculate surface areas of all six sides and place them in a table.
3. Assume two different versions of the enclosure: (a) made from 100 mm concrete and (b) made from 100 mm hardwood.
4. Use Embodied Carbon Footprint Database (C. Jones & Hammond, 2019) to calculate embodied emissions for versions (a) and (b) of the enclosure.
5. Create version (c) where carbon storage is included in the material from version (b) using information from Section 2.2.1.
6. Organise your calculations using the following table format or similar for all three versions (Table 2):
7. Compare the totals from all three versions and discuss with your colleagues.
8. Repeat the process with OneClickLCA.
9. Compare the results between the manual and OneClickLCA calculations and discuss with your colleagues.

TABLE 2 TABLE FORMAT FOR SIMPLE EMBODIED EMISSIONS CALCULATION							
Surface	Area (m²)	Thickness (m)	Volume (m³)	Density (kg/m³)	Mass (kg)	Embodied emissions factor (kgCO$_2$/kg)	Embodied emissions (kgCO$_2$)
Floor							
Ceiling							
Larger wall-front							
Larger wall-rear							
Smaller wall-left							
Smaller wall-right							
						Total	

CHAPTER 3

METHODS AND TOOLS FOR DETERMINING OPERATIONAL EMISSIONS

Before a simulation of building performance can be carried out, a computer model of the building needs to be developed. The term **modelling** is defined as making a logic machine that represents the material properties of the building and physics processes in it. **Simulation** is then defined as numerical experimentation with the model to investigate its response to changing conditions inside and outside the building. Models that are based on first principles and can replicate dynamic heat transfer in a building in response to external and internal influences on the time scale of one hour or less, are called **dynamic simulation models (DSMs)**.

Modelling involves a certain degree of abstraction. A simulation model is not a detailed representation of all geometry and all processes in a building but just of those aspects of geometry and processes that are important for objectives of our analysis. According to John Holland, one of the early pioneers and giants of Computer Science, '*the art of model building turns on selecting a level of detail*' (Holland, 2000, p. 46). In the creation of a dynamic simulation model, the importance of abstraction through setting a level of detail is comparable with the process of making architectural scale models. It '*means taking away any unnecessary components or detail that will not aid the understanding of the design being communicated*' (Dunn, 2014, p. 28).

Computer modelling is however different from architectural modelling as it changes with time. Generally, it involves capturing a definition of a real-world system (in our case a building) at a time 't' through observation and measurement, applying abstraction and transferring this definition into a model domain at time 't', running the model for one time step till time 't+1', and transferring the results into real-world domain and interpreting their meaning at time 't+1' (Holland, 2000). Abstraction and interpretation are therefore two critical aspects of dynamic simulation modelling.

3.1 DESIGN APPROACH

The main design principle that will be adopted in this book will be that of experimentation with a DSM. In order to design zero operational emissions, we first need to reduce energy demand by improving building energy efficiency. After the energy efficiency has been maximised, we then need to consider the efficiency of various systems and various renewable energy options that are suitable and feasible for the building we are designing. Energy efficiency and carbon emissions analysis, together with economic analysis and comfort analysis, are essential ingredients of this method, which ensure the overall success of a design. Details of this design approach will be presented throughout this book. A simple example of improving energy efficiency using multiple simulations is described in the optimisation section later in this chapter.

3.2 DESIGN VARIABLES

The variables that we will be able to affect by design are as follows:

- Response to climate context: climatic conditions for a particular location, taking into account predicted climate change
- Response to site context: solar radiation, building orientation, prevailing winds, site configuration, overshadowing by the land configuration or existing objects

 DOI: 10.4324/9781003342342-4

- Building geometry
- Thermal insulation
- Airtightness
- Passive solar gain
- Thermal mass
- Natural ventilation
- Natural daylight
- Electrical lighting
- Renewable energy systems
- Internal heat gains
- Additional heating or cooling

These variables will be discussed in detail in corresponding chapters in this book.

3.3 DESIGN TOOLS AND EVALUATION TOOLS: DSM VERSUS SAP, SBEM, AND PHPP

We need to differentiate very clearly between what is and what is not a design tool. As we can see from the introductory part of this chapter, DSMs are based on first principles, running on time scales of one hour or less, and are for that reason capable of replicating dynamic performance of the building that is close to the performance of its real-world equivalent – a building constructed on the basis of DSM design. Therefore, using a DSM, a designer can establish relative merits of different design options. This makes DSM a design tool.

Methods that are used to investigate building performance in response to much larger time scales of monthly changing conditions or based on one calculation per year are not DSM and cannot be used as design tools. These methods, which we consider merely as performance **evaluation tools**, such as SAP – Standard Assessment Procedure (BRE Group, 2018), PHPP – Passive House Planning Package (Passive House Institute, 2023), and other similar evaluation tools are outside of the scope of this book.

It is important to understand that the information content offered by evaluation tools is not sufficient to fully evaluate building performance. I experienced this first hand when running a DSM model and an evaluation tool of the same building. The DSM model revealed a significant overheating problem in the building, but no overheating was reported by the evaluation tool. This is easily understood when considering the difference in resolution between evaluation tools and DSM tools. Evaluation tools use 12 sets of numbers if they are based on monthly average calculations, and DSM tools use at least 8,760 sets of numbers equal to the number of hours in the year, taking even sub-hourly time steps for more detailed calculations, and applying dynamic heat transfer principles. Let us draw an analogy: if we are given a choice of two computer screen resolutions: 12 pixels and 8,760 pixels, it is intuitive that we will get much more information from the higher resolution than from the lower resolution screen. Taking this analogy back into the domain of building performance calculation, it is apparent that most of the information generated by DSM will not even be on the 'radar' of the evaluation tools – it will simply not be visible there.

3.4 PHYSICAL SCALE MODELS VERSUS DYNAMIC SIMULATION MODELS

Physical scale models (Figure 10) are suitable for investigation of certain aspects of building behaviour. Because of a difference in dimensions between the actual building and the scale model, physical models are less suitable for investigation of thermal comfort. To scale a building down 10 times (a scale of 1:10) to a size of a physical model, thicknesses of all materials need to be scaled down proportionally. We will show throughout this book how scale models can be used to investigate certain aspects of building performance, such as thermal insulation, thermal mass, natural ventilation, and natural daylight.

However, scale models are hard to be built using real materials; instrumentation systems are required to monitor their behaviour; and we cannot put people into them to tell us how they feel. DSMs overcome these disadvantages and have a much wider application scope than physical scale models. Whereas we can use a physical scale model to investigate a particular aspect of building behaviour, a simulation model can be used to investigate multiple aspects of behaviour simultaneously.

Figure 10   Physical scale models of various aspects of building performance created by my students

3.5 OVERVIEW OF DYNAMIC SIMULATION TOOLS SPECIFICALLY USED IN THIS BOOK

BEST – 'Building Energy Software Tools' Directory[1] contains information about hundreds of software tools. Some of these tools are suitable for whole building analysis, and some are specialised for demonstrating compliance with codes and standards; lighting; ventilation; HVAC components and systems; and other various aspects of building energy performance.

It is clearly not practical to review all of these software tools in this book. Instead, I will focus on the tools that I have used in this book, in alphabetical order. These are overviewed in the next section.

3.5.1 DesignBuilder

DesignBuilder (DesignBuilder Software Ltd, 2022) is an advanced modelling tool. A dashboard user interface contains the majority of controls available from a single screen, which is reconfigured according to the type of editing operation being carried out.

In addition to energy simulation, DesignBuilder can carry out heating and cooling design calculations, computational fluid dynamics (CFD) simulation, daylighting simulation, and it can calculate construction cost and embodied carbon.

A particularly useful and well-developed aspect of energy simulation in DesignBuilder is multi-objective optimisation, which is becoming one of the major directions in which the building simulation field is developing. Optimisation will be discussed in detail elsewhere in this book, and will be used as one of the main methods for zero carbon design in Chapter 18.

A simulation model is created in 'Edit' mode (Figure 11). The user first needs to create a site, and define its layout, location and region, where the simulation weather data file is also selected. DesignBuilder uses EnergyPlus weather (EPW) data files in native format, as EnergyPlus is its main simulation engine. After setting the site parameters, the user needs to create a building and define the following:

• Layout: creation of building geometry;
• Activity: specification of activity templates, internal heat gains, environmental control settings, etc.;
• Construction: specification of wall/floor/roof constructions, airtightness, and costs, and including material choices from conventional to PCM;
• Openings: specification of windows, doors, and vents;
• Lighting: specification of electrical lighting power density, lighting controls, and costs;
• HVAC: specification of heating, cooling, DHW, natural and mechanical ventilation, earth tubes, and costs.

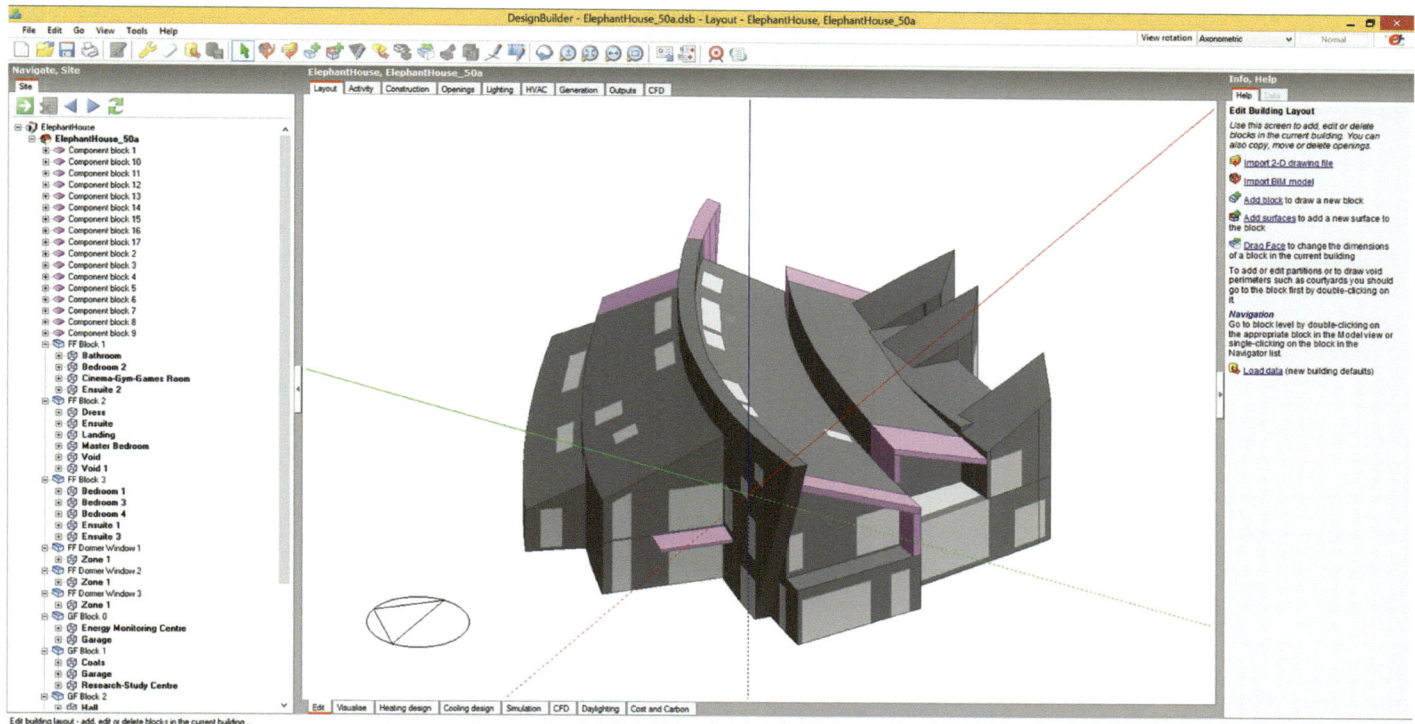

Figure 11   DesignBuilder 'Edit' dashboard (model based on design of Elephant House by Vivid Architects Ltd)

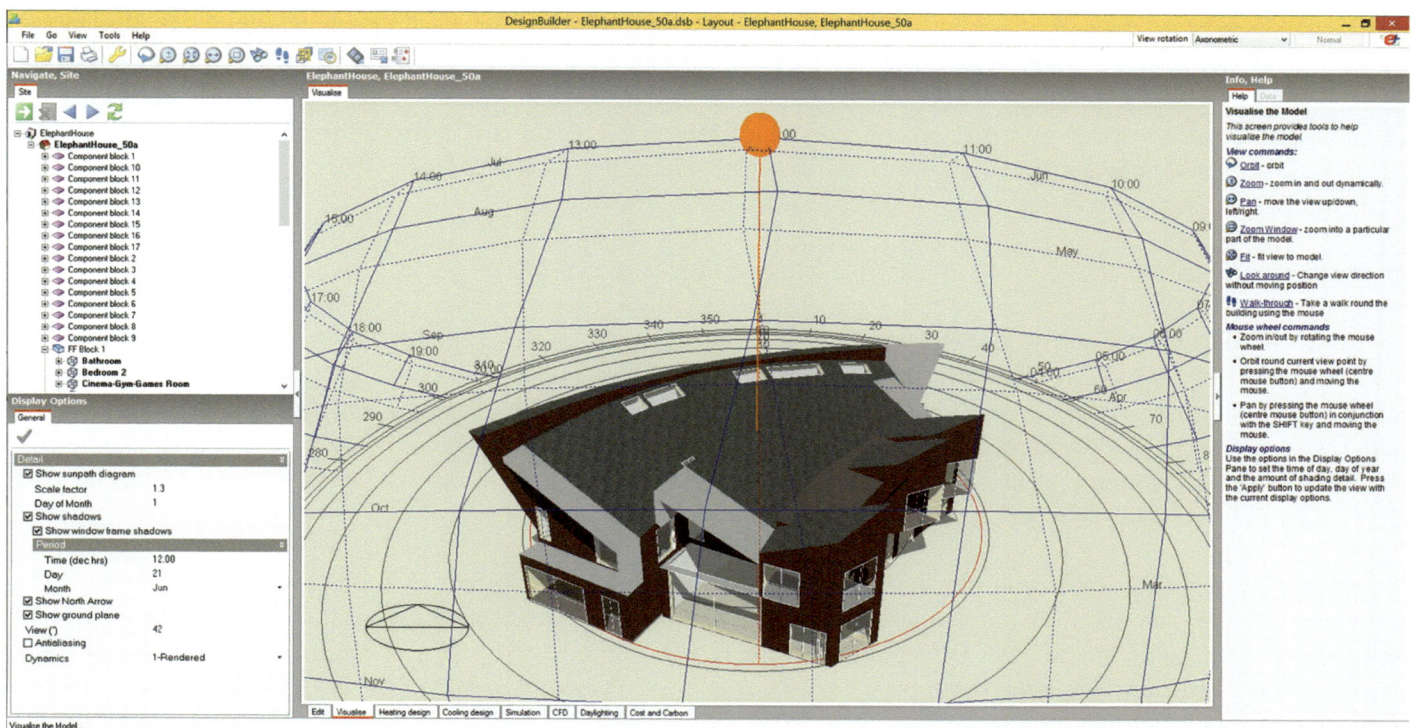

Figure 12   DesignBuilder rendered geometry (model based on design of Elephant House by Vivid Architects Ltd)

In addition to the above settings which are necessary for energy simulation, the following optional settings are also available:

• Generation: renewable energy generation settings if applicable;
• Outputs: heating and cooling design and simulation output options, if non-default outputs are required;
• CFD: applicable only if computational fluid dynamics analysis is carried out.

Building geometry is created under 'Layout' editing, either created using 3D solid modelling tools, or imported from a BIM model in gbXML (Green Building XML) format. Geometry of simulation models has an architectural look and feel (Figures 11 and 12).

All dialogues give comprehensive written and visual access to various controls. For instance, setting a construction type opens a dialogue box with several different tabs, which enable the user to select construction layers and choose materials and their properties (Figure 13a). Another tab then gives the corresponding U-value calculation (Figure 13b), reminding the user that this steady state value is not used in dynamic simulation. This will be elaborated upon in Chapter 8, where it will be explained which construction parameters are used in dynamic simulation. The same dialogue for setting the construction layers enables the user to choose if any of the layers are thermally bridged, and by which material and percentage (Figure 13c). This enables the user to quickly evaluate the effect of a thermal bridge, as shown in Figure 13d. As can be seen from Figure 13c, thermal bridging of the insulation layer with 1% of steel doubles the U-value of the entire construction in Figure 13d. Constructions are also represented visually (Figure 13e), and condensation analysis is carried out for each construction (Figure 13f).

Other features include the capability to set a target U-value (Figure 13a and 13c) and choose which layer to vary (changing its thickness) in order to achieve the target. All constructions have a cost per square metre setting, which enables overall construction cost calculation.

Simulations are run in the 'Simulation' tab, which opens a simulation setting dialogue box to enable the user to set the simulation period, time step resolution of outputs, output details, and an optional use of an external simulation manager (Figure 14). The latter feature is particularly useful in 'Optimisation' mode, accessible from the 'Simulation' tab, where numerous simulations are sent to an external simulation server and thus do not use CPU time of the local machine.

Simulation results are all on one page within the same dashboard (Figure 15), from where different types of outputs (graphical or tabular) and different variables can be selected.

After a single annual simulation is completed, the user can proceed to the 'Parametric' tab to carry out multiple parametric simulations (single simulations that go through multiple values of one or more design variables), or to the 'Optimisation' tab to carry out multi-objective optimisation (Figure 15).

In the 'Optimisation' tab, the user can set optimisation objectives (Figure 16a), constraints, and design variables (Figure 16b). The objectives can be set as comfort, cost, daylight, energy/loads, environmental impact (including CO_2 emissions), heat gains, unmet loads, or other user-defined objectives. The design variables can be set to vary within a certain range of airtightness, wall constructions, thermal mass, renewable energy generation options, and numerous others. Optimisation can be set for two objectives only. Constraints are used to exclude solutions where one or more simulation outputs exceed a particular value. For instance, if cost and carbon emissions are set as objectives, a constraint could be set as a number of discomfort hours not exceeding a certain maximum value. An example of optimisation results is shown in Figure 17, from where the user can find a combination of parameters that generated Pareto points in the scatter plot.

Before starting the optimisation the user needs to set initial population size, maximum number of generations, and other parameters. There is an option to use a simulation manager, which can be set to utilise an external simulation server for optimisation. As optimisation can be very CPU-intensive and can render a laptop or a personal computer unusable for anything else for many hours whilst optimisation is in progress, it is advisable to use an external simulation server for this purpose. In my particular case, a multicore simulation server running up to 48 parallel threads has proved to be essential hardware for extensive optimisation work.

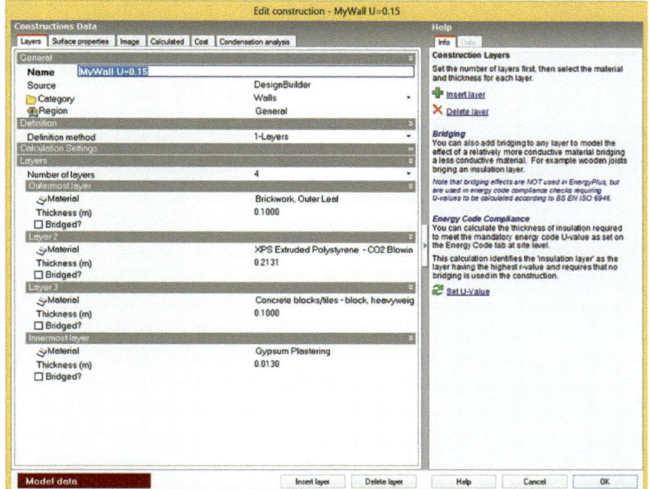

a) Construction layers with no thermal bridging

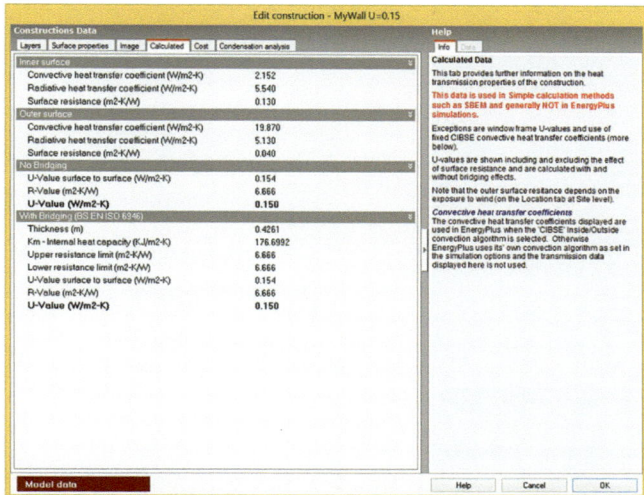

b) Corresponding U-value calculations with no thermal bridging

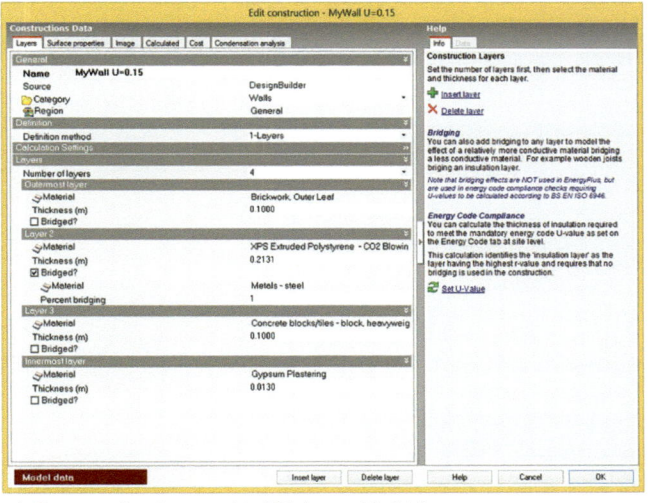

c) Construction layers with 1% thermal bridging

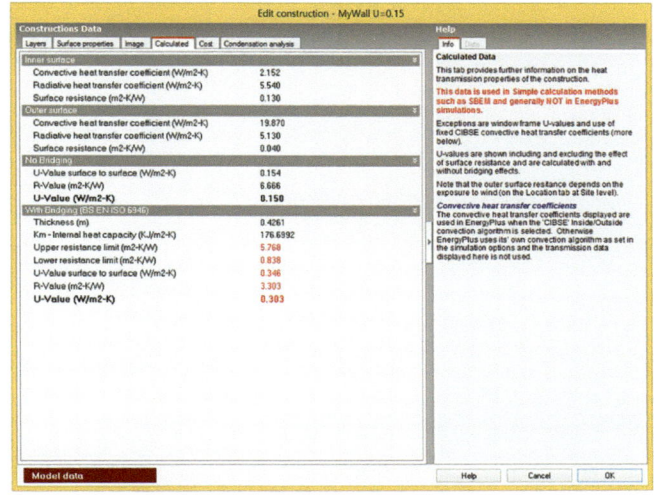

d) Corresponding U-value calculation with thermal bridging

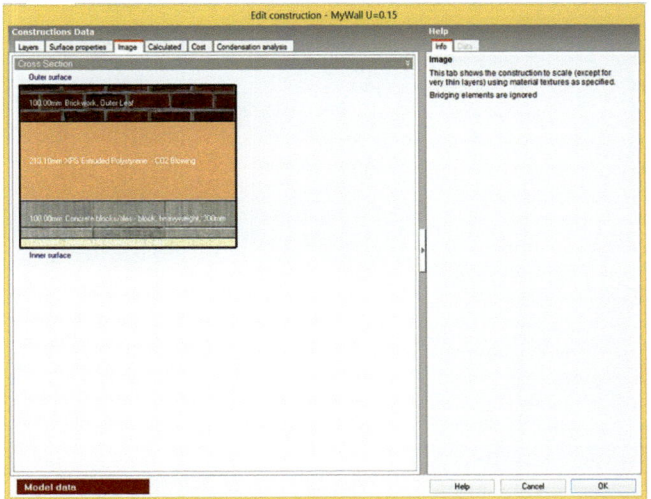

e) Visual representation of a construction

f) Condensation analysis for a construction

Figure 13   DesignBuilder constructions setting

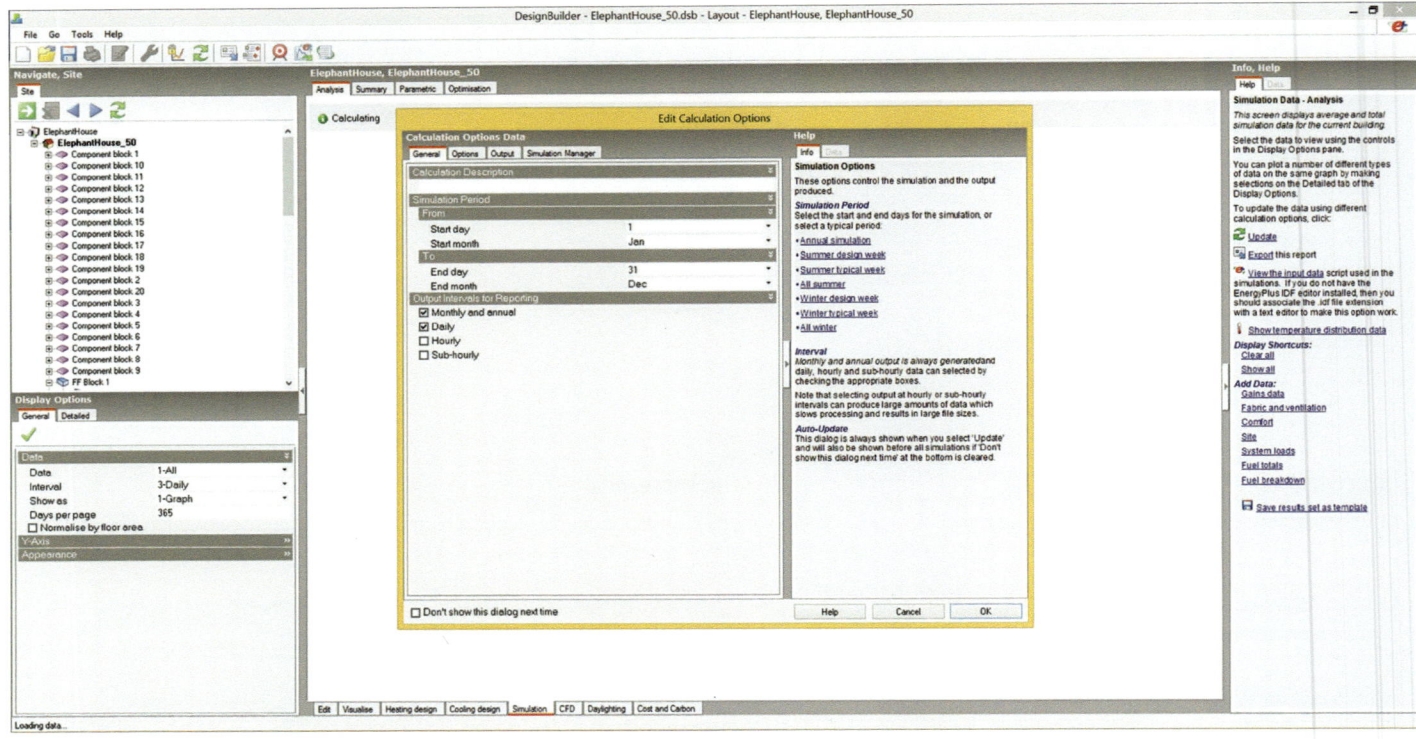

Figure 14   DesignBuilder simulation setting dialogue

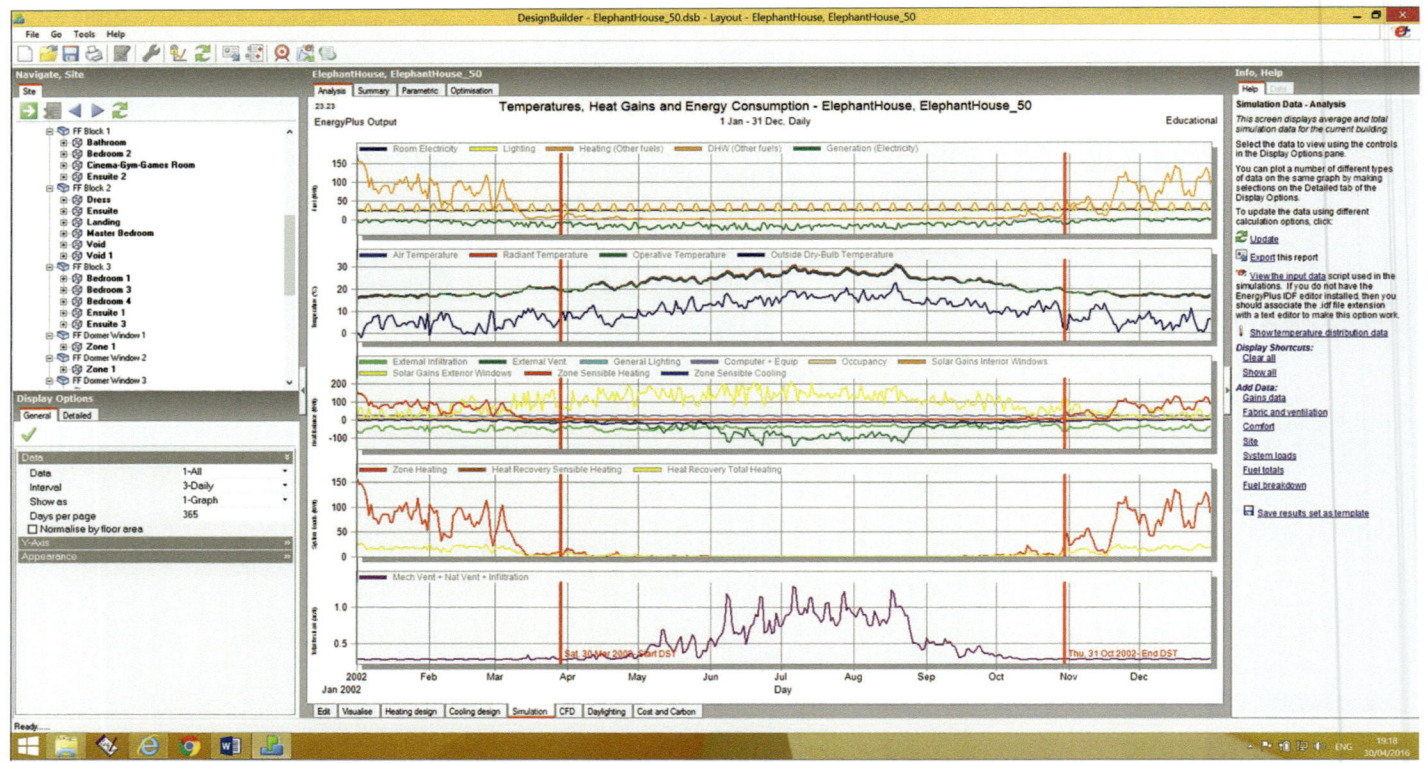

Figure 15   DesignBuilder single simulation results

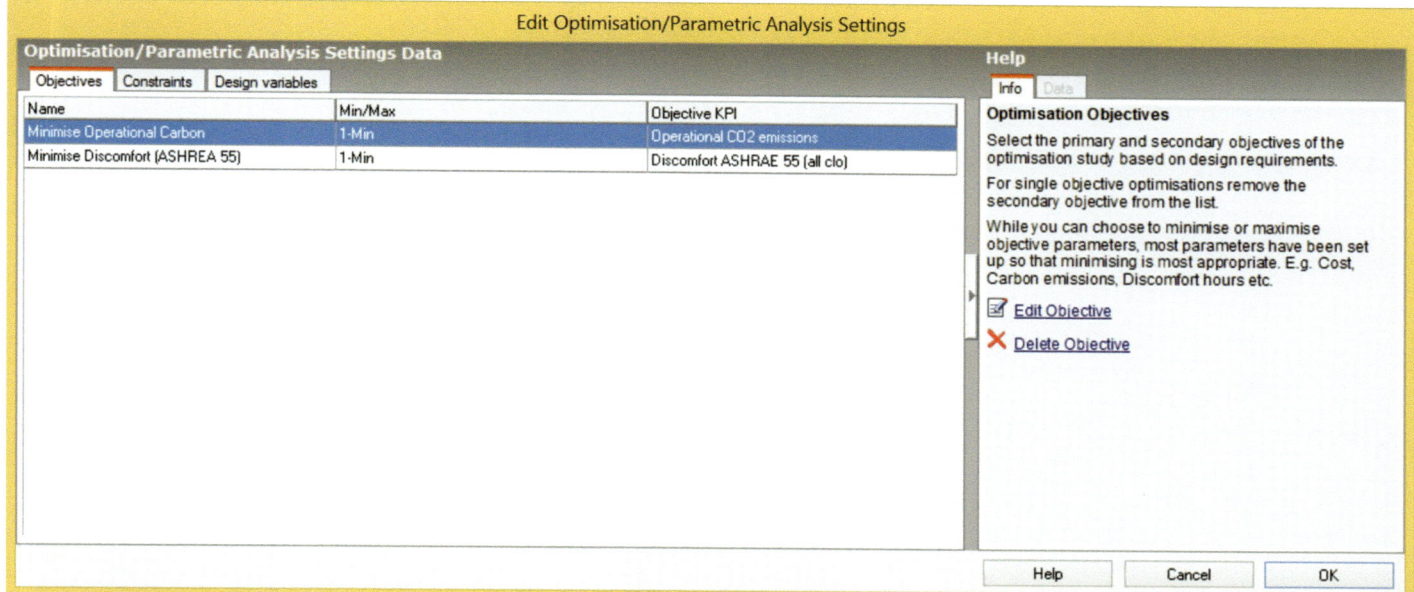

a) Setting optimisation objectives

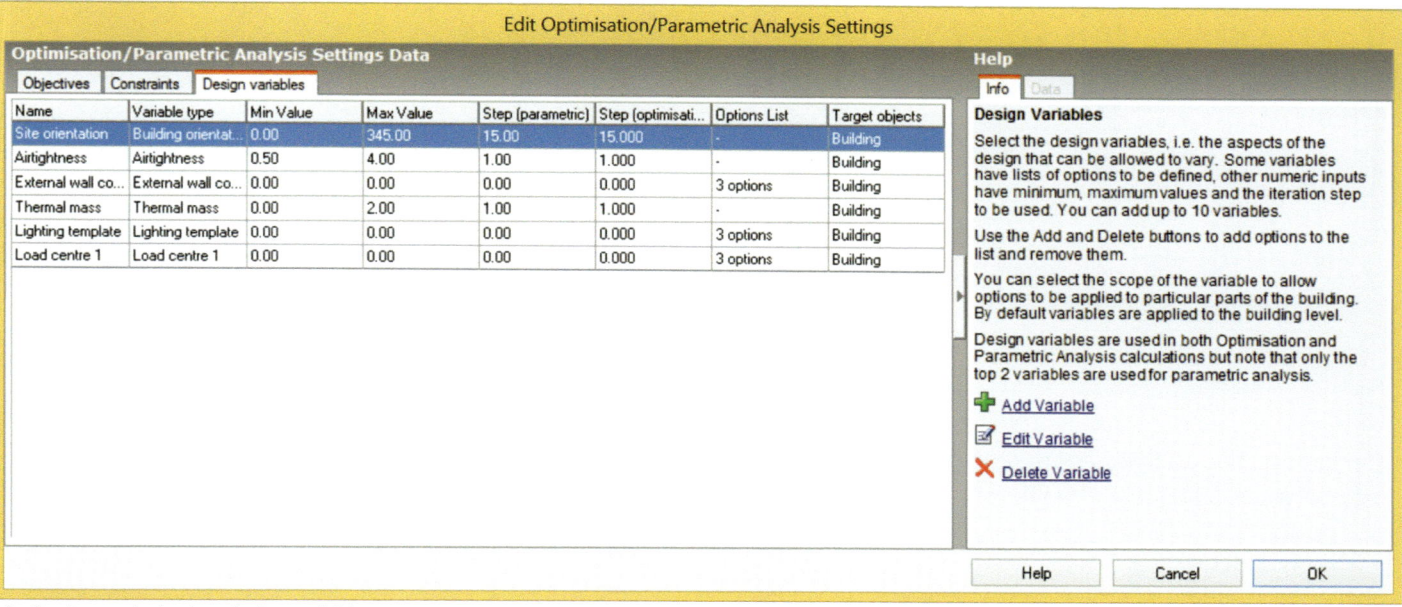

b) Setting optimisation design variables

Figure 16   Optimisation settings in DesignBuilder

Strengths. DesignBuilder combines a dashboard-like user interface with the power of EnergyPlus as the main simulation engine. Its workflow follows the way designers use simulation tools, which led to the success of winning the ASHRAE Modelling Challenge Award for 'Best Innovative Workflow' in 2015. The well-developed optimisation process enables designers to use multi-objective optimisation routinely, thus pushing the boundaries of traditional design based on a small number of single simulations. It also has extensive online documentation and excellent training videos, which enable novice users to learn and start using building simulation quickly.

Weaknesses. Although DesignBuilder provides a number of different simulation outputs, the range of these outputs is not as extensive as in some other simulation tools. The multi-objective optimisation is limited to two objective functions for practical reasons, although there are cases where three or more simultaneous objective functions would be desirable.

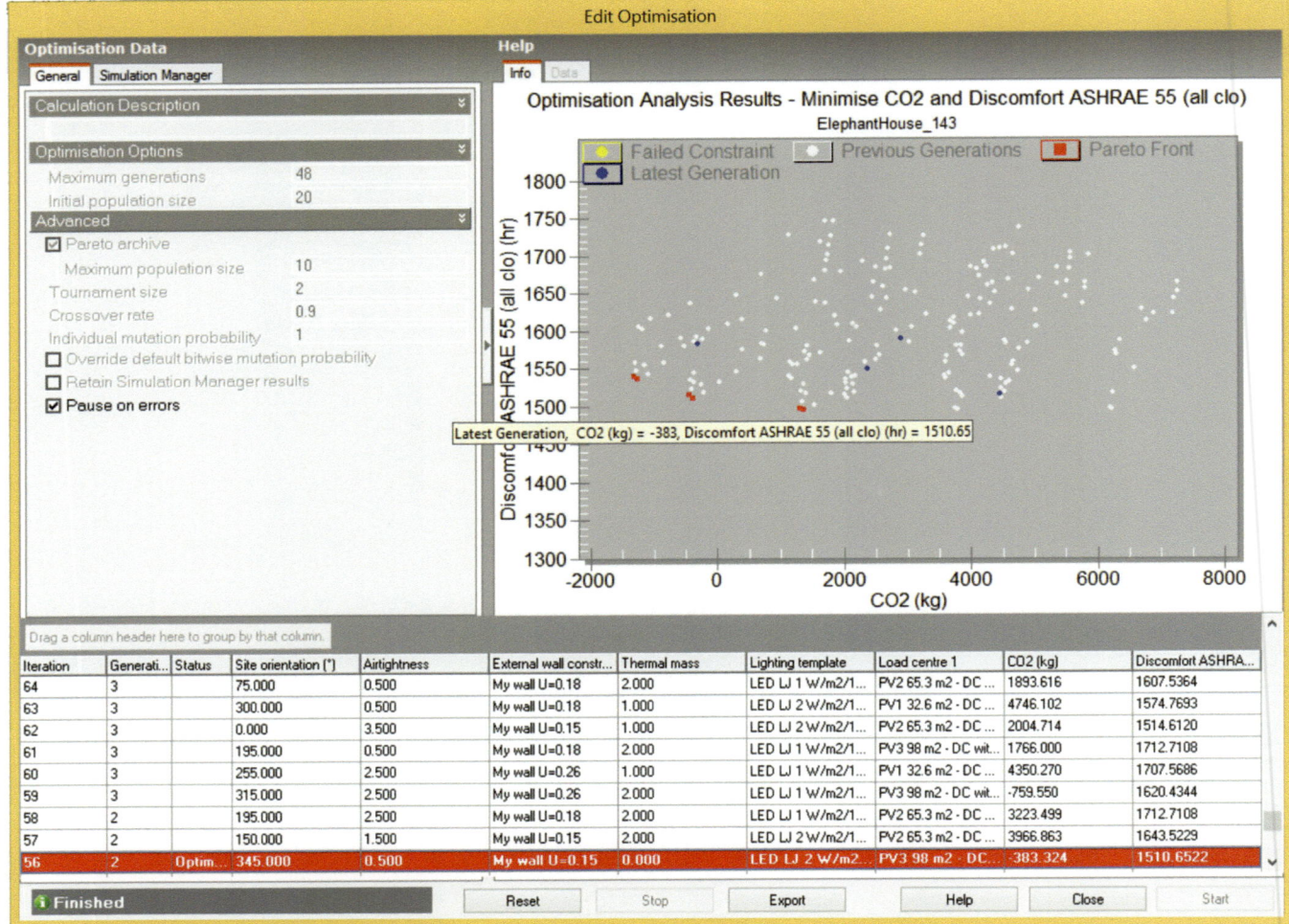

Figure 17   *DesignBuilder optimisation results dialogue*

On balance, the weaknesses are not significant in comparison with the strengths, making DesignBuilder one of the most powerful, versatile, and intuitive simulation tools available.

3.5.2 EnergyPlus

EnergyPlus (DOE, 2022) is a dynamic simulation software tool developed on the basis of BLAST and DOE-2 simulation programs, which were originally developed in the US in the late 1970s and early 1980s. It has the most comprehensive list of heat transfer and HVAC system models than any other building simulation software. However, it is a simulation engine only, and thus it only has a basic user interface. It will therefore not correct user input errors, although it will report them. In case of severe errors, the simulation will terminate, and the user will need to go through an error finding and correcting process, commonly called 'debugging' in relation to software matters, and this may increase the lead time before a dynamic simulation of a building is ready for use.

EnergyPlus interfaces well with third-party software tools, so that graphical outputs of the building geometry are created in either DXF format that can be read by many CAD programs, or in WRL format that can be viewed in by VRML viewers in web browsers with an appropriate plug-in. Numerical outputs are generated in CSV format (comma separated values) compatible with many spreadsheet programs.

EnergyPlus is an open system that encourages development contributions from individuals. It contains comprehensive developer guides for modules, user interfaces, and guidance on programming standards. This ensures engagement from the building simulation community and expansion of the program's capabilities driven by individual users. The software is developed in

FORTRAN programming language using a modular approach. New modules can be developed independently of the overall system, enabling individual developers to focus on key features of their modules, without the need for detailed knowledge of the overall program. This open framework, combined with free download, ensures a continuously increasing scientific and technical capability and superiority in building simulation.

A starting point in the preparation of an EnergyPlus model is a creation of an input definition file (IDF file). This is done with an IDF editor (Figure 18) that enables the user to set simulation parameters, building geometry, materials, composition of multi-layered walls and other components, and make a choice and specification of systems, controls, and simulation outputs. Each new definition is handled with entities called objects, which are inserted, populated with required input parameters, and in most cases linked to other objects, thus collectively specifying the dynamic simulation model.

To run the model, the user needs to run a launcher called EP-Launch, in which the IDF and the weather data files are specified (Figure 19).

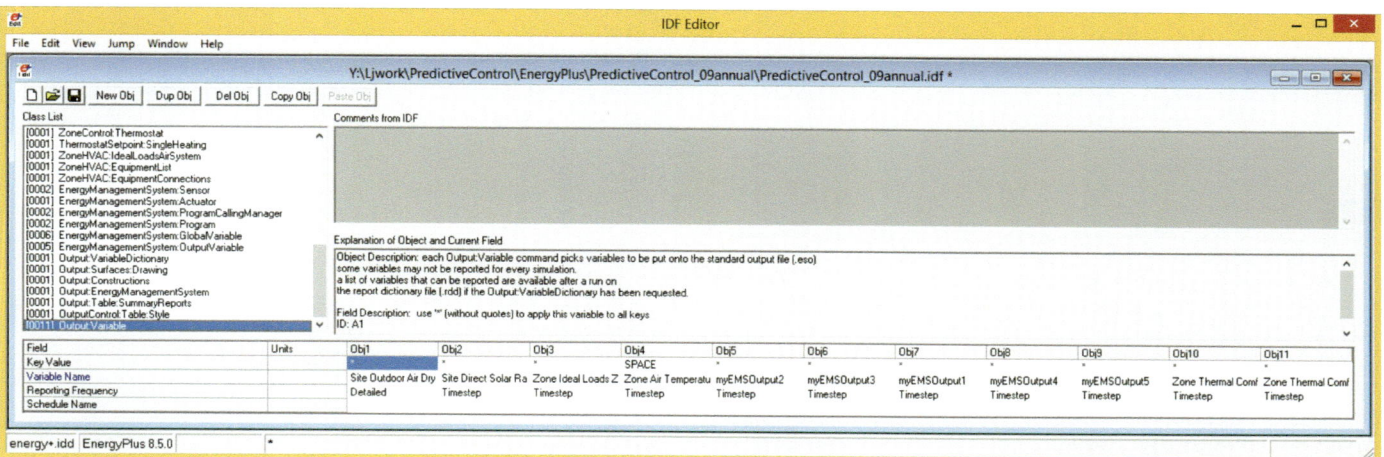

Figure 18   EnergyPlus input definition file editor

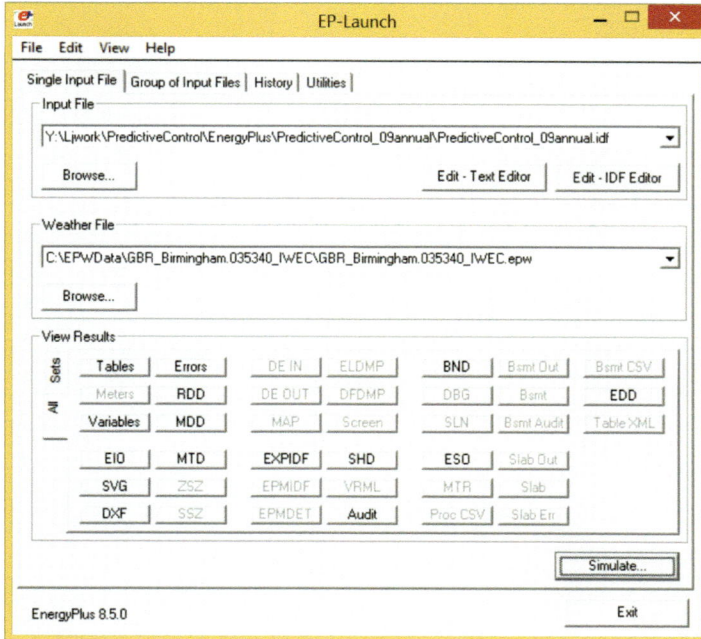

Figure 19   EnergyPlus launcher

There is a range of outputs that can be generated during the simulation. An example of a numerical output in CSV format loaded into Excel spreadsheet is shown in Figure 20. In this particular example only two output parameters were specified in the IDF file, and thus the output file shows the time step and the two parameters in hourly time steps. An example of a DXF drawing file is shown in Figure 50 in Chapter 4. The user can also display an error log in case the simulation has been terminated as result of an error, and this will contain clues as to how to resolve the error.

Energy Management System (EMS) Scripting is an advanced feature in EnergyPlus, running under EnergyPlus Runtime Language (Erl). It enables users to develop their own functionality and implement it in an EnergyPlus simulation model via an IDF file. A family of EMS classes are available for that purpose within the IDF editor. A simple user application might involve Energy Management System: Program, which defines bespoke functionality required by the user; Energy Management System:Sensor, which links Output variables to Program; and EnergyManagementSystem:Actuator, which overrides classes, such as Schedule, using values calculated by Program. An example of a simple room thermostat model in the EMS Program is shown in Figure 21.

Coding in EMS is recommended for experienced programmers only. As the language is still under development and with minimum debugging facilities, coding in EMS can be a hard task, but ultimately it provides bespoke functionality not found in many other simulation tools.

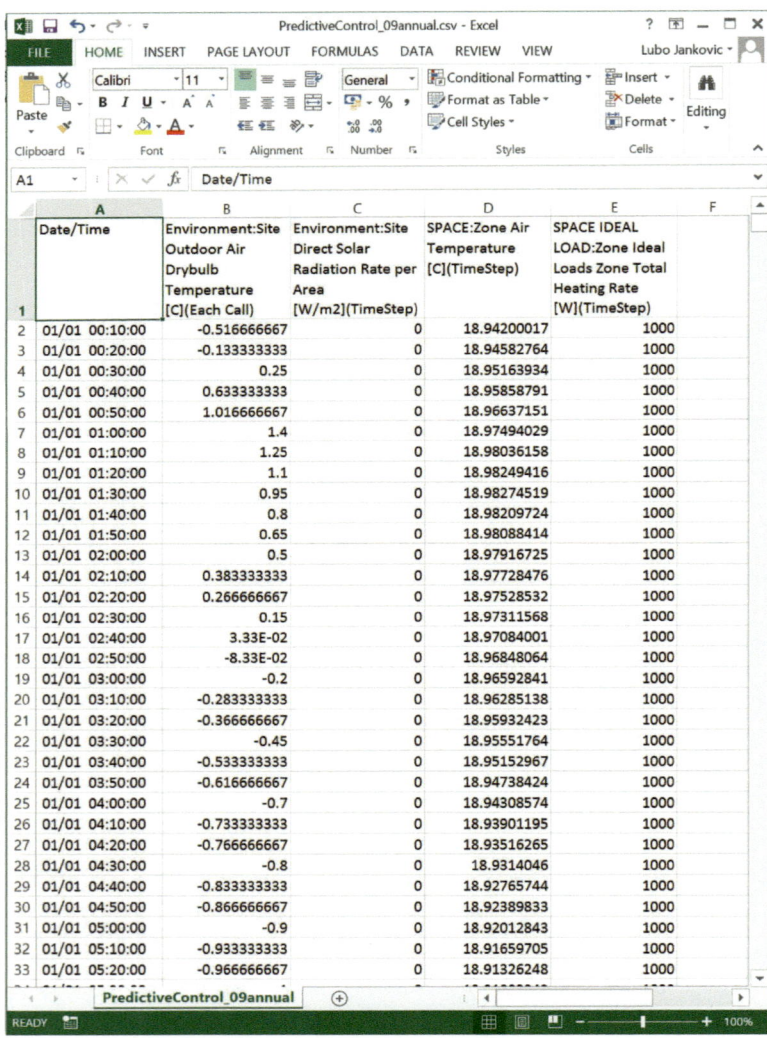

Figure 20   EnergyPlus output in comma separated values (CSV) format viewed in Excel spreadsheet

```
EnergyManagementSystem:Program,
 Heating_Manager ,              !- Name
 SET deltaT = 2.0,             !- Program Line 1
 SET Tz = Tzone,               !- Program Line 2
 SET Tmax  = 19.0,            !- A3
 SET Tmin = Tmax - deltaT,!- A4
 IF (Tz > Tmax),              !- A5
  SET Heating_Status = 0, !- A6
  SET SetTEmperatureAdjustment = 5.0 !- A7
 ELSEIF (Tz < Tmin),         !- A8
  SET Heating_Status = 1, !- A9
  SET SetTEmperatureAdjustment=25.0, !- A10
 ENDIF;                       !- A11
```

Figure 21   Room thermostat code in EnergyPlus EMS

Strengths. EnergyPlus has the most comprehensive coverage of materials and systems. Detailed warning and error reporting during the run time will help the user overcome the absence of a more sophisticated user interface. Fast execution times of simulation are measured in seconds. The program has been conceived with expansion in mind, making it easy for third parties to develop their own modules and thus contribute to increasing strengths of EnergyPlus. The addition of runtime language Erl which enables CMS coding, EnergyPlus can be used for bespoke applications and experimentation. EPW data format has become the industry standard, compatible with most mainstream simulation software tools. EnergyPlus is available for free use, under a specific end user licence agreement.

Weaknesses. The absence of a comprehensive graphical user interface can sometimes lead to tedious model preparation process. Geometry needs to be entered numerically, vertex by vertex. The program is prone to human error in the model development stage, and these errors are only revealed during the run time. This extends to EMS coding, which can be very hard to develop by users not experienced in coding in other programming languages.

On balance, however, the strengths of EnergyPlus by far outweigh the weaknesses, making this program a truly superior simulation tool under continuous development.

3.5.3 IES Virtual Environment
IES Virtual Environment (IES Ltd, 2022) is a dynamic simulation modelling system that originated in the UK in the mid-1990s and is built around an idea of shared content between different simulation tools. The shared database enables the model specification to be entered only once and facilitates the integration between different applications within the simulation modelling system from which the database can be further expanded. The software has an extensive graphical user interface and a range of modules for the simulation of energy, air movement, lighting, HVAC systems, and others.

A modelling project would typically start by entering geometry using a tool called ModelIT (Figure 22). The user can view geometry in 3D (Figure 23) as it develops, and this function remains active in other modules, such as Apache for thermal simulation and RadianceIES for lighting simulation. Geometry can also be imported from Sketchup or Revit, or using a gbXML file.

The same geometry is subsequently used for solar shading and solar illumination analysis. Figure 24 shows a building viewed from the sun's position (top) and in a shadow casting mode (bottom). Solar shading information generated by SunCast is subsequently used in thermal simulation in IES Apache (Figure 25).

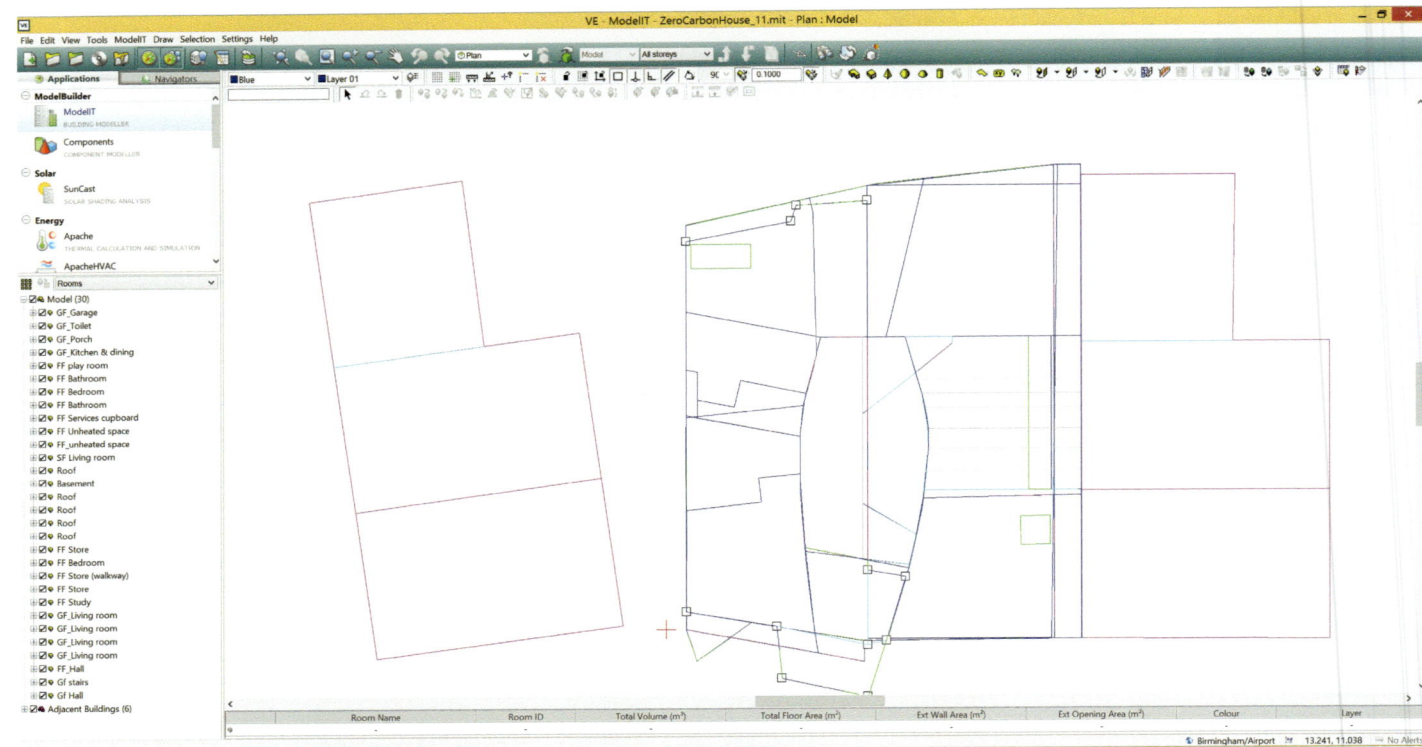

Figure 22   IES ModelIT tool for entering 3D geometry

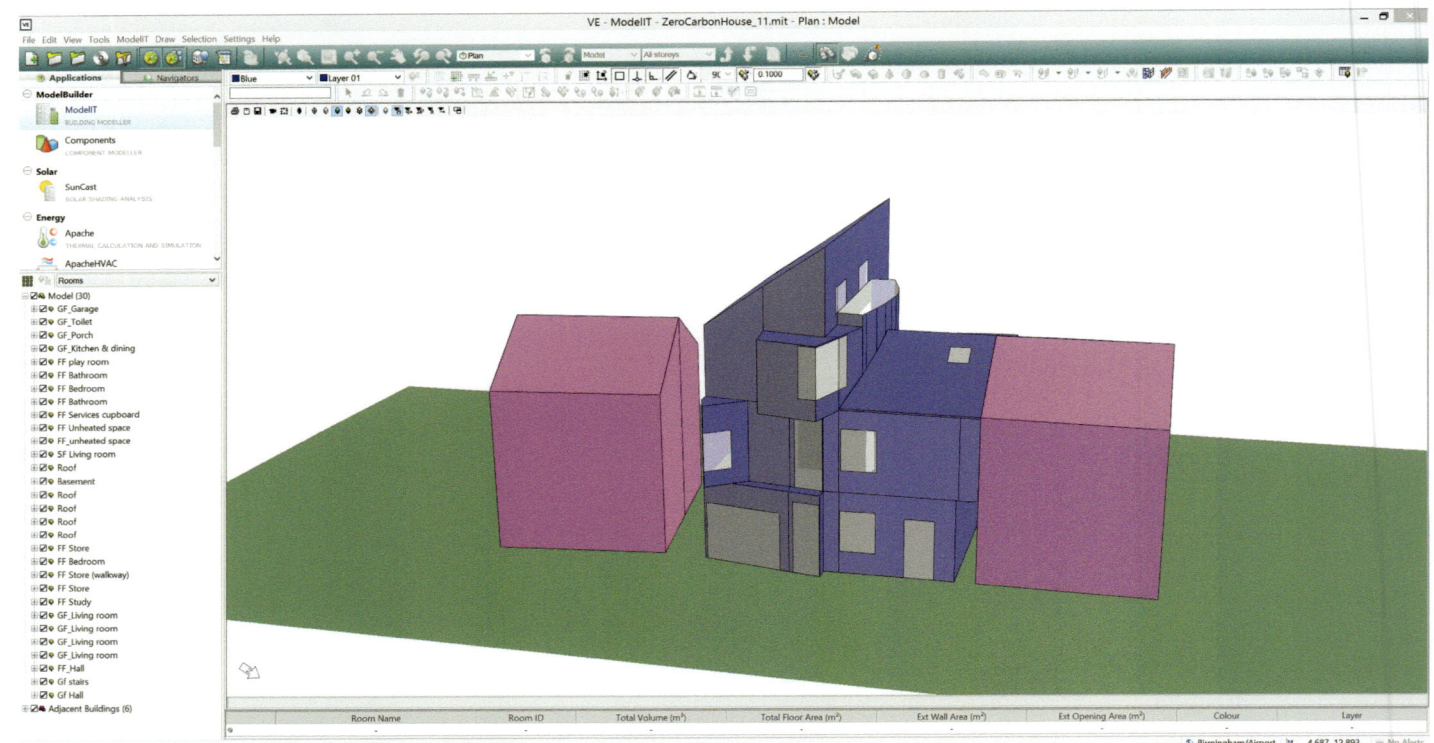

Figure 23   Viewing 3D geometry in IES ModelIT

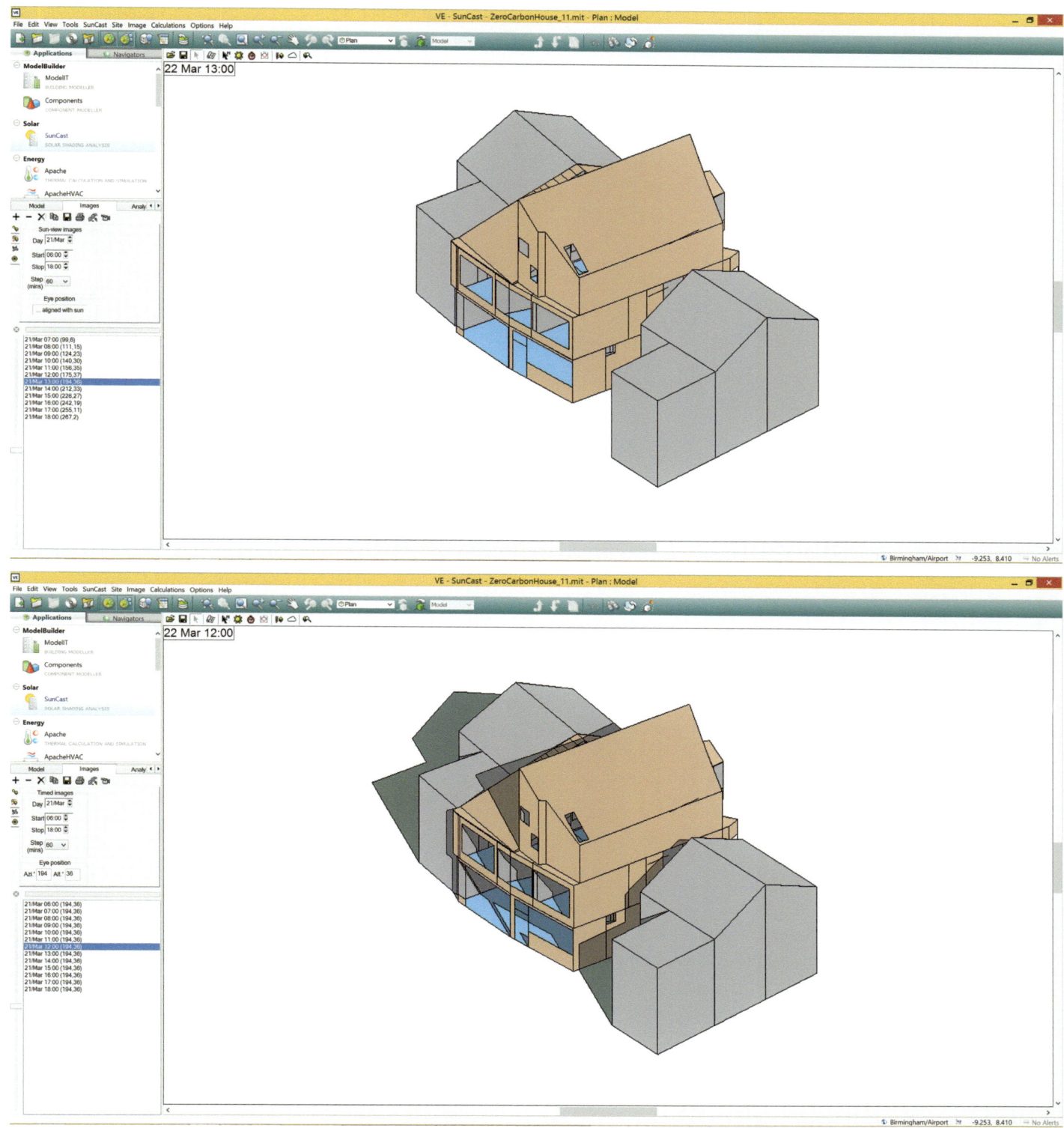

Figure 24   *Sun view image (top) and shadow casting image (bottom) in IES SunCast*

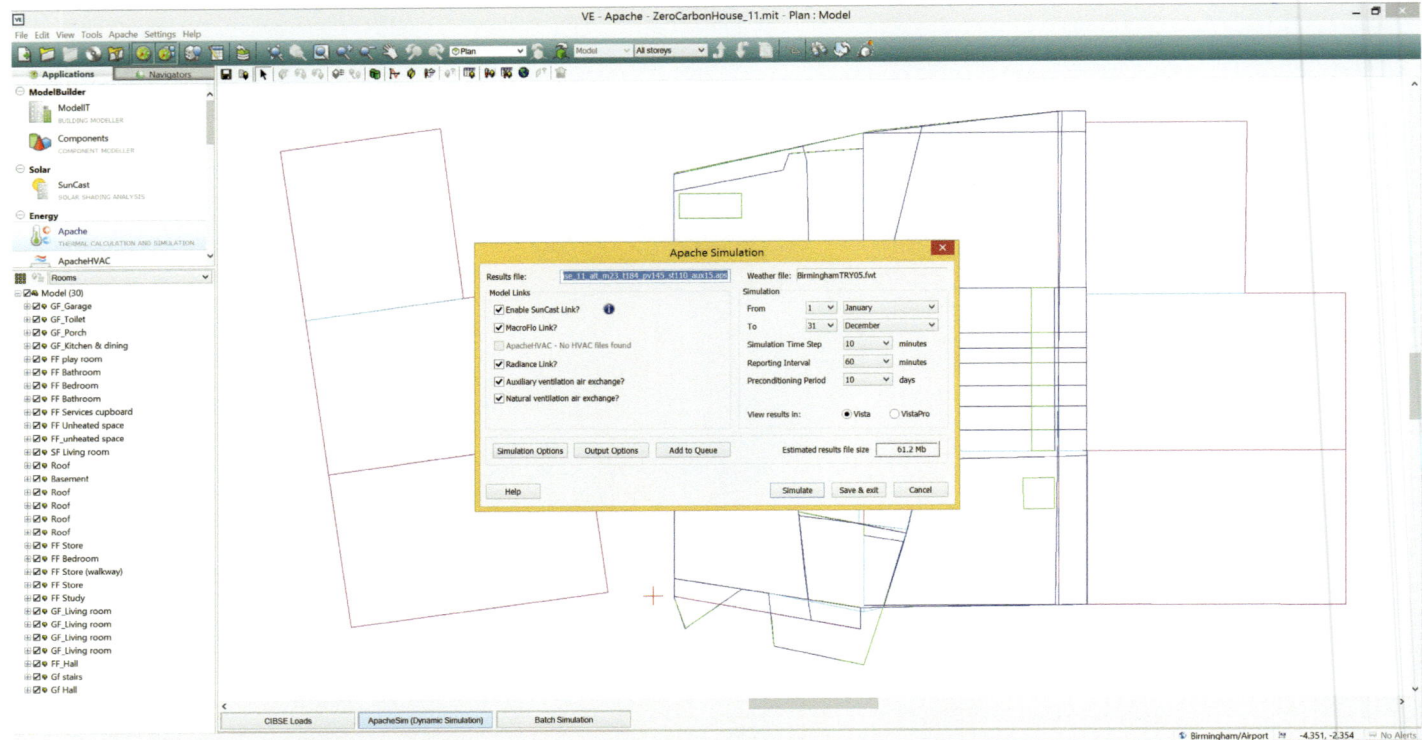

Figure 25   IES Apache module for thermal simulation

Prior to setting up the thermal simulation the user needs to set the heating and cooling systems, and templates for room conditions, including heat gains, applicable heating and cooling, internal gains, and air exchanges (Figure 26).

Operation profiles for occupancy, lighting, window opening, and others are set using APpro, a profiles database (Figure 27). This enables the creation of daily, weekly, and annual profiles (schedules), which are then used in various parts of the simulation model.

After various settings have been specified, the user is ready to conduct thermal simulation. In the Apache Simulation control dialogue (Figure 25) the user can enable links to other modules, such as SunCast, to take solar shading into account; MacroFlo to take into account bulk air flow movement; HVAC link to connect to an ApacheHVAC system; and Radiance link to take into account daylight sensor readings separately generated in a RadianceIES module. The Apache Simulation dialogue also enables the user to set simulation start and stop dates and times, simulation time step, a reporting interval and a preconditioning period during which the simulation is run in order to eliminate the influence of any arbitrarily chosen starting conditions on the simulation results, such as initial temperatures.

After the simulation is completed, the results are accessed through Vista (Figure 28), now superseded by VistaPro, from where numerous reports can be obtained as charts or tables.

In addition to thermal simulation, the IES VE can be used for lighting simulations in Radiance IES (see Chapter 13), computational fluid dynamics simulation of external air flow (Chapter 6), internal air flow (Chapter 12), and others.

Strengths. IES VE provides a number of simulation tools that share the same information and thus increase the efficiency of simulation projects as well as time efficiency of designers. For instance, integration of Radiance, that exports daylight sensor information into thermal simulation and allows the thermal simulation to control electrical lights using daylight

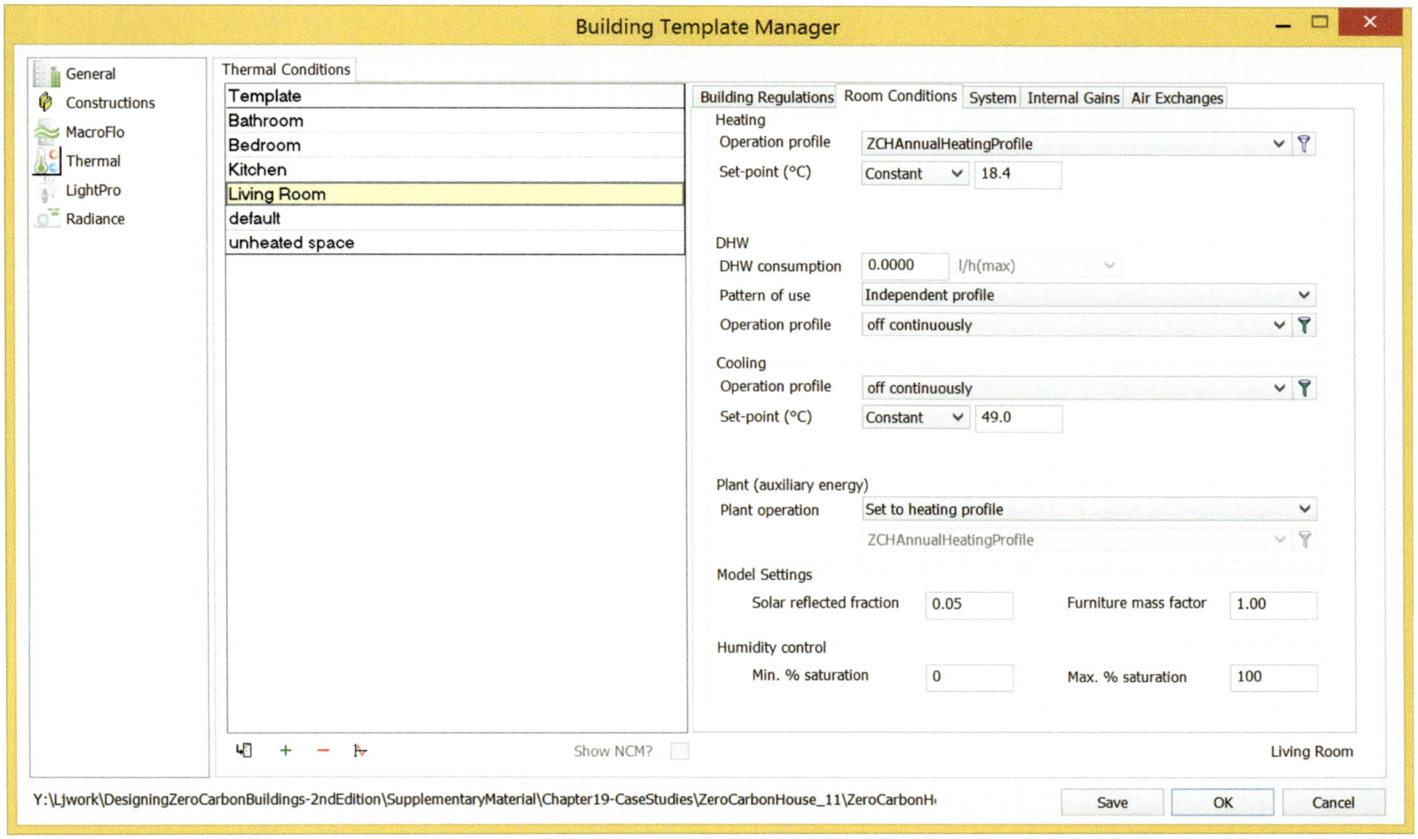

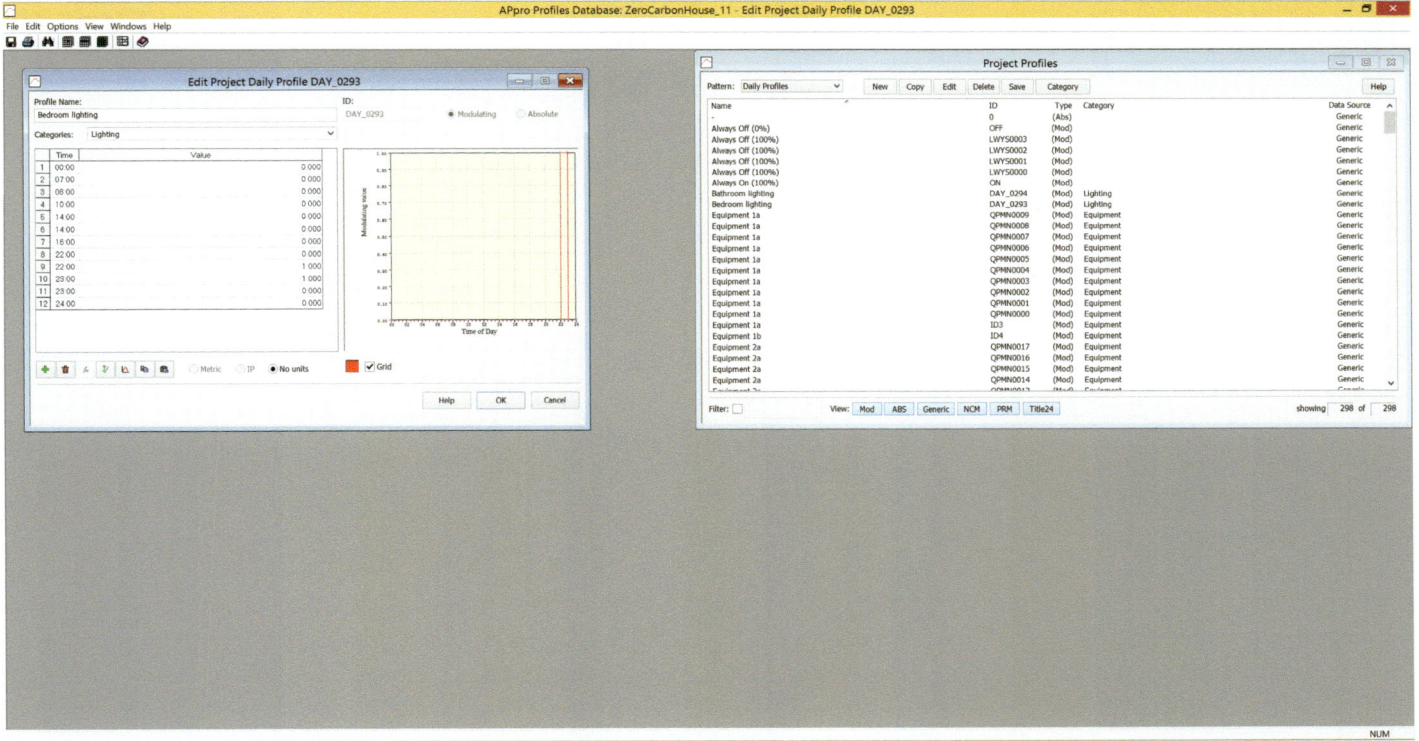

Figure 26 Building Template Manager used for setting building regulations, internal conditions, heating and cooling systems, heat gains, and air exchanges

Figure 27 IES APpro – Profiles database

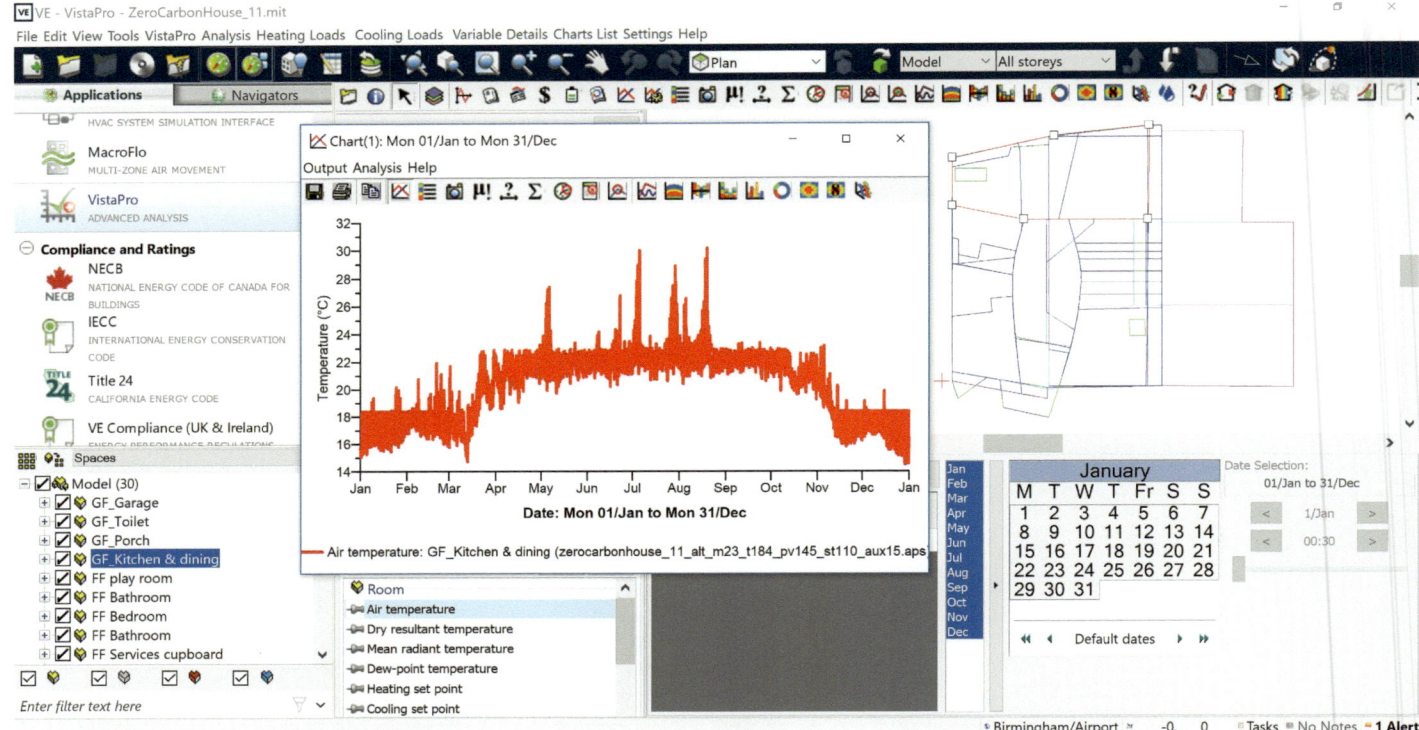

Figure 28   Access to simulation results through IES VistaPro

sensitivity, enables the designer to optimise the building within the same suite of simulation tools, thus improving design productivity.

Weaknesses. IES VE is very extensive and requires considerable experience to use it effectively. Unlike EnergyPlus, this is a closed system and no third-party modules can be easily incorporated in it. At the time of writing the software did not directly support transparent insulation, phase change materials, evaporative cooling and some other less usual but increasingly essential materials and systems that can be particularly useful for zero carbon design.

On the whole, IES VE has an excellent combination of modelling capabilities and the user interface, making it one of industry standard building simulation tools worldwide.

3.5.4 JEPlus/JEPlus+EA

JEPlus is a parametric analysis tool for EnergyPlus, developed for the audience of scientists, researchers, engineers, and other advanced end users (Zhang, 2020b). As explained earlier in this chapter, parametric simulations are single simulations that go through multiple values of one or more design variables. Parametric simulations provide results on each simultaneous combination of these values, representing points in the solution space. As it can be seen from Table 4, the solution space arising from parameterising a relatively small number of design variables can run into millions, and simulations of all points in this solution space can take a long time.

In comparison, it is worth noting that optimisation tools do not carry out simulations of all combinations of values of design variables – they do not do exhaustive search of the solution space. Instead, they use 'smart' shortcuts that can in turn lock them onto a path that leads to optimum solutions. Depending on the algorithm used, however, the global optimum solution solutions are not guaranteed to be found. Inversely, parametric simulations do exhaustive search of the solution space, as they simulate every single combination of the values of design variables. This guarantees to hit a global optimum solution when the steps between discrete values are sufficiently small, but it will take longer than optimisation analysis.

A parametric simulation project in JEPlus starts by selecting an EPW file, an EnergyPlus IDF file that has been previously tested in EnergyPlus, and EnergyPlus RVI file, which chooses the outputs of EnergyPlus simulations (Figure 29). The user also needs to set up parameter definitions: names, variation range, and search strings (see Parameter Tree in Figure 29). The variation range is set with a list of arbitrary values or use the shorthand syntaxes such as '[start value: step: end value]', and the parameter search string is enclosed with '@@' notation. In the particular example in Figure 29, the parameters were set as Building Orientation and Insulation Thickness.

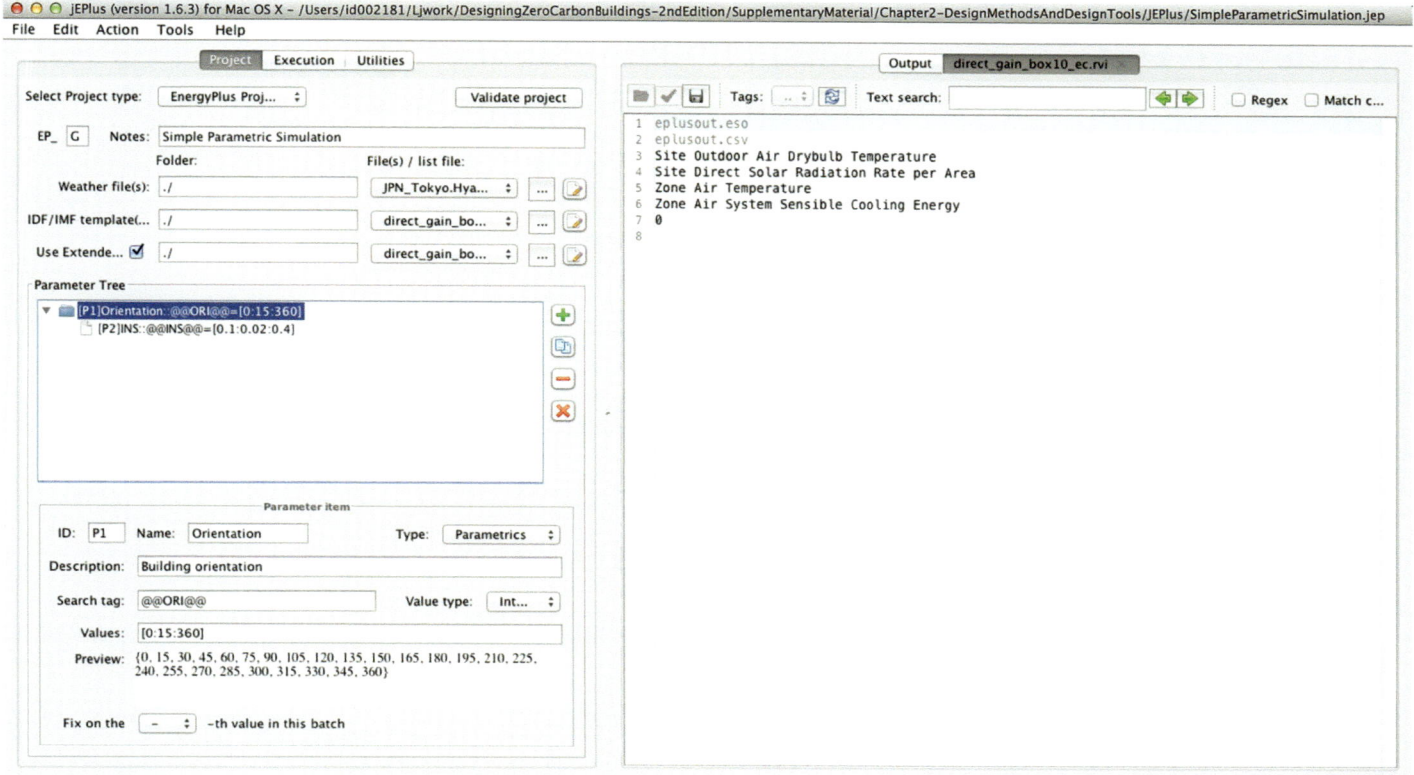

Figure 29   Setting up a JEPlus project

After the parameters and their variation ranges have been defined, the corresponding single values in the IDF file are replaced by the parameter search strings. Hence, a fixed value of 'North Axis' in the IDF file (Figure 30) is replaced with the parameter search string '@@ORI@@' (Figure 31), and the fixed value of wall insulation thickness is replaced with the parameter search string '@@INS@@'.

After all file paths and parameters have been defined, and parameter search strings placed in the IDF file, the user needs to save the modified IDF file and initiate the project validation. Upon successful validation, a message showing the total number of parametric simulation jobs will appear (Figure 32).

The user can then go to 'Execution' tab, start the parametric simulation, and follow its progress (Figure 33).
Upon the completion of the parametric simulation, the results are available in a comma separated values (CSV) file (Figure 34).

Each line of the output file contains the simulation results together with the parameters that define the case to which these results correspond. Therefore, upon selecting the output value, such as 'Sensible Cooling Energy' in this particular example, the user has the 'recipe' for this building design in the form of the corresponding design parameter set. Global optimum can then be found by performing calculations in the spreadsheet file, such as ordering the outputs, calculating minimum values, and finding parameters that correspond to the minimum.

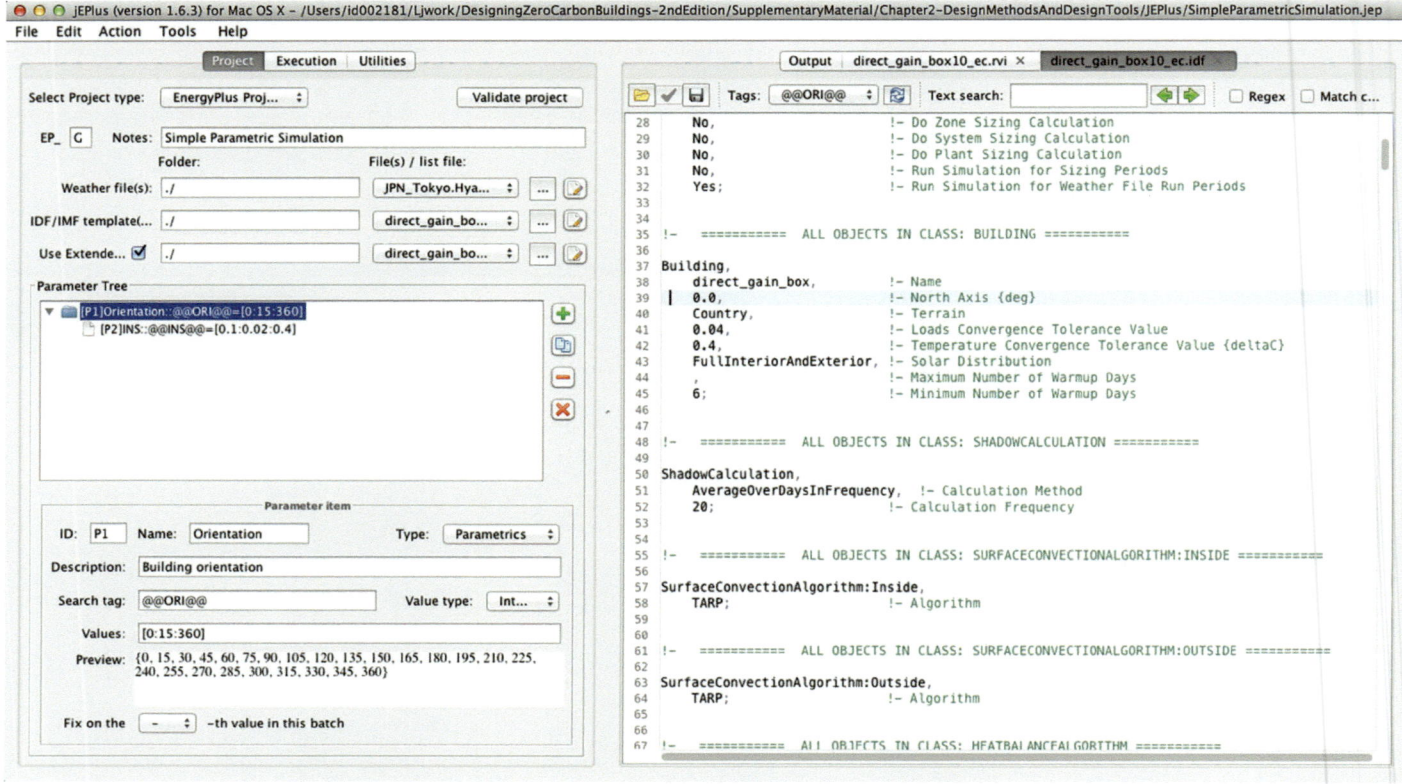

Figure 30   Finding a fixed value to replace in the IDF file

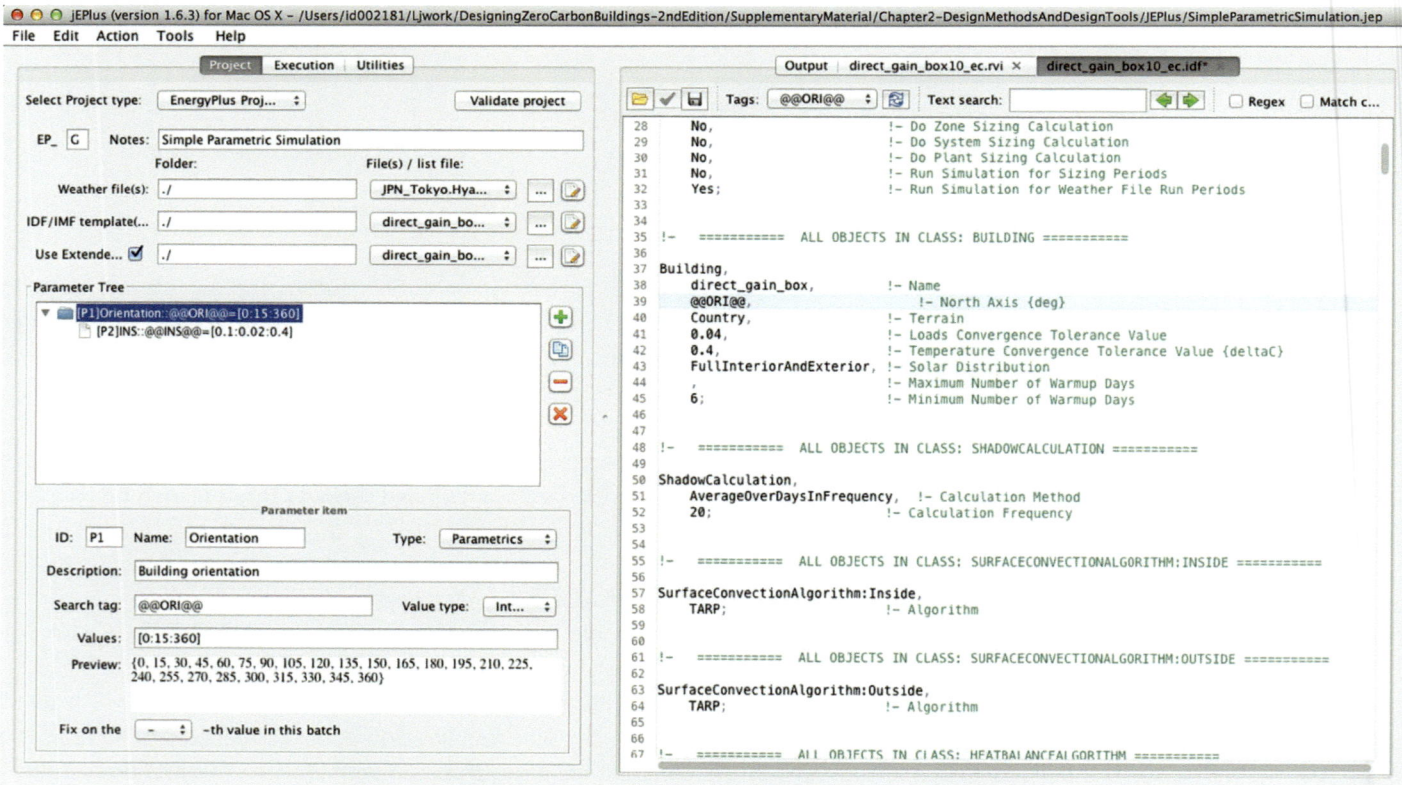

Figure 31   Replacing fixed parameters with corresponding search strings in the IDF file

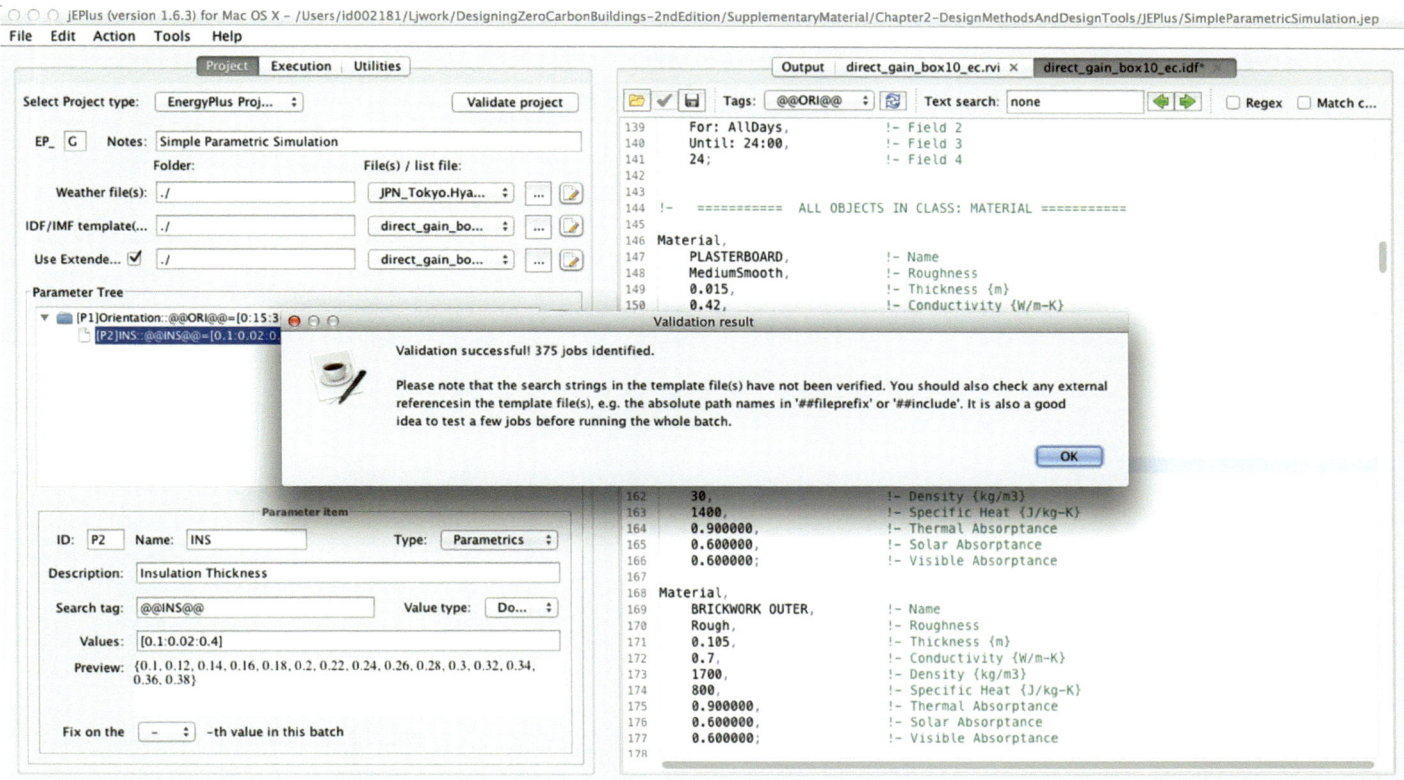

Figure 32   JEPlus project after successful validation

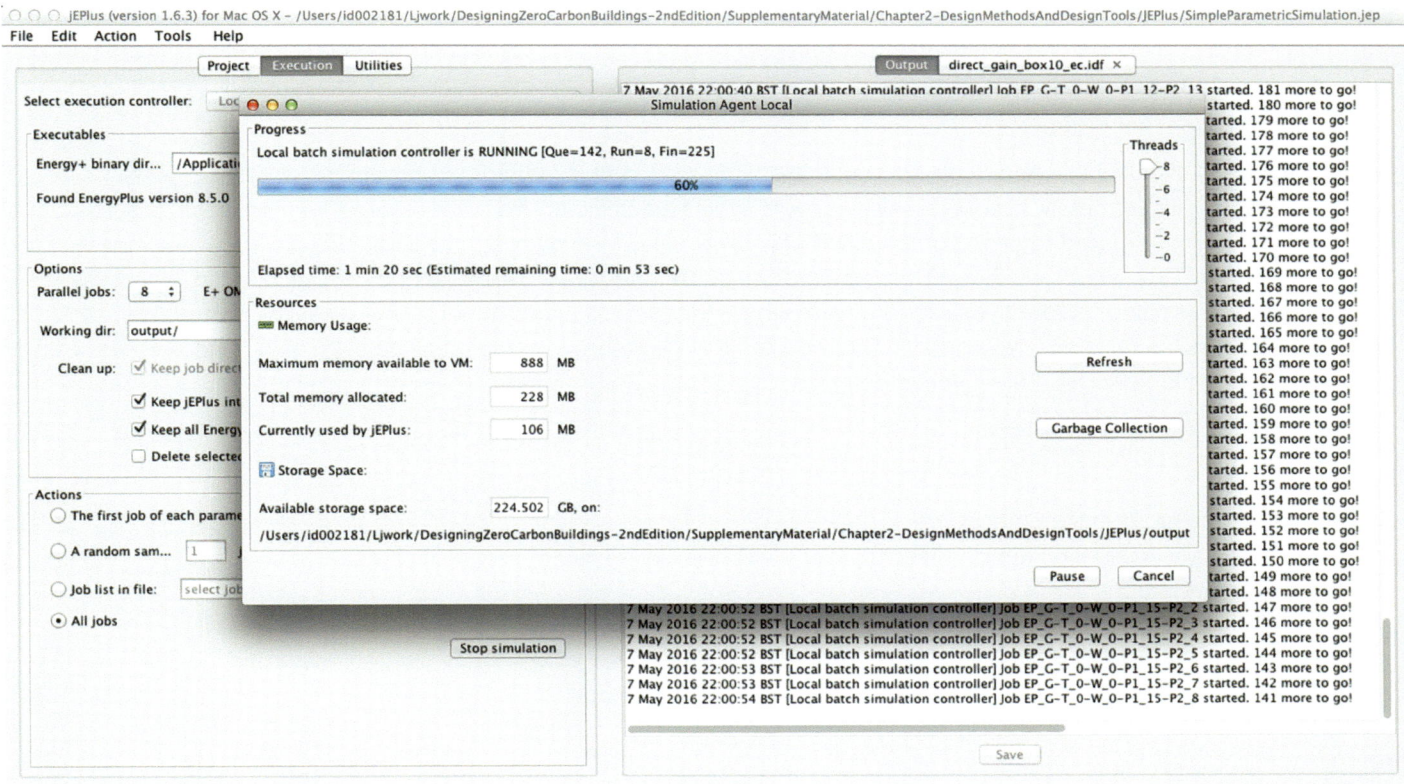

Figure 33   JEPlus parametric simulation in progress

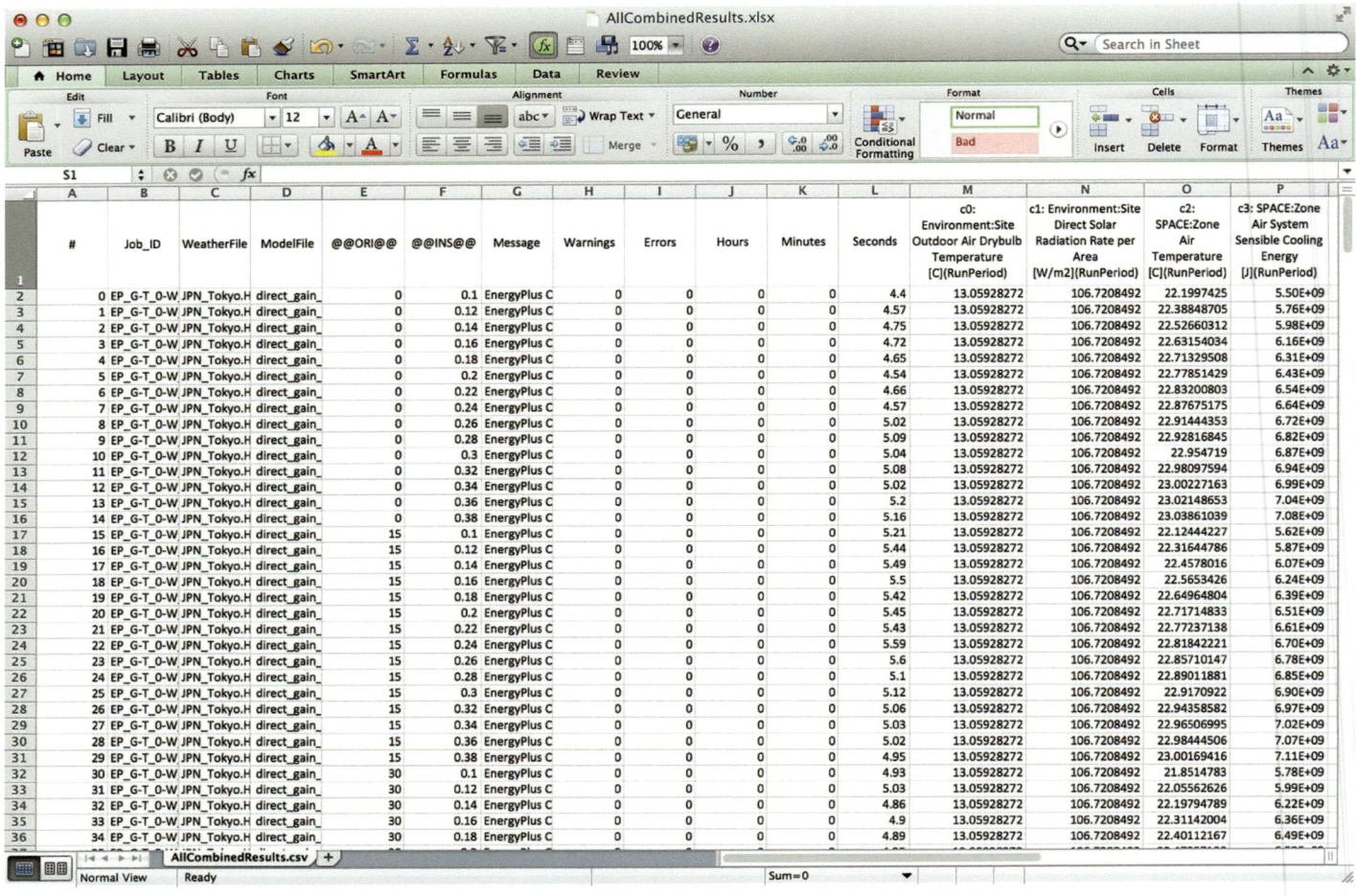

Figure 34 JEPlus results in a CSV file

JEPlus+EA uses evolutionary algorithms to conduct optimisation over a solution space defined by JEPlus (Zhang, 2020a). Whilst RVI file in JEPlus merely specified which outputs would be selected from the EnergyPlus simulation (Figure 29), in JEPlus+EA that file is also used to define optimisation objectives. Thus, in addition to the lines in Figure 29, we now extend the RVI file to include CO_2 emissions (line 7) and discomfort hours (line 8), as well as to define optimisation objectives (lines 11–14), using the syntax of <objective; unit; formula>, where the formula either refers to individual columns of JEPlus outputs, or combines these in a functional relationship. The modified RVI file is shown in the listing below (line numbers are not part of the original file and are inserted for reference only):

1   eplusout.eso
2   eplusout.csv
3   Site Outdoor Air Drybulb Temperature
4   Site Direct Solar Radiation Rate per Area
5   Zone Air Temperature
6   Zone Air System Sensible Cooling Energy
7   Environmental Impact Total CO_2 Emissions Carbon Equivalent Mass
8   Zone Thermal Comfort ASHRAE 55 Simple Model Summer or Winter
9   Clothes Not Comfortable Time
10 0
11 !-objectives
12 ASHRAE 55 discomfort hours; hrs; c5

13 CO$_2$ emissions; kg; c4

14 !-end objectives

The objectives block in RVI is replaced by the new RVX file, which gives additional flexibility in defining evaluation metrics.

When the JEPlus project is modified using the RVI file above and loaded into JEPlus+EA, the original IDF file and the parameter search and replacement strings are shown as design parameters, and the optimisation settings from the RVI file are recognised as optimisation objectives in the JEPlus+EA project (Figure 35). The tool is designed to work with an external simulation server, with user login dialogue contained in the project dashboard (Figure 35).

After the optimisation gets started, the progress is displayed textually (Figure 36 top) and graphically (Figure 36 bottom), creating a Pareto chart gradually until the process is completed (Figure 37). The final Pareto chart then becomes a design decision-making tool, giving a trade-off relationship between the objectives and cross-referencing each point to the parametric case that has created it. Choosing a desired point on the Pareto chart and clicking on it creates a cross-reference to the combination of parameters that created that point, effectively representing a recipe for a corresponding design.

Despite the inherent risk of the genetic optimisation of potentially locking in a less optimum region of the solution space, the advantages over parametric optimisation become obvious by comparing the outputs of JEPlus and JEPlus+EA. Whilst the former produces an exhaustive list of outputs from each individual parametric simulation, multi-objective optimisation takes the analysis of building performance to a completely new level. Instead of leaving the user to search for the optimum manually through the parametric simulation outputs, multi-objective optimisation searches the solution space using simultaneous influence of the functional relationship between design variables and design objectives. In addition to providing an automated search process of the solution space, multi-objective optimisation also provides the functional relationship between the objectives in the form of a Pareto chart. The chart is very useful as an enhanced design decision-making tool, as it enables the user to visually explore the optimum solutions, and the trade-offs between design objectives.

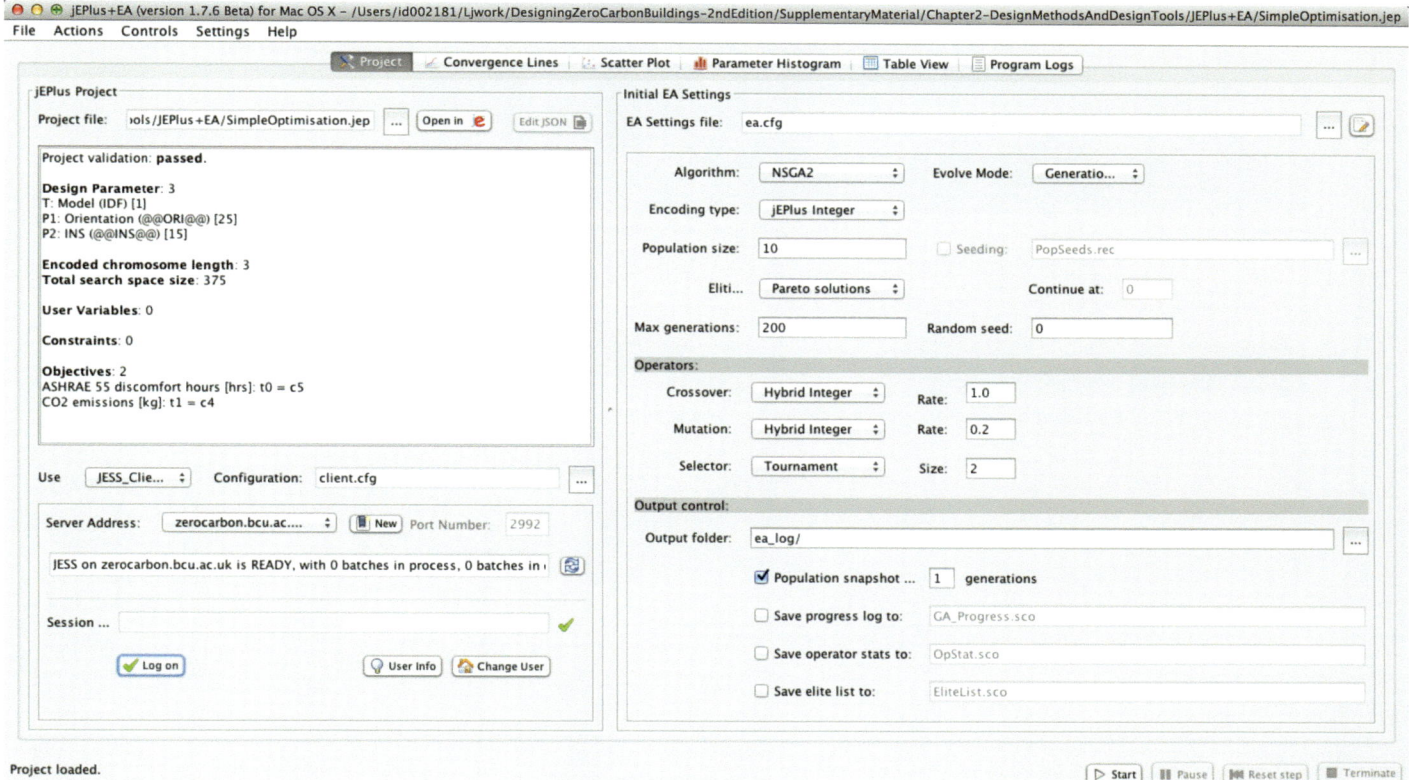

Figure 35   JEPlus+EA optimisation project using modified RVI file to define optimisation objectives

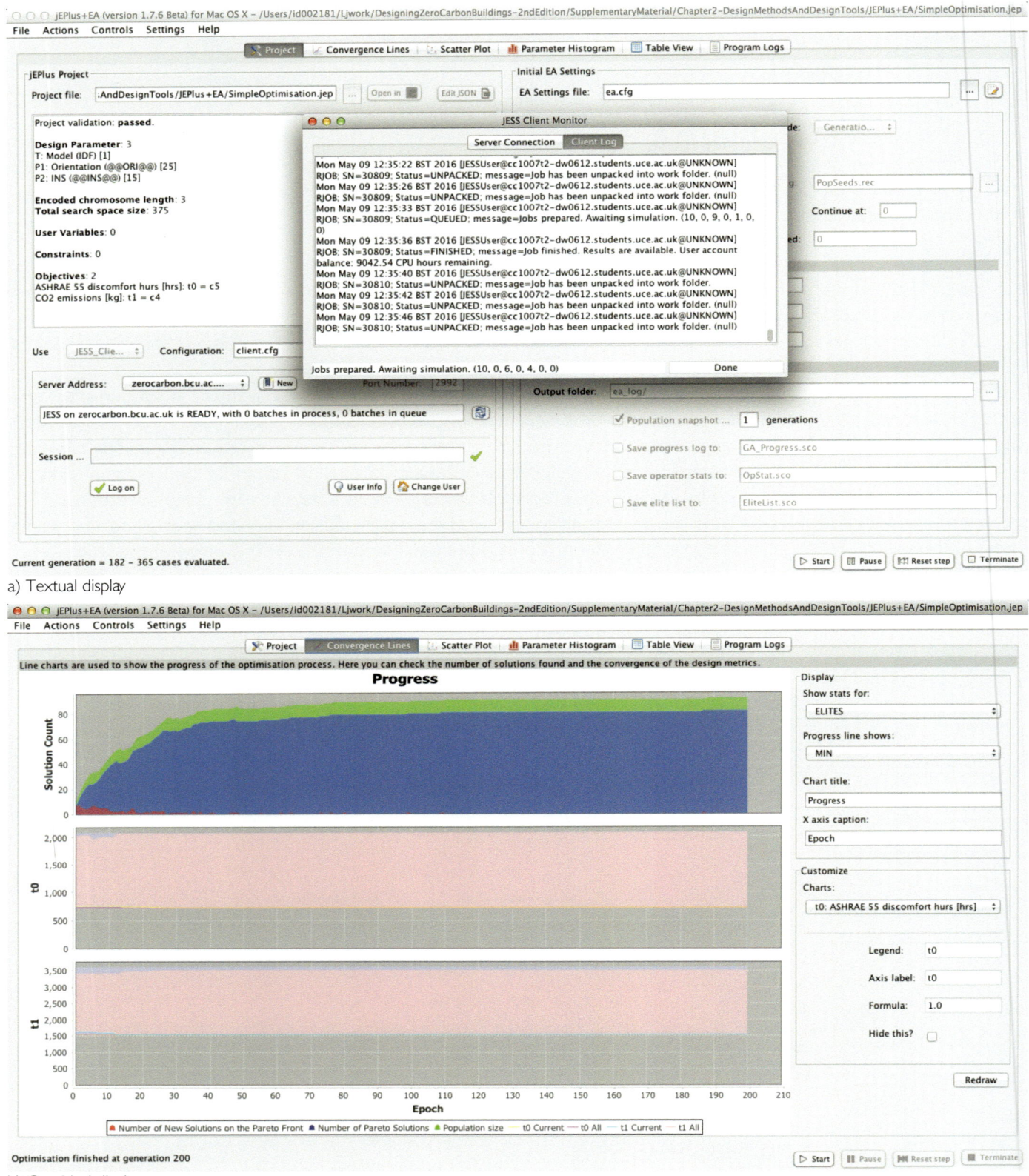

a) Textual display

b) Graphical display

Figure 36   Progress of JEPlus+EA optimisation

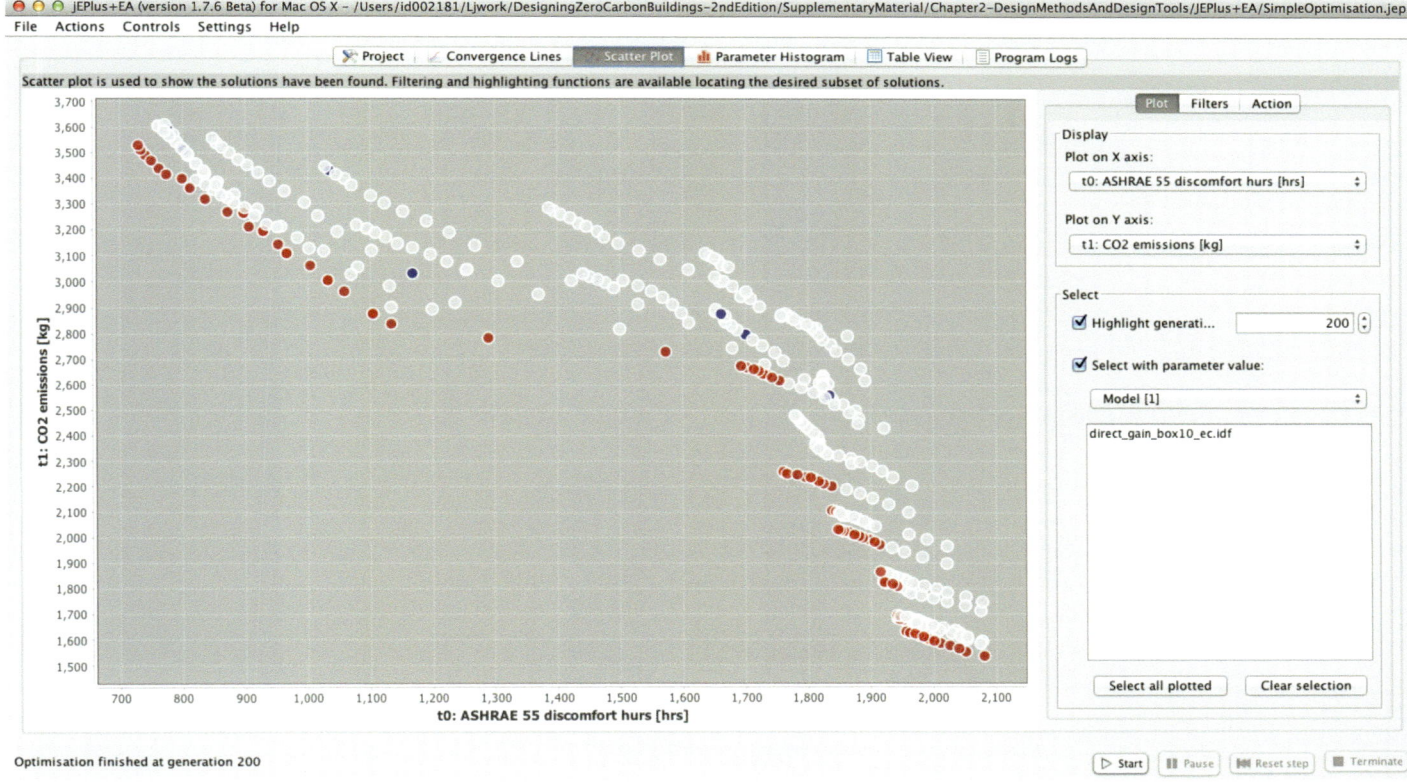

Figure 37   Pareto chart as one of the final outputs of JEPlus+EA optimisation

Strengths. JEPlus and JEPlus+EA are very powerful analysis tools, which enable the user to run parametric simulations, and multi-objective optimisation using evolutionary computation. They provide the user with a considerable flexibility of analysis that is not easily attainable in other software packages. This includes the capability to create bespoke objective functions using formulae. The optimisation process is designed to be used with external computing resources including dedicated simulation servers and cloud computing services, thus minimising the load on the local CPU. The tools are provided under a free licence.

Weaknesses. JEPlus and JEPlus+EA are designed for an expert audience. The users need to have experience of EnergyPlus and some coding capability in order to be able to use these tools effectively.

On balance, JEPlus and JEPlus+EA provide unprecedented parametric simulation and optimisation capability that enable analysis and design of building performance to be extended beyond the current practice.

3.6 CREATING A DYNAMIC SIMULATION MODEL

The creation of a dynamic simulation model may appear to be a complicated process; however it can be considerably simplified by breaking it down into several tasks. First, the user needs to define the geometry of the building in three dimensions. Second, the site location needs to be specified together with the associated weather data file. Third, building construction types need to be defined, including all building components such as walls, floors, ceilings, windows, and doors, and these need to be assigned to the corresponding building geometry. Fourth, building use patterns need to be defined, including room conditions, internal gains, and infiltration and ventilation air exchange. Fifth, heating and cooling systems need to be defined and associated with corresponding parts of the building.

Within these five tasks, other subtasks need to be performed, depending on specific requirements related to individual buildings. We will address these subtasks in the context of individual subject areas in the chapters of this book.

Commercial DSMs contain ready-made generic models of buildings. Modelling therefore involves filling in the 'blanks' in the generic model, to make it specific for the analysed building. However, any remaining 'blanks' filled in by the software as default parameters can considerably affect the results.

Despite the simplified process of the creation of a dynamic simulation model described here, and in the context of the default parameters used by some models, the complexity of modelling should not be underestimated. Dynamic simulation is to some extent comparable to playing the violin: one needs to practice a lot, learn a lot, and use all of one's skills to perform well. This book is intended to point the reader some way towards achieving this capability.

3.7 PLANNING AND RUNNING SIMULATIONS

Before running any simulations, we must first define the design objectives, and then define design variables and their values which will be investigated in order to achieve these objectives. Depending on the number of simulation cases to be investigated and the approach to this investigation, we can either carry out a single simulation, or a series of parametric simulations, or optimisation that takes into account a large number of simulation cases.

3.8 DESIGN OPTIMISATION

3.8.1 Single simulations and parametric simulations

One of the best ways to plan simulations is to list all design variables that need to be investigated during the simulation into a table. These variables are assigned discrete values within a specified range during the simulation. These discrete values effectively represent what-if scenarios, and assigning a single value to a single variable at a time will enable us to find out the effect of that variable on building performance. An example of a plan for a simple simulation analysis from a real-life project is shown in Table 3.

TABLE 3 SIMULATION TABLE – EACH CASE FROM EACH ROW IS SIMULATED WITH EACH OTHER CASE FROM EACH OTHER ROW					
Design variable	**Values assigned to design variable**				**No. of cases**
Building type	Datum type: masonry, cavity wall, insulation as per building regulations	Heavy weight type: all masonry, super insulated	Light weight type: timber frame, masonry core, super insulated	Hybrid type: masonry core, timber frame skin, super insulated	4
South glazing increase by (%)	50	100	150	200	4
Density of masonry (kg/m³)	650	1,300	2,000		3
Mechanical ventilation (m³/day)	11,000 no heat recovery	27,000 no heat recovery	27,000 plus heat recovery		3
			Total number of cases (product of the above) =		144

A **single simulation** run is carried out with a single set of design variables and single values for these design variables. This is useful if no comparison between different design options is required, although this is very rarely the case.

Multiple simulation runs are carried out when a comparison between design options is required. In such cases, each value assigned to each design variable, as in Table 3 for instance, is considered to be a parameter, and combinations of all these parameters through several single simulations is called **parametric simulation**.

Here is an example of investigation of energy performance of four building types at the design stage, with an objective to select the best building type and related parameters. Figure 38 shows a simulation analysis of four different construction types. One type needs to be chosen to build an estate of houses. The simulations of each construction type were conducted using an annual weather data set for the location in which the houses were to be built, and with a number of assumptions, as set out in Table 3. The results of the simulations were then interpreted and supplied to a design team. From the results of this analysis (Figure 38), the advice given to the design team was to use the hybrid construction type.

Another simulation analysis taken from Table 3 had the objective to establish the optimum size of the south-facing windows (Figure 39). When the surface area of the windows is too small, the building does not receive enough solar energy, and therefore the annual heating energy consumption is higher than the optimum. If the window sizes are increased in small steps, the building will be receiving more solar energy than the amount of heat losses through the windows, and the conventional heating energy consumption will be decreasing. After the optimum window size is reached, increasing the surface area of windows will make the heat losses through windows greater than the heat gains from solar radiation, and the conventional heating energy consumption will be greater. From the results in Figure 39, it appears that the optimum window size is 150% of the original size, and therefore the advice given to the design team was to increase the window size by 50%.

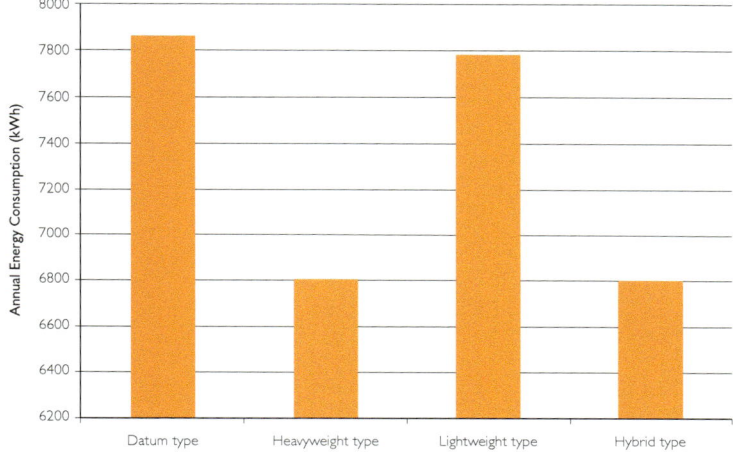

Figure 38   Simulation analysis of four construction types

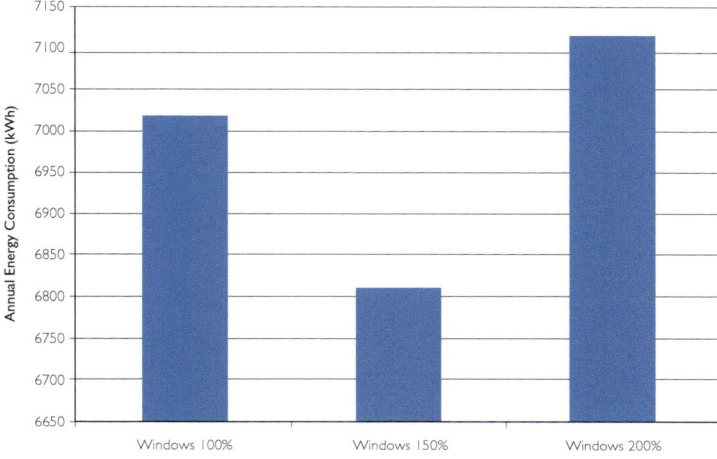

Figure 39   Simulation analysis of south-facing window sizes

3.8.2 Single-objective and multi-objective optimisation

Depending on the number of combinations of different parameters, parametric simulations can produce a very large number of single simulation outputs, and searching through these outputs manually may become very time-consuming and in many cases an impossible task. This is where optimisation search methods become very useful. There are two types of optimisation: single-objective and multi-objective optimisation, where the objective is a function that needs to be optimised.

Optimisation is a process of finding either a minimum or a maximum of a function, by finding values of independent variables for which the objective function reaches its maximum or minimum value. In the context of building simulation, the independent variables are design variables. Optimisation always seeks a global minimum ($y = y_{min, G}$) or maximum, as opposed to a local minimum ($y = y_{min, L}$) or maximum that represents a sub-optimum, in other words 'not so good' performance (Figure 40). The simplest case of optimisation is a single-objective optimisation in which the objective function has only one independent variable. For instance, Figure 39 shows the process of finding the best solution for energy consumption of a building, depending on the south-facing window size. The best solution in that particular case, the minimum energy consumption, is achieved for a south-facing window size that is half way between the initial window size and the window size that is doubled. In that example, the single-objective function is the annual energy consumption, and the independent variable is the window size. The result of the single-objective optimisation, which is carried out manually in that example, is the value of the independent variable, the window size that achieves the optimum, in that case the minimum energy consumption.

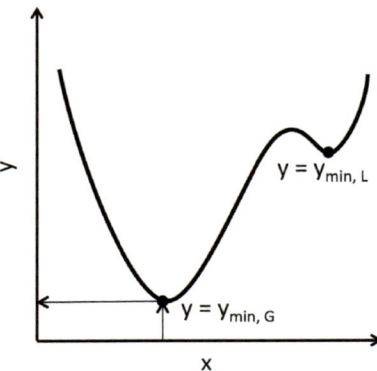

Figure 40   Local and global minimum of a function with one independent variable

Single-objective optimisation with a single independent variable is relatively straightforward. Unless there are any local minima where the optimisation process could get locked into (Figure 40), it will quickly proceed to finding the best solution.

Multi-objective optimisation will be typically looking at two or more objective functions, such as energy consumption and cost, and each of these objective functions will have its own or shared independent variables. These multiple objectives and corresponding independent variables will create a multi-dimensional solution space, in which optimisation is far from trivial.

As different combinations of design variables can amplify each other (for instance volume to surface ratio and glazing surface area) or oppose each other (for instance thermal insulation and glazing surface area), optimisation will analyse their combined effect, thus giving an insight into building performance that cannot be achieved by single individual simulations.

3.8.3 Multi-dimensional solution space: 'God's view' and 'Pedestrian's view'

Let us consider two objective functions: (1) to minimise operational carbon emissions and (2) to minimise discomfort. The independent variables shared between these two objectives could include those shown in Table 4.

This large solution space containing over eight million variations of a single design will be optimised using evolutionary computation methods that search such large spaces quickly, in order to achieve minimum CO_2 emissions and minimum life cycle cost normalised to the present value.

As there are ten independent variables in this example, the solution space therefore becomes ten-dimensional. In order to visually illustrate the problem, and as it is difficult to visualise the ten-dimensional space, a three-dimensional analogy will be used for explanation. In this analogy, the solution space consists of hills and valleys (Figure 41) and the highest hill or the deepest valley will be the best solution. If we had 'God's' view (Figure 41a), it would be easy to see where the highest hill or the deepest valley is.

TABLE 4 AN EXAMPLE OF DESIGN VARIABLES INVESTIGATED IN MULTI-OBJECTIVE OPTIMISATION	
Design variables and values assigned to these	**No. of cases**
Site orientation (clockwise declination from north): min = 140°, max = 220°, step = 10°	9
External wall construction: U-value (W/m²·K): 0.26, 0.18, 0.10	3
Roof construction: Green roof U-value (W/m²·K): 0.10, 0.16, 0.25	3
Thermal mass: insulated ground floor concrete slab U-value 0.10, 0.16, 0.22; internal slab: timber floor, 100 concrete, 150 mm concrete	6
Airtightness (m³/h/m²): min = 0.5, max = 4, step = 0.5	8
Glazing: double, triple, quadruple	3
Electrical lighting power density (W/m²/100 lux): 3, 2, 1 uncontrolled, plus 3, 2, 1 linear dimming control (daylight following)	6
Window to wall ratio (%): min = 20, max = 80, step = 20	4
External window opening (%): min = 0, max = 80, step = 20	5
PV array: Six different sizes to facilitate optimisation	6
Total number of cases (product of the above) =	8,398,080

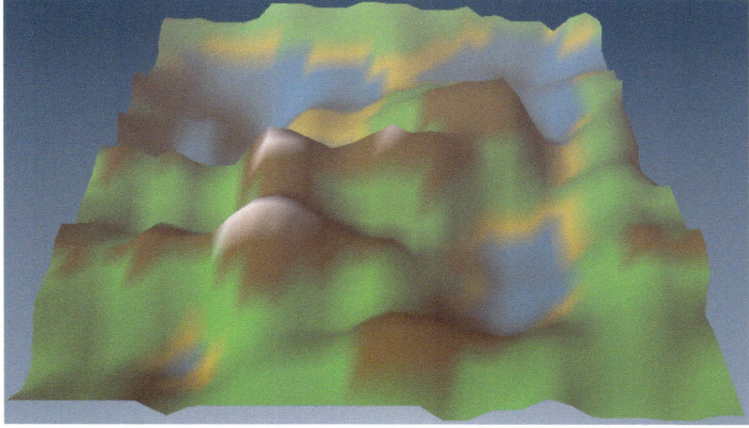

a)   'God's view'

b)   'Pedestrian's view'

Figure 41   Conceptual multi-dimensional solution space

However, when starting from the surface of the solution space, we in fact have a 'pedestrian's' view (Figure 41b), and the horizon might look the same in all directions, so that we will not know where to go to find the highest hill or the deepest valley. Following an initially promising path, our search for the optimum solution might lock into a local minimum, and result in model parameters that give sub-optimum performance.

There are many well-established numerical methods for dealing with this kind of problem of multi-dimensional optimisation, and one of the most efficient of these methods is based on evolutionary computation, namely on genetic algorithms. Building simulation tools with multi-objective optimisation capabilities primarily use a method by Deb et al. (2002) named NSGA-II: A Fast and Elitist Multiobjective Genetic Algorithm. This method will give us 'God's' view that we need in order to find the optimum solution in the multi-dimensional solution space.

In typical practice, designers only consider a small number of variations of design parameters, and carry out a small number of single simulations, perhaps not more than half a dozen or so, and a single-objective function, potentially leading to sub-optimum building performance and significant missed opportunities. In contrast, the above approach that potentially considers over eight million possibilities for a single design will give a range of solutions and an opportunity for trade-off choices between cost and CO_2 emissions, whilst ensuring minimum discomfort levels.

3.8.4 Tools for parametric simulation and multi-objective optimisation

The need to run building simulation with different values of design variables and to find optimum building performance from a combination of a number of alternative values of parameters has created a new range of tools, or new functionality within existing simulation tools. Thus DesignBuilder, JEPlus/JEPlus+EA, TRNSYS, IES Virtual Environment and others now have capabilities for parametric simulation and multi-objective optimisation.

3.9 VALIDITY OF SIMULATION RESULTS FOR DESIGN DECISION-MAKING

Until simulation models are experimentally validated, they cannot give absolute answers. The results can be completely wrong, as a number of default parameters in the simulation model may have not been set by the user. So why simulate? Because simulation has a great value as a comparative analysis tool. It is easy to tell from simulation analysis which design option is better, when compared with other design options from the same simulation model. What is not easy to tell is how good these individual options are in absolute terms. In this way the design team can be advised on the relative significance of design parameters and use the simulation results as a decision-making tool for trade-offs between various design options.

3.9.1 Simulation performance gap and how to deal with it

Differences between theoretical values and user assumptions in the simulation model on the one hand and actual properties of the building and the conditions in and around it on the other hand cause a discrepancy called performance gap.

The performance gap in existing buildings can be eliminated by a process of calibration of the simulation model using actual performance data. However, as simulations are predominantly conducted for design purposes, when the building does not yet exist, the calibration is not possible.

Findings by Menezes et al. (2012) suggest that some of the reasons for performance gap are shortcomings of simulation tools and poor assumptions made and implemented by their users. Studies by Carbon Trust, RIBA, CIBSE, and others (The Construction Wiki, 2022) divide the causes of performance gap into regulated energy (from fixed building services), unregulated energy (from plug loads, lifts, etc.), and poor commissioning, maintenance, and control.

As one of the possible ways of reducing the performance gap, Menezes et al. (2012) suggested a development of realistic building energy performance benchmarks on the basis of post-occupancy monitoring data. However, this approach would only provide information for the calibration of simulation models using single annual figures for heating and electricity energy consumptions. As explained in the previous section, that would help with improving accuracy of annual performance of simulation models, but it would not help with improving dynamic response to time-dependent inputs on an hourly basis.

I worked on an alternative approach, which creates a relationship between the simulated and monitored building performance of an existing building on an annual basis in the form of a 'digital filter'. After the filter has been created, its application to a non-existing building of a similar type 'morphs' the output of the simulation model into values that are equivalent to results of monitoring of that non-existing building, something that sounds intuitively impossible. This approach offers opportunities for development of dynamic benchmarks in the form of digital filters applicable on an hourly basis throughout the simulation year. As performance gap is particularly significant in the cases of buildings made from photosynthetic materials, such as lime-bonded hemp, there is experimental evidence that this approach can lead to significant improvements of accuracy of design analysis, leading to significant capital and operational savings in buildings (Jankovic, 2016). I explain details of this approach in Chapter 22, Section 22.1.1, entitled 'Using reverse modelling to reduce simulation performance gap'.

3.9.2 Model calibration

Calibration is a process of increasing the accuracy of a simulation model by comparing its output with measured data and adjusting its parameters in order to reduce the discrepancy between the two. The existence of 'measured data' suggests that this process applies to existing buildings, which either have energy bills, or are monitored with instruments in regular time intervals.

Calibration can be described as a process analogous to bracketing in artillery fire (Figure 42), so that the error 'bracket' is reduced by changing parameters of the model until desired accuracy is achieved.

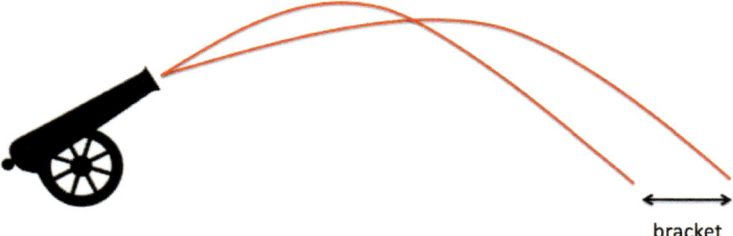

bracket

Figure 42   Calibration as analogous process to bracketing in artillery fire

The simplest way of calibrating the simulation model is to use measured annual energy consumption figures and adjust the model to produce outputs that closely match these figures. In the absence of any detailed performance monitoring, these measured figures can come from energy bills.

Whilst this type of calibration guarantees annual performance accuracy, it does not guarantee the accuracy of dynamic response of the simulation model to time-dependent inputs. In order to overcome this limitation, a second calibration is required, in which hourly outputs are compared with hourly data from instrumental monitoring, and root mean squared error between the two is minimised over the entire simulation period.

However, in cases of calibration of time-dependent simulation outputs, such as hourly room air temperatures, it is no good using library weather data files. Instead, the weather data file needs to be synthesised using hourly data from monitoring. In other words, actual weather data that created the building response needs to be used to drive the simulation model during the calibration process from within the weather data file used by the simulation tool.

EPW data files are suitable for this process, as they are well documented (Crawley et al., 1999), available for numerous locations around the world, and available in plain text format. The process of synthesising the weather data file takes the following steps:

1   Convert <weather_file_name>.EPW for the location of the simulated building into <weather_file_name>.CSV, where extension CSV means 'Comma Separated Values'. The file then becomes readable and editable by spreadsheet programs. Open the CSV file in a spreadsheet program.

2 Referring to weather data specification developed by Crawley et al. (1999), locate the columns in the weather data file to be replaced by data from monitoring. In the case of dry bulb temperature and relative humidity, the replacements are straightforward. However, in the case of certain other parameters, such as dew point temperature, direct normal radiation and others, these need to be calculated from available monitored data before the replacement, and that will require relevant expertise.

3 After all intended replacements of data columns are completed, save the file as <modified_weather_file_name>.CSV, rename it into <modified_weather_file_name>.EPW, and place the file in the directory of weather data files for the corresponding simulation tool.

This makes the weather data file ready for the purpose of calibration of the simulation model with reference to hourly time-dependent building performance parameters.

3.10 EXPERIMENTS WITH DYNAMIC SIMULATION IN THIS BOOK

We will use several test case models to demonstrate relevant principles in this book. One of the models chosen to be used in this book is based on a Greenpower Centre building, which opened in July 2010 (Figure 43). This building is a home for Greenpower Education Trust, an organisation promoting sustainable engineering to young people.

Figure 43   Greenpower Centre near Portsmouth, UK

As I was a member of a design team that created an initial concept and initial simulation model for this building, this enabled me to create new experimental simulation models based on this building and to use these to demonstrate steps towards zero carbon design. Other simple models will also be used throughout the book where appropriate, in order to demonstrate certain principles of modelling and simulation.

3.11 USEFULNESS OF DIFFERENT SIMULATION TOOLS

Which of these simulation tools is the best? There is no simple answer. Each system has strengths and weaknesses as described above. In order to have a comprehensive simulation capability, we might need to use all of them, sometimes even at the same time. For instance, I found it useful to use a geometry input tool from IES Virtual Environment when specifying coordinates of vertices of the same building in EnergyPlus. Most of my early work was in TRNSYS and most of my recent work has been in IES VE and DesignBuilder, but I also used EnergyPlus to investigate less usual experimental buildings. DesignBuilder is capable of multi-objective optimisation in addition to single simulations, and can fulfil a range of analyses required in a design project. JEPlus and JEPlus+EA offer extensive flexibility in defining and running parametric simulations and multi-objective optimisation using EnergyPlus as a base model.

So, there is no single tool that will fulfil all user requirements. These need to be used as part of a simulation toolbox, in which different tools fit different jobs, just like in a toolbox with spanners, pliers, and screwdrivers one tool cannot be easily replaced by another.

3.12 TASKS FOR SIMPLE EXERCISES

1 Download a free version of EnergyPlus.
2 Create a model of a box with dimensions of 10 m × 6 m × 3 m. Point the larger wall towards the south and install 40% glazing on it. Use materials of your own choice and thickness for walls, floor and roof.
3 Download an EPW file for a location of your choice.
4 Run an annual EnergyPlus simulation.
5 Change the window size to 60% of glazing on the south wall, rerun the simulation and compare the results with the first simulation.
6 Download a free version of JEPlus and JEPlus+EA.
7 Use EnergyPlus IDF file to create and run single-objective simulation with JEPlus and JEPlus+EA in order to optimise the glazing size, setting up the '@@...@@' search strings in the IDE file and setting the range for glazing size from 20% to 100%.
8 Discuss the findings with colleagues.

NOTE

1 https://www.buildingenergysoftwaretools.com/ (Accessed: 23 May 2023).

CHAPTER 4

WEATHER AND CLIMATE

The analysis of global issues is essential for understanding the context of the world we are living in and we are designing buildings in. This chapter will give insights into the world as it is, and as it will be.

4.1 SOLAR RADIATION

In this and the next section, we will show how we can use data from a DSM software to get a feel about the climate we are designing in. We will first look at solar radiation in this section and prevailing winds in the next section.

Solar radiation changes daily as a result of the earth's rotation around its axis and yearly as a result of its revolution around the sun. On 21st June, the day of the summer solstice, solar radiation is perpendicular on the ground located on the Tropic of Cancer, at 23°26' northern latitude, marking the beginning of summer in the northern hemisphere (Figure 44). On 21st December, solar radiation will be perpendicular on the ground located on the Tropic of Capricorn, at 23°26' southern latitude, when summer starts in the southern hemisphere. In addition to daily and seasonal variation, cloud cover and other atmospheric conditions will contribute to a further variation of solar radiation. We can find out about the solar radiation that reaches individual locations on earth, taking into account astronomic and atmospheric influences from weather data files used in DSM software tools.

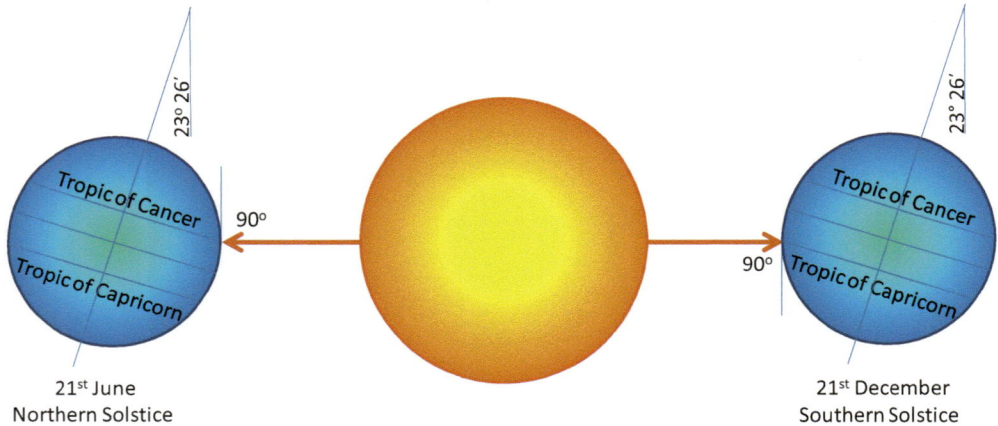

Figure 44   Sun–earth relationship

For this purpose, we will use a simple box model, as in Figure 45, to run annual simulation with hourly time steps, specifying external incident solar flux in detailed output options in IES Virtual Environment. The model dimensions are 10 m × 6 m × 3 m with two south-facing windows, 3 m × 2.1 m each. After the simulation is completed, we are able to click on individual surfaces (south wall, west wall, north wall, east wall, and horizontal flat roof) and obtain hourly values of the incident solar flux in W/m².

DOI: 10.4324/9781003342342-5

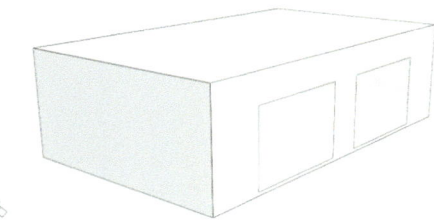

Figure 45   A simple box model used to investigate solar radiation incident on various surfaces in IES Virtual Environment

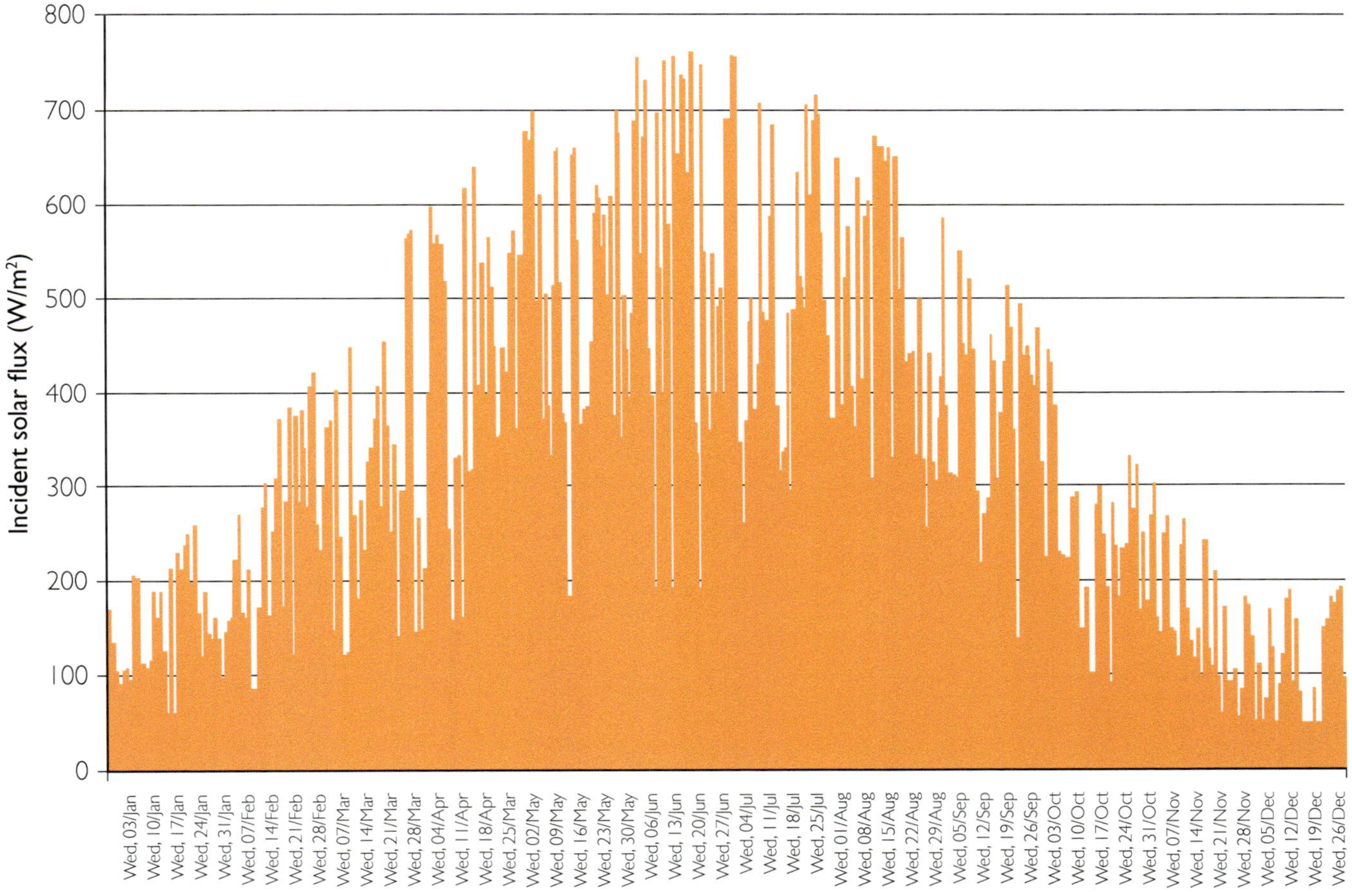

Figure 46   Hourly solar radiation on horizontal surface for an entire year based on Birmingham weather data

An example of hourly solar flux obtained in this way is shown in Figure 46, while hourly fluctuations resulting from the earth's rotation, cloud cover, and other atmospheric influences can be better seen in Figure 47.

In order to get a quick overview of radiation on various surface orientations, we can create monthly totals as shown in Figure 48. We can see from this figure that different surfaces receive different amounts of solar radiation. South surfaces will receive more than horizontal surfaces in winter months but less in summer months. East and west surfaces will receive a similar amount of solar radiation throughout the year, and, contrary to our intuition, north surfaces will receive certain amount of solar radiation in winter (mostly diffuse) and a lot more in summer (mostly from direct radiation in early morning and evening hours).

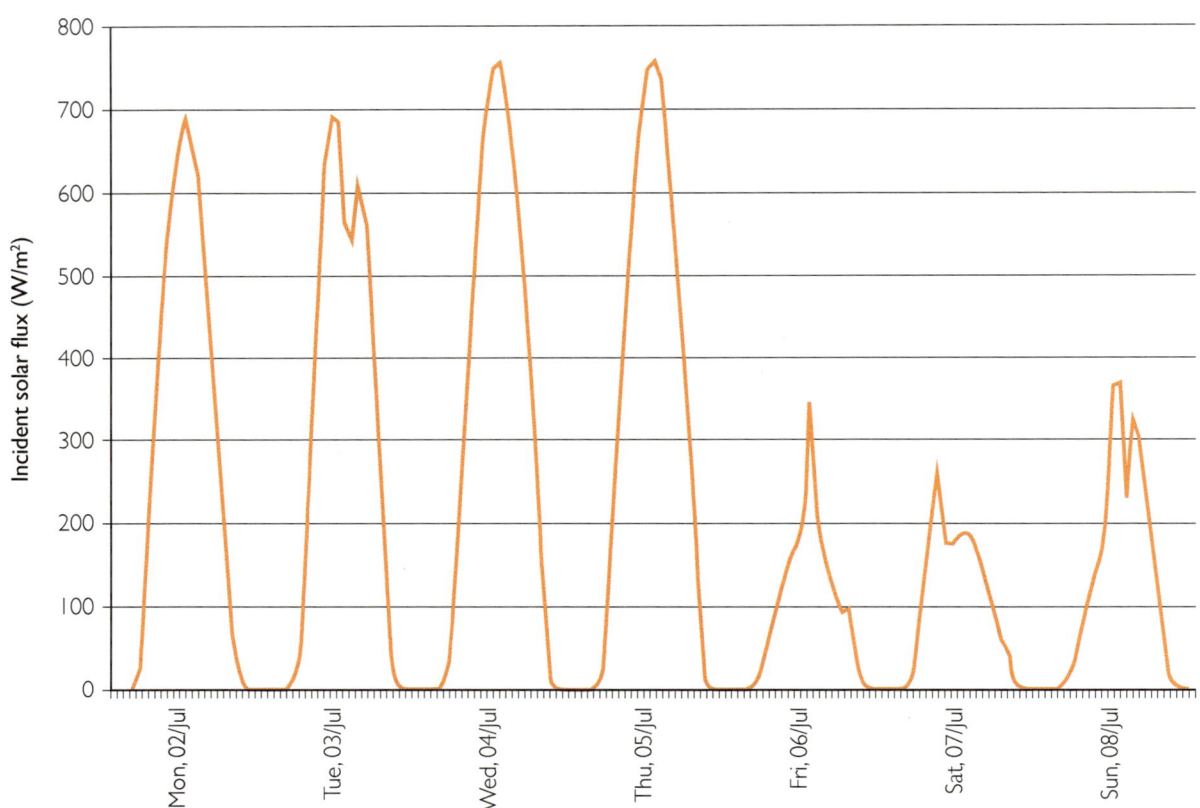

Figure 47   Hourly solar radiation data on horizontal surface from Figure 46 magnified for a week in July

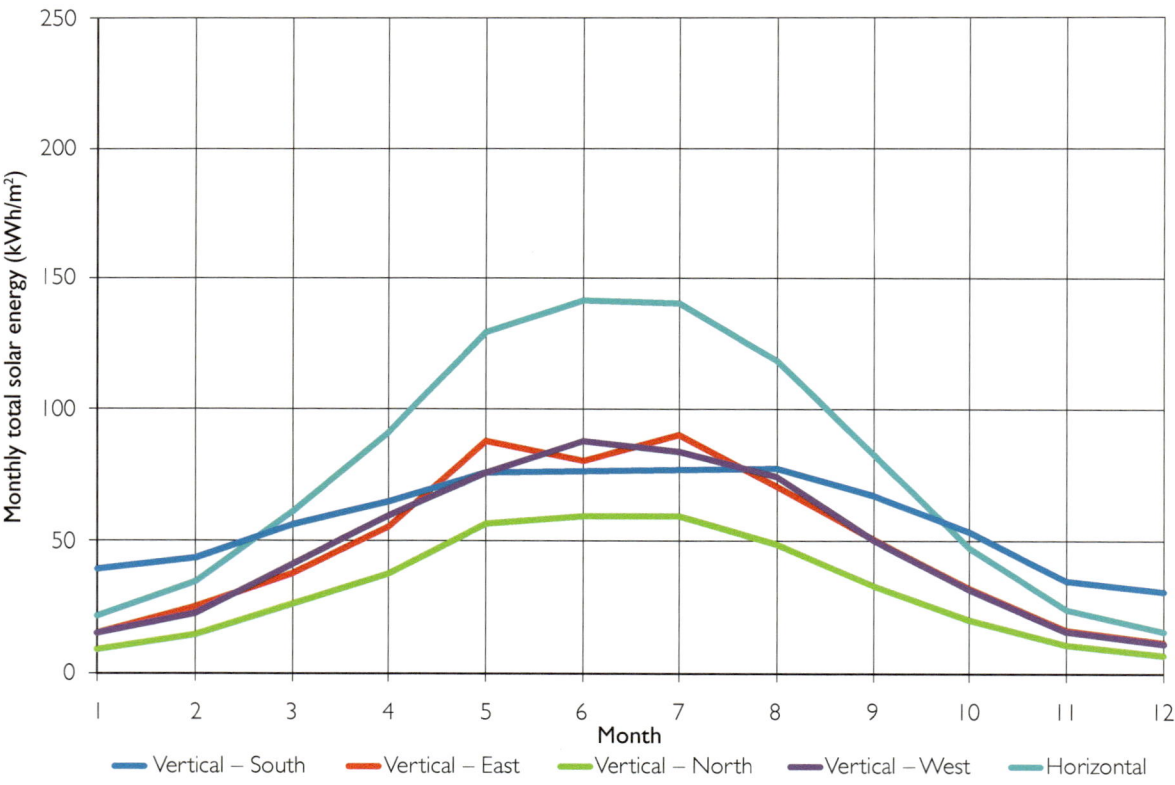

Figure 48   Monthly total solar radiation on various orientations from Birmingham data

This demonstrates the need for careful choice of the building orientation. If we want to get as much solar radiation into the building at this latitude (52.45°), we will need to choose south orientation and possibly use roof lights to increase solar gain. If we, however, wish to protect the building from solar gain, we will need to choose north orientation.

4.2 PREVAILING WINDS

Global winds are driven by temperature differences and corresponding densities of air. Warm air has a lower density than cold air and as a result of this rises as it is replaced by colder air at lower altitudes. The most intensive heating occurs near the equator and this results in upward air currents (Figure 49). As the air gets to higher altitudes, it cools down and drops to the surface level, providing the source for horizontal circulation and completing the circulation cycle. This air movement then influences the rest of the global air circulation and manifests itself as a pattern of permanent global winds.

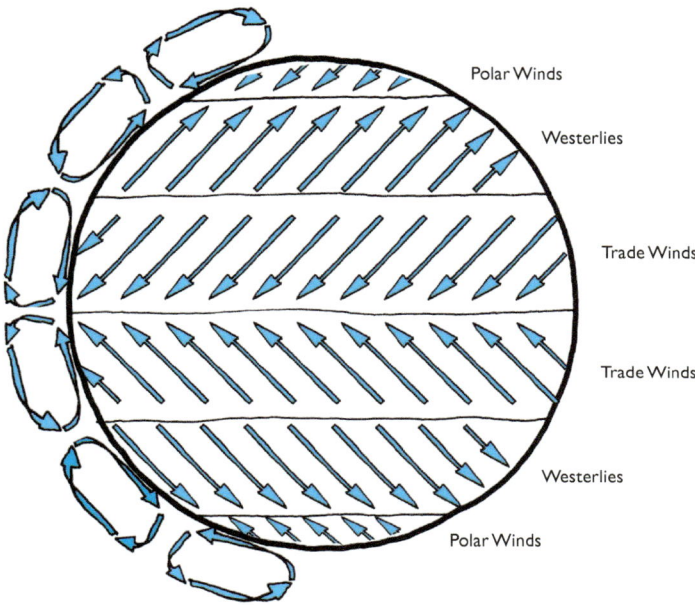

Figure 49   Global winds (illustration by Holly Doron)

This pattern is reflected in simulation weather data for different locations, so that each location will have a corresponding annual wind pattern and prevailing wind. This will be discussed further in the next chapter.

4.3 SOURCES AND CONTENTS OF WEATHER DATA FILES

An excellent overview of different types of data and consequences on dynamic simulation results was published by Drury Crawley (Crawley, 1998). He found that different weather data file types had caused considerable variations in simulated annual energy consumption (between −11% and +7%); annual peak cooling loads (between −11% and +30%); and annual peak heating loads (between −48% and +3%). These variations were mainly due to the use of Typical Meteorological Year (TMY) data for a single year. Based on this evidence, his advice is to avoid using TMY and to use synthetic years aggregated from calendar years over a period of several decades.

In the above figures, we used data from a Test Reference Year (TRY). TRY is one of the types of weather data used in dynamic simulation analysis. A TRY is assembled by choosing months with lowest values from a range of different years, and aggregating them into a full year. Another type of data called Design Summer Year (DSY) is also used in simulations. A DSY is obtained by choosing near-extreme data occurring once within a 5–10-year window and ranking these values before aggregation. Due to the way the TRY and DSY are assembled, these file types seem to address the issues raised by Crawley (1998). A large number of weather data for various locations across the world are available from EnergyPlus Weather site (DOE, 2023). The data is in EPW (EnergyPlus Weather) format and can be used with a wide range of DSM software tools, including IES VE, DesignBuilder, and others.

Details of weather data format definition for EnergyPlus can be found in a paper by Crawley et al. (1999). Each EPW weather data file contains location details, design conditions, a list of typical and extreme periods, ground temperatures, holiday/daylight saving details, etc., plus annual hourly data for numerous variables, such as dry bulb temperature, dew point temperature, relative humidity, direct normal radiation, diffuse radiation, global horizontal radiation, a range of illuminance categories, wind speed and direction, snow cover, and others. With an individual EPW data file open either in a plain text editor or as comma-separated values in a spreadsheet application, it is necessary to refer to the weather data definition in the paper by Crawley et al. (1999) in order to understand what is in the data file.

4.4 CLIMATE CHANGE AND FUTURE WEATHER DATA

Climate change occurs as a consequence of the Greenhouse Effect, explained in Section 1.5.1 and Figure 6. As greenhouse gases, including carbon dioxide, methane, nitrous oxide, and others, reduce the optical transparency of the atmosphere, more heat is retained within the atmosphere and less heat dissipates into space.

Climate change predictions have been used to morph current weather data into future data for use in dynamic simulation. The UK Engineering and Physical Sciences Research Council funded a PROMETHEUS project to develop a methodology for the creation of future weather data, based on the work by the UK Climate Change Projections. Future weather data for over 40 UK locations generated by the PROMETHEUS project were thus produced (Eames et al., 2011) and made available in the EPW EnergyPlus format on the project website (University of Exeter, 2012).

What will be the actual impact of climate change on buildings? We will investigate this in the next section.

4.4.1 Simulation experiments with future weather data
To estimate the actual impact of climate change on buildings, we will run several experiments on a simple test building using the EnergyPlus dynamic simulation model. The test building, almost identical to the one used in the section entitled 'Solar Radiation' in this chapter, is shown in Figure 50. The dimensions of the model are 10 m × 6 m × 3 m with two large double-glazed windows on the south-facing side, dimensions 3 m × 2.1 m, and a wooden door on the north-facing side, dimensions 1 m × 2.1 m.

This model was subjected to four DSY weather data sets: control set (1961–1990); 2030; 2050; and 2080. The weather files used were for a median estimate of future change (50th percentile) and for a medium emissions scenario. Eames et al. (2011) explain in detail how these probabilistic data sets were generated.

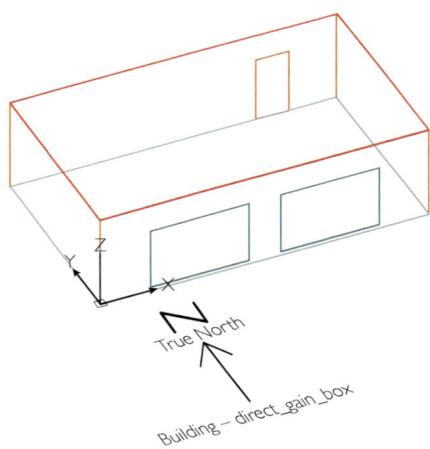

Figure 50   A simple EnergyPlus box model used to investigate the influence of future weather

The model was set to work in a free-floating mode, without any heating or cooling, so that the internal air temperature was solely the consequence of the balance between heat gains from solar energy and heat losses through the building fabric. There was no intention to optimise this model in terms of reducing overheating, but the main aim was only to compare the influence of future weather data sets with reference to the control data set. A working version of this model is available on the supplementary material website: www.ljankovic.com.

The summary of average internal temperatures is shown in Table 5 and the cumulative frequency of occurrence of temperatures in Figure 51. As we can see from Table 5, there is a gradual increase of average internal air temperatures, by 2°C between the control data set and 2030, and by 1°C from 2030 through 2050 to 2080. Although the investigated model shows overheating even in the control case, the comparison between the control case and future weather cases is quite revealing.

The intersection between the vertical line at 28°C and the cumulative frequency curves (Figure 51) can help us estimate the increase in annual overheating hours. These estimates are summarised in Table 6. We can see from this table that there will be a 9% increase in annual hours above 28°C between the control set and the 2030 data set, and a 16% increase between the control set and the 2080 data set.

TABLE 5 AVERAGE INTERNAL AIR TEMPERATURES IN THE ENERGYPLUS MODEL RESULTING FROM FUTURE WEATHER	
Weather data set	Average internal temperature (°C)
Control (1961–1990)	22.0
2030	24.2
2050	25.2
2080	26.4

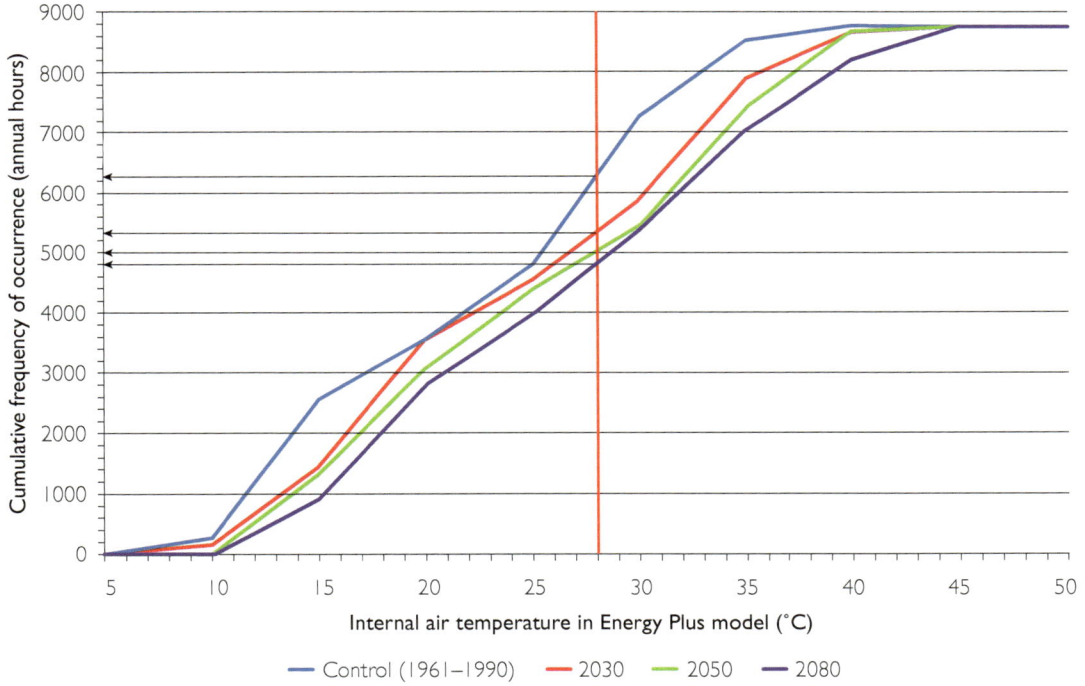

Figure 51   Cumulative frequency of occurrence of internal air temperatures in EnergyPlus model resulting from future weather

TABLE 6 INCREASE OF HOURS ABOVE 28°C RESULTING FROM FUTURE WEATHER		
Weather data set	**Increase of hours above 28°C**	**Percentage of total annual hours**
Control (1961–1990)	–	–
2030	800	9%
2050	1200	14%
2080	1400	16%

This is a significant increase that will require careful attention to design, as well as the development of new methods, materials, systems, and tools.

Although a similar analysis with a different probabilistic data set (10th or 90th percentile) would have reached different results, this analysis based on the 50th percentile data gives us realistic numbers that hopefully increase our understanding of possible consequences of climate change. However, the reader is encouraged to investigate the behaviour of the EnergyPlus box model, available on the supplementary material website, using different probabilistic data sets from the PROMETHEUS project.

4.5 RESPONDING TO THE CLIMATE CHANGE WITH ADAPTABLE DESIGN

In order to respond effectively to climate change, we recommend the following design measures, which will be explained in detail in various chapters in this book:

- Consider building geometry and shallow plan early in the design stage (Chapter 7).
- Use thermal mass to smooth out fluctuations of internal temperatures (Chapter 11).
- Use natural ventilation methods and nighttime ventilation to condition thermal mass and mitigate the effects of external temperature increase (Chapter 12).
- Use correctly designed solar shading to mitigate increased solar gain in summer (Chapter 10).
- Improve thermal insulation to mitigate heat gains in summer (Chapter 8).
- Use reflective coating on the outside of the building to reduce solar absorption (Chapter 9).
- Design for use of natural daylight (Chapter 13) and integrate electrical lighting with natural daylight using daylight sensitive controls (Chapter 14).
- Reduce internal heat gains (Chapter 15).
- Use dynamic simulation methods to optimise the design and achieve zero carbon performance (Chapter 19).

4.6 THE BIG PICTURE

Climate change and its predicted consequences can make us think that the planet is on a self-destruction trajectory, and that when it dies we will all die with it. However, let us consider some facts about energy supply and demand that could make us think differently. To do that, we will compare the total world energy demand with energy supplied from the sun. Table 7 shows a simple calculation of the time that is needed to get the total world energy demand supplied by the sun. The world energy consumption is based on a report by International Energy Agency (2022, p. 436), and the energy intercepted by the earth is based on a simple calculation using a solar radiation intensity at the top of the atmosphere, also called the solar constant (Duffie & Beckman, 2013), and simplified earth geometry in Figure 52.

We can see from this calculation that the entire annual world energy consumption could have been met in 42 minutes of solar radiation in 2021! As world energy demand increases by 24% by 2050, the total annual energy consumption will be equivalent to 53 minutes of solar radiation.

Description	Units	2021	2050
World energy consumption (International Energy Agency, 2022, p. 436)	EJ (ExaJoule = 10^{18} Joule)	439	544
	J	4.39×10^{20}	5.44×10^{20}
Earth radius r	m	6.37×10^{6}	6.37×10^{6}
Earth area that intercepts radiation $= \pi \times r^2$	m²	1.28×10^{14}	1.28×10^{14}
Solar constant – solar radiation at the top of the atmosphere (Duffie & Beckman, 2013)	W/m²	1,353	1,353
Total energy from the sun	W	1.73×10^{17}	1.73×10^{17}
Time in which solar energy equals to world energy consumption	seconds	2,544	3,153
	minutes	42	53

TABLE 7 THE TIME IT TAKES TO GET THE WORLD ANNUAL ENERGY DEMAND SUPPLIED BY THE SUN

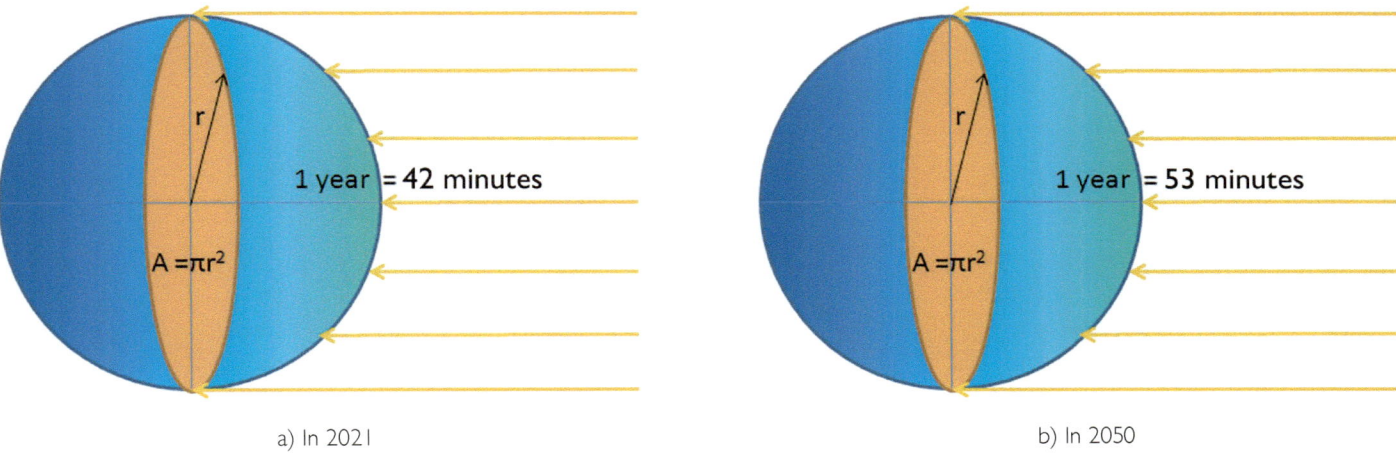

a) In 2021　　　　　　　　　　　　　　b) In 2050

Figure 52   Solar radiation intercepted by the earth

Another abundant source of energy is just below our feet: the earth's magma. Driven by radioactive decay and residual heat from the formation of the planet, there is an increasing temperature gradient below the surface (Johnson et al., 2021). At 100 km depth, this temperature is estimated to be around 1,500 K. The temperature progressively rises as the depth increases towards the earth's core, where it reaches 7,000 K at 6,400 km depth. Magma therefore provides sufficient renewable heat that could fulfil worldwide energy demand indefinitely. But our civilisation does not yet have the technology capable of a wide-ranging use of this geothermal heat.

There are several compelling arguments that we can develop from here. First, there is enough energy around, above our heads and below our feet, but we need to increase the level of our technological capability to be able to capture it. Today's technology that we call 'advanced' is not yet capable of capturing this abundant energy from the sun and from earth's magma. Second, change of behaviour alone will just slow down energy consumption but it will not enable us to tap into this vast energy resource. Tapping into this energy source would result in the total abandonment of fossil fuels and a drastic reduction of CO_2 emissions.

Designing zero carbon buildings is only one of the steps that we must take in order to make our planet's resources sustainable. This should not only involve the balancing of energy production and consumption, but also the substantial reuse of materials that introduce strong feedback loops into our processes. It should not only involve the design of new buildings, but even more so the retrofit of existing buildings.

We need more cutting-edge research and development work and focused effort to give us the capability to use the abundant solar energy and earth's energy, which is available but currently not easily exploitable due to technological and social limitations of our civilisation. We need a united world, reviewing its priorities, eliminating conflicts, and focusing its effort in this direction. We cannot afford to have continuous economic growth on a finite planet. We need to change our technological, social, and economic ways. I hope that this book will make a contribution, however small, towards achieving that goal in going forward towards a zero carbon future.

4.7 SUMMARY OF DESIGN PRINCIPLES

Use of resources

• Balance production and consumption.
• Introduce feedback loops into societal processes through the reuse of materials and other resources.

Response to global context

• Respond to solar radiation by choosing appropriate orientation.
• Respond to prevailing wind by sheltering the building from its direction.

Response to climate change

• Use shallow plan and curved surfaces.
• Use increased amount of thermal mass.
• Use natural ventilation methods.
• Use correctly designed solar shading.
• Improve thermal insulation.
• Use natural daylight and daylight sensitive controls of electrical light.
• Use DSM to optimise the design and achieve zero carbon performance.

PART 2
EMBODIED EMISSIONS CONTEXT

CHAPTER 5

WORKING OUT EMBODIED EMISSIONS IN A BUILDING

This chapter introduces a workflow of calculations of embodied emissions in a building, using Birmingham Zero Carbon House as an example. Zero Carbon House was retrofitted and extended in 2010 from an 1840s end-of-terrace house in Birmingham, UK. The building therefore consists of an original building that was retrofitted and a new extension. A detailed case study of this house is in Chapter 21.

5.1 EMISSIONS EMBODIED IN CONSTRUCTION MATERIALS

We will first calculate embodied emissions in construction materials using the Embodied Carbon Footprint Database (C. Jones & Hammond, 2019). As the database generally provides values in the units of emissions per mass of the corresponding material, this means that we need to calculate the mass of each building component. The process can be made easier if a simulation model of the building can be used, such as Virtual Environment 2022 (IES, 2022), as explained below.

In order to calculate the mass of a building construction, such as wall, floor or slab, we need to know its surface area, thickness, and density. Thus, from the simulation model of this building (Figure 53), we first select the surface of interest from a wireframe axonometric model with selectable individual spaces (Figure 54). After selecting the spaces of the original building on the street side, for instance, we read external wall surface area from the model (Figure 54). We then open the

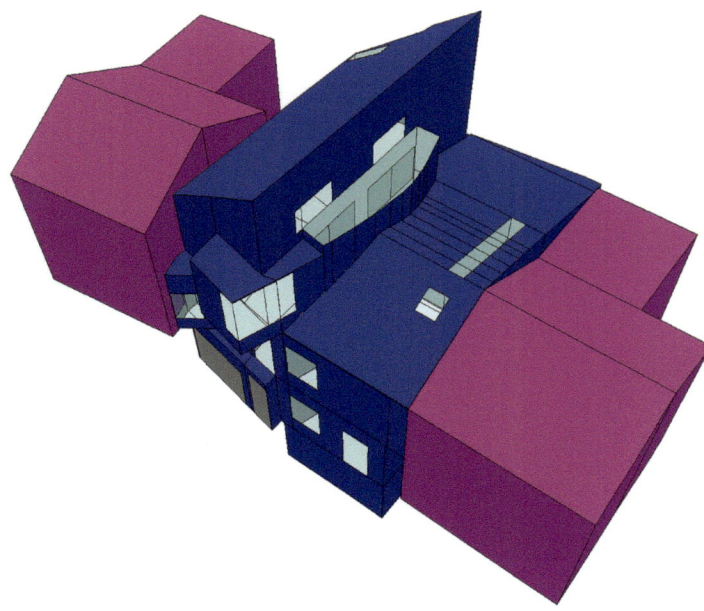

Figure 53   Rendered axonometric view of the building model

construction database of this simulation model, from where we obtain the thicknesses of individual layers of the wall and the corresponding material densities (Figure 55). Multiplying the surface area and thickness values, we obtain the volume of the building components (Table 8, column 4). Subsequently, multiplying the volume with density, we obtain the mass of each layer (Table 8, column 6), and multiplying the mass by embodied emissions per unit of mass, we obtain embodied emissions in each layer (Table 8, column 8). The sum of embodied emissions in each layer represents embodied emissions in that construction (Table 8, column 8, last row). This is then repeated for each set of constructions and each different material within these constructions to obtain total embodied emissions in building materials. The result for the entire building is summarised in Table 9.

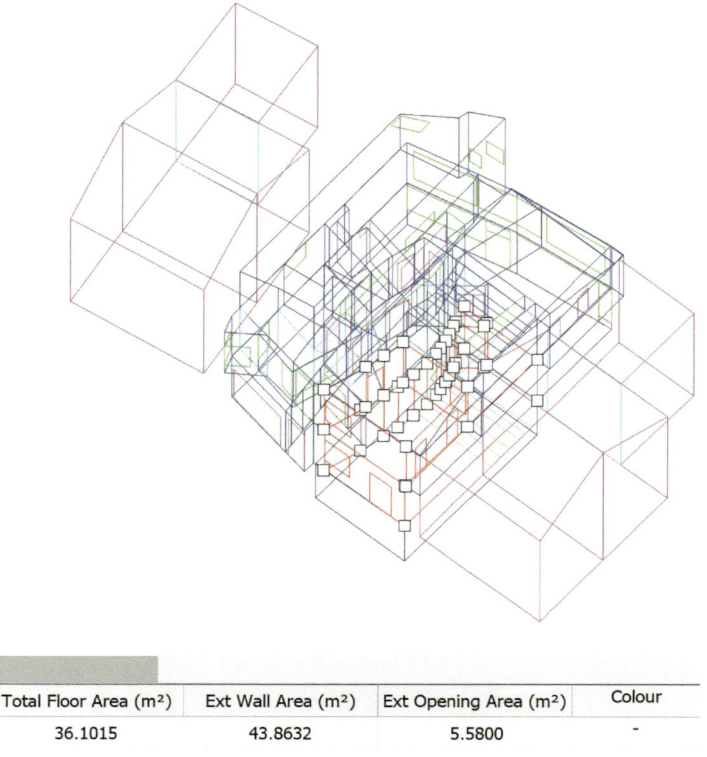

Total Floor Area (m²)	Ext Wall Area (m²)	Ext Opening Area (m²)	Colour
36.1015	43.8632	5.5800	-

Figure 54   Wireframe axonometric view of the building model with selected individual spaces on the street side

Construction Layers (Outside To Inside)

Material	Thickness mm	Conductivity W/(m·K)	Density kg/m³
[BRO1] Brickwork	215.0	0.8400	1700.0
[SFOAM11] Warmcell	350.0	0.0380	39.0
[GPL1] Glaster 12mm (4:1 glass to lime putty) - recycled crushed glass part 80%	9.6	1.0600	2500.0
[GPL11] Glaster 12mm (4:1 glass to lime putty) - lime putty part 20%	2.4	0.2700	1350.0

Figure 55   Thickness of material layers and material density obtained from the simulation tool database

TABLE 8　EMBODIED EMISSIONS CALCULATION FOR EXTERNAL WALL ON THE STREET SIDE

Materials in wall layers	Area (m²)	Thickness (m)	Volume (m³)	Density (kg/m³)	Mass (kg)	Material embodied emissions factor (kgCO$_2$/kg)	Construction embodied emissions (kgCO$_2$)
(1)	(2)	(3)	(4) = (2) · (3)	(5)	(6) = (4) · (5)	(7)	(8) = (6) · (7)
External wall on the street side							
Brickwork	43.86	0.2150	9.43	1700	16030.83	Existing material – zero emissions	0.00
Warmcel (cellulose insulation reprocessed from recycled paper)	43.86	0.3500	15.35	39	598.69	0.630	377.17
Glaster 12 mm (4:1 glass to lime putty) – recycled crushed glass part 80%	43.86	0.0096	0.42	2500	1052.64	Recycled material – delivery emissions only	0.00
Glaster 12 mm (4:1 glass to lime putty) - lime putty part 20%	43.86	0.0024	0.11	1350	142.11	0.780	110.84
Embodied emissions in external wall on the street side (kgCO$_2$) =							488.01

TABLE 9　SUMMARY OF EMBODIED EMISSIONS CALCULATION IN BUILDING MATERIALS AND GLAZING

Construction	Mass (kg)	Material embodied emissions factor (kgCO$_2$/kg)	Construction embodied emissions (kgCO$_2$)
Triple glazing windows – glass	3057.75	0.91	2782.55
Triple glazing windows – timber	317.10	0.31	97.03
Original building – External wall			
Brickwork	16030.83	Existing material – zero emissions	0.00
Warmcel (cellulose insulation reprocessed from recycled paper)	598.69	0.63	377.17
Glaster 12 mm (4:1 glass to lime putty) – glass part	1052.64	Recycled material – delivery emissions only	0.00
Glaster 12 mm (4:1 glass to lime putty) – lime putty part	142.11	0.78	110.84
Extension building – External wall			
Sto (acrylic) render 3 mm	1135.70	0.78	885.85
Neopor-expanded polystyrene with infrared reflecting additive	1727.39	3.29	5683.10
Unfired clay block	113850.52	See Q19	3347.21
Glaster 12 mm (4:1 glass to lime putty) – glass part	6730.08	Recycled material – delivery emissions only	0.00
Glaster 12 mm (4:1 glass to lime putty) – lime putty part	908.56	0.78	708.68
Roof type 1			
Warmcel (cellulose insulation reprocessed from recycled paper)	220.17	0.63	138.71
Plasterboard 12 mm	183.88	0.39	71.71

TABLE 9 CONTINUED

Construction	Mass (kg)	Material embodied emissions factor (kgCO$_2$/kg)	Construction embodied emissions (kgCO$_2$)
Roof type 2			
Single ply membrane 1.2 mm	118.90	0.97	115.33
Pavatex wood fibre board 100 mm	1249.32	0.72	893.26
Warmcel (cellulose insulation reprocessed from recycled paper)	1176.08	0.63	740.93
Joists 360 mm 6.5% of the roof surface	1411.30	0.44	616.74
Plasterboard 12 mm	982.22	0.39	383.07
Warmcel (cellulose insulation reprocessed from recycled paper)	129.40	0.63	81.52
Plasterboard 12 mm	108.07	0.39	42.15
Floors			
Ground floor – rammed earth 100 mm	17188.20	0.02	412.52
Ground floor – bees wax 0.01 mm	0.91	0.00	0.00
First floor – rammed earth 100 mm	15280.20	0.02	366.72
First floor – bees wax 0.01 mm	0.81	0.00	0.00
Second floor – rammed earth 100 mm	8357.40	0.02	200.58
Second floor – bees wax 0.01 mm	0.44	0.00	0.00
Ceilings			
Ground floor – plasterboard 12 mm	1088.59	0.39	424.55
Ground floor – paint 0.1 mm	11.94	0.44	5.25
First floor – plasterboard 12 mm	967.75	0.39	377.42
First floor – paint 0.1 mm	10.61	0.44	4.67
Second floor – plasterboard 12 mm	529.30	0.39	206.43
Second floor – paint 0.1 mm	5.80	0.44	2.55
Other building elements			
Internal partition walls (where new) are 100 mm clay blocks	11440.00	0.03	336.34
Ground floor new build part only, 200 mm insulation boards, 100 mm limecrete floor slab			
Insulation board 200 mm	167.20	3.29	550.09
Limecrete slab 100 mm	0.00	0.00	0.00
1 part NHL5 glaster screed	886.67	0.15	128.57
2 parts fine sharp sand	5269.33	0.01	39.36
Newbuild foundations, limecrete trench fill 450 × 600, 19 linear metres			
1 part NHL5 glaster screed	1197.00	0.15	173.57
2 parts fine sharp sand	7113.60	0.01	53.14
Internal doors, 13 nos, mostly standard timber doors + frames	674.69	0.31	206.46

TABLE 9   CONTINUED

Construction	Mass (kg)	Material embodied emissions factor (kgCO$_2$/kg)	Construction embodied emissions (kgCO$_2$)
Kitchen, recycled glass worktops	300.00	0.22	64.80
Kitchen, open shelves from reclaimed timber	140.00	0.31	42.84
Stairs: Reclaimed timber + hemp rope handrail			
0.018 × 0.8 × 0.3 × 32 nos	96.77	0.00	0.00
2 nos Canvas doors (top floor cupboards)	50.40	0.31	15.42
Reclaimed bricks for front elevation (from a local school)			
0.103 × 0.553 × 8 m	774.64	0.00	0.00
Total mass of construction materials (kg) =	222,682.97	Embodied emissions in construction materials (kgCO$_2$) =	20,687.13

The house has three energy systems that also need to be taken into account: a solar photovoltaic system consisting of 35.6 m² corresponding to 5.04 kW$_p$, a solar thermal system consisting of 8.8 m² evacuated tube solar thermal collectors, and a wood-burning stove.

Emissions from the solar photovoltaic system can be calculated on the basis of kW$_p$ rating. Guidance from Circular Ecology suggests that 2,560 kgCO$_2$ is attributed to each kW$_p$ (kilo-Watt-peak) (Circular Ecology, 2021). However, there are forecasts that this will change with time (Worboys, 2021), so that 325 kgCO$_2$ will be attributable to each kW$_p$ in 2040. The calculated embodied emissions in the solar PV system are shown in Table 10. The typical value will be used in the calculation of the initial embodied emissions and the forecast value for 2040 will be used in the calculation of maintenance emissions.

Row	Description	Typical	Forecast for 2040
	TABLE 10 EMBODIED EMISSIONS IN SOLAR PHOTOVOLTAIC SYSTEM		
(1)	kW$_p$ rating of solar PV system	5.04	5.04
(2)	Embodied emissions factor (kgCO$_2$/kW$_p$)	2,560	325
(3)	Embodied emissions in solar PV system (kgCO$_2$) = (1) · (2)	12,902	1,638

Emissions from the solar thermal system are calculated on the basis of guidance from Menzies and Roderick (2010). They break down embodied emissions per functional unit representing one solar thermal collector. The calculation of embodied emissions in the solar thermal system is shown in Table 11.

TABLE 11   EMBODIED EMISSIONS IN SOLAR THERMAL SYSTEM

	Mass per functional unit (collector) (kg)	Embodied emissions factor per functional unit (collector) (kgCO$_2$)	Number of functional units (collectors)	Total mass (kg)	Embodied emissions in solar thermal system (kgCO$_2$)
Solar thermal collector	57.42	240.10	8	459.36	1920.80
Solar cylinder	63.74	344.47	8	509.92	2755.76
Support and external pipework	23.60	143.02	8	188.8	1144.16
			Totals	1,158.08	5,820.72

Embodied emissions in the wood-burning stove are calculated on the basis of guidance from CIBSE TM65 – a calculation methocology for embodied carbon in building services (Hamot et al., 2021). The manufacturer's specification of the wood-burning stove was used to apportion different materials in it, and Embodied Carbon Footprint Database (C. Jones & Hammond, 2019) was used to assign embodied emissions to these materials, as shown in Table 12, based on the total stove mass of 219 kg.

TABLE 12 EMBODIED EMISSIONS IN WOOD-BURNING STOVE				
Materials	Percentage of total mass (%)	Mass of each material fraction (kg)	Material embodied emissions factor (kgCO$_2$/kg)	Embodied emissions in each material fraction (kgCO$_2$)
Steel 10%	10%	21.90	2.97	65.04
Glass 1%	2%	4.38	1.44	6.31
Iron 60%	78%	170.82	1.52	259.65
Copper tube/sheet 10%	10%	21.90	3.81	83.44
Totals	100%	219.00		414.44

The total initial embodied emissions in the materials are therefore obtained from the totals in Tables 9–12, as summarisec in Table 13.

TABLE 13 TOTAL EMBODIED EMISSIONS IN MATERIALS	
Total embodied emissions in materials	kgCO$_2$
Constructions	20,687.13
Solar PV	12,902.40
Solar thermal	5,820.72
Wood-burning stove	414.44
Total embodied emissions in materials (kgCO$_2$) =	39,824.69

5.2 EMISSIONS FROM THE CONSTRUCTION PROCESS: CONSTRUCTION WORKERS TRAVEL, MATERIAL DELIVERIES, OPERATION OF SITE MACHINES

Retrofitting and extending Zero Carbon House lasted 200 days, or approximately 29 weeks. This is divided into five construction phases, as shown in Table 14. This now enables informed estimates of construction workers' travel, material deliveries, and operation of site machines.

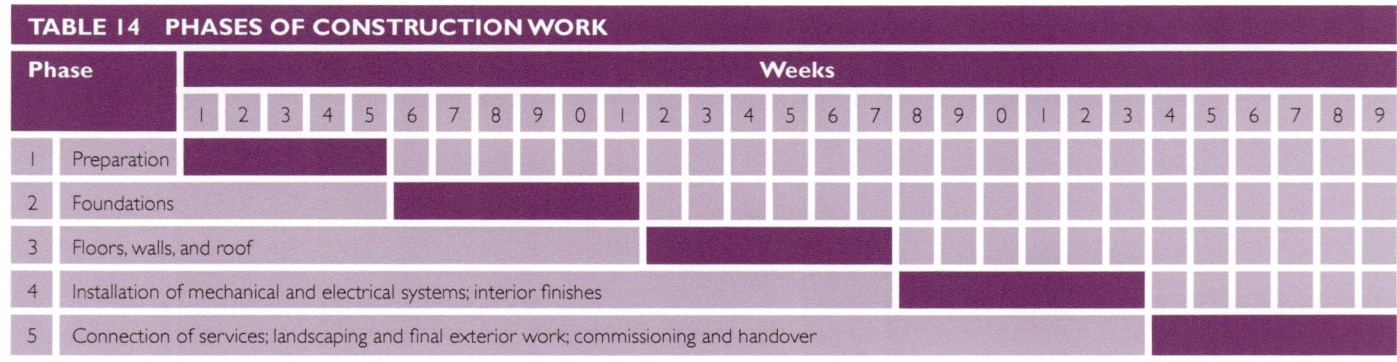

TABLE 14 PHASES OF CONSTRUCTION WORK	

Phase		Weeks (1–29)
1	Preparation	Weeks 1–5
2	Foundations	Weeks 6–12
3	Floors, walls, and roof	Weeks 13–17
4	Installation of mechanical and electrical systems; interior finishes	Weeks 18–24
5	Connection of services; landscaping and final exterior work; commissioning and handover	Weeks 25–29

Over the 200 days, up to four construction workers were on site. It is assumed that they travelled in three-person vans, and this means that between one and two vans were used. The project was delivered by a local construction company, and therefore it is assumed that the construction workers travelled only locally within a radius of ten miles, on 20-mile day return trips. Whilst at the time of writing there were more recent sources of carbon emissions from vehicles, the source used here (VCA, 2016) was chosen to more closely coincide with 2010 when Zero Carbon House was retrofitted and extended. This now enables the calculation of emissions from construction workers' travel, as shown in Table 15.

TABLE 15   CARBON EMISSIONS FROM TRAVEL OF CONSTRUCTION WORKERS

Phase (from Table 14)	No. of people	No. of vans	No. of days	Miles travelled per day	Miles travelled	km travelled	Emissions factor (gCO$_2$/km)	Total emissions – people travel (kgCO$_2$)
1–5	4	1.3	200	20	5,333	8,583	156.5	1,343
							Total emissions from people travel (kgCO$_2$) =	1,343

Emissions from material deliveries are calculated using the total mass of construction materials from Table 9, subtracting the mass of existing brickwork in the original building, and adding the mass of the wood-burning stove, solar thermal system, and solar PV system. This is summarised in Table 16.

TABLE 16   MASS FOR MATERIAL DELIVERIES

Description	Source	Mass (kg)
Total mass of construction materials	Table 9	222682.97
Less mass of brickwork in the original building	Table 9	−16030.83
Plus mass of the PV system	PV data sheets and installer's invoice	444
Plus mass of the solar thermal system	Table 11	1158.08
Plus mass of wood-burning stove	Table 12	219
	Total mass for material deliveries	208,254.22

This now enables us to calculate emissions from material deliveries, as summarised in Table 17. We first apportion deliveries across the construction phases (Table 17, column 2), and apply this to the total mass for material deliveries from Table 16. Deliveries are assumed to be done by 7.5 metric-ton trucks. The number of trucks is obtained by dividing the mass of materials delivered by 7,500 kg (Table 17, column 4). Another assumption is that all deliveries are local within the range of 15 miles, and thus each delivery consists of a 30-mile round trip (Table 17, column 5). The total number of miles travelled is obtained by multiplying the number of trucks by the number of miles travelled per day (Table 17, column 6). As data for truck emissions is available in gCO$_2$/km, the trips in miles are converted into kilometres (Table 17, column 7). Total emissions from material deliveries are then obtained by multiplying the kilometres travelled by the corresponding emissions per kilometre (Table 17, column 9).

Emissions from the operation of site machines are calculated using the power rating of these machines and days spent on site, and applying published emissions factors (Heidari & Marr, 2015). The source of the emissions factors is chosen to coincide more closely with the time when Zero Carbon House was retrofitted and extended. Thus, the emissions from the operation of site machines are summarised in Table 18.

Total emissions from construction of 1956 kgCO$_2$ are obtained by adding the totals from Tables 15, 17, and 18.

TABLE 17 CARBON EMISSIONS FROM MATERIAL DELIVERIES

Phase (Table 14)	Percentage of materials per phase	Materials delivered (kg)	No of 7.5 t trucks	Miles travelled per day	Total miles travelled	Total kilometres travelled	Emissions factor (gCO$_2$/km)	Total emissions - material deliveries (kgCO$_2$)
(1)	(2)	(3) = (2) · 208,254	(4) = (3)/7500	(5)	(6) = (4) · (5)	(7) = (5) · 1.609	(8) (Source: (TheyWork-ForYou, 2013)	(9) = (7) · (8)/1000
1	0%	0	0	0	0	0	327	0.00
2	40%	83,302	12	30	360	579.36	327	189.45
3	48%	99,962	14	30	420	675.92	327	221.03
4	12%	24,991	4	30	120	193.12	327	63.15
5	0%	0	0	30	0	0	327	0.00
	Total =	208,254					Total =	473.63

TABLE 18 CARBON EMISSIONS FROM THE OPERATION OF SITE MACHINES

Phase	Hours of operation per day (hr/day)	Days of operation	Power (kW)	Emissions factor (g CO$_2$/(kW.hr))	Total emissions – site machines (kgCO$_2$)
(1)	(2)	(3)	(4)	(5)	(6) = (2) · (3) · (4) · (5) /1000
2	8	2	42	138	93
3	8	1	42	138	46
Total emissions from the operation of site machines (kgCO$_2$) =					139

5.3 EMISSIONS FROM MAINTENANCE

RICS (2017) provides guidance on scenarios for maintenance, repair, and replacement during the building use stage. In the particular case, these scenarios were developed in consultation with the owner of the building, and the corresponding carbon emissions were obtained from detailed calculations. This is summarised in Table 19.

TABLE 19 CARBON EMISSIONS FROM MAINTENANCE

Description	Material	Embodied emissions (kgCO$_2$)	Source	Frequency in the 60-year cycle	Total emissions (kgCO$_2$)
Repaint the house once every 20 years	Sto (acrylic) renderer	885.85		2	1,772
Re-wax floors	Bees wax	0.00	Table 9	59	0
Repaint ceilings	Paint	12.47		2	25
Replace solar PV system		1,638	Table 10	1	1,638
Replace solar thermal system		5,821	Table 11	1	5,821
			Total emissions from maintenance (kgCO$_2$) =		9,256

In buildings with conventional walls, repainting of the walls would also be a part of regular maintenance. However, as all internal walls in this building are finished in glaster, there is no need for periodic repainting.

5.4 END-OF-LIFE EMISSIONS: DECONSTRUCTION, DEMOLITION, TRANSPORT, AND DISPOSAL

The end-of-life emissions are calculated on the basis of RICS guidance (RICS, 2017). A life cycle of 60 years was adopted, and calculations were carried out for deconstruction and demolition emissions, and transport and disposal emissions.

Deconstruction and demolition emissions are calculated using gross internal floor area multiplied by the demolition emissions factor (RICS, 2017, sec. 3.5.4.1 [C1]), as shown in Table 20.

TABLE 20 DECONSTRUCTION AND DEMOLITION EMISSIONS		
Gross internal floor area (m²)	Demolition emissions factor (kgCO$_2$/m²)	Total emissions – deconstruction and demolition (kgCO$_2$)
226.81	3.4	771.15
Total deconstruction and demolition emissions (kgCO$_2$) =		771.15

Transport emissions in relation to deconstruction and demolition are calculated using RICS guidance (RICS, 2017, sec. 3.5.4.2 [C2]), as shown in Table 21. The total mass for deconstruction and demolition is taken from Table 16, where existing brickwork from the same table was added to the total, as it needs to be included in the end-of-life treatment. Transport of the material from deconstruction and demolition is assumed to be carried by 7.5 ton trucks, and the total number of trucks is calculated by dividing the amount of material by the truck capacity. A longer range of 50 miles is assumed. The total miles travelled is calculated using the number of trucks and the number of miles travelled. This is then converted into kilometres travelled and emissions factor applied to obtain the total emissions from transport in relation to deconstruction and demolition (Table 21, column 7).

TABLE 21 TRANSPORT EMISSIONS – DECONSTRUCTION AND DEMOLITION						
Mass (kg)	No of 7.5 t trucks	Miles travelled per day	Total miles travelled	Total kilometres travelled	Emissions (gCO$_2$/km)	Total emissions – transport (kgCO$_2$)
(1)= (Table 16 plus brickwork)	(2) = (1)/7500	(3)	(4) = (2) · (3)	(5) = (4) · × 1.609	(6) (source: (TheyWorkForYou, 2013)	(7) = (5) · (6)
224,285	30	50	1,500	2414.02	327	789.38
Total transport emissions - deconstruction and demolition (kgCO$_2$) =						789.38

Disposal emissions are also calculated on the basis of RICS guidance, as shown in Table 22, and assuming that disposal will be into a landfill with landfill gas recovery. These are based using the same mass of material as in Table 21 and applying the recommended emissions factor for disposal per each kilogram of waste (RICS, 2017, sec. 3.5.3.4 [C4]).

TABLE 22 DISPOSAL EMISSIONS		
Mass (kg) (Table 16 plus brickwork)	Disposal emissions factor (kgCO$_2$/kg)	Total emissions – disposal (kgCO$_2$)
224,285	0.013	2915.71
Total disposal emissions (kgCO$_2$) =		2915.71

Combining the totals from Tables 20–22, a total of the end-of-life emissions is obtained as 4476.24 kgCO$_2$.

5.5 TOTAL EMBODIED EMISSIONS

The total embodied emissions are the sum of all of the emissions calculated in this chapter, as summarised in Table 23. As it can be seen from this table, the materials have the greatest impact on embodied emissions of 71.7%. The second largest impact of 16.7% is from maintenance, whilst end of life is responsible for 8.1% of embodied emissions and the construction process is responsible for 3.6% of embodied emissions. This applies to the specific example introduced in this chapter. Different buildings under different circumstances will make a different impact under these headings.

TABLE 23 TOTAL EMBODIED EMISSIONS	Embodied emissions (kgCO$_2$)	Embodied emissions (kgCO$_2$)	Percentage of the total (%)	Percentage of the total (%)
Embodied emissions in materials				
Embodied emissions in building constructions	20,687		37.3%	
Embodied emissions in solar photovoltaic system	12,902		23.2%	
Embodied emissions in solar thermal system	5,821		10.5%	
Embodied emissions in wood-burning stove	414		0.7%	
Sub-total: Embodied emissions in materials		**39,825**		**71.7%**
Emissions from the construction process				
Emissions from the travel of construction workers	1,343		2.4%	
Emissions from material deliveries	474		0.9%	
Emissions from the operation of site machines	139		0.3%	
Sub-total: Emissions from the construction process		**1,956**		**3.5%**
Emissions from maintenance				
Emissions from maintenance	9,256		16.7%	
Sub-total: Emissions from maintenance		**9,256**		**16.7%**
End-of-life emissions				
Deconstruction and demolition emissions	771		1.4%	
Transport emissions – deconstruction and demolition	789		1.4%	
Disposal emissions	2,916		5.3%	
Sub-total: End-of-life emissions		**4,476**		**8.1%**
Totals	**55,512**	**55,512**	**100%**	**100%**

5.6 DISCUSSION

Now, when we have the above calculations we can investigate the sensitivity of these calculations in relation to local sourcing, transportation and construction methods, and materials reuse. This will help to answer the questions below.

How locally sourced materials can influence embodied emissions? This can be analysed by making changes to the range of miles travelled per day in Table 17, column 5. If, for instance, the currently assumed range of 15 miles radius (30 miles round trip per delivery) is reduced to zero so that all materials are sourced from the construction site, that would eliminate 473.6 kgCO$_2$ of emissions. If however the delivery range is increased by a factor of 10 so that materials come from 150 miles distance (300 miles round trip per delivery), the delivery emissions will also increase by a factor of 10 and become 4736 $kgCO_2$. That would increase emissions from the construction process from 3.5% to 10.4% of the total embodied emissions.

How less intensive transportation method can influence embodied emissions? The calculations in the paragraph above are based on transportation by 7.5-ton trucks running on diesel fuel. If, however, all transportation could be achieved by electric vehicles, that would eliminate 473.63 $kgCO_2$ of emissions from Table 17.

How different construction methods can influence embodied emissions? This can be analysed by making changes to the number of days of operation of site machines in Table 18, column 3. If days of operation are set to zero, in other words, no site machines are used, that would eliminate 139 $kgCO_2$ from this table. Using electric construction machines, such as diggers, would have the same effect. If, however, the days of operation are increased by a factor of 10, that would increase the total in Table 18 also by a factor of 10 to 1,391 $kgCO_2$ and it would increase emissions from the construction process from 3.5% to 5.7% of the total embodied emissions.

How can reuse of materials reduce embodied emissions? End-of-life deconstruction, demolition, transport, and disposal as calculated in Tables 20–22, amount to 4476.24 $kgCO_2$, or 8.1% of the total embodied emissions. This could be eliminated if the building is retrofitted again after the current lifecycle period of 60 years. That was done previously in the case of Zero Carbon House, which was retrofitted in 2010, and it was made possible by using durable construction material when the building was first constructed in the 1840s. If, however, it turns out that deconstruction and demolition are inevitable, but the material from deconstruction and demolition could be completely reused, then disposal emissions from Table 22 would not apply, thus reducing the total embodied emissions by 2916 $kgCO_2$ and thus reducing the end-of-life emissions from 8.1% to 3%. Disposal emissions into landfill as calculated in Table 22 assumed that landfill gas recovery would be in operation. However, if landfill disposal does not have a gas recovery facility, then according to RICS, a much higher disposal emissions factor of 2.15 $kgCO_2$ would be applicable for bio-based materials, such as timber. If all timber is reused, then these much higher emissions would be prevented. Further emissions benefits and burdens apply for repurposed elements or energy recovered beyond the project lifecycle, as outlined in the RICS guidance (RICS, 2017, sec. 3.5.5 Module [D]).

However, not all the emissions calculated in this chapter occur on the first day when the building is completed. The maintenance emissions occur during the building use stage and the end-of-life emissions occur when the building is deconstructed and disposed or recycled. These embodied emissions calculated here will be used in Chapter 21, Section 21.1.11 within a corresponding timeline when they occur, in combination with operational emissions in Zero Carbon House, in order to calculate the lifecycle emissions and the time when the total emissions from the house reach zero.

How accurate are embodied emissions calculations? Although the examples of embodied emissions calculations in this chapter, in Chapter 19 and elsewhere in this book are introduced with significant attention to detail, there will be some discrepancies between the calculations and the actual embodied emissions. These will occur for two reasons: (1) a possibility that some materials have been missed out and therefore not accounted for; and (2) databases of embodied emissions of materials do not correspond to live changes of grid carbon intensities in the countries where the materials and products originate from. Item (1) will increase the calculation inaccuracy and item (2) will reduce it because the grid carbon intensities are getting better around the world. This means that at the time of writing, our calculations were close but could not be identical to actual embodied emissions.

5.7 TASKS FOR SIMPLE EXERCISES

I Create a model of a box with dimensions of 10 m × 6 m × 3 m. Install 40% glazing on one of the larger walls and 1 m wide by 2 m tall timber door on the opposite wall. The ground floor slab consists of 100 mm thermal insulation and 100 mm cast concrete. The walls are 100 mm brick and 100 mm block with 100 mm cavity insulation. The roof is 50 mm concrete deck with 100 mm internal insulation. The windows are double glazed with 10% of the window area used by a timber frame.

2 Replicate embodied emissions calculation from this chapter, including emissions in construction materials, construction process, maintenance, and end of life, using information sources referenced in this chapter.

3 Change the wall and roof construction material to 300 mm and 200 mm hemp-lime bio-composite material and re-run your calculations.

4 Discuss the results and differences with your colleagues.

PART 3

OPERATIONAL EMISSIONS CONTEXT

CHAPTER 6

SITE ISSUES

In this chapter, we will look at local environment or site issues. We will learn how to find information about the intensity and direction of prevailing wind from weather data, and how to mitigate its effect on a building. We will also learn about a sun path diagram and how to use it to design site geometry. We will then investigate how to control solar gain using orientation and reflective surfaces, and how to analyse overshadowing and a potential for daylighting. These investigations will be based on dynamic simulation experiments and comparisons between different simulation cases.

6.1 PREVAILING WIND AND SHELTER PLANTING

We have seen from the previous chapter that there are permanent winds that define global air circulation. These winds are reflected in local conditions as prevailing wind. A chart that displays the annual wind pattern for a particular location is called a 'wind rose'. Some DSM tools can display the wind rose either as wind intensity chart or wind frequency chart using information from the selected weather data file (Figure 56). The prevailing wind can be found by combining the highest magnitude and the highest frequency from these charts.

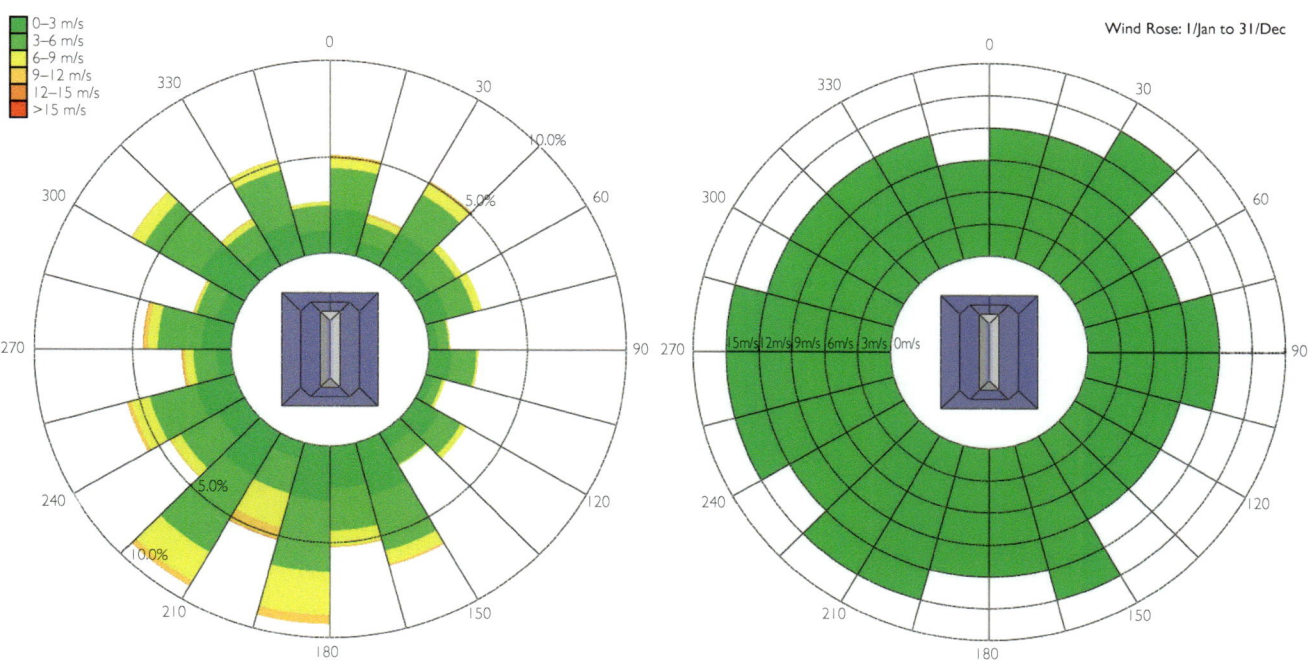

Figure 56   Wind rose for Birmingham in IES Virtual Environment: left – wind magnitude; right – wind frequency

DOI: 10.4324/9781003342342-9

To mitigate the effect of prevailing wind we can use shelter planting, which consists of trees and bushes planted on the prevailing wind side. The recommended distance from the building is between three and four heights of the shelter planting, which leaves enough space from the south-facing side so that the building can receive solar radiation (Figure 57). In three-dimensional space, the shelter planting may take the form as shown in Figure 58a. To convert it into a dynamic simulation model, trees could be replaced with solid objects as in Figure 58b. From the north-facing side, the plants can be located quite near to the building, creating a cushion of stagnant air that prevents heat losses from the building (Figure 58a). In order to ensure that desired conditions are achieved, the distance and configuration of the shelter planting need to be investigated in detail using a shadow casting module and a Computational Fluid Dynamics (CFD) module in a DSM.

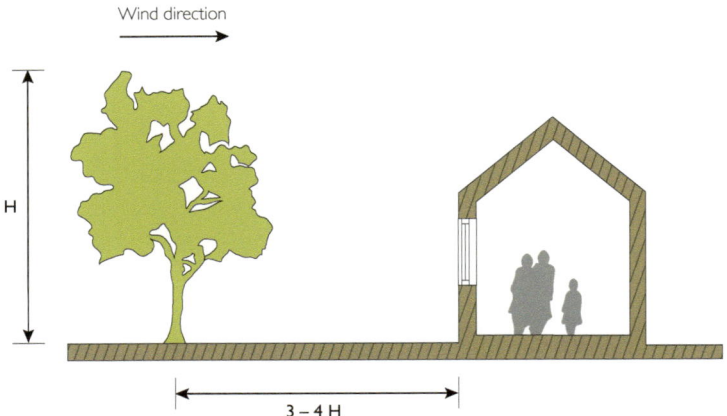

Figure 57   *Recommended distance for shelter planting (illustration by Oliver Mould)*

We will first investigate the effect of wind using a CFD module. Results of CFD analysis of an unprotected building are shown in Figure 59. The diagram on the left shows wind velocities on the windward side up to and in excess of 14 m/s, dropping down to less than 2 m/s on the leeward side immediately after the building. The diagram on the right shows pressure build-up on the windward side, with pressures exceeding 66 Pa in a small region of the roof.

The CFD analysis is repeated after the introduction of shelter planting, and the results are shown in Figure 60. The diagram on the left shows that wind velocities approaching the building are now lower, down to 10 m/s or less, but the diagram on the right now shows two high-pressure points on different sides of the roof. Iterative repositioning of the shelter planting and CFD analysis is therefore required in order to eliminate the high-pressure spots as well as to achieve lower velocities on the windward side. If needed, we ought to be able to make a trade-off between solar gains and infiltration losses, and perhaps reduce the distance between the shelter planting and the building on the windward side accordingly.

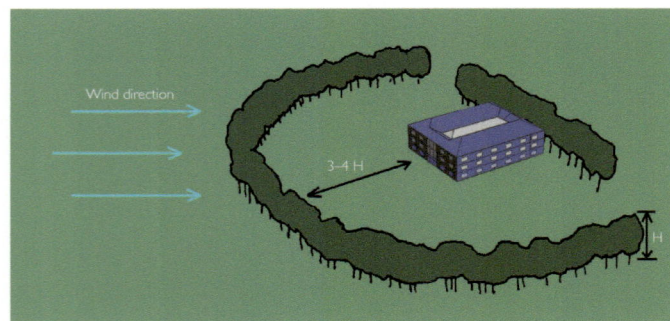

a) shelter planting in axonometric view (illustration
by Holly Doron)

b) solid objects replace trees in dynamic simulation
model

Figure 58   *Conceptual shelter planting in DSM*

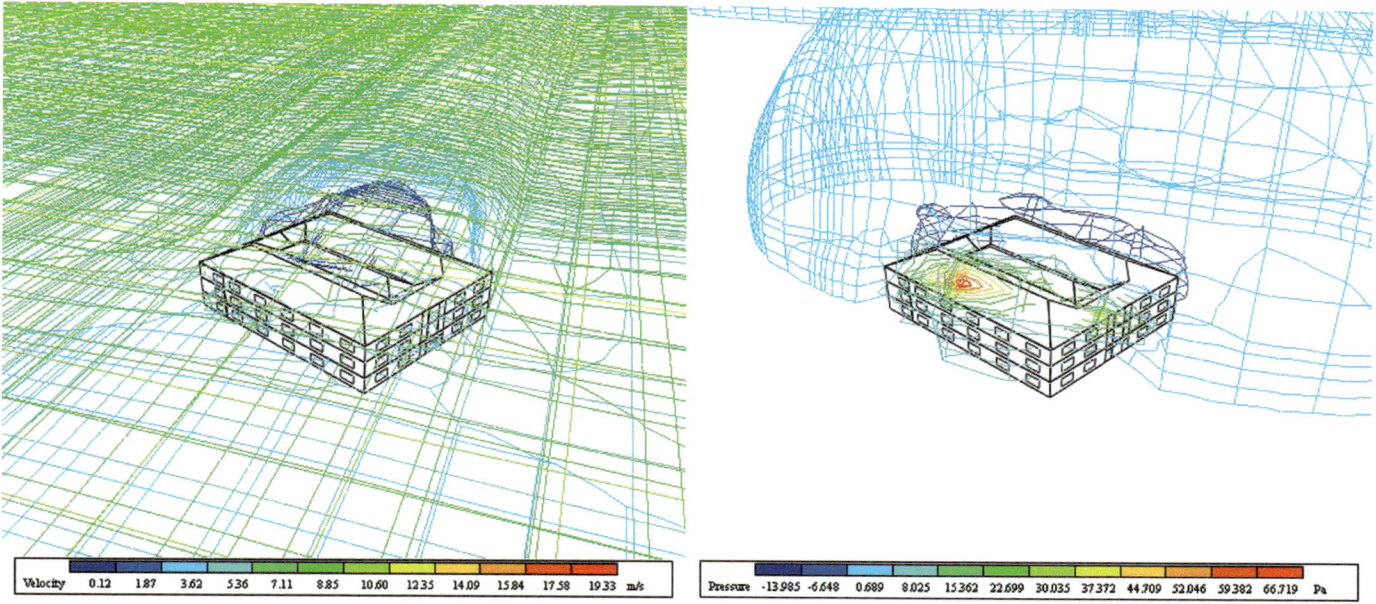

Figure 59   CFD analysis of wind velocities (left) and air pressures (right) without shelter planting

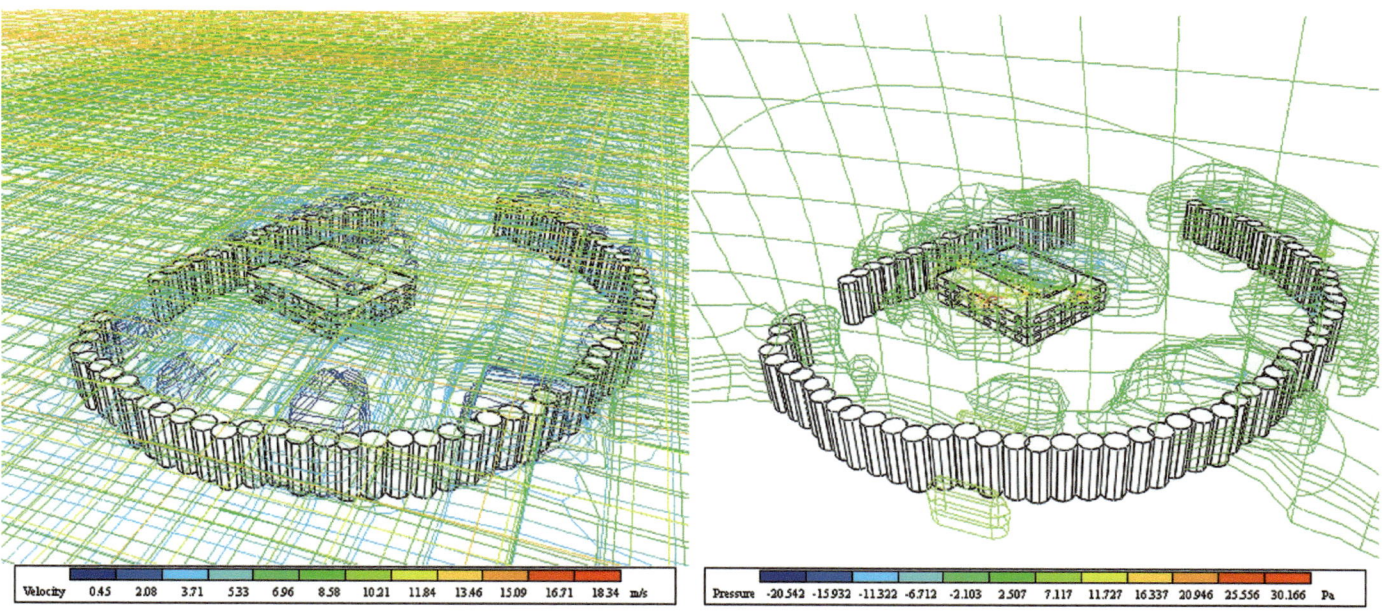

Figure 60   CFD analysis of wind velocities (left) and air pressures (right) influenced by partially open shelter planting

Having analysed wind velocities and pressures with shelter planting in place, we now need to analyse overshadowing from shelter planting. We can do this using a shadow casting module in a DSM, as shown in Figure 61.

We can see from this figure that the building is fully exposed to solar gain between 11:00 and 13:00 on 21st December, with partial exposure between 10:00 and 11:00 hours and between 13:00 and 14:00 hours. From the overshadowing point of view, we can conclude that the position and height of the shelter planting are appropriate.

If we decide to make changes to shelter planting to further mitigate wind velocities and pressures, we need to repeat both the CFD and shade casting analysis.

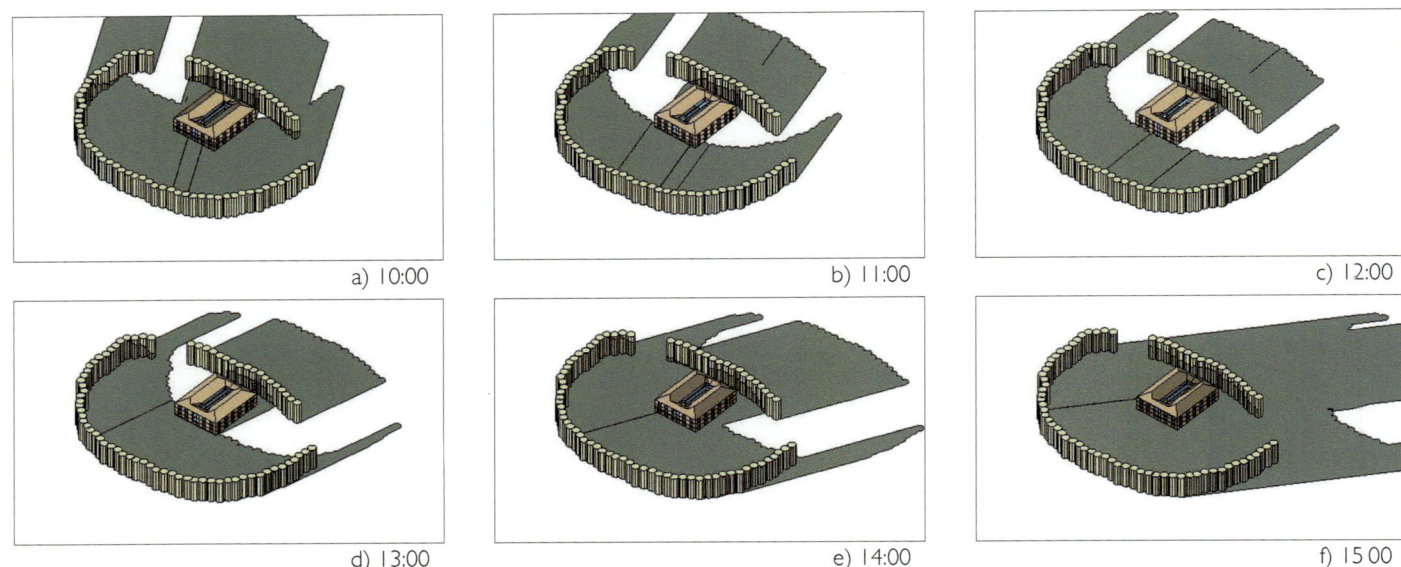

a) 10:00   b) 11:00   c) 12:00

d) 13:00   e) 14:00   f) 15:00

Figure 61   Sun cast analysis of overshadowing on 21st December resulting from shelter planting

6.2 SUN PATH DIAGRAM

The sun path diagram gives information on solar angles for different times of the day and different months of the year for a particular latitude. These are important parameters for site analysis, however, we will also use the sun path diagram for solar shading design in Chapter 10.

Some DSM tools can create sun path diagrams for different locations around the world. The diagram in Figure 62 was obtained from IES Virtual Environment. Here we will first explain the main features of the sun path diagram and a simple way of determining solar altitude angles. Further use of the sun path diagram will be explained in examples in the remainder of this chapter.

The outer circle in the sun path diagram represents the horizon. The scale marked on this circle represents azimuth angles – angles of deviation from the north direction. Azimuth angles are shown in degrees and are increasing in a clockwise direction.

Date (or month) lines run as arcs from left to right in the diagram. Figure 62 shows four date lines: summer and winter solstice lines (21 June and 21 December) and two overlapping equinox lines (21 March and 21 September). Lines perpendicular to the date lines are time (hour) lines. The intersection between each date and timeline corresponds to a particular hour on the corresponding date.

The concentric circles in the diagram represent a sky grid and correspond to solar altitude angles. These are the angles between the solar ray and its projection on the horizontal plane.

Example 1: To determine the solar altitude angle at 15:00 hours on the day of the summer solstice (21st June) in Birmingham, UK, we locate the intersection of the corresponding date and timelines (marked with a red dot in Figure 62), and follow the corresponding circle line to the scale of solar altitude angles (marked with a red arc). The arrowhead of the red arc shows the solar altitude angle of 50°.

Example 2: We would like to find solar altitude angles and times when a south-facing building receives solar radiation on 21st June. We represent the south facade as a straight line drawn through the centre of the sun path diagram (Figure 63). We first mark intersections between this line and the dateline for 21st June, as shown with red dots on the diagram. We then follow each red dot along the concentric circles that represent the sky grid until we get to the angle scale, and read the two angles.

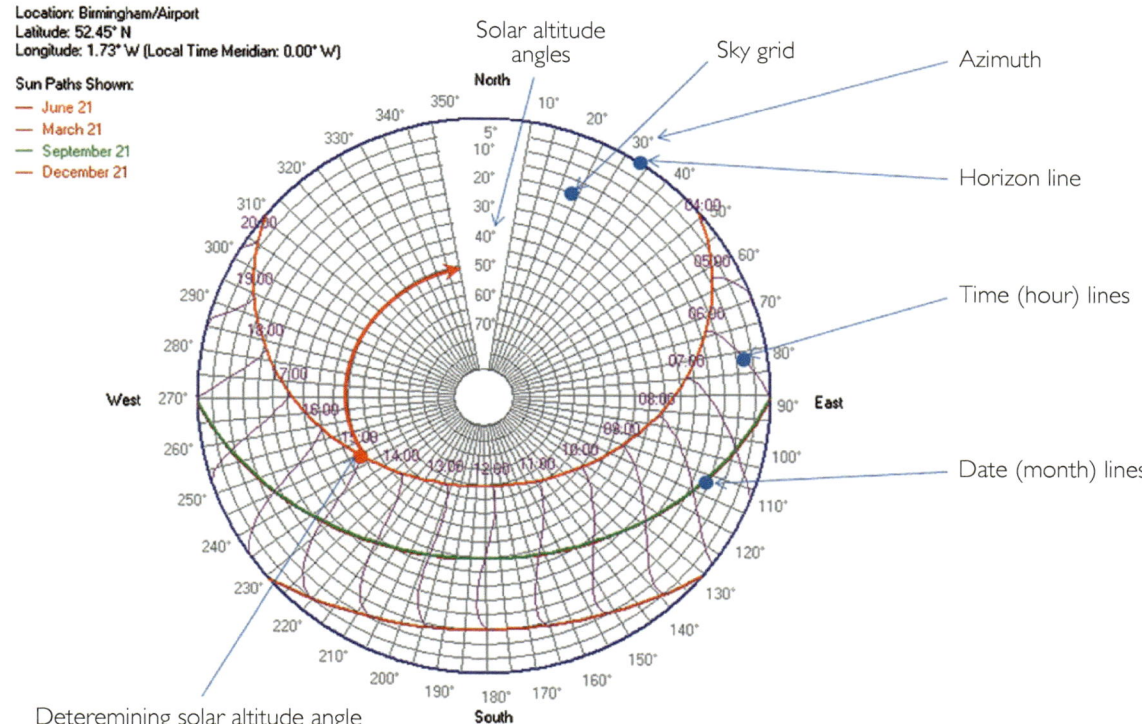

Figure 62   Sun path diagram for Birmingham, UK

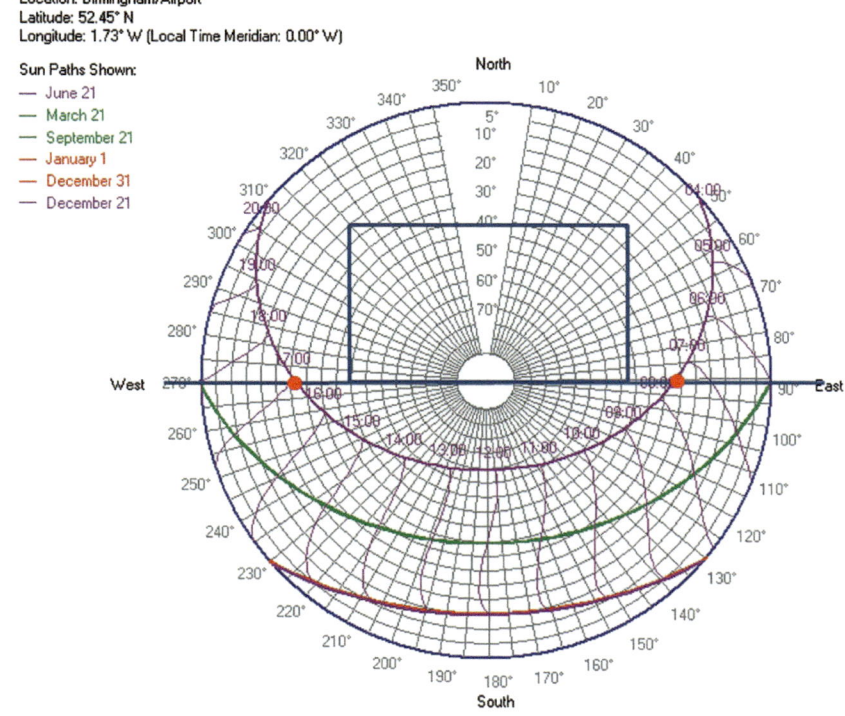

Figure 63   Determining solar angles for a south-facing building

Location: Birmingham/Airport
Latitude: 52.45° N
Longitude: 1.73° W (Local Time Meridian: 0.00° W)

Sun Paths Shown:
— June 21
— March 21
— September 21
— January 1
— December 31
— December 21

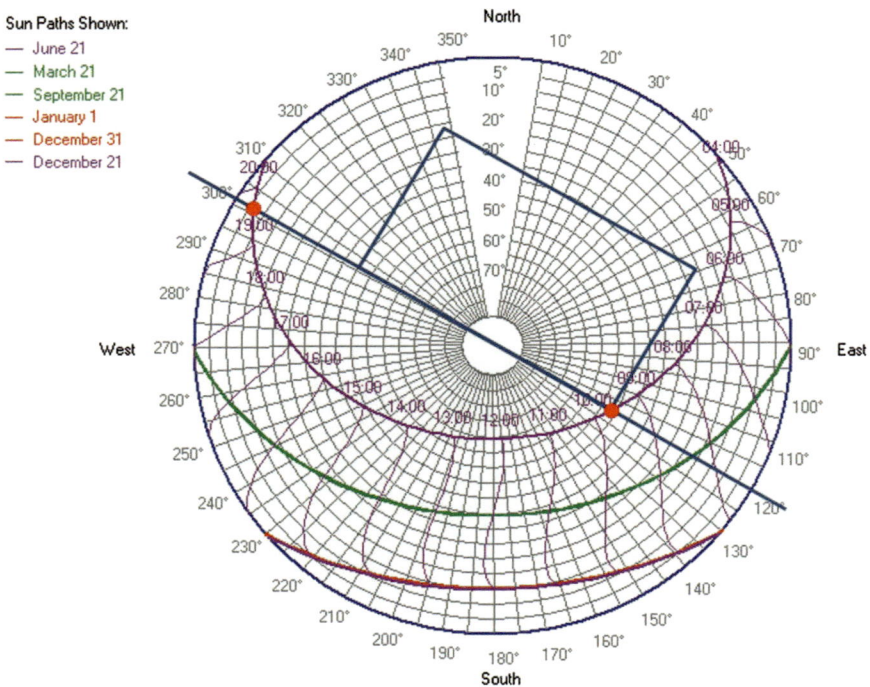

Figure 64   *Determining solar angles for a building facing 30° west of south*

They are exactly the same, approximately 32°, and occur at 7:30 and 16:30. The south facade will therefore be receiving solar gain between these times on 21st June.

Example 3: The building faces 30° west of south, and we would like to find angles and times when this facade receives solar radiation on 21st June. We draw the line representing the facade through the centre of the sun path diagram, and mark the intersection points with the date line for 21st June (Figure 64). We then follow the sky grid lines from the intersection points until we get to the angle scale. The corresponding angles and times read from the diagram are: approximately 52° just after 9:30 in the morning, and about 7° just after 19:30 in the evening.

6.3 MAXIMISING/MINIMISING SOLAR GAIN

6.3.1 Orientation

We saw in the previous chapter that solar radiation varies with the orientation, time of the year and latitude. A rule of thumb says that the optimum orientation in the northern hemisphere is ±30° due south. We are now going to investigate optimum orientation using a dynamic simulation experiment and check whether this rule of thumb is correct. We will use a model of a simple building located in London, with large south-facing windows as shown in Figure 65. We will vary its orientation from east-facing to south-facing in steps of 10 degrees.

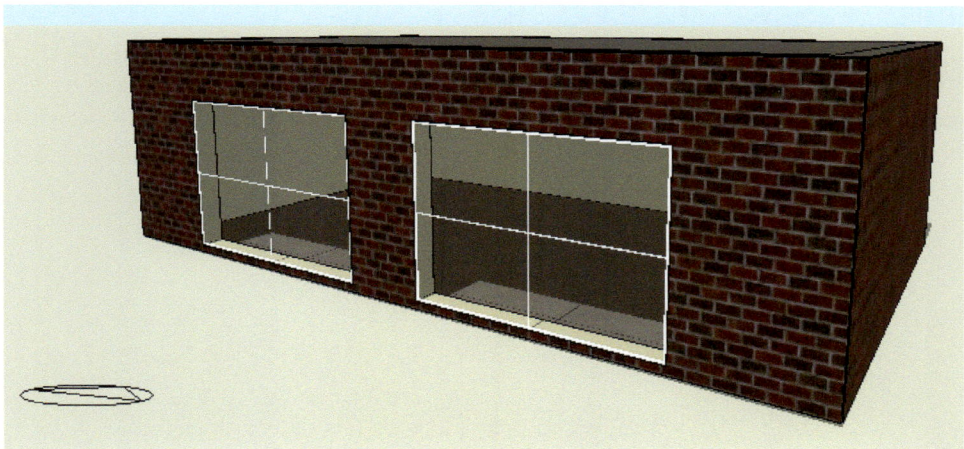

Figure 65   A simple model in DesignBuilder for orientation analysis

The results of this analysis are shown in Table 24, where azimuth angle denotes a clockwise angle from the direction of north, and 180° minus azimuth denotes the angle from the direction of south in the anticlockwise direction. Looking from the last row of the table, solar gain changes very slowly as the deviation from the south starts increasing. There is less than 3% change as the deviation from the south increases to 30°. Thereafter, the rate of change increases, quickly approaching and exceeding 3% for each 10° of the deviation from south. The analysis confirms that for this latitude the ±30° rule of thumb is generally correct. However, choosing the orientation based on simulation experiments rather than rules of thumb is highly recommended, as different latitudes and different site configurations will lead to different results.

TABLE 24 VARIATION OF SOLAR GAINS WITH CHANGING ORIENTATION			
Azimuth (°)	180-Azimuth (°)	Solar gain (kWh)	% of south solar gain
90	90	3524	70%
100	80	3791	75%
110	70	4051	80%
120	60	4296	85%
130	50	4516	89%
140	40	4702	93%
150	30	4846	96%
160	20	4948	98%
170	10	5014	99%
180	0	5055	100%

6.3.2 Use of reflective surfaces

We can increase solar gain on the south-facing vertical surface by creating reflective horizontal surface in front of the building (Figure 66). This could be a pond; or a light-coloured solid surface; or a highly reflective material; or this could be a consequence of a snow cover. This kind of surface will increase the amount of radiation falling on the vertical surface of the building.

This kind of effect is well known in various fruit-growing regions around the world, where land overlooks a significant amount of water on the south-facing side, such as a river or a lake. Solar radiation reflects from the water and delivers an increased amount of energy to the fruit plantation, making the fruit sweeter and of better quality in comparison with those regions that do not benefit from the same effect. Nature created these favourable conditions and people discovered them through many years of experience. We can replicate these conditions around buildings and increase the amount of solar radiation received on the south-facing surface.

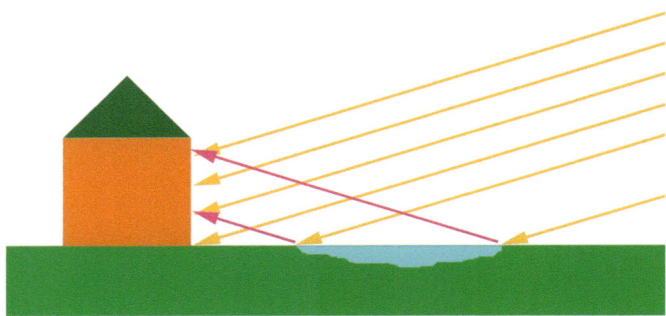

Figure 66   Reflective horizontal surface on the south side increases solar gain

Two simulations were conducted in order to establish the magnitude of change of solar gains resulting from increased ground reflectance, using the same model as in Figure 65. Table 25 shows that increasing ground reflectance from a typical 0.2 to a highly reflective 0.9 increases solar gain by over 43%. This is a significant increase, and we can therefore use ground reflectance as one of the control mechanisms for regulating solar gain.

TABLE 25 INFLUENCE OF GROUND REFLECTANCE ON SOLAR GAIN		
Ground reflectance	**Solar gain (kWh)**	**% increase**
0.2	5055	
0.9	7236	43.1%

But where exactly we should position the reflective surface? We can calculate the extent of the reflective surface as shown in Figure 67.

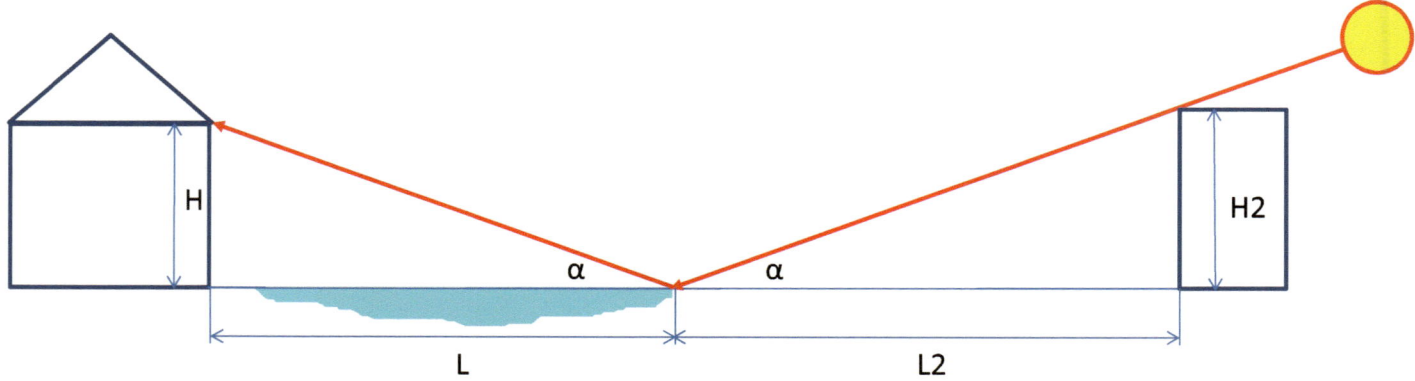

Figure 67   Geometry of solar angles on site: positioning a reflective surface (left) and calculating shade length (right)

Using the sun height angle α obtained from the sun path diagram we can write the following equation for the outer extent of the reflective surface L:

$$L = H \times \tan(\alpha) \tag{5}$$

from which we calculate distance L from the building by substituting actual values for building height H and sun height angle α. Note that distance L is aligned with azimuth angles in the sun path diagram (see Figure 62), rather than being perpendicular to the building facade.

We need to decide on the range of solar height angles to be taken into account, taking into consideration dates and hours during the year when we want the reflective surface to work. By repeating the above calculation for the corresponding range of angles we will obtain the outline of the reflective surface on the south side of the building.

6.4 OVERSHADOWING

Continuously changing angles of the sun will create shadow regions behind buildings throughout the year. These regions, called shadow prints, are shown in the plan and side elevation in Figure 68. We can determine the outline of the shadow print using a similar method as in the previous section (see Figure 67) as follows:

$$L2 = H2 \times \tan(\alpha) \tag{6}$$

where

$H2$ – obstacle height
$L2$ – length of the shadow

We can determine the outline of the shadow print by repeating this process for dates, times, and corresponding sun height angles of interest, bearing in mind that $L2$ is aligned with azimuth angles in the sun path diagram in Figure 62.

If constraints of a particular site allow us to choose, we need to ensure that new buildings are not positioned inside shadow prints of existing buildings or other topographical objects. A building inside a shadow print will receive a reduced amount of

Figure 68   Shadow prints and overshadowing

solar energy, and this will need to be compensated by another source of energy in order to maintain comfort conditions throughout the year, most likely leading to higher energy consumption and higher carbon emissions.

We can evaluate overshadowing using a shadow casting module of DSM software tools. An example is shown in Figure 69 where a taller building on the south-facing side casts shadow and causes increased energy consumption in the analysed building.

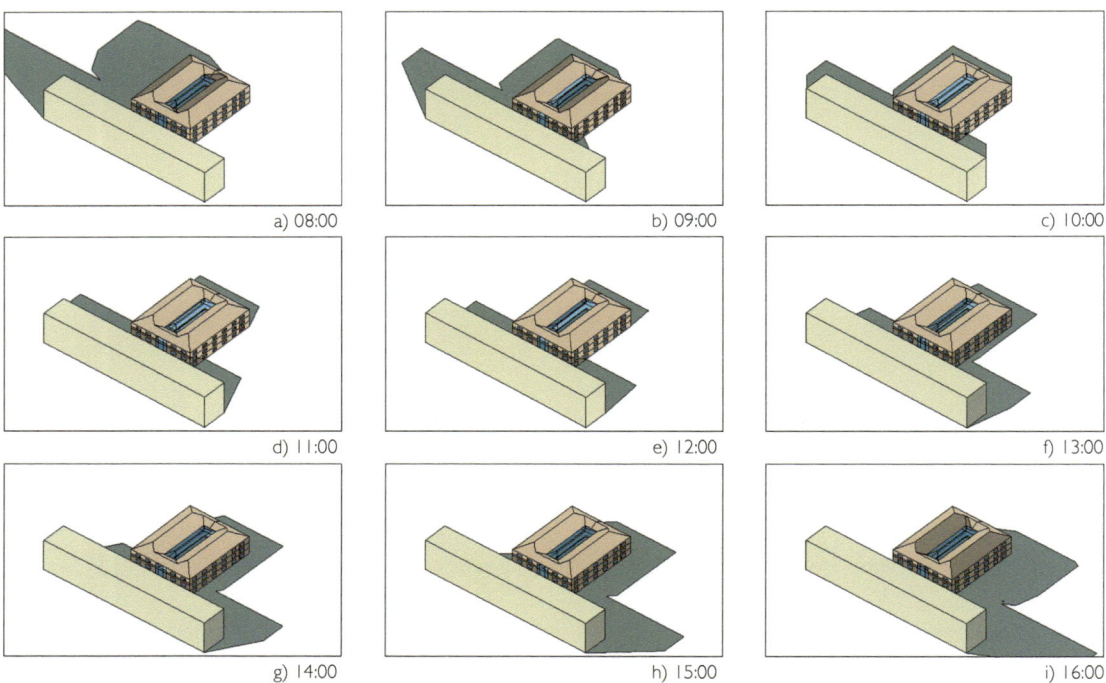

a) 08:00	b) 09:00	c) 10:00
d) 11:00	e) 12:00	f) 13:00
g) 14:00	h) 15:00	i) 16:00

Figure 69   Overshadowing from a taller building on 21st March

To mitigate this problem, we need to carry out site analysis and explore various positions for the new building so as to maximise the amount of solar radiation available throughout the year.

6.5 DAYLIGHTING POTENTIAL

Daylighting potential is a site parameter expressed in terms of a vertical sky component (VSC). This is a ratio between daylight received at the centre of a vertical window plane and daylight received at the same point on a horizontal plane from unobstructed sky.

Using the notation from Figure 70, we can define the VSC as

$$\text{VSC} = \frac{E_v}{E_{out}}$$

(7)

where

E_v – illuminance at the centre of a vertical window plane
E_{out} – illuminance at the horizontal plane from unobstructed sky

We can measure the VSC from an image generated by a lighting simulation program, such as RadianceIES (Figure 71), using the same model with a large obstruction from Figure 69. The image on the left corresponds to noon on 21st March and the image on the right corresponds to noon on 21st June. As windows in this software are represented as clear surfaces, we need to convert them into opaque surfaces (in this case doors) before we are able to do the measurements on the outside glazing plane. For comparison, the lower right window on the south-facing side is left in its original form, where illuminance measurement would show internal rather than external values.

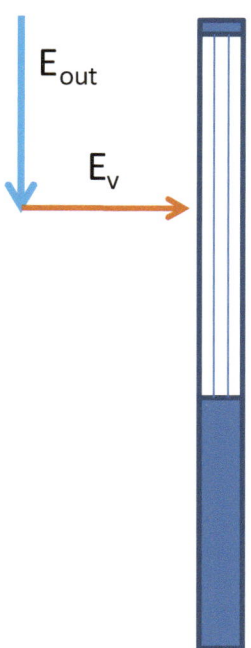

Figure 70   Definition of the Vertical Sky Component

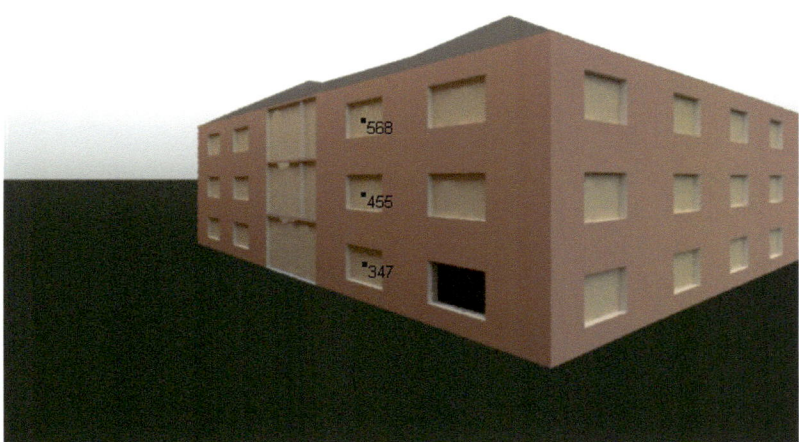

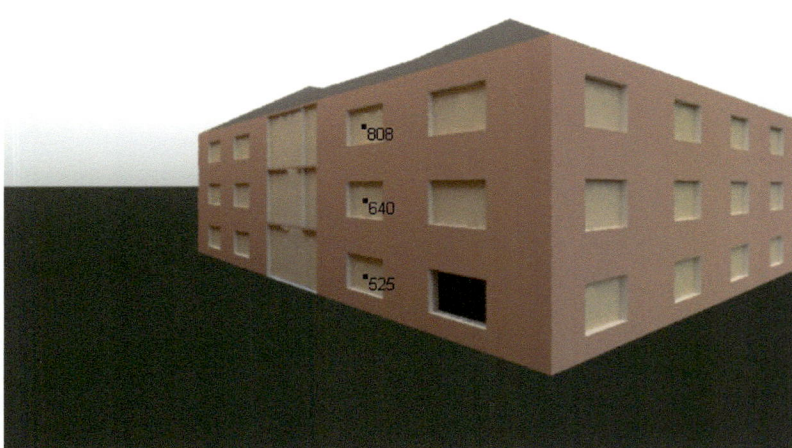

Figure 71   Vertical sky component on 21st March (top) and 21st June (bottom) in Birmingham

The numbers in Figure 71 are in lux and represent illuminances at the centre of the corresponding window surfaces. As the simulation was carried out using the Standard CIE Overcast Sky with an illuminance of 10,000 lux, we can obtain the VSC by dividing the illuminance values measured at each individual window by the illuminance of the standard sky. Thus, for the top window on 21st March, the VSC is equal to 568/10000 = 0.0568 or 5.68%. VSC values obtained in this way are shown in Table 26. We can see that the VSC increases with height above the ground, proportionally to the visible sky. However, as the standard sky model used was the 'overcast sky' rather than the 'sunny sky', we do not see as much variation between the windows on different floors as we might expect from the images in Figure 69.

TABLE 26 CALCULATION OF THE VERTICAL SKY COMPONENT (VSC)				
	Vertical illuminance		VSC	
Window location	21st March (lux)	21st June (lux)	21st March (%)	21st June (%)
Ground floor	347	525	3.47	5.25
1st floor	456	640	4.56	6.40
2nd floor	568	808	5.68	8.08

6.6 TASKS FOR SIMPLE SIMULATION EXPERIMENTS

1   Create a sun path diagram and wind rose for a location of your choice using IES VE. What can you tell about summer and winter solar angles and prevailing winds by observing these diagrams?
2   Create a massing model of a building and position shelter planting using the guidelines provided in this chapter. Run external CFD analysis with and without shelter planting and observe air pressures and velocities. Are there any high-pressure points in the model before or after the insertion of shelter planting? What kind of problems can high-pressure points cause if they are not eliminated by design?
3   Create a box of 10 m × 6 m × 3 m with a longer side facing south. Insert 40% of glazing on the south-facing side. Copy the box vertically ten times, placing it the copies on the most recently created box below, thus creating a ten storey building block. Surround the model created in this way with four shading blocks of varying sizes placed 10 m away from each side of the central ten storey box.

   a)   Carry out shading analysis of the central ten storey block for a location of your choice on 21st December, 21st March, and 21st June.
   b)   Calculate daylighting potential for each south-facing window on each storey of the central block.
   c)   Set the ground reflectance to 0.2 and calculate the annual solar gain on each storey of the central block.
   d)   Change the ground reflectance from 0.2 to 0.9 and repeat the above sub-task.
   e)   Move the surrounding four blocks from 10 to 20 m away from the central block and repeat all of the above steps from this task.
   f)   Discuss the results with your colleagues.

6.7 SUMMARY OF DESIGN PRINCIPLES

Prevailing wind:

• Use DSM wind rose to identify the prevailing wind direction and intensity.
• Specify shelter planting to protect building exposure to the prevailing wind.
• Use CFD and shadow casting modules in DSM to evaluate the effects of shelter planting.
• Make adjustments to shelter planting if necessary and repeat the above step until desired conditions are achieved.

Solar gain:

• Use building orientation and external horizontal surface reflectance to control solar gain.
• To maximise solar gain, use south-facing orientation with no more than 30° deviation from south.
• Use a highly reflective external horizontal surface on the south side to increase solar gain.
• Determine the outline and the position of the reflective surface using the sun path diagram and the method described in this chapter.
• Use DSM to quantify the effects of the chosen orientation and ground surface reflectance.

Overshadowing:

• Use the sun path diagram and the method described in this chapter to calculate the outline of a shadow print.
• Conduct shadow casting analysis for different site configurations so as to maximise the amount of solar radiation available for the building throughout the year.

Daylighting potential:

• Replace windows in a DSM model with opaque surfaces such as doors.
• Run a Radiance simulation and obtain measurements of illuminance at the centre of a vertical window plane.
• Obtain daylighting potential by calculating VSC as the ratio between the measured illuminance and the standard sky illuminance.

BUILDING GEOMETRY

In this chapter, we will set out early foundations for zero carbon design by making early design decisions about building geometry.

7.1 SHALLOW PLAN VERSUS DEEP PLAN

One of the most important design decisions to be made at an early design stage, before the first line is drawn on paper is the plan depth. This will influence several aspects of zero carbon design, including natural daylight, natural ventilation, energy consumption, as well as the amenity of internal spaces, availability of views to the outside, and others.

There is one of two choices to be made: shallow plan (Figures 72 and 73) or deep plan (Figure 74). In a single-sided building (a building with windows on only one facade) with a typical floor-to-ceiling height of about 3 m, the shallow plan means that

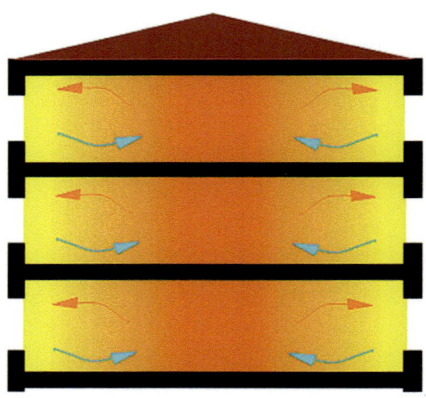

Figure 72   Shallow plan

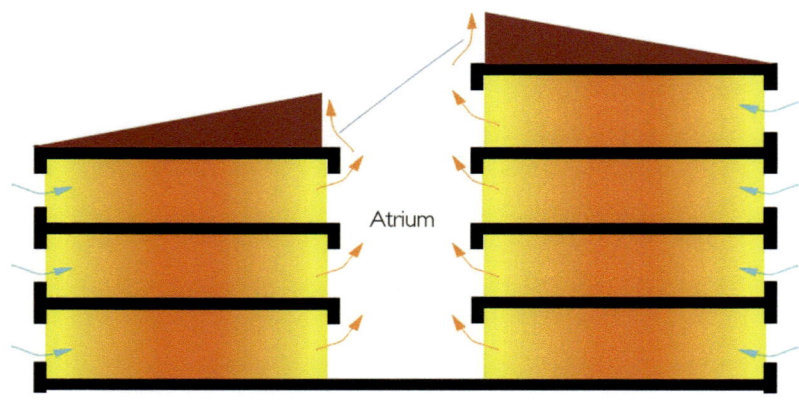

Figure 73   Shallow plan with atrium

the building is not deeper than 6m from the window to the back wall. In a double-sided building (a building with windows on two opposite facades) with the same typical floor-to-ceiling height of 3 metres, the shallow plan means that the building is not deeper than 12 m between the opposite facades (Figure 72).

One of the consequences of the shallow plan is a good potential for natural ventilation. This will reduce the need for mechanical ventilation or air conditioning, unless specifically required by the work process in the building, or dictated by a hot climate,

DOI: 10.4324/9781003342342-10

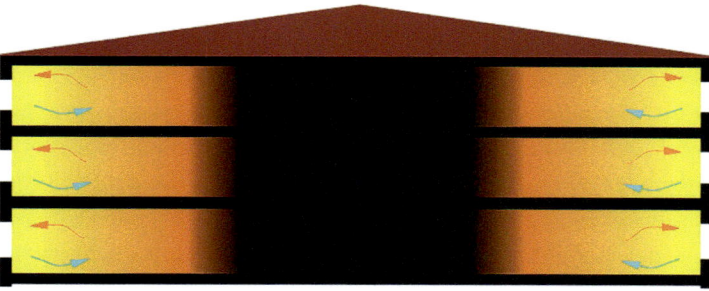

Figure 74 Deep plan

and will have a significant impact on energy consumption and carbon emissions. We will discuss natural ventilation in detail in Chapter 12.

Another consequence of the shallow plan is the opportunity to use natural daylight to its full potential. If electrical lights are controlled in response to the available daylight, either manually or automatically, this will have a consequence of a reduced use of electricity. We will discuss the issues of natural daylight and electrical light in Chapters 13 and 14. In Chapter 13, we will explain that natural daylight also has physiological effects on the human body through suppression of the secretion of a hormone called melatonin.

For a different floor-to-ceiling height, the depth of the plan will need to be established through experimentation with dynamic simulation models, so as to ensure good potential for natural ventilation and natural daylight.

There are other less tangible advantages of the shallow plan which are equally important. In a shallow-plan building, all occupants have a good chance to have views to the outside. This will have a psychological effect on them and will reduce the occurrence of fatigue. The visual contact with the outside environment will enable the change of the focal distance between the inside and outside environment, and reduce the occurrence of fatigue. The reduced need for electrical lighting will also work in the same direction, as seemingly unnoticeable flickering of electrical light is known to cause fatigue in office workers. The overall result will be a more pleasant and healthy building to live or work in and happier building occupants.

A variation of the shallow plan described above is a building with a central atrium and spaces which are not deeper than 12m from the outside facade to the atrium (Figure 73). In addition to all advantages of the shallow plan already described, the atrium will provide an additional amenity for the building's occupants, and an internal circulation space protected from external influences.

The other choice we can make is the deep plan (Figure 74). The meaning of deep plan is that the horizontal distance from the window to the back wall is more than 6 m in a single-sided building, or more than 12m in a double-sided building. In this type of plan it will not be possible to naturally ventilate the core of the building, and hence a mechanical ventilation or air conditioning will be required, having consequences on increased energy consumption and increased carbon emissions. The deep plan will also have a detrimental effect on the availability of natural daylight in the building, and will result in an increased need for artificial lighting. A number of occupants will not have views to the outside, and the increased requirement for mechanical and electrical services will reduce the amenity and quality of the internal space.

Although the deep plan will have a better site space utilisation, as it will be possible to put more people per square metre of site, the price to pay will not just be increased energy consumption and increased carbon emissions but also reduced quality of the internal environment and negative consequences on health and productivity of building users.

7.2 VOLUME-TO-SURFACE RATIO

Volume-to-surface ratio is another important consideration to make at an early design stage. The context for this consideration is that heat is generated and stored within the volume of the building and exchanged through its surface. There are three typical geometric forms that we will use for explanation: a sphere, a cube, and a pyramid. In Figure 75 the basic dimension is diameter in the case of sphere, edge length in the case of cube, and base side length in the case of pyramid. As we increase the basic dimension whilst keeping the surface areas of three objects equal by adjusting their dimensions accordingly, we see the fastest increase of volume-to-surface ratio in the case of sphere, the slowest in the case of pyramid, and in between in the case of cube. This means that the volume of the sphere will increase faster with reference to its surface area, making it good at retaining heat. In the case of pyramid, the volume will be increasing slower with reference to its surface area, making it good at losing heat.

My intuition tells me that it is not a coincidence that traditional buildings made by the indigenous peoples in the Arctic region resemble the shape of a sphere (or hemisphere to be more precise), and that monumental buildings made by ancient Egyptians resemble the shape of a pyramid. The former building needed to retain heat, and the latter needed to dissipate heat. Most of the buildings in Western civilisation are in between these two extremes and resemble the shape of a cube and various variations of it. The real reasons for different building shapes in different regions are more likely to be structural properties of locally available materials, corresponding construction methods and space utilisation, with heat transfer properties potentially being of secondary importance. We can, however, use this knowledge of volume-to-surface ratio in modern buildings, using curved shapes to retain more heat, or shapes with numerous corners to lose more heat, depending on whether our building predominantly needs heating or cooling.

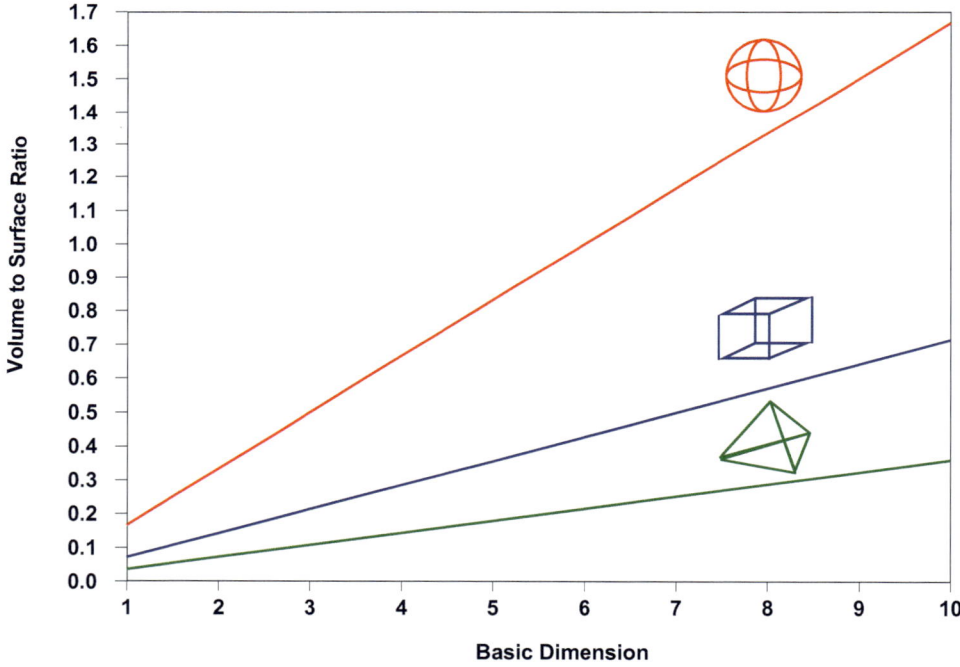

Figure 75   Volume-to-surface ratio

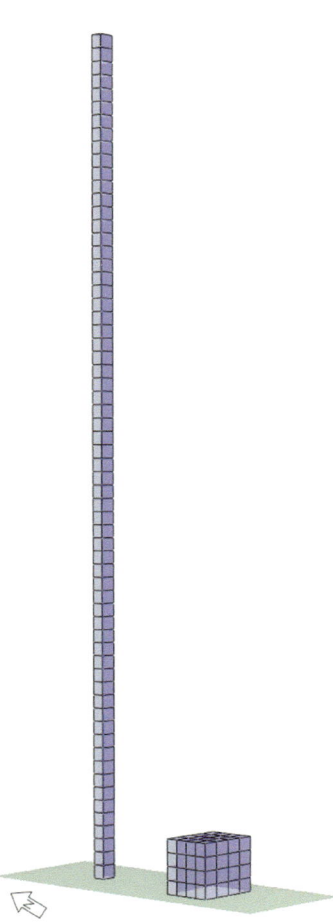

Figure 76   Geometry for volume-to-surface simulation test

In order to make this discussion more tangible, we will use an extreme example shown in Figure 76. The two volumes have been made from 64 blocks of 1 m³. The taller volume (Tall Box) consists of 64 blocks placed vertically, and the 'smaller' compact volume (Compact Box) consists of 4 × 4 × 4 blocks. The total exposed surface area of the Tall Box is 257 m² (walls 256 m², roof 1 m²), and the total exposed surface area of the Compact Box is 80 m² (walls 64 m², roof 16 m²). Therefore, the volume-to-surface ratio of the Tall Box is 64/257 = 0.249 m and volume-to-surface ratio of the Compact Box is 64/80 = 0.8 m.

The walls, ground floor, and roof are made of concrete blocks or concrete, and are uninsulated. The models are without glazing and are set to be unoccupied with no internal gains. Infiltration rate is set to 0.167 air changes per hour. The heating operates continuously with a set point of 19°C, and cooling is permanently off. The location is set to Birmingham.

The result of an annual simulation is shown in Figure 77. As can be seen from this figure, the heating load is much higher in the Tall Box than in the Compact Box. The annual totals are 14.411 MWh and 5.989 MWh, respectively, therefore 141% higher in the Tall Box, calculated as (14.411–5.989)/5.989. This illustrates that volume-to-surface ratio is an important design parameter, and it should be given careful consideration.

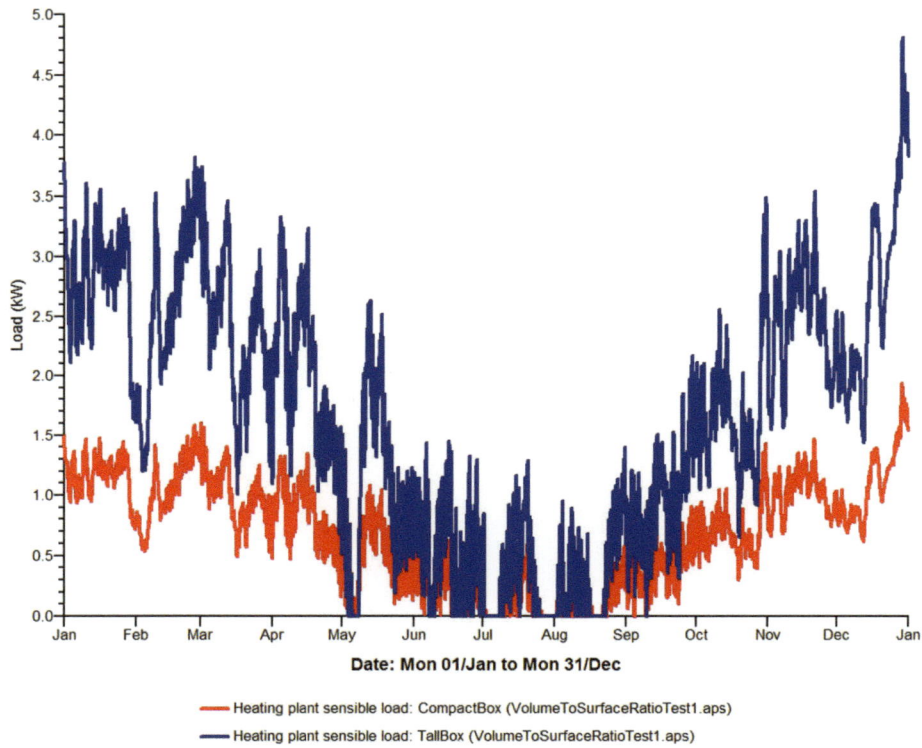

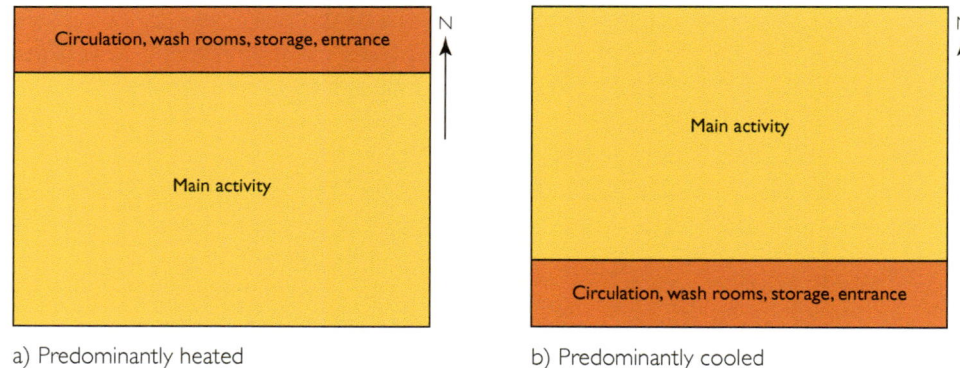

Figure 77   Volume-to-surface ratio simulation result

7.3 INTERNAL BUILDING LAYOUT

Buildings will require less heating energy and will generate fewer emissions if the main activity is positioned towards the sun-facing side. Buildings that are predominantly cooled will require less cooling energy and will generate fewer emissions if the main activity is positioned to face away from the sun. Figure 78 illustrates this for the northern hemisphere, whilst an inverse layout will be applicable for the southern hemisphere.

a) Predominantly heated   b) Predominantly cooled

Figure 78   Internal building layout for northern hemisphere (inverse layout applies to southern hemisphere)

7.4 TASKS FOR SIMPLE SIMULATION EXPERIMENTS

I   Investigate the effect of the volume-to-surface ratio on three different configurations of 64 blocks of 1 m³ each as follows: (1) 1 wide × 1 deep × 64 tall, (2) 2 wide × 1 deep × 32 tall, and (3) 4 wide × 1 deep × 16 tall. Calculate the percentage difference between respective annual heating loads.

2 Create models of two buildings using two sets of 64 blocks of 1 m³ each as follows: (1) 4 wide, 4 deep, and 4 tall and (2) 5 wide, 5 deep, and 4 tall, with 3 × 3 × 4 blocks taken out from the middle, thus creating a courtyard shape. Run annual simulations and compare annual and hourly simulation loads. Note the conflicting constraint between the plan depth and the volume-to-surface ratio. What steps would you take to resolve this conflicting constraint in a design project?

3 Discuss the results with colleagues.

7.5 SUMMARY OF DESIGN PRINCIPLES

Building geometry:

• Use shallow-plan depth to maximise natural ventilation and natural daylight.
• Use curved surfaces to reduce heat losses of sharp edge surfaces to increase heat losses.
• Investigate different configurations of volume-to-surface ratio of design options and select the most favourable option in the context of the conflicting constraint of plan depth.
• Position high activity spaces towards high solar gains if the building is predominantly heated or the other way round if the building is predominantly cooled.

CHAPTER 8

THERMAL INSULATION AND AIRTIGHTNESS

In this chapter, elementary principles of heat transfer associated with thermal insulation and airtightness will be explained and their implementation in dynamic simulation will be demonstrated. Simulation experiments will then be conducted to evaluate effects of the following: high versus low insulation; external versus internal insulation; and high versus low airtightness. Consequences on building energy efficiency and carbon emissions will be discussed on the basis of these simulation experiments.

8.1 HEAT TRANSFER PROPERTIES OF MATERIALS AND THERMAL DIFFUSIVITY

When we touch a piece of metal at room temperature, we feel that it is cold, and when we touch a piece of polyurethane foam also at room temperature, we feel it is warm. How can there be such a difference in our perception of two different materials, both at room temperature? What are the parameters that cause this difference? Three parameters will be introduced to help us answer this question and understand some fundamental principles of heat transfer: specific heat, thermal conductivity, and density.

The first parameter to discuss is the specific heat. It tells us how much heat in Joules (J) we need to add to a unit mass of a material in order to increase its temperature by 1°C. The specific heat is a property of materials as it applies to unit mass. For instance, the unit mass of material of higher specific heat will absorb more heat than the unit mass of material of lower specific heat.

Whilst specific heat is a property of materials, heat capacity is a property of bodies

$$C = m \times c \tag{8}$$

where

C – heat capacity (J/K)
m – mass (kg)
c – specific heat (J/(kg·K))

Heat capacity effectively tells us how much heat should be added to a body, rather than to a unit mass, in order to increase its temperature by 1°C.

If we add Q amount of heat to a body, its temperature will change according to the following equation:

$$Q = C \times (T_{end} - T_{start}) \tag{9}$$

where

T_{start} – is initial temperature of the body equal to room temperature from our example above
T_{end} – is the final temperature that the body reaches, having received an amount of heat Q.

To find out how different values of specific heat and heat capacity affect the final body temperature, we solve the equation above for T_{end} as follows:

$$T_{end} = \frac{Q}{C} + T_{start}$$ (10)

or if we substitute C with $m \times c$, we get

$$T_{end} = \frac{Q}{m \times c} + T_{start}$$ (11)

This means that the higher specific heat c, the lower the end temperature T_{end}, and vice versa. Specific heat will help a bit further below to explain why we feel that a piece of metal is cold and a piece of thermal insulation is warm when both are at room temperature.

The second parameter to discuss is thermal conductivity, expressed in units of W/(m·K). This parameter tells us how easily heat goes through material. As metals have higher thermal conductivity than insulation materials, heat from our hand when touching a piece of metal will be drawn away from the hand. Conversely, as thermal insulation materials have low thermal conductivity, heat from our hand when touching a piece of polyurethane will not be drawn away from the hand. Although the answer to the question why a piece of metal feels cold and a piece of polyurethane feels warm when we touch both at room temperature is now beginning to emerge, we still have one parameter missing to get the full answer. This parameter is material density.

Density of materials is expressed in kg/m³. It tells us how much mass of material is found in a unit volume. It effectively tells us how compact the material is, and this will affect its capability to transmit heat. Higher-density materials will transmit heat quicker because of denser 'packing' of material in the unit volume.

Using values for these three properties of materials (specific heat, thermal conductivity, and density) will now help to analyse why we feel that metal is cold and thermal insulation is warm when both are at room temperature. Faster transmission of heat will make it possible for material to remove heat from another body, such as from our hand, but only if it can quickly absorb heat as well. We will therefore feel that an object is cold at room temperature if it has high conductivity and high capacity to store heat that it has removed. Comparing the properties of steel and polyurethane in Table 27, it now becomes clear why steel feels colder.

The ratio between the conductivity and the specific heat-density product shown in the rightmost column in Table 27, called 'thermal diffusivity', measures the rate of heat removal from a heat source. Metals have high thermal diffusivity and will remove heat from our hand quickly, making us feel that the material is cold. Insulation materials have low diffusivity and will remove heat from our hand slowly, making us feel that the material is warm. As the thermal diffusivity of steel is almost 19 times higher, we will feel that steel is much colder than polystyrene, even though both are at the same room temperature.

8.1.1 Thermal insulation materials

There are numerous materials characterised by low thermal conductivity which can be used for thermal insulation. For example, natural or recycled materials, in addition to saving on operational carbon emissions, also save on embodied carbon. The purpose of this section is not to give a detailed overview of these materials, but to highlight certain materials and their applications. Detailed information about common building materials and their properties can be found in CIBSE Guide A (Butcher et al., 2015), and Suhr et al. (2019) give an excellent overview of different types of insulation materials and their application, especially in retrofit situations.

A particular natural insulation material has received considerable interest in the UK recently: hemp-lime bio-composite, also called hempcrete, is a building material based on naturally grown hemp bonded with lime (Figure 79) and it is therefore renewable. Its growth captures carbon dioxide from the atmosphere and thus reverses carbon emissions. Hemp-lime material is characterised by very low thermal diffusivity (Table 27), and it therefore removes heat very slowly from heat sources. The internal structure of the material is characterised by interconnected capillary tubes. This makes the material breathable. Heating and cooling cause evaporation and condensation of moisture in the capillary tubes, thus adding latent heat transfer properties to the material. For this reason, heat gains and losses differ from those predicted on the basis of conductive heat transfer alone. The combination of its sensible and latent heat characteristics gives hemp-lime material good heat insulation properties, as well as good heat storage properties, a combination that is not easily found in many other materials. Examples of two hemp-lime housing projects completed in the UK in 2011 are shown in Figure 80.

a) Long Meadow development at Diss by C-Zero

b) The Triangle development at Swindon by Hab Oakus

Figure 79   Hemp-lime material (the sample courtesy of Greencore Construction)

a) 300 mm block with hemp leaf on top

b) Hemp-lime block close-up

Figure 80   Recent housing projects developed with hemp-lime bio-composite material

As hemp-lime bio-composite is a relatively new material, mainstream simulation software tools did not have the capability to model it at the time of writing this text. The best software candidates for creating models are those that have built-in capabilities for new materials, such as EnergyPlus and TRNSYS (see Chapter 3). As these two simulation tools are built for easy expansion, this would facilitate development of modules that can simulate new materials, such as hemp-lime material.

TABLE 27 COMPARATIVE ILLUSTRATION OF THERMAL PROPERTIES OF MATERIALS					
	Conductivity	Density	Specific heat	Diffusivity	Relative to steel
Material	W/(m·K)	kg/m³	J/(kg·K)	m²/s × 10^{-7}	
Steel	45	7800	480	120.19	
Polyurethane	0.028	30	1470	6.35	18.93
Hemp-lime	0.05	240	1750	1.19	100.96

8.2 THERMAL RESISTANCE AND U-VALUE

Having analysed heat transfer properties of materials in the previous section, we are now ready to define some derived parameters, such as thermal resistance R and thermal transmittance U.

In a single-layered wall, thermal resistance is calculated as

$$R = \frac{d}{k} \tag{12}$$

where

d — wall thickness in metres
k — thermal conductivity in W/(m·K).

The unit of measure for R is therefore derived as m²K/W. Thermal transmittance U for a single-layered wall is then defined as a reciprocal of thermal resistance

$$U = \frac{1}{R} \tag{13}$$

and therefore the unit of measure for U is W/(m²K).

Most constructions in buildings however have more than one layer. We will now find out how to calculate thermal resistance and U-value for a multi-layered construction as in Figure 81.

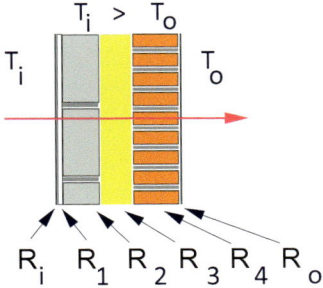

Figure 81   Multi-layered wall construction

The layers that we must take into account are not only the physical layers but also internal and external imaginary surface layers R_i and R_o.

The total thermal resistance of the above multi-layered construction is calculated as:

$$\sum R = R_i + R_1 + R_2 + R_3 + R_4 + R_o \tag{14}$$

Thermal transmittance is then calculated as a reciprocal of the sum of all thermal resistances as

$$U = \frac{1}{\sum R} \tag{15}$$

8.3 THERMAL DIFFUSIVITY VERSUS THERMAL TRANSMITTANCE

Having introduced both thermal diffusivity and thermal transmittance (U-value) in this chapter, it is important to put both into context. Thermal diffusivity is calculated from thermal conductivity, density, and specific heat. U-value is calculated from thermal conductivity only, and therefore it has much lower information contents regarding heat transfer, in comparison with thermal diffusivity.

Although DSMs report on U-values, they do not use U-values for heat transfer calculations. Instead, they use thermal diffusivity, which is why they are capable of representing dynamic heat transfer. Tools that use U-values only are not capable of representing dynamic heat transfer, and can therefore be only used as evaluation tools for steady-state heat transfer.

8.4 SURFACE RESISTANCES AND AIR GAPS

Whilst resistances of physical layers are calculated on the basis of conductive heat transfer in the same way as the resistance of the single-layered wall described above, resistances of internal and external surface layers in Equation (14) are not calculated. Instead, they are defined in relevant standards and available from engineering handbooks, for instance from CIBSE Guide A (Butcher et al., 2015). But what do these resistances represent considering that they are different from resistances of physical layers?

Heat transfer on the internal and external wall surfaces is a complex combination of radiation and convection heat transfer. This is described in detail in the 'Heat transfer' chapter in CIBSE Guide C (W. P. Jones et al., 2007). Radiation heat transfer depends on wall surface temperature and temperatures and geometry of the surrounding surfaces, distance and temperature difference as well as surface emissivity, surface inclination, direction of heat transfer, and sky conditions for external surfaces, along with other factors. Replacing this complexity with a single number is a significant simplification, but nevertheless, it enables a reasonable representation of the processes at the wall surfaces.

CIBSE Guide A (Butcher et al., 2015) as well as other engineering handbooks contain tables of internal and external resistances, applicable to different conditions, including whether the surface is exposed, sheltered, or subject to horizontal or vertical, upward, or downward heat transfer. Sheltered external surfaces have higher surface resistance and exposed surfaces have lower surface resistance. Internal surfaces have higher resistance than external surfaces.

Heat transfer in air gaps is also manifested by radiation and convection heat transfer. It is influenced by the thickness and geometry of the air gap, and by the emissivity of internal surfaces. It also depends on whether the gap is ventilated, and whether it is horizontal, vertical, or inclined.

Air flow near the surface is laminar, characterised by parallel layers of moving air. As the distance from the surface increases, air flow becomes less regular and turbulence starts occurring. Heat transfer from the surface is slower under laminar flow and faster under turbulent flow. Hence increasing the air gap beyond certain thickness will not lead to a proportional increase of thermal insulation properties of the gap.

Due to their underlying complexity, resistances of air gaps are represented in a simplified way as single numbers and are made available in engineering handbooks, such as CIBSE Guide A (Butcher et al., 2015).

8.5 CONDUCTIVE HEAT LOSS

Summing up the products of U-values and corresponding surface areas for all building elements results in the conductive heat loss coefficient for the building expressed as

$$H_c = \sum_{i=1}^{i=N} U_i A_i \tag{16}$$

where

H_c – conductive heat loss coefficient (W/K)
U_i – thermal conductance of the i-th building element (W/(m²K))
A_i – surface area of the *i*-th building element (m²)
N – number of building elements

Conductive heat loss rate for a building is proportional to the conductive heat loss coefficient and temperature difference:

$$Q_c = H_c \times (T_i - T_o)$$

(17)

where

Q_c – conductive heat loss rate (W)
T_i – internal air temperature (K or °C)
T_o – external air temperature (K or °C)

As part of preparation for every dynamic simulation we will need to define construction types of all building components, and U-values will be automatically calculated as we define each component. An example will be shown later in this chapter.

8.6 HEAT LOSS DUE TO THERMAL BRIDGING

A thermal bridge is an area in a building element that is characterised with a higher U-value than the one calculated using Equations (10) and (11). Thermal bridges occur at junctions between building elements, through joists, through insulation mounting pins, or around openings.

Linear thermal bridges at material junctions are characterised with a ψ-value, expressed in W/(m·K). The heat loss through a linear thermal bridge is then calculated as

$$H_{tb} = \sum_{i=1}^{i=N} L_i \Psi_i$$

(18)

where

H_{tb} – thermal bridging heat loss coefficient (W/K)
L_i – length of the *i*-th linear thermal bridge (m)
Ψ_i – linear thermal transmittance of the *i*-th thermal bridge (W/(m·K))

If thermal bridging details are not known, then a slightly different calculation based on a surface factor and a total external area is used, also resulting in a thermal bridge heat loss coefficient H_{tb}.

Heat loss rate due to thermal bridging is proportional to the thermal bridging heat loss coefficient and temperature difference expressed as follows:

$$Q_{tb} = H_{tb} \times (T_i - T_o)$$

(19)

The effect of a thermal bridge as result of materials of different properties within wall construction is illustrated in Figure 13d, where DesignBuilder construction specification is shown. As it can be seen from that figure, 1% of steel protrusion through insulation material results in a 102% increase of the U-value, from 0.15 to 0.303 W/(m²K). Therefore, thermal bridging can considerably degenerate the effect of thermal insulation and should be taken into careful consideration at the design stage. When calculated in this way, thermal bridge is accounted for through the U-value, in which case Equations (16) and (17) apply.

If thermal bridging details are not known, then a slightly different calculation based on a surface factor and a total external area is used. The reader is advised to consult BRE (BRE Group, 2018) for further details on thermal bridging calculations and accredited construction details.

8.7 VENTILATION AND INFILTRATION HEAT LOSS

Ventilation and infiltration are essentially the same type of process of replacing internal air with external air. However, infiltration occurs spontaneously through cracks and gaps in the building structure, and ventilation occurs as result of planned and intentional replacement of internal air. We can therefore say that infiltration is unwanted ventilation.

In order to define ventilation and infiltration heat loss, we first need to define a concept of volume air change per hour. If we imagine a room of certain dimensions, for example, 10 m × 6 m × 3 m, then the total room volume is 180 m³. If we, through ventilation, replace the entire volume of 180 m³ in one hour, we call this one volume air change per hour. If we say that the volume air change rate is 0.5, that means that half of the volume of air within a room (i.e. 90 m³ from the above example) is replaced by external air within one hour.

We can now define ventilation heat loss rate as

$$Q_v = N \times V \times (T_i - T_o) \times \frac{c \times \rho}{3600}$$
(20)

where

Q_v – ventilation heat loss rate (W)
N – volume air change per hour (h⁻¹)
T_i – internal air temperature (K or °C)
T_o – external air temperature (K or °C)
c – specific heat of air (J/(kg·K))
ρ – density of air (kg/m³)
3600 – number of seconds in an hour

As $c \times \rho / 3600$ is approximately 1/3, the above equation becomes

$$Q_v = \frac{N \times V}{3} \times (T_i - T_o)$$
(21)

As we have seen from Equations (17) and (19), the term next to temperature difference is the heat loss coefficient. In the case of Equation (21), the term next to temperature difference is the heat loss coefficient due to ventilation and infiltration, which is expressed as

$$H_v = \frac{N \times V}{3}$$
(22)

Volume air change rate N will need to be defined in every dynamic simulation model in order to account for ventilation heat loss. We will show this in an example later in this chapter.

8.7.1 Volume air change rate versus air permeability

In some cases, we will need to convert volume air change rate into air permeability or vice versa. Air permeability is defined as volume flow rate of air (m³/h) through a unit area of the external building envelope (in m²). Hence the unit for air permeability is (m³/h)/m².

Example 1. How do we convert air change rate into air permeability? We need to know the volume of the space in question and the area of the building envelope. For instance, in the case of a space with the internal dimensions of 10 m × 6 m × 3 m, the volume is 180 m³, and the area of the envelope is 10 × 6 × 2 + 10 × 3 × 2 + 6 × 3 × 2 = 216 m². If the volume air change rate due to infiltration is 0.25, we can calculate air permeability AP as

$$AP = \left(180 \text{ m}^3 \times 0.25/\text{hr}\right) / 216 \text{ m}^2 = 0.2083 \left(\text{m}^3/\text{h}\right)/\text{m}^2.$$

In other words, we need to multiply the building volume with the air change rate and divide it by the area of the building envelope to obtain air permeability.

If we establish through airtightness testing that a building has air permeability of 5 (m³/h)/m² under a standard testing pressure of 50 Pa, this means that our room with the above dimensions will have approximately 5/0.2083 = 24 times more volume air changes per hour than in the above calculation, therefore six volume air changes per hour under the test conditions.

Example 2. How do we convert air permeability into air change rate? If we are given air permeability for a building of 5 (m³/h)/m² and we need air changes per hour N for the simulation model, we can calculate that as follows:

$$N = 5(\text{m}^3/\text{h})/\text{m}^2 \times 216 \text{ m}^2 / 180 \text{ m}^3 / 24 = 0.25/\text{hr}.$$

Effectively, we need to multiply air permeability with the area of the envelope and divide it by the building volume and the volume air change factor, to obtain the number of air changes per hour.

CIBSE Guide A (Butcher et al., 2015) gives the empirical relationship between air permeability at 50 Pascal in (m³/h)/m² and infiltration rate in h⁻¹ for a range of different building sizes and types. The reader is advised to consult Table 4.13 in this guide for a detailed relationship between these two parameters.

For a number of cases in this table, values appear to be closely approximated by the following simple relationship for air permeability up to 10 (m³/h)/m² and for building sizes of up to 8000 m² of floor area:

$$N = AP \times 0.033$$

where

N – volume air change rate in h⁻¹
AP – air permeability at 50 Pascal in (m³/h)/m²

The above relationship is only an approximation, but it will give us a quick start when entering initial values into a simulation program.

8.7.2 How to achieve airtight buildings in practice
Having personally measured airtightness in buildings, and having come across many buildings that underperformed in terms of airtightness, the following points emerged as ways to achieving good airtightness in buildings:

• Design: Attention to detail in specifying building construction elements and other components is essential in achieving good airtightness. This includes specifying vapour-permeable airtight membranes and durable airtight tapes, and continuous coverage across the building envelope, especially between different building components. Unsealed service cavity behind the plasterboard can be a major source of air leakage. This especially occurs through unsealed parts of the internal surfaces, such as electricity sockets. Leaving the openings for cables unspecified at the design stage and at the discretion of construction site personnel can result in post-construction holes being drilled and being left unsealed, causing air leakage. Therefore, every interface between different building components, every cable run and hole, as well as the sealing methods, should be specified at the design stage.
• Construction: Awareness raising and training of the workforce are essential in implementing good airtightness design into airtight buildings.
• Installation of building services: In addition to awareness raising and training, it is essential that installers of building services are given clear instructions on where they can and where they cannot drill, specifying diameters of drills and how to seal the holes.

Airtightness tests during the construction phase are essential to reveal and rectify any imperfections in the building envelope. Airtightness tests should be carried out in every single building. In housing projects, it is a common practice to test only a sample of the total number of houses being built, which leaves a majority of houses untested and potentially with worse airtightness than the tested sample. Designs that go into the very detailed specification of airtightness and leave very little to the uncertainty of ad-hoc decisions on site have a better chance of achieving good airtightness in buildings.

See 'Airtightness tests' in Chapter 20 for sources of air leakage found during airtightness tests.

8.8 OVERALL BUILDING HEAT LOSS

On the basis of the previous three sections, we can now calculate the overall heat loss from a building as follows:

$$Q = (H_c + H_{tb} + H_v) \times (T_i - T_o) \tag{23}$$

where

Q – overall heat loss rate (W)
H_c – conductive heat loss coefficient (W/K)
H_{tb} – thermal bridging heat loss coefficient (W/K)
H_v – ventilation and infiltration heat loss coefficient (W/K)
T_i – internal air temperature (K or °C)
T_o – external air temperature (K or °C)

8.9 SCALE MODEL EXPERIMENTS

Scale model experiments with thermal insulation were conducted by my students as part of coursework in Environmental Design. The brief for insulation experiments was as follows:

> **External versus internal insulation.** Create a cardboard model of 20 cm × 12 cm × 6 cm (box 1). Create two internal boxes so that box 2 is 2 cm smaller in all directions than box 1, and box 3 is 2 cm smaller than box 2 in all directions. Run the following experiments: **Experiment 1** - Fill the gap between box 1 and 2 with sand and the gap between box 2 and 3 with thermal insulation, such as mineral wool or polystyrene. Place a teacup of boiling water into the box, and record internal temperature and time in 1 minute intervals until the box has reached its maximum temperature. Continue recording the temperature until the box has cooled down to room temperature. **Experiment 2** – The same as Experiment 1, except the gap between box 1 and box 2 is filled with thermal insulation, and the gap between box 2 and box 3 is filled with sand. Compare the results from the two experiments and comment on the findings.

This experiment is shown in Figure 82. The model has two cavities. In the first 'insulation inside' case the outer cavity is filled with sand and the inner cavity with thermal insulation (Figure 82a). In the second 'insulation outside' case, the outer cavity is filled with thermal insulation and the inner cavity with sand (Figure 82b). A temperature probe is suspended in the inner box with a cable clip. A container with boiling water is then placed into the model and a transparent lid is secured on top of it using adhesive tape. Internal and external air temperatures were recorded in one-minute intervals over a period of 60 minutes using a data logger.

Results of monitoring are shown in Figure 83, from where we can see that the model with internal insulation heats up much faster than the model with external insulation. After reaching the peak temperature, the slope of the cooling curve for the model with inside insulation is steeper, showing that it cools down faster than the model with external insulation. This means that the two temperature lines will cross over after a while so that the external insulation model temperature will become higher than the internal insulation model temperature. This behaviour can be explained with higher heat storage capacity of sand in comparison with thermal insulation. Effectively, external insulation with internal high-density material increases thermal mass in buildings, and this will be discussed in more detail in Chapter 11. The behaviour observed in this experiment is consistent with dynamic simulation experiments in the next section.

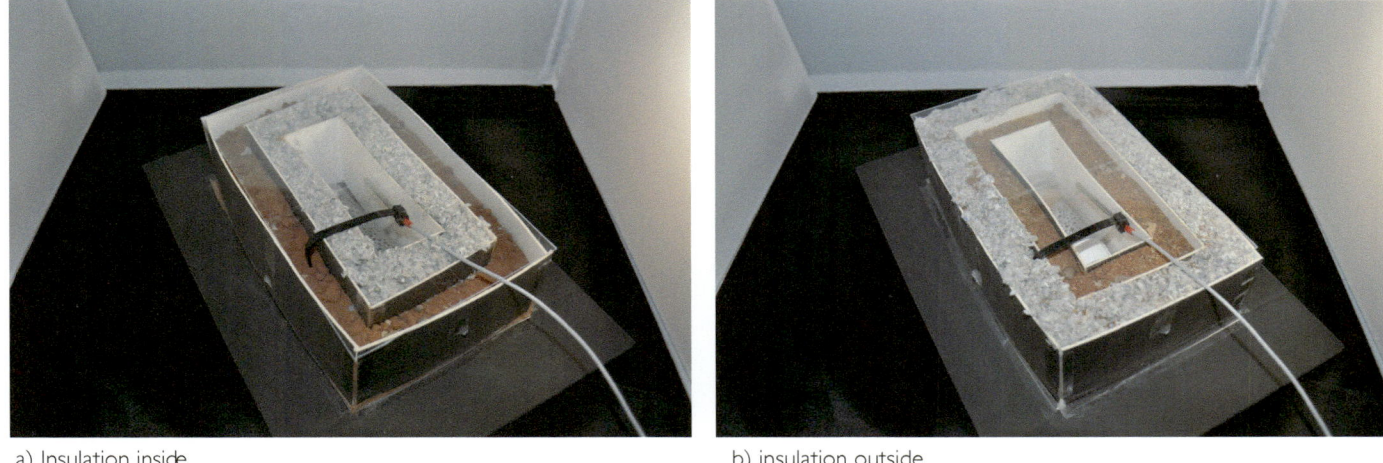

a) Insulation inside   b) insulation outside

Figure 82   Thermal insulation scale model experiment (model by my former students Simon Pope, Parminder Dhillon, Etienne Amion, Antonios Papanastasiou, and Pierre Arnou)

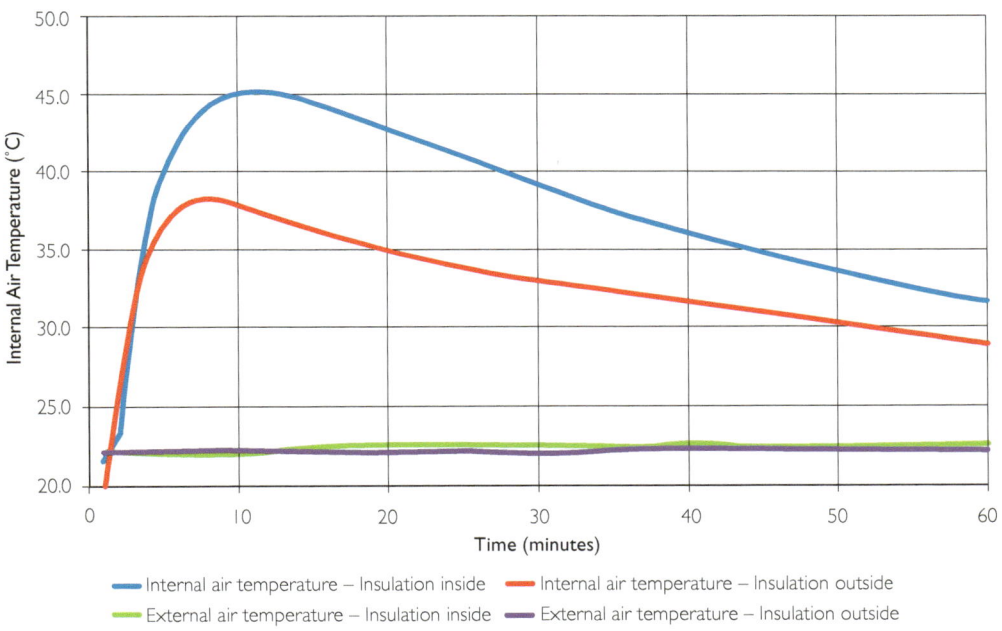

Figure 83   Results of monitoring thermal insulation scale model experiment

8.10 DYNAMIC SIMULATION EXPERIMENTS

Several simulation experiments were conducted in order to increase the understanding of the effects of insulation thickness, position, and type, as well as the influence of building airtightness. The simulations were conducted using DesignBuilder and were based on a simple 'shoe box' model in Figure 65, with dimensions of 10 m × 6 m × 3 m.

The climate location was London, Heathrow. The construction types of walls which were varied between different simulations are shown in Table 28. Details of the model and its variations are available on the supplementary material website www.ljanko-vic.com.

TABLE 28   CONSTRUCTION TYPES IN THERMAL INSULATION COMPARISON CASES

Low insulation wall

Outer surface

100.00mm Brickwork Outer

113.20mm XPS Extruded Polystyrene - CO2 Blowing

100.00mm Concrete Block (Medium)

Inner surface

U-Value = 0.26 W/m²·K

High insulation wall

Outer surface

100.00mm Brickwork Outer

322.40mm XPS Extruded Polystyrene - CO2 Blowing

100.00mm Concrete Block (Medium)

Inner surface

U-Value = 0.10 W/m²·K

Insulation inside wall

Outer surface

100.00mm Brickwork Outer

100.00mm Concrete Block (Medium)

171.30mm XPS Extruded Polystyrene - CO2 Blowing

Inner surface

U-Value = 0.18 W/m²·K

Insulation outside wall

Outer surface

100.00mm Brickwork Outer

171.30mm XPS Extruded Polystyrene - CO2 Blowing

100.00mm Concrete Block (Medium)

Inner surface

U-Value = 0.18 W/m²·K

8.10.1 Insulation thickness

In this experiment, we used high and low insulation walls from Table 28 and the results are shown in Figure 84.

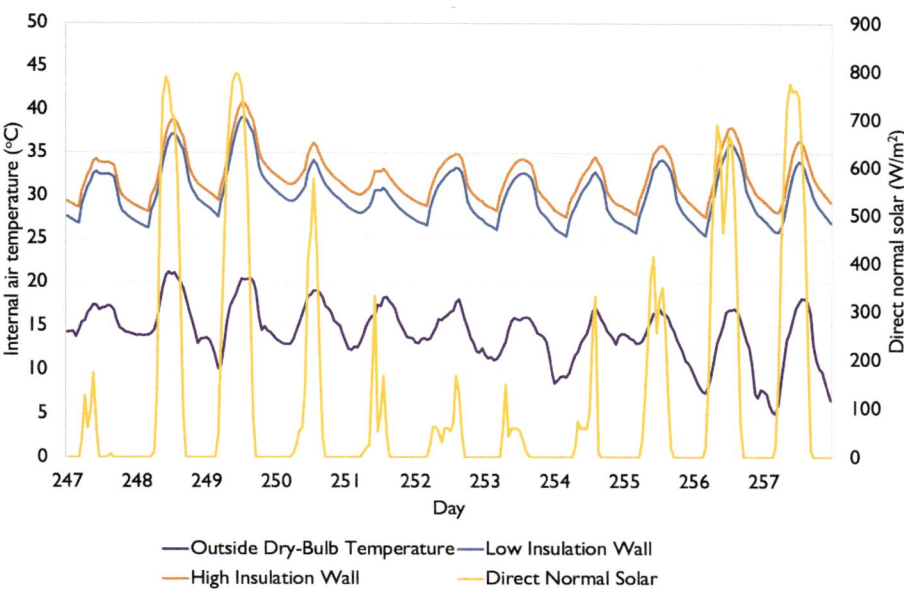

Figure 84   Low versus high thermal insulation

We can see from this figure that a higher level of thermal insulation generally elevates the internal temperature profile. This means that there will be lower requirement for heating in winter but an increased risk of overheating in summer. The latter can be mitigated using solar shading discussed in Chapter 10.

8.10.2 Internal versus external insulation

In this experiment, we used walls with inside and outside insulation from Table 28 and the results are shown in Figure 85.

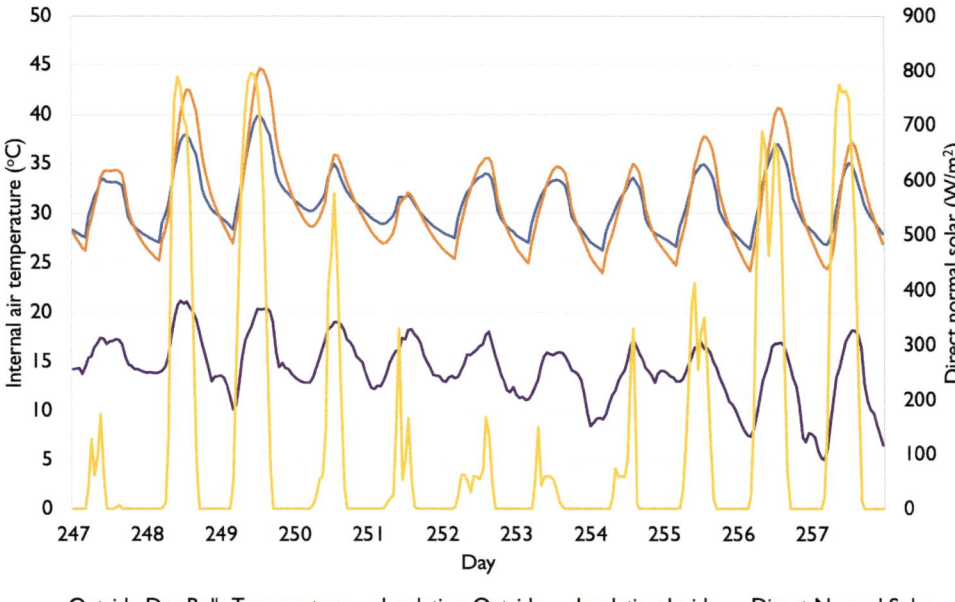

Figure 85   Internal versus external insulation

We can see from this figure that external insulation results in less extreme temperatures and internal insulation results in more extreme temperatures. External insulation therefore smooths out internal temperature fluctuations, makes the internal environment more thermally stable, and results in lower heating and cooling requirements. The reason for this behaviour is that external insulation makes internal high-density materials such as brick, blockwork, and concrete slabs exposed to absorption of internal heat, and thus results in the higher thermal mass of the building. We will discuss thermal mass in detail in Chapter 11. External insulation will however result in a slower response to heat input whilst the response of internal insulation to heat input will be faster. It is important to note that both wall constructions used in this experiment have the same U-value, and the only difference is the order of individual layers.

8.10.3 Airtightness

In this experiment we used the 'insulation outside wall' from Table 28, and a low and high airtightness. The low airtightness was represented with 0.7 air changes per hour and high airtightness by 0.1 air changes per hour. Therefore, we refer to a lower number of air changes per hour as 'higher airtightness' and vice versa. The results are shown in Figure 86.

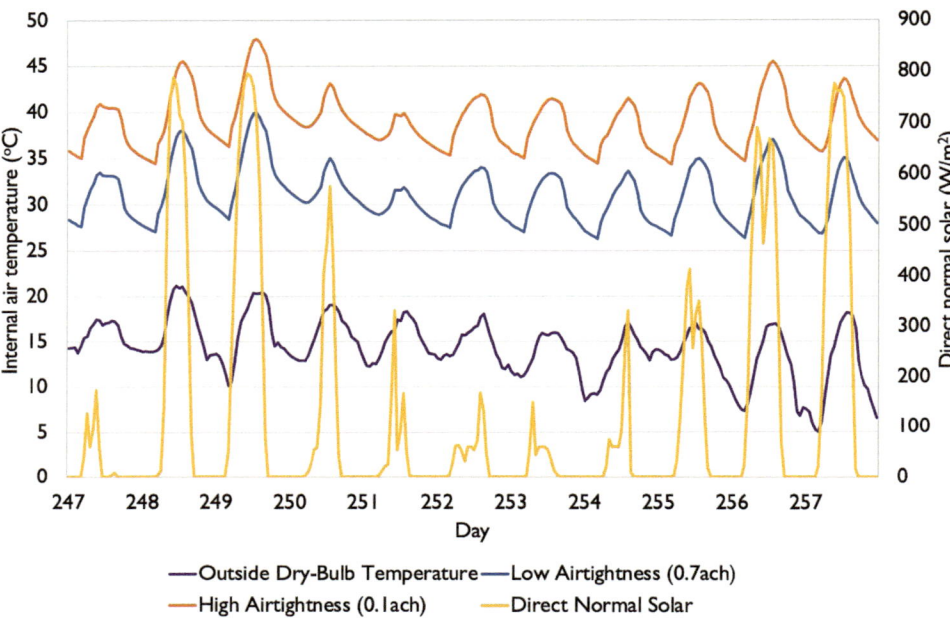

Figure 86   High airtightness (0.1 ach) versus low airtightness (0.7 ach)

As can be seen from this figure, higher airtightness will generally elevate the internal temperature profile in comparison with lower airtightness. The increase of airtightness will therefore have a similar effect as the increase of thermal insulation in Figure 84, and will lead to lower energy consumption and lower carbon emissions. However, in winter months, high airtightness will result in a requirement for mechanical ventilation and consequent higher electricity consumption. These conflicting constraints and trade-offs between airtightness and mechanical ventilation need to be investigated and resolved using dynamic simulation methods.

8.11 TASKS FOR SIMPLE SIMULATION EXPERIMENTS

1   Create a model of a box with dimensions of 10 m × 6 m × 3 m. Point the larger wall towards the south and install 40% glazing on it. Set the following U-values in W/m²·K for the model constructions: Walls 0.35; Windows 1.98; Floor slab 0.13; Roof 0.25. The ground floor slab consists of a minimum of 100 mm thermal insulation and 100 mm cast concrete. The walls are brick and block with cavity insulation. The roof is a concrete deck with internal insulation. The windows are double glazed. Calculate the overall UA value as a sum of the products of individual U-values and corresponding surface areas.

2 Simulate a co-heating test of the model in the example above, as follows:

   a) Set a daily temperature profile for heating control to increase from an initial unheated temperature to 25°C in a single step at 8 am and stay on the rest of the day.

   b) Set another daily temperature profile for heating control that is on at 25°C throughout the day.

   c) Set a weekly heating control profile with the first day controlled by the temperature profile from a. and the rest of the weekdays controlled by the temperature profile from b.

   d) Set another weekly temperature profile with all days controlled by the temperature profile from b.

   e) Set an annual temperature profile with the first seven days controlled by the weekly temperature profile from c and the rest of the year controlled by the temperature profile from d.

   f) Set electricity heating with 100% efficiency and limit it to 3 kW. Switch off cooling and domestic hot water.

   g) Exclude all internal heat gains from people and appliances, and set infiltration air exchange to 0.167 ach.

   h) Install internal blinds in the glazing that is lowered at zero W/m² and risen at 1300 W/m² of solar radiation, and therefore practically always on. This is to exclude the solar gain interfering with the test.

   i) Set the initial internal air temperature to the initial external air temperature and the simulation preconditioning period to zero days.

   j) Carry out dynamic simulation with an hourly time step over 15 days, starting on 1st January.

   k) Export the internal room air temperature, external air temperature, and heating plant sensible load into a spreadsheet.

   l) Calculate daily average temperature differences between internal and external air temperatures, from the point when internal air temperature has reached 25°C. Also calculate daily average heating plant sensible load from the same point.

   m) Calculate averages of all values from step l. above.

   n) Convert the average heating plant sensible load into Watts and divide it by the average temperature difference in °C. The result is the overall UA value of the model in W/K.

   o) How does this value compare with the manually calculated value in example 1. above? How would you explain any discrepancies between the two values? How would you carry out this test in practice in a real building?

   p) Discuss the process and the outcome with your colleagues.

This example was developed for and tested in IES VE, but it can be customised for other simulation tools.

8.12 SUMMARY OF DESIGN PRINCIPLES

Insulation type and thickness
• Thermal insulation with higher thermal resistance will elevate internal temperature profile.
External versus internal insulation

• External thermal insulation will smooth out internal temperature fluctuations, in comparison with internal thermal insulation with equal thermal resistance.
• External thermal insulation will reduce peaks and troughs of internal temperatures and will make the internal environment more thermally stable.
• External thermal insulation will result in higher level of thermal mass.
• Internal thermal insulation will respond to heat input quickly.
• External thermal insulation will respond to heat input slowly.

Airtightness

• Higher airtightness resulting in lower infiltration rates will elevate internal temperature profile in comparison with lower airtightness (higher infiltration rates).
• The effect of high airtightness is similar to the effect of higher thermal insulation levels – both will elevate internal temperature profile.
• Higher airtightness may require mechanical ventilation in order to provide sufficient amount of air for building users.
• Conflicting constraints between high airtightness and the need for mechanical ventilation need to be resolved through dynamic simulation.

CHAPTER 9

SOLAR GAIN

This chapter discusses fundamental properties of solar radiation and the ways of capturing solar radiation in buildings.

9.1 SOLAR ENERGY SPECTRUM AND SOLAR ABSORPTION

Solar radiation that reaches the earth has a spectral distribution of wavelengths as shown in Figure 87. The wavelength is inversely proportional to temperature, so that for instance the wavelength of 400 nm corresponds to 7500 Kelvin anc the wavelength of 2400 nm corresponds to 1250 Kelvin.

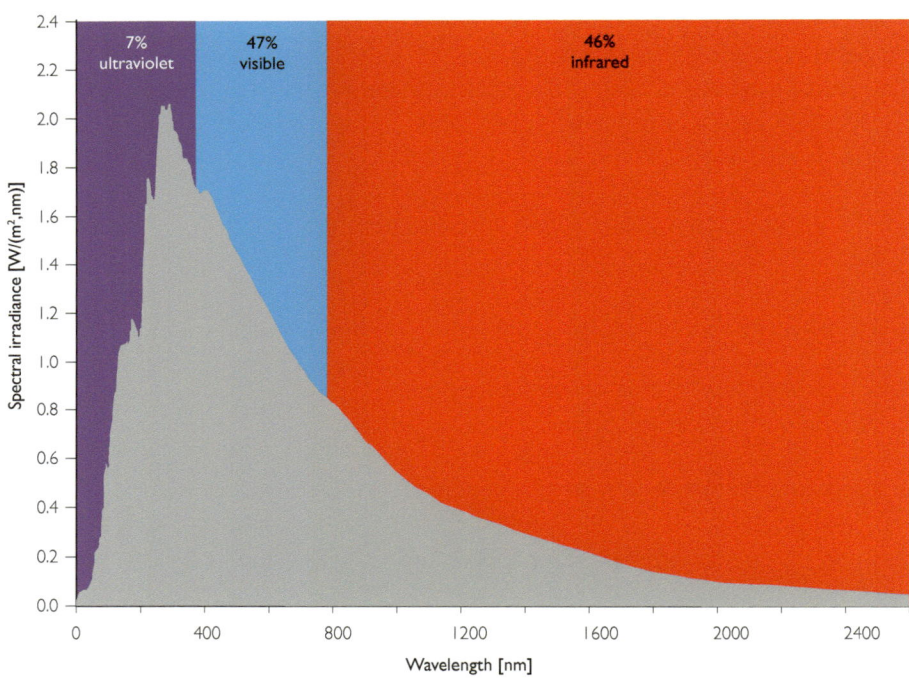

Figure 87   Solar spectrum (adapted from Duffie and Beckman (2013))

Material surface properties called absorptance, reflectance, and emittance determine how individual materials respond to solar radiation. Absorptance (α) is defined as a fraction of absorbed radiation at the surface relative to radiation absorbed by the black body. Reflectance (ρ) is a fraction of reflected radiation from the material surface. Emittance (ε) is a fraction of emitted radiation, relative to radiation emitted by the black body.

9.1.1 Selective surfaces

The spectral distribution of solar radiation, the corresponding temperature, and the surface properties of materials can now be used to define types of absorption surfaces.

DOI: 10.4324/9781003342342-12

Absorptance is equal to emittance in materials with conventional surfaces. In other words, materials with conventional surfaces gain and lose energy at the same rate, and they are not the best materials to use for solar absorption. Materials with selective surface properties have high absorptance at high radiation temperatures (low wavelengths) and low emittance at low radiation temperatures (high wavelengths). For instance, Szokolay (2014) defines the relationship for selective surfaces as $\alpha_{6000} > \varepsilon_{60}$, meaning that absorptance at radiation temperatures of 6000°C is higher than emittance at temperatures of 60°C that are likely to occur at material surfaces as a result of heating by solar radiation. On this basis, selective surface materials are better suited for the absorption of solar radiation, as they absorb more than they emit.

9.2 SOLAR GAIN PROCESSES AT GLAZING

When reaching the outer surface of the glazing, incoming solar radiation I_o will partially get reflected to the outside (I_r), and partially refracted through the glazing. The refracted part will get partially absorbed in the mass of the glazing (I_a) and partially transmitted into the internal environment (I_t). The absorbed fraction will be dissipated through heat losses to the outside (I_{co}) and inside (I_{ci}). The total amount of heat gain I_g will therefore consist of the transmitted solar radiation and the absorbed solar radiation that is transferred by convection to the inside ($I_g = I_t + I_{ci}$). Figure 88 shows this process for a single layer of glazing. For multiple layers of glazing, the process is repeated as shown in Figure 89.

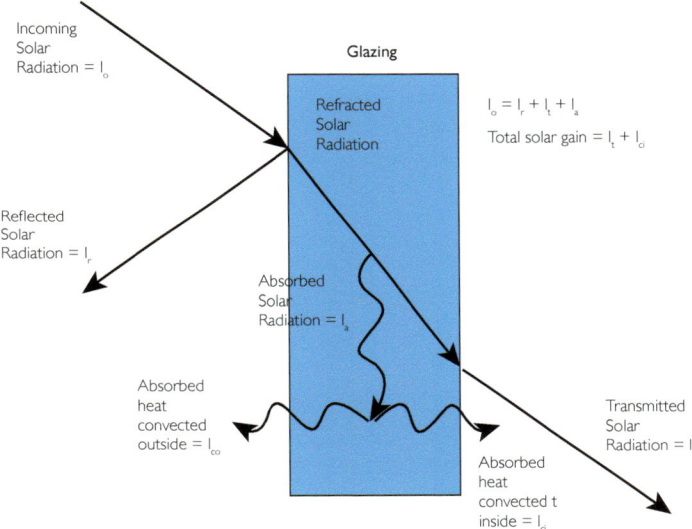

Figure 88   Solar gain processes at single glazing

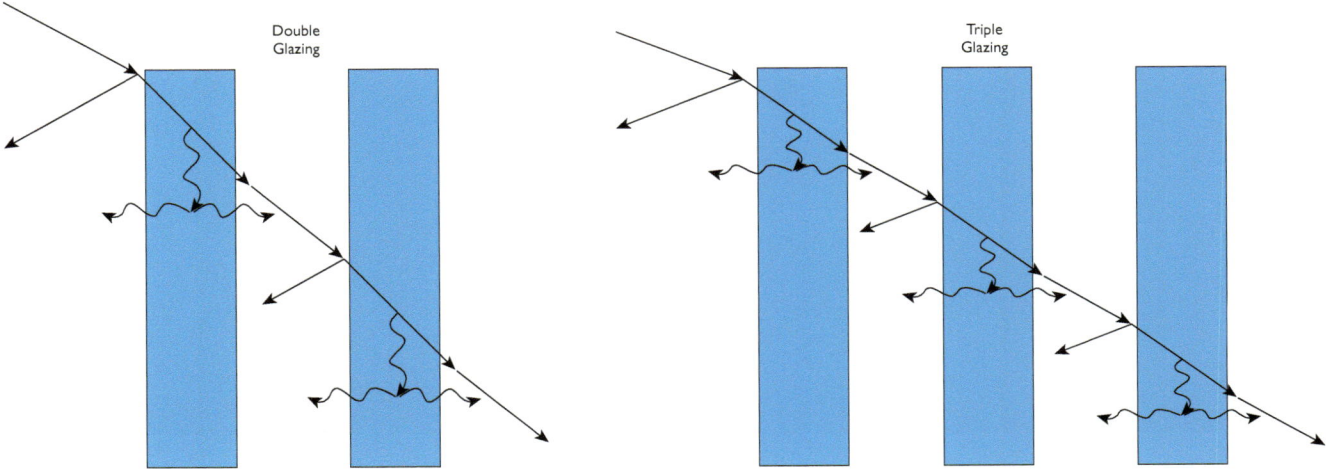

Figure 89   Solar gain processes at double and triple glazing

The amount of reflected, transmitted, and absorbed solar radiation will vary between different types of glass. Reflective glass will reflect more than clear glass, and tinted glass will absorb more than clear glass. Depending on the type of glass, the amount of heat dissipated to the outside may be greater or smaller than the amount of heat dissipated to the inside.

Examples of how to set properties of different types of glazing in a dynamic simulation model will be shown later in this chapter.

9.3 SOLAR TRANSMISSION COMPONENTS

9.3.1 Glazing

Clear glazing is the simplest and earliest type of glazing. The amount of transmitted radiation through clear glazing will diminish with the addition of glass layers. Following the diagram in Figure 89, and in the case of the clear glass transmittance of 0.8, the amount of incoming radiation transmitted through double glazing will be 0.8 × 0.8 = 0.64, and the amount transmitted through triple glazing will be 0.8 × 0.8 × 0.8 = 0.51. This indicates a considerable reduction of solar gain; however, heat losses from double and triple glazing would also be considerably reduced.

Low emissivity glazing, or low-e glazing, has a low emissivity metal or metal oxide coating applied to the glass surface, either inside the air cavity, or on the outside or inside the glass pane. Whilst not affecting light transmission properties or conductive heat transfer, the coating reduces radiation heat transfer through the glazing and thus increases thermal insulation properties. Depending on whether the glazing is required to keep the outside heat out, or the inside heat in, the low-e coating is applied to the outside or the inside glass pane respectively.

Gas-filled glazing contains an inert gas in the cavity. Gases such as krypton or argon have lower conductivity than air, and will improve thermal insulation properties of glazing whilst not reducing light and radiation transmission properties.

Photo-chromic glazing changes its transmittance as the intensity of solar radiation increases. It is used commonly in eye care, for glasses that change colour and transmittance in high daylight conditions, but it can also be used on buildings.

Thermo-chromic glazing is based on a thin layer of chemical compounds that form molecular chains of variable length, depending on temperature. At room temperatures, the molecular chains are shorter and the thermo-chromic layer allows full radiation and light transmission. At temperatures higher than room temperatures (the exact temperature is programmed by a combination of chemicals in the manufacturing process) longer molecular chains are formed, restricting the transmission of solar radiation and light.

Electro-chromic glazing can be based on several different types of materials that change transparency in response to electric voltage. One of these types of glazing is based on a thin layer of liquid crystal, the material commonly used in wristwatches,

a) In opaque mode

b) In transparent mode

Figure 90   The effect of electro-chromic glazing

applied to the glass surface through a special manufacturing process. Liquid crystals are randomly oriented until a voltage is applied across the layer, when they take uniform orientation. In the random orientation mode, the crystals scatter the light passing through the layer, and the layer appears nearly opaque. In the uniform orientation mode, the layer becomes transparent and it allows transmission of solar radiation and light (Figure 90).

9.3.2 Transparent insulation

The main principle behind transparent insulation materials is the air capture in small volumes of transparent materials, such as polycarbonate or silica gel. In cavities larger than 20 mm, turbulent flow of air occurs naturally as a result of heat transfer and that makes thermal insulation properties of air obsolete. By capturing air in polycarbonate capillary tubes or in silica gel granules, good thermal insulation properties of air are preserved, but the overall optical transmittance is reduced.

Figure 91a shows light transmission through aerogel – a nano-structure of interconnected molecules of silicon dioxide. A nervous system-like structure of molecular chains captures air and creates one of the best thermal insulators known today. Its conductivity is lower than any other insulation material, making it possible to achieve excellent insulation properties of buildings with a lower thickness of insulation layer. The material is translucent and thus lends itself to natural daylight applications in buildings which at the same time provide excellent thermal insulation.

a) Aerogel -a nano-structure of SiO$_2$ molecules (shown inside a polycarbonate panel)

b) Perpendicular view of polycarbonate capillary tubes

c) Oblique view of polycarbonate capillary tubes

Figure 91   Transparent insulation materials

Figure 91 also shows polycarbonate capillary tubes viewed from a perpendicular angle (Figure 91b) and from an oblique angle (Figure 91c). These two images demonstrate that there is an angular dependency of light transmission through the material. This results in a degree of self-regulation of solar gain, which is higher at lower solar angles and vice versa.

Figure 92 shows an example of application of transparent insulation material on an external wall surface. The material is based on capillary tubes of 3 mm in diameter, positioned perpendicularly to the wall surface. As solar radiation reaches the outer surface of the material, it undergoes multiple reflections inside the tube until it reaches the wall surface. This surface is then heated up, whilst the air captured in the capillary tubes prevents heat losses to the outside. The absorbed heat has only one way to travel, through the wall and into the internal space.

9.3.3 Double skin

A double skin is a system consisting of large cavity glazing, ventilation openings at the base and top on the outside and inside glass pane to control air flow and a solar absorber in the form of adjustable louvre shading inside the cavity (Figure 93). The double skin has four modes of operation, depending on the open/closed status of the ventilation louvres:

• Heating – absorbed heat is recirculated between the cavity and the internal space
• Heat rejection – the cavity is ventilated to the outside, rejecting the absorbed heat
• Preheating – absorbed heat in the cavity results in natural buoyancy of air, which draws external air and supplies it preheated into the internal space
• Ventilation – absorbed heat in the cavity draws air from the internal space and rejects it to the outside.

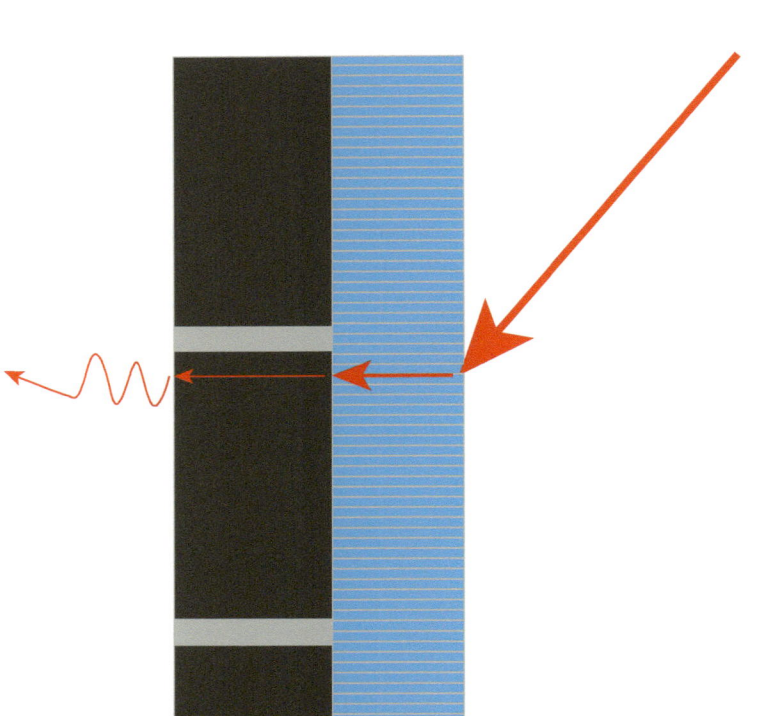

a) How it works

b) Practical application

Figure 92   Polycarbonate capillary tubes attached to external wall surface

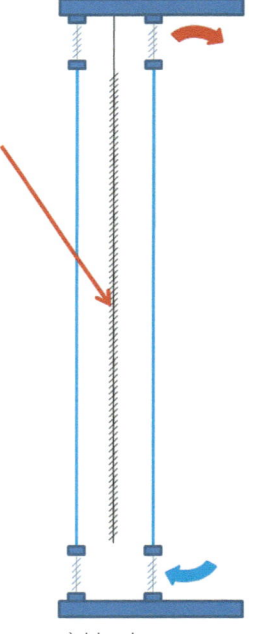

a) Heating

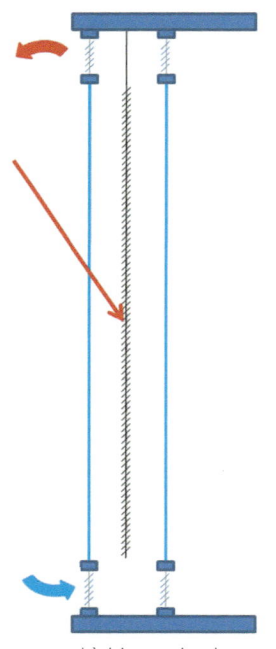

b) Heat rejection

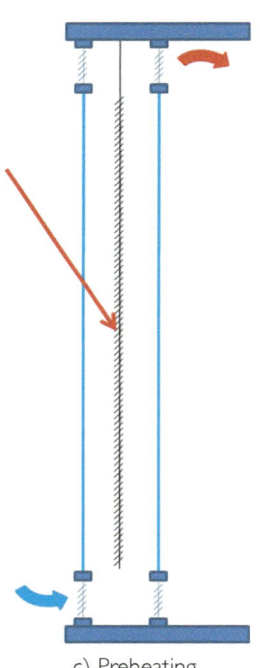

c) Preheating

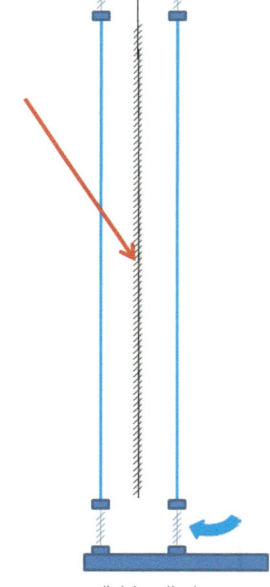

d) Ventilation

Figure 93   Double skin

9.4 PASSIVE SOLAR SYSTEMS

9.4.1 Direct gain systems

A direct gain system is one of the simplest passive solar systems (Figure 94). The main characteristics of the system are a large sun-facing glazing panel (south-facing in northern latitudes, north-facing in southern latitudes) and high-density material inside, such as concrete, stone, and brick. The glazing transmits solar radiation into the internal space, where it is absorbed by the high-density material (the absorber). Darker floor surface colours will increase the amount of absorbed heat and the energy efficiency of this system. The absorbed heat is re-radiated into internal space with a time delay. The absorber is protected from heat losses to the outside with thermal insulation.

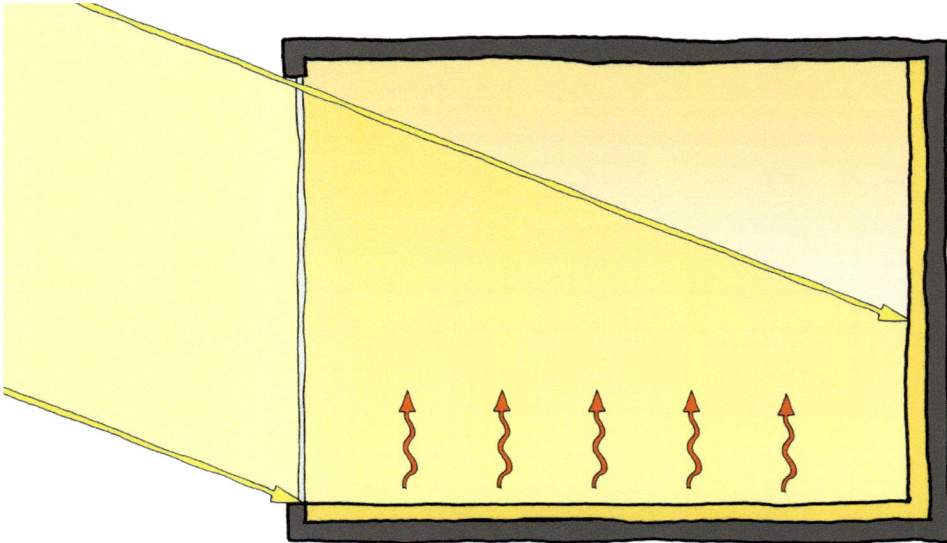

Figure 94   Direct gain system (illustration by Holly Doron)

9.4.2 Isolated gain systems

Isolated gain systems are greenhouses attached to the sun-facing side of the building (Figure 95). We also call them sunspaces. As solar radiation enters through glazing, it is absorbed in the high-density material on the floor of the sunspace. As in the direct gain system, a darker floor surface will result in higher absorption and higher energy efficiency. In order to minimise heat losses, the high-density material is protected with thermal insulation from underneath. The internal space receives heat from the sunspace by natural circulation of air through top and bottom openings. These openings can be controlled with shutters. However, if air circulation fans and automated shutters are installed in the sunspace, it becomes less of a passive system.

Figure 95   Isolated gain system (illustration by Holly Doron)

9.4.3 Trombe-Michel wall systems

Trombe-Michel wall system is named after two Frenchmen, Félix Trombe and Jacques Michel We will also refer to t as a 'Trombe wall'. It consists of a high-density wall covered with absorptive surface and protected by glazing from heat losses on the sun-facing side (Figure 96). The heat absorbed at the sun-facing surface is slowly transmitted to the inside surface of the wall and released into the inside space. Selective surfaces discussed in the section with the same name in this chapter, are used to increase the energy efficiency of this system. Top and bottom openings in the wall allow natural circulation of air through the air gap between the glazing and the wall. Hot air gathers at the top of the air gap and enters the internal space as the cooler air from it enters the bottom of the air gap. The air circulation between the air gap and internal space can be controlled with shutters. Similarly to isolated gains systems, the air circulation can be regulated by a fan and auto-mated shutters.

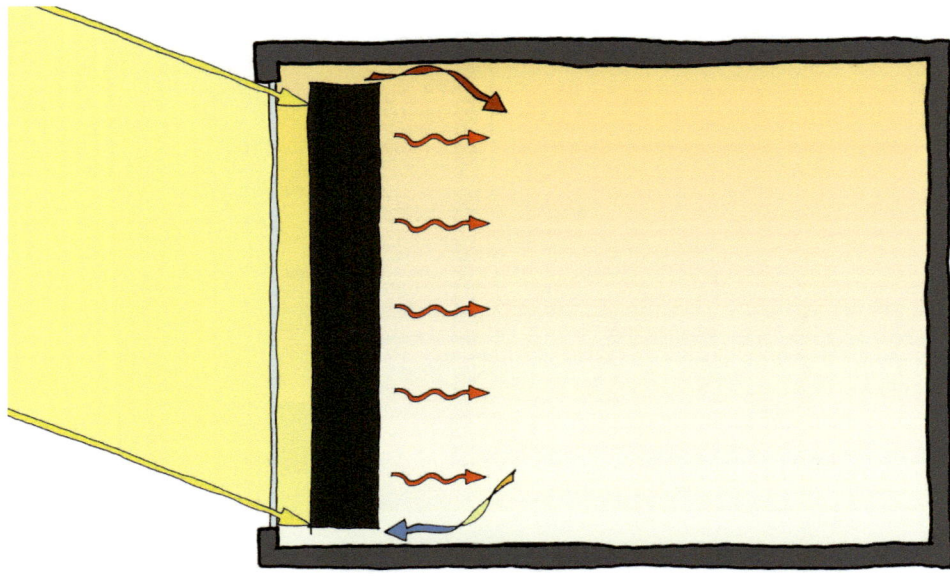

Figure 96   Trombe-Michel wall system (illustration by Holly Doron)

9.4.4 Main design parameters for passive systems

The Trombe wall, as well as the other passive solar systems described above, is characterised by several design parameters. These are:

• Time lag
• Decrement factor
• Type and characteristics of the absorption surface
• Type and characteristics of glazing
• Surface area exposed to the sun.

The time lag and decrement factor are dynamic heat transfer characteristics of the solar absorber, as shown in Figure 97. These parameters are defined on the basis of differences of temperature fluctuations at the external and internal surface of the absorber. Figure 97 shows a conceptual temperature curve at the outside (red) and inside (blue) surface of the absorber.

When the external surface is exposed to a heat input, this input will not be felt at the internal surface instantaneously, but only some considerable time later. The time difference between the peaks of these temperature curves is called the 'time lag'. The time lag therefore represents the time delay between a heat input at the outside surface and the occurrence of this heat input at the inside surface. As a very rough guide, the time lag in a 300 mm lightweight construction (1200 kg/m³) will be 8 hours, and in a heavyweight construction (2400 kg/m³), it will be 12 hours (Figure 98).

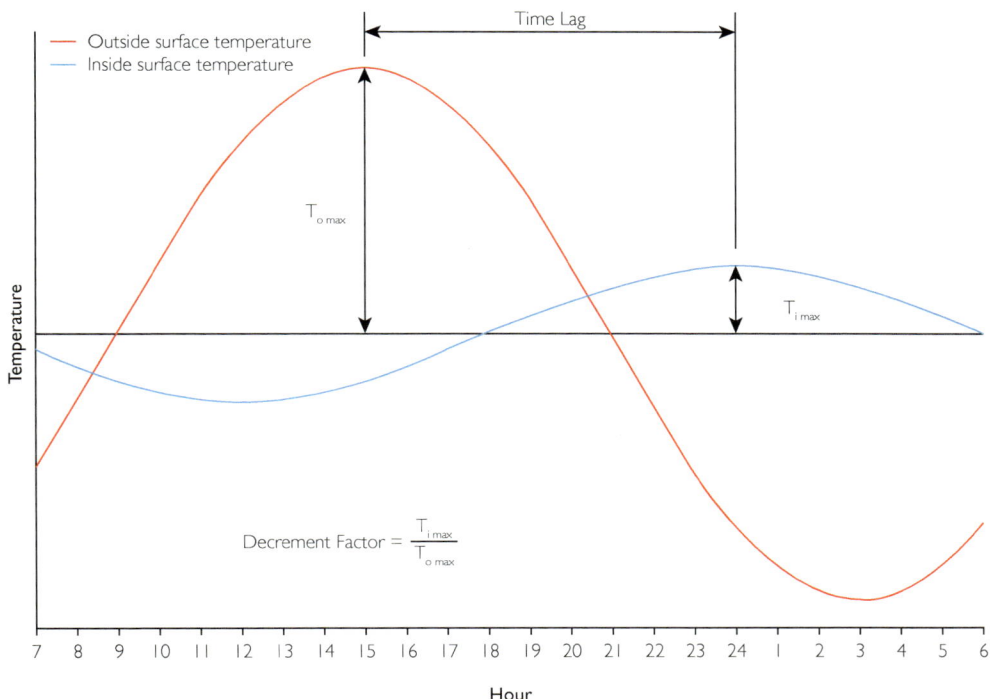

Figure 97   Time lag and decrement factor

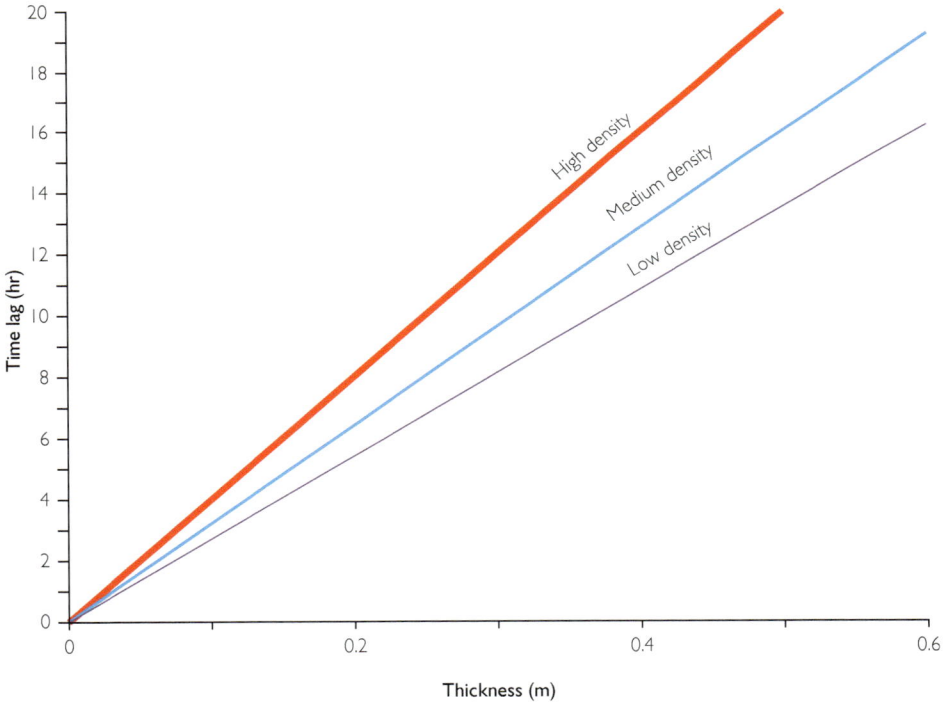

Figure 98   Time lag for different material densities

The magnitude of the heat input received at the internal surface will only be partially transferred to the internal surface. This is referred to as a decrement factor D, and is numerically equal to the ratio between the maximum temperature on the inside and outside surfaces ($D = t_{i\,max}/t_{o\,max}$). The decrement factor is a dimensionless number, and Figure 99 shows a rough guide to how it varies with thickness in a high-density wall. At the inside surface of a 200 mm thick wall, the decrement factor is 0.43,

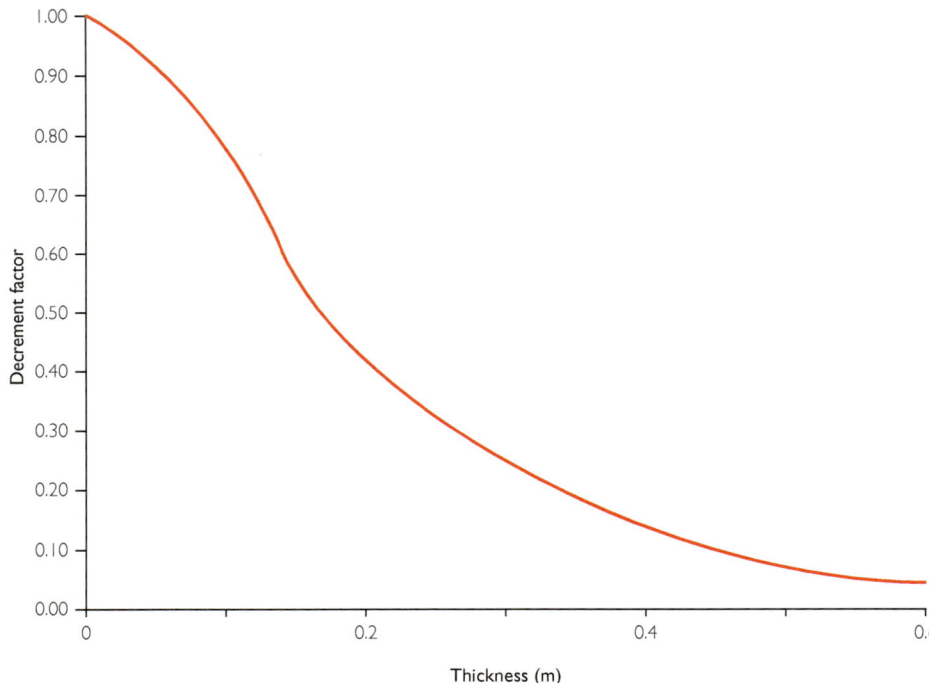

Figure 99   Decrement factor as a function of wall thickness

compared with 0.26 in a 300 mm wall (Figure 99). This means that the thicker the wall, the less heat input will be transmitted to the inside surface.

In the case of the Trombe wall, it is clear that the external surface faces the glazed air gap and the internal surface faces the internal space. In the case of direct gain and isolated gain systems, the external surface faces the internal space, and the internal surface faces the thermal insulation underneath. In these two cases, the absorbed heat will 'bounce' off the thermal insulation before propagating back and reaching the internal space.

Dynamic heat flow in buildings will be further explained in Chapter 11, but for now, it is sufficient to say that the time lag and decrement factor are derived from the following parameters that need to be specified in the dynamic simulation model:

• Thermal conductivity – ability of material to transmit heat
• Density – mass of a unit volume of the material
• Specific heat capacity – the amount of heat required to change a unit mass of a material by one °C
• Total volume of the material.

The first three parameters will be entered directly into the specification of building construction types in the DSM. The fourth parameter will be entered indirectly by a combination of specifying the building geometry and material thicknesses, therefore specifying volumes of individual building components in the DSM.

Regarding the other three design parameters for passive solar systems (absorption surface, glazing, and exposed surface area), we have already discussed the first two in 'Selective surfaces' and 'Solar transmission components' in this chapter. These two parameters need to be chosen by the designer at the outset, and the third one, the surface area, will be obtained as a result of dynamic simulation and design optimisation.

9.5 SCALE MODEL EXPERIMENTS

Scale model experiments of passive solar systems were conducted by my students. The brief was as follows:

Direct gain system and Trombe wall system. Create two cardboard models: Model 1—20 cm × 12 cm × 6 cm, with 2 × 6 cm × 4 cm openings on the larger wall (wall with dimensions of 20 cm × 6 cm). Model 2 – the same as model 1, except the south wall has an opening as large as the wall itself. Create internal boxes for both models, so that they are by 2 cm smaller than the external boxes in all directions. In model 1, fill the walls and ceiling with thermal insulation, and the floor with sand. In model 2, fill the south wall with sand, and all other walls, the floor and the ceiling with thermal insulation. Expose the south side to solar radiation. Record temperatures in the inside space over 1 hour at intervals of 1 minute. Compare the results from the two experiments and comment on the findings.

A model created by one of the groups of students is shown in Figure 100. The figure shows a Trombe wall model and a direct gain model both exposed to solar radiation on a windowsill between 8:30 and 18:30 on a day in early May. Temperature sensors were inserted through the centre of the back wall and connected to a data logger (Figure 101), monitoring temperatures in one minute intervals.

Results of the monitoring of the behaviour of these two models under partial cloud and varying solar radiation are shown in Figure 102. We can see that the direct gain model heats up and cools down faster than the Trombe wall model. Energy accumulated in the Trombe wall model contributes to higher temperatures towards the end of the monitoring period.

Figure 100   Trombe wall (left) and direct gain (right) models exposed to solar radiation on a window sill (models by my former students Anil Bandha, Aishah Afzal, and Robert Shaw)

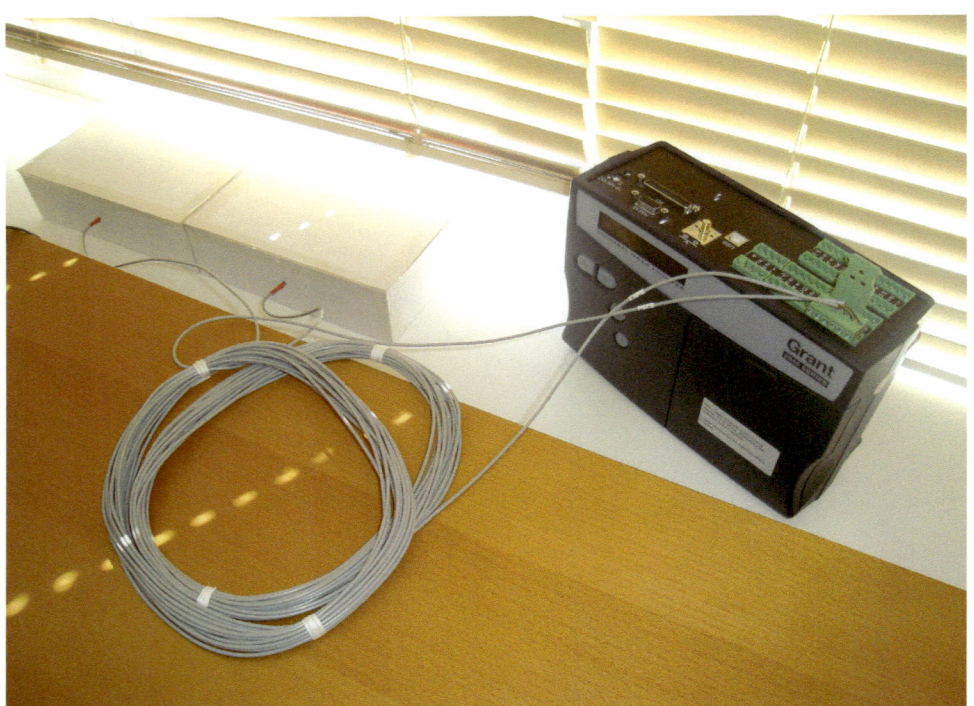

Figure 101   Temperature monitoring of the two solar gain scale models

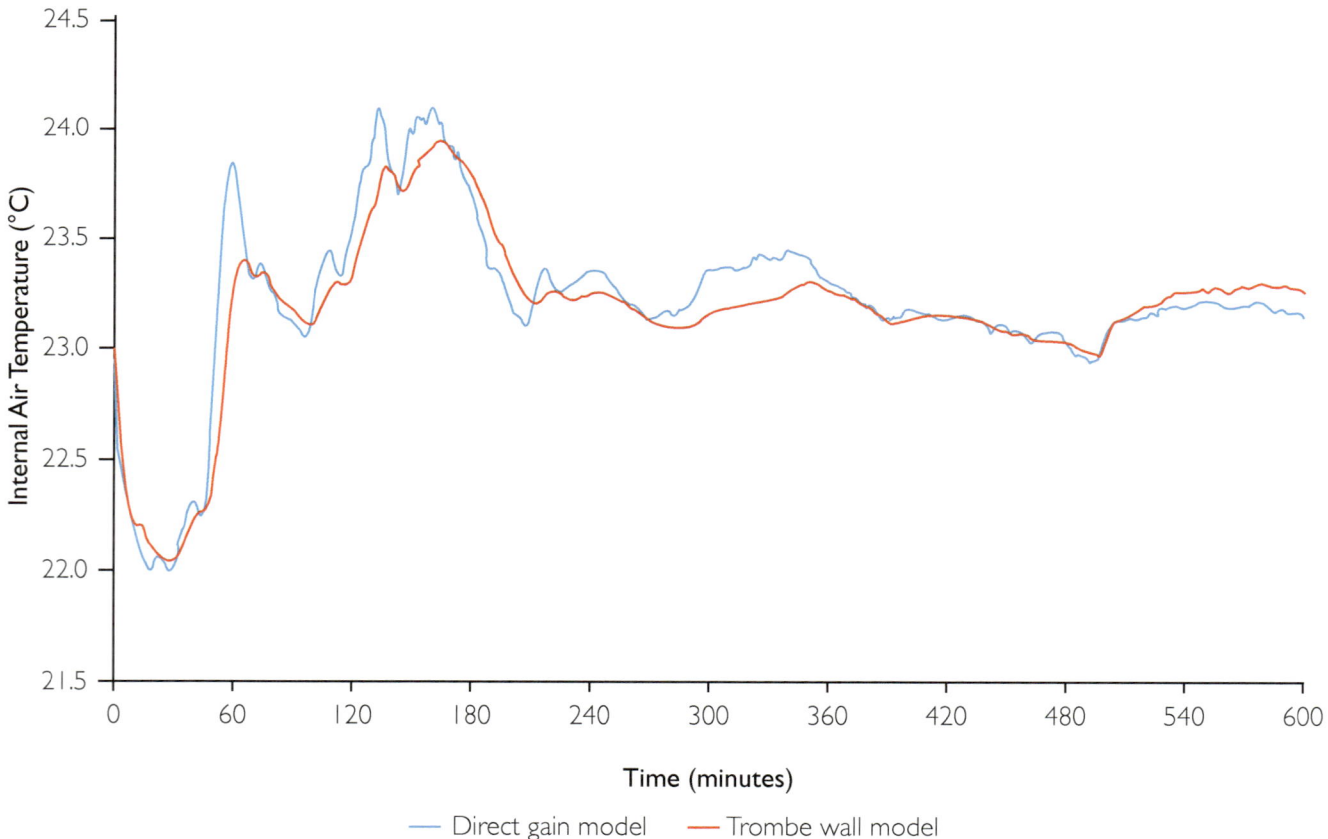

Figure 102   Results of monitoring of two solar gain scale models

9.6 SIMULATION EXPERIMENTS WITH PASSIVE SOLAR SYSTEMS

This section explains how to build models of various passive solar systems in the IES Virtual Environment. The performance of the models will be compared with a reference case. The weather data location for the above models was set to Birmingham (UK), and the activity type was set to open plan office. The models used natural ventilation, with a control profile for windows to open above 23°C internal air temperature. Construction types used in all models are shown in Table 29. The reference case is a simple box-type building with north-facing windows (Figure 103). Working models of all simulation cases are available for download from the supplementary material web site www.ljankovic.com.

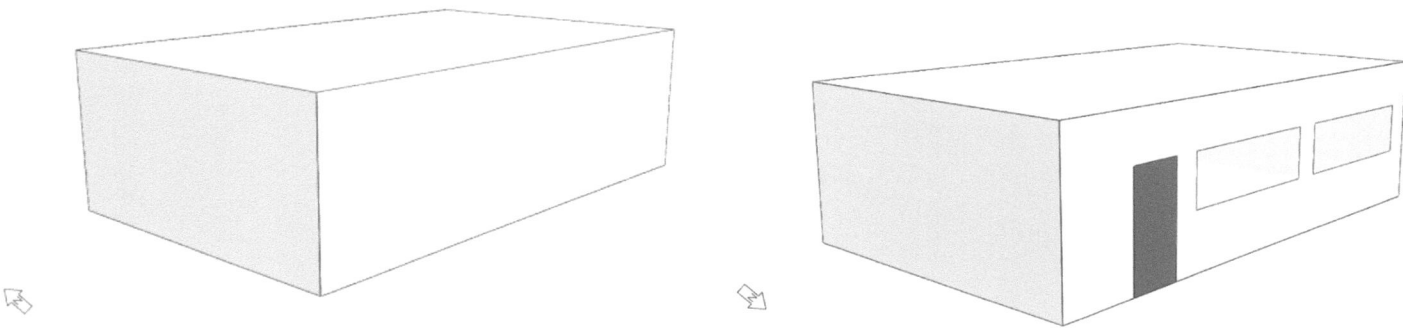

Figure 103   Geometry of the simulation reference model

The direct gain system is modelled using the same overall geometry, with large windows on the south-facing side (Figure 104). The isolated gain system is modelled using two adjacent spaces: the direct gain model and an attached glazed space on the south-facing side (Figure 105). There are four of 1 m × 0.4 m and two of 1 m × 0.2 m controllable openings between the sunspace and

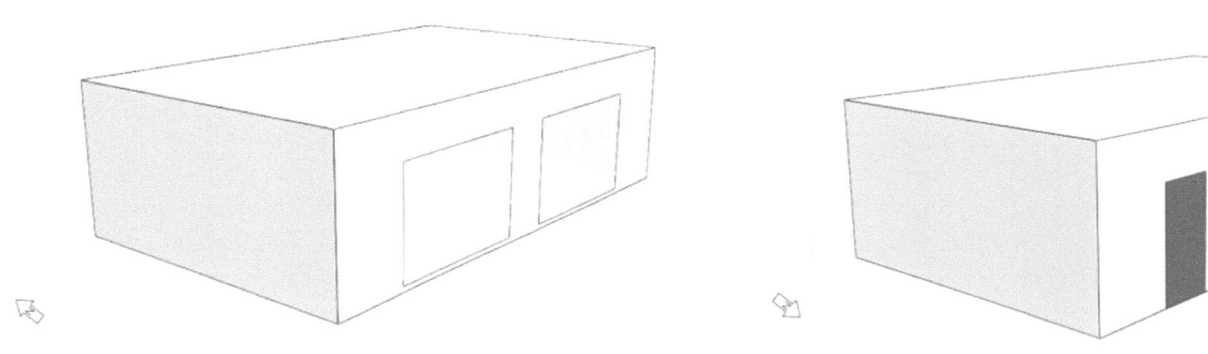

Figure 104   Direct gain model geometry

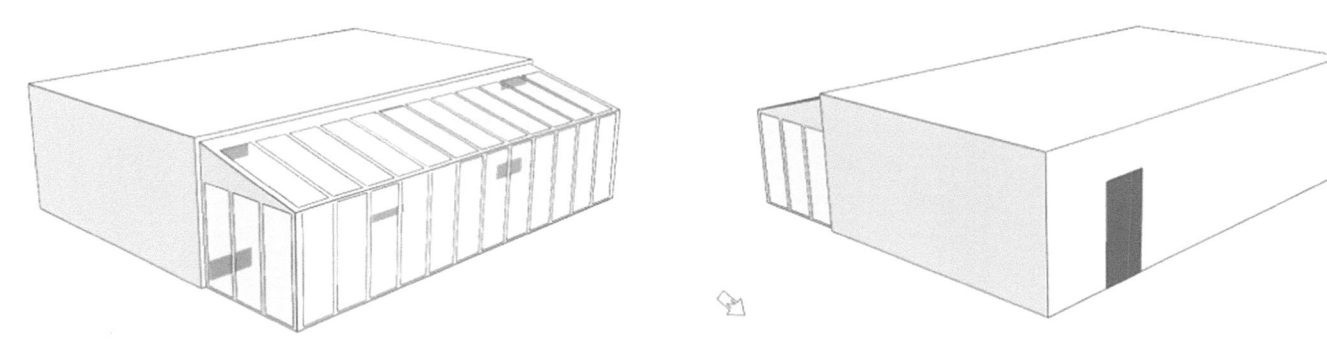

Figure 105   Isolated gain model geometry

the room, at the floor and ceiling level, for air circulation. These are modelled as small-size doors and are controlled by a formula-based profile, which makes them 100% open when the sunspace temperature is greater than the room temperature, and closes them when the room temperature reaches 23°C. The sunspace is ventilated to the outside when this condition is reached

The Trombe wall model (Figure 106) uses a high-density wall and an adjacent 150 mm deep space, fully glazed on the south-facing side. The selective absorbing surface of the Trombe wall is based on black chrome properties, with a solar absorptance of 0.87 and an emittance of 0.09. The north-facing wall is the same as in the reference case model. There are ten of 1 m × 0.2 m controllable openings at the floor and ceiling level of the Trombe wall, modelled as doors and controlled by a formula that opens them 100% when the air gap temperature is greater than room temperature and closes them when the room temperature reaches 23°C. The room is ventilated above this temperature using the north-facing windows.

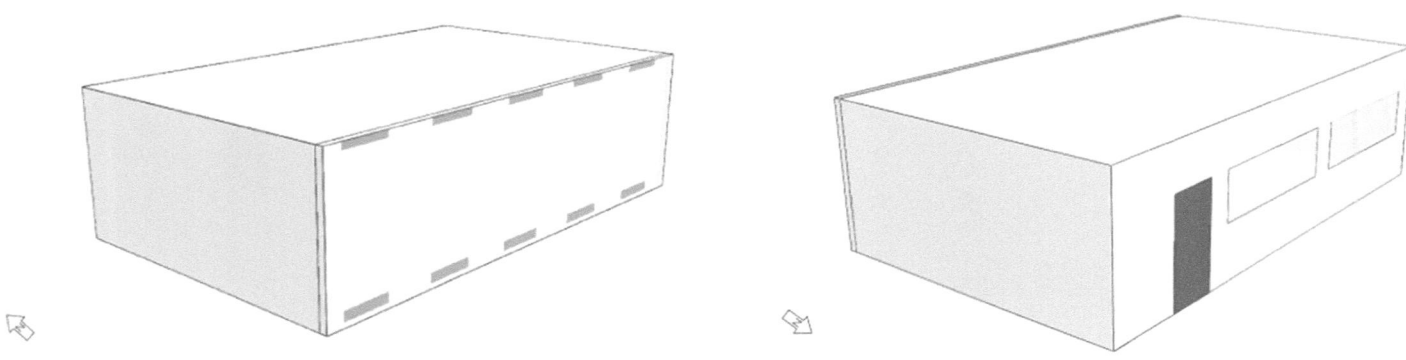

Figure 106   Trombe wall model geometry

There are also several examples of models of solar transmission components, as described below.

A double-skin model is shown in Figure 107. It consists of a 0.6 m deep fully glazed air space on the south-facing side of the standard room geometry identical to the above models. There are twenty of 1 m × 0.2 m controllable openings, both on the outer and inner skin, positioned at the floor and ceiling level, and controlled by a formula. There are four modes of operation of the double skin, as shown in Figure 93. In the heating mode (mode 1), the inner skin flaps open when temperature in the air space is greater than room temperature, and are closed when room temperature reaches 23°C and the room carbon dioxide level is below 1000 ppm. In the heat rejection mode (mode 2), the outer flaps are open when both air space temperature and room temperature exceed 23°C and the room carbon dioxide level is below 1000 ppm. In the preheating mode (mode 3), the

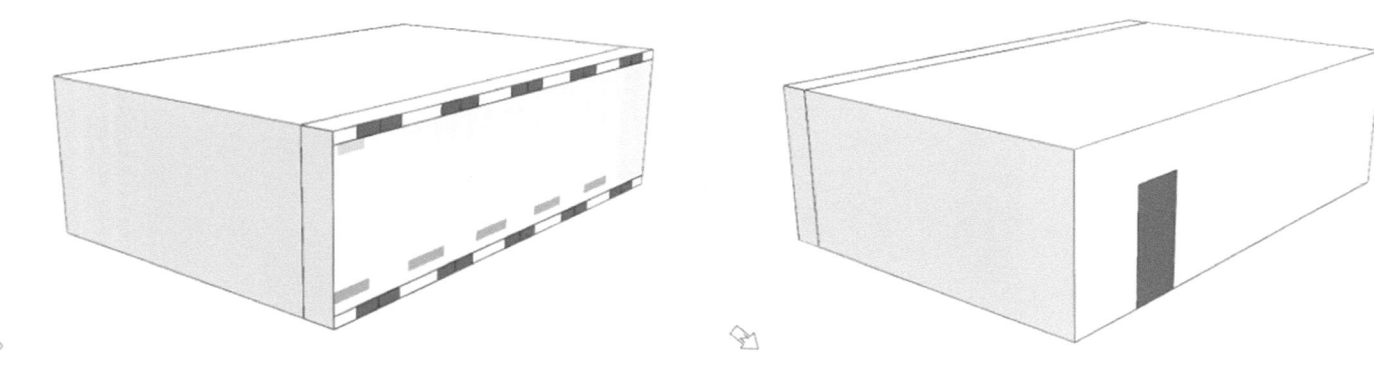

Figure 107   Double-skin model geometry

outer bottom flaps and inner top flaps are open under the same temperature conditions as in the heating mode (mode 1), but only when the room carbon dioxide level exceeds 1000 ppm. In the ventilation mode (mode 4), the inner bottom and the outer top flaps are open under the same conditions as in the heat rejection mode (mode 2), but only when the room carbon dioxide level exceeds 1000 ppm.

A glazing model is shown in Figure 108. It has a fully glazed south-facing surface, and will be used to demonstrate four different types of glazing. The first two types, clear, and low emissivity glazing, are chosen directly from the DSM construction database. The third type, transparent insulation, is modelled by modifying single clear glazing characteristics using manufacturer's data, as shown in Table 29. The fourth type, switchable glazing, is modelled using clear glazing and a control profile that activates an internal blind as shown in Table 29, thus giving it desired transmittance characteristics.

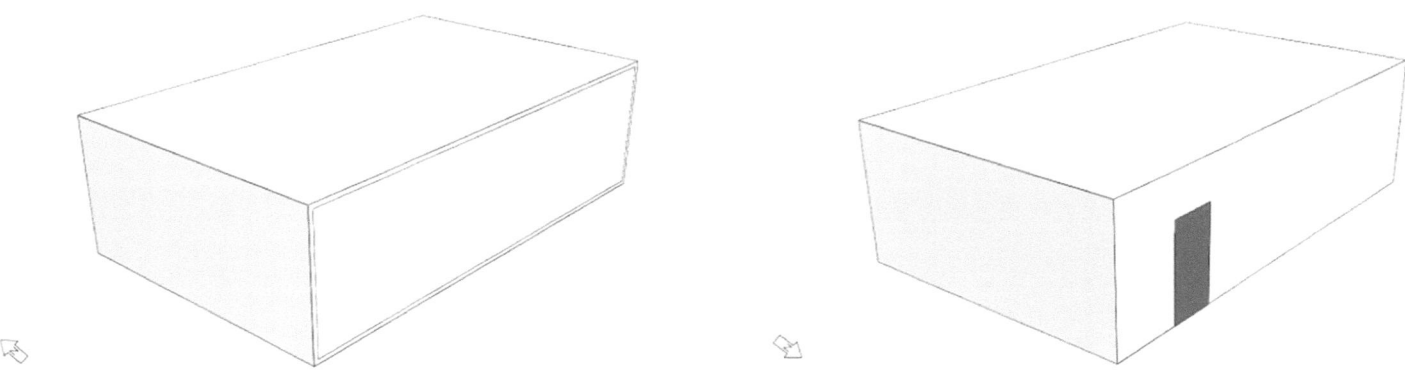

Figure 108   Glazing model geometry

TABLE 29   CONSTRUCTION TYPES IN THE SIMULATION MODELS. UNLESS SPECIFIED, THE CONSTRUCTION TYPES OF OTHER MODELS ARE THE SAME AS THOSE IN THE REFERENCE MODEL.

Reference model

Exterior walls

Layer Description	Thickness m	Conductivity W/(m·K)	Density kg/m³	Capacity J/(kg·K)	Resistance m²K/W
BRICKWORK (OUTER LEAF)	0.1000	0.840	1700.0	800.0	0.1190
DENSE EPS SLAB INSULATION – LIKE STYROFOAM	0.0585	0.025	30.0	1400.0	2.3400
CONCRETE BLOCK (MEDIUM)	0.1000	0.510	1400.0	1000.0	0.1961
GYPSUM PLASTERING	0.0150	0.420	1200.0	837.0	0.0357
Inside Surface					0.130
Outside Surface					0.040
Total Resistance					2.861
CIBSE Net U-Value W/m²·K	0.3487				
EN-ISO U-Value W/m²·K	0.3495				
Outside surface absorptance	0.7000				
Inside surface absorptance	0.5500				
Inside Emissivity	0.9000				
Outside Emissivity	0.9000				

Floor

Layer Description	Thickness m	Conductivity W/(m·K)	Density kg/m³	Capacity J/(kg·K)	Resistance m²K/W
LONDON CLAY	0.7500	1.410	1900.0	1000.0	0.5319
BRICKWORK (OUTER LEAF)	0.2500	0.840	1700.0	800.0	0.2976
CAST CONCRETE	0.1000	1.130	2000.0	1000.0	0.0885

TABLE 29 CONTINUED

Reference model

Exterior walls

Layer Description	Thickness m	Conductivity W/(m·K)	Density kg/m³	Capacity J/(kg·K)	Resistance m²K/W
DENSE EPS SLAB INSULATION — LIKE STYROFOAM	0.0635	0.025	30.0	1400.0	2.5400
CHIPBOARD	0.0250	0.150	800.0	2093.0	0.1667
SYNTHETIC CARPET	0.0100	0.060	160.0	2500.0	0.1667
Inside Surface					0.170
Outside Surface					0.040
Total Resistance					4.001
CIBSE Net U-Value W/m²·K	0.2533				
EN-ISO U-Value W/m²·K	0.2499				
Outside surface absorptance	0.7000				
Inside surface absorptance	0.5500				
Inside Emissivity	0.9000				
Outside Emissivity	0.9000				

Roof

Layer Description	Thickness m	Conductivity W/(m·K)	Density kg/m³	Capacity J/(kg·K)	Resistance m²K/W
STONE CHIPPINGS	0.0100	0.960	1800.0	1000.0	0.0104
FELT/BITUMEN LAYERS	0.0050	0.500	1700.0	1000.0	0.0100
CAST CONCRETE	0.1500	1.130	2000.0	1000.0	0.1327
GLASS-FIBRE QUILT	0.1345	0.040	12.0	840.0	3.3625
Cavity	0.1000				0.1700
CEILING TILES	0.0100	0.056	380.0	1000.0	0.1786
Inside Surface					0.100
Outside Surface					0.040
Total Resistance					4.004
CIBSE Net U-Value W/m²·K	0.2487				
EN-ISO U-Value W/m²·K	0.2497				
Outside surface absorptance	0.5000				
Inside surface absorptance	0.5500				
Inside Emissivity	0.9000				
Outside Emissivity	0.9000				

Glazing
low-e double glazing (6 mm + 6 mm)

Layer Description	Thick. m	Cond. W/(m·K)	Res. m²K/W	Solar Trans.	Out. Solar Refl.	Ins. Solar Refl.	Visible Trans.	Out. Vis. Refl.	Ins. Vis. Refl.	Ref. Index	Outside Emiss.	Inside Emiss.
PILKINGTON K 6 MM	0.0060	1.0600		0.6900	0.0900	0.0900	0.0000	0.0000	0.0000	1.5260	0.0000	0.0000
Cavity	0.0120		0.3247									
CLEAR FLOAT 6 MM	0.0060	1.0600		0.7800	0.0700	0.0700	0.0000	0.0000	0.0000	1.5260	0.0000	0.0000

Frame occupies 10.00% of area		
Inside Surface Resistance	0.13000	
Outside Surface Resistance	0.04000	
Outside Emissivity	0.90000	
Inside Emissivity	0.90000	
CIBSE U-value (glass only)	1.94894	
CIBSE net U-value	1.95002	
EN ISO U-value (glass only)	1.97620	
EN ISO net U-value	1.97731	

TABLE 29 CONTINUED

Reference model

Exterior walls

Layer Description	Thickness m	Conductivity W/(m·K)	Density kg/m³	Capacity J/(kg·K)	Resistance m²K/W

Door
wooden door

Reference ID: DOOR
Construction is from the project database

Layer Description Vapour res. GN·s/(kg·m)	Thickness m	Conductivity W/(m·K)	Density kg/m³	Capacity J/(kg·K)	Resistance m²K/W
PINE (20% MOIST)	0.0400	0.140	419.0	2720.0	0.2857
Inside Surface					0.130
Outside Surface					0.040
Total Resistance					0.456
CIBSE Net U-Value W/m²·K	2.1608				
EN-ISO U-Value W/m²·K	2.1944				
Outside surface absorptance	0.7000				
Inside surface absorptance	0.5500				
Inside Emissivity	0.9000				
Outside Emissivity	0.9000				

Direct gain model

Floor

Layer Description Vapour res. GN·s/(kg·m)	Thickness m	Conductivity W/(m·K)	Density kg/m³	Capacity J/(kg·K)	Resistance m²K/W
SAND	1.0000	0.329	1515.0	796.0	3.0395
CRUSHED BRICK	0.0750	0.550	1580.0	1057.0	0.1364
CAST CONCRETE	0.1250	0.870	1800.0	1920.0	0.1437
POLYURETHANE BOARD	0.0990	0.025	30.0	1400.0	3.9600
CAST CONCRETE (DENSE)	0.1000	1.400	2100.0	840.0	0.0714
PLASTIC	0.0050	0.500	1050.0	837.0	0.0100
Inside Surface					0.170
Outside Surface					0.040
Total Resistance					7.571
CIBSE Net U-Value W/m² K	0.1330				
EN-ISO U-Value W/m²·K	0.1321				
Outside surface absorptance	0.9000				
Inside surface absorptance	0.9000				
Inside Emissivity	0.9000				
Outside Emissivity	0.9000				

Isolated gain model

Floor — the same as Direct gain model

Trombe wall model

Partition wall between the air space and the room
Trombe Wall — 8 In. Heavy Weight Concrete

Layer Description	Thickness m	Conductivity W/(m·K)	Density kg/m³	Capacity J/(kg·K)	Resistance m²K/W
HW CONCRETE UNDRIED AGGREGATE — HF-C12	0.2032	1.730	2243.0	837.0	0.1175
Inside Surface					0.130
Outside Surface					0.130
Total Resistance					0.377

TABLE 29 CONTINUED

Reference model

Exterior walls

Layer Description	Thickness m	Conductivity W/(m·K)	Density kg/m³	Capacity J/(kg·K)	Resistance m²K/W
CIBSE Net U-Value W/m²·K	1.4702				
EN-ISO U-Value W/m²·K	2.6493				
Outside surface absorptance	0.8700				
Inside surface absorptance	0.8700				
Inside Emissivity	0.0900				
Outside Emissivity	0.0900				

Air space — 150 mm
Air space glazing — the same as reference model

Double skin model

Floor — the same as direct gain model
Glazing — the same as reference model

Clear glazing model

Glazing

Layer Description	Thick. m	Cond. W/(m·K)	Res. m²K/W	Solar Trans.	Out. Solar Refl.	Ins. Solar Refl.	Visible Trans.	Out. Vis. Refl.	Ins. Vis. Refl.	Ref. Index	Outside Emiss.	Inside Emiss.
CLEAR FLOAT 6 MM	0.0060	1.0600		0.7800	0.0700	0.0700	0.0000	0.0000	0.0000	1.5260	0.0000	0.0000
Cavity	0.0120		0.1688									
CLEAR FLOAT 6 MM	0.0060	1.0600		0.7800	0.0700	0.0700	0.0000	0.0000	0.0000	1.5260	0.0000	0.0000

Frame occupies 5.00% of area

Inside Surface Resistance	0.13000
Outside Surface Resistance	0.04000
Outside Emissivity	0.90000
Inside Emissivity	0.90000
CIBSE U-value (glass only)	2.79956
CIBSE net U-value	2.79976
EN ISO U-value (glass only)	2.85616
EN ISO net U-value	2.85637

Low e-glazing model

Glazing — the same as reference model

Transparent insulation glazing model

Nanogel panel

Layer Description	Thick. m	Cond. W/(m·K)	Res. m²K/W	Solar Trans.	Out. Solar Refl.	Ins. Solar Refl.	Visible Trans.	Out. Vis. Refl.	Ins. Vis. Refl.	Ref. Index	Outside Emiss.	Inside Emiss.
KAPPA-FLOAT 6 MM (CHAMPAGNE)	0.0060	1.0600		0.5800	0.1600	0.1600	0.0000	0.0000	0.0000	1.5260	0.0000	0.0000
Cavity	0.0100		1.4650									
Nanogel material	0.0200	0.0180		0.2100	0.0800	0.0800	0.0000	0.0000	0.0000	1.5900	0.0000	0.0000
Cavity	0.0100		1.4650									
KAPPA-FLOAT 6 MM (CHAMPAGNE)	0.0060											

TABLE 29 CONTINUED

Reference model

Exterior walls

Layer Description					Thickness m	Conductivity W/(m·K)	Density kg/m³	Capacity J/(kg·K)	Resistance m²K/W
Sharing Device information Local Device:none External Device:none Internal Device:none									

Frame occupies 5.00% of area

Inside Surface Resistance	0.13000								
Outside Surface Resistance	0.04000								
Outside Emissivity	0.90000								
Inside Emissivity	0.90000								
CIBSE U-value (glass only)	0.23643								
CIBSE net U-value	0.36479								
EN ISO U-value (glass only)	0.23683								
EN ISO net U-value	0.36801								

Switchable glazing model

Glazing

Switchable glazing

Layer Description	Thick. m	Cond. W/(m·K)	Gas Conv. Coeff. W/m²·K	Res. m²K/W	Solar Trans.	Out. Solar Refl.	Ins. Solar Refl.	Vis-ible Trans.	Out. Vis. Refl.	Ins. Vis. Refl.	Ref. Index	Outside Emiss.	Inside Emiss.
Switchable glass	0.0092	1.0600			0.4900	0.0700	0.0700	0.0000	0.0000	0.0000	1.5260	0.0000	0.0000
Cavity	0.0120		AIR 2.0800	0.1730									
PERFECTLY CLEAR	0.0060	1.0600			1.0000	0.0000	0.0000	0.0000	0.0000	0.0000	1.0000	0.0000	0.0000

GLAZING MATERIAL
Sharing Device information
Internal:
Device:Blinds
Percentage profile group:NONE
Incident radiation to lower device:500.00

TABLE 29 CONTINUED

Reference model

Exterior walls

Layer Description		Thickness m	Conductivity W/(m·K)	Density kg/m³	Capacity J/(kg·K)	Resistance m²K/W
Incident radiation to raise device:400.00						
Nighttime resistance:0.00						
Daytime resistance:0.00						
Shading coefficient:0.49						
Short wave radiant fraction:0.55						
Fraction of daylight hour closed:0.10						
Frame occupies 5.00% of area						
Inside Surface Resistance	0.13000					
Outside Surface Resistance	0.04000					
Outside Emissivity	0.90000					
Inside Emissivity	0.90000					
CIBSE U-value (glass only)	2.74395					
CIBSE net U-value	2.85406					
EN ISO U-value (glass only)	2.79830					
EN ISO net U-value	2.91467					

Properties of the transparent insulation material have been chosen to closely match similar material from Table 28, however due to the maximum thickness limitation of 20 mm for transparent materials in the particular software tool used in this chapter, these properties had to be enhanced by increasing the air cavity resistance.

The comparison of annual simulations of all nine models is shown in Figure 109. We can see that the direct gain system performs better than the reference case, and in this particular case, it is almost identical to the Trombe wall system. The combined isolated and direct gain system is worse than the direct gain and Trombe wall, but better than the reference case. This is because of lower solar gain in the room due to the two sets of double-glazing panels (room and sunspace). The double-skin system appears to be very similar to direct gain and Trombe wall systems.

The comparison of four different glazing systems in the same figure shows that the clear glazing case is worse than the reference case, as the former has a smaller amount of low-e glazing and the latter has a larger amount of clear glazing. The low-e glazing case is better than the reference case as its glazing is larger and south-facing. The case with the transparent insulation material (tim_glazing), based on 30 mm nanogel, outperforms all other systems, whilst the switchable glazing system is better than the reference case but worse than all other cases due to lower transmittance in the clear mode and consequent reduction of solar gain.

Whilst specific differences between these results are not to be generalised, as they only serve the purpose of comparison between different simulation models presented in this chapter, there are two outcomes of this analysis that are worth emphasising: simulation results are meaningful only if at least two different simulation cases are compared; and the explanation of performance differences between different models is essential for making design decisions with confidence. If differences between some models cannot be explained, these models should be looked at again in detail and simulations re-run until we are satisfied that we have eliminated any possible errors or inconsistencies.

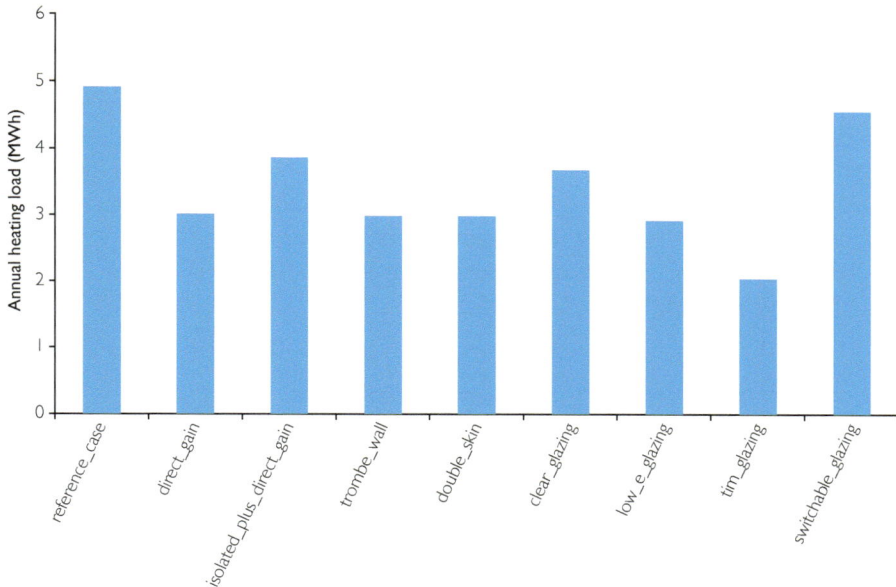

Figure 109   Comparison of performance of nine simulation models

9.7 NORTH-FACING GLAZING

Judging by the inferior performance of the reference case in Figure 109, north-facing glazing is less desirable than south-facing glazing. As we need daylight in north-facing spaces, the surface area of glazing will be typically smaller and will need to be optimised so as to achieve the balance between the heat losses through glazing and the energy used for electrical lighting due to reduced glazing area. A method that can be used for optimisation of glazing surface area is described in the section entitled 'Design optimisation' in Chapter 3.

9.8 INTERNAL BUILDING LAYOUT

From the analysis in the previous section, we have seen that the north-facing orientation (Figure 109- reference case) requires more energy than various systems with south-facing orientation. In order to make the internal building layout energy responsive, the main activity areas should be located on the sun-facing side, and less frequent activities on the opposite side, such as shown in Figure 110a. This layout corresponds to buildings that predominantly need heating. Buildings that predominantly require cooling need to have a reversed layout as shown in Figure 110b.

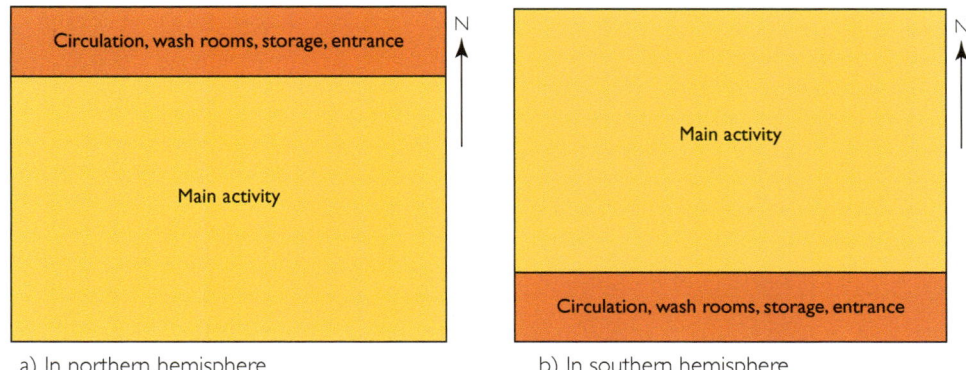

a) In northern hemisphere

b) In southern hemisphere

Figure 110   Energy-responsive layout for buildings that predominantly require heating

9.9 TASKS FOR SIMPLE SIMULATION EXPERIMENTS

1 Create a model of a dome with 10 m base diameter and 5 m height. Apply default construction type and double glazing. Set the location to a cool climate. Set the model to run in free-floating mode, without any heating, or cooling, and without any internal gains from people, lights and appliances.

2 Run annual simulations of the dome without any glazing. View the graph of internal air temperatures and record the maximum temperature. In a cool climate, this temperature will typically not exceed 20°C.

3 Glaze the dome, using minimum and maximum azimuth of 0° and 360° respectively, and minimum and maximum tilt of 0° and 90° respectively, and specifying 40% of the glazed area. Run the annual simulation, view the graph of internal room temperatures, and record the maximum temperature. In a cool climate, this temperature could easily exceed 50°C.

4 Scale down the glazing progressively until the maximum internal temperature does not exceed 25°C.

5 Experiment with increased north and reduced south glazing and the other way round.

6 Discuss the results with your colleagues.

9.10 SUMMARY OF DESIGN PRINCIPLES

Solar transmission components:

• Choose the optimum type of solar transmission components, balancing between thermal insulation properties and solar transmission properties using DSM.
• Optimise the size of the transmission component by balancing heat gains and losses using DSM.
   Passive solar systems:

• Choose an appropriate passive system, taking into consideration their main characteristics.
• Test the total annual heating and cooling load and summertime temperatures for at least two design options using DSM.
   Internal building layout:

• Position high-activity spaces towards high solar gains and low-activity spaces towards the opposite side in predominantly heated buildings.

CHAPTER 10

SOLAR SHADING DESIGN

Whilst the previous chapter discusses the ways of capturing solar radiation in buildings, this chapter discusses the ways of preventing solar radiation entering the building. Methods for solar shading design are discussed, to be used in conjunction with systems from the previous chapter. The sun path diagram, explained in Chapter 6, will be instrumental in designing solar shading. The focus of this chapter will be on the design of overhang geometry and louvre geometry and on testing designs using a DSM sun-casting facility.

10.1 CALCULATING THE GEOMETRY OF AN OVERHANG

In Chapter 6, we learnt how to determine solar altitude angles. With this knowledge, we can now calculate the geometry of an overhang. Calculation spreadsheets from this chapter are available for download from the supplementary material website www.ljankovic.com.

Let us assume that we have already determined the solar altitude angle α as in Example 1 in the section entitled 'Sun path diagram' in Chapter 6, referring to Figure 62. Then from Figure 111, we find that the relationship between the glazing height H and overhang length L is

$$\tan(90 - \alpha) = \frac{L}{H} \tag{24}$$

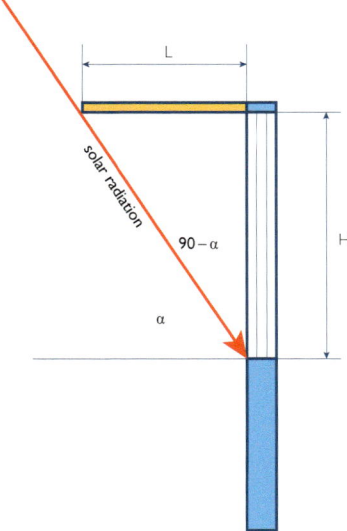

Figure 111   Geometry of an overhang

Note: We must ensure that the angles in the above equation are in the right units for the calculator or spreadsheet that we are using. If conversion into radians is required, we need to multiply the angle in degrees by π/180.

From this, it follows that

$$L = H \times \tan(90 - \alpha) \tag{25}$$

As both H and α are known, we can calculate the length of the overhang by substituting these known values in Equation (25).

10.2 CALCULATING GEOMETRY OF EXTERNAL LOUVRE SHADING

The geometry of louvre shading is shown in Figure 112. We again assume that we have already obtained solar altitude angle α, as in Example 1 in the section entitled 'Sun path diagram' in Chapter 6, and that the louvre is fixed and with tilt angle β that is different from solar altitude angle α.

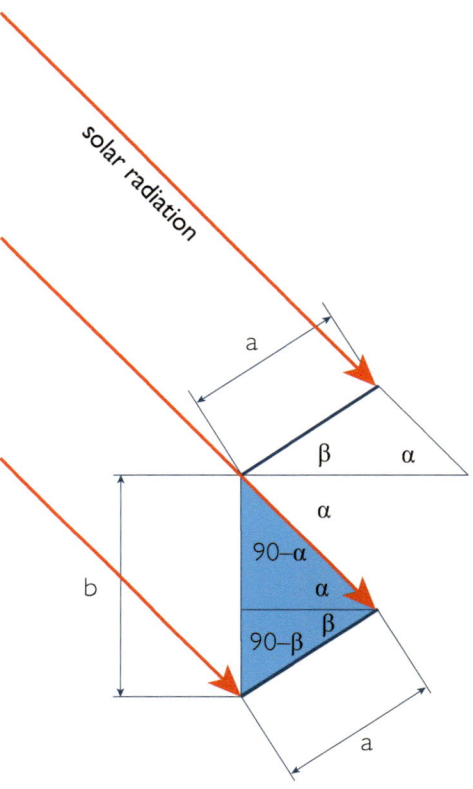

Figure 112   Geometry of louvre shading

The relationship that links relevant parameters in the louvre geometry is based on the following trigonometric relationship between sides and angles in the highlighted scalene triangle:

$$\frac{a}{\sin(90 - \alpha)} = \frac{b}{\sin(\alpha + \beta)} \tag{26}$$

From here, it follows that

$$b = a \times \frac{\sin(\alpha + \beta)}{\sin(90 - \alpha)} \tag{27}$$

In order to specify the louvre geometry, we need to determine the width of the louvre blade a, the distance between louvre blades b, and the louvre tilt angle β. The easiest way to do this is by setting up the above relationship in a spreadsheet, and experimenting with different parameter values until we find a suitable combination that satisfies our design requirements. An example of this calculation is shown in the spreadsheet in Table 30.

TABLE 30	EXAMPLE CALCULATION OF LOUVRE GEOMETRY	
	Parameter (units)	Value
1	α (°)	50.00
2	β (°)	30.00
3	$\sin((90-\alpha) \cdot \pi/180)$	0.64
4	$\sin((\alpha+\beta) \cdot \pi/180)$	0.98
5	a (m)	0.50
6	b (m)	0.77

First, the value of the solar altitude angle α in Table 30 is determined from the sun path diagram (Figure 62). Second, values of the louvre tilt angle β and the louvre blade width a are chosen arbitrarily and the distance between the louvre blades b is calculated. The process is repeated until values that satisfy the specific design objectives of an individual project are obtained. Note that angles in the sine functions in Table 30 are converted from degrees into radians, as per spreadsheet syntax requirements.

10.3 COMBINED LOUVRE AND OVERHANG GEOMETRY

In many instances, it is not practical to design an overhang that is sufficiently long to shade the entire glazed surface. This is especially the case with full-height glazing. In such case, a combination of an overhang and a louvre can provide a suitable solution (Figure 113). The design of this combination is based on the examples we have already covered in this chapter. The overhang length L is calculated in a similar way as in Equation (25), except the glazing height H is now replaced with $H2 - H1$ as follows:

$$L = (H2 - H1) \times \tan(90 - \alpha) \qquad (28)$$

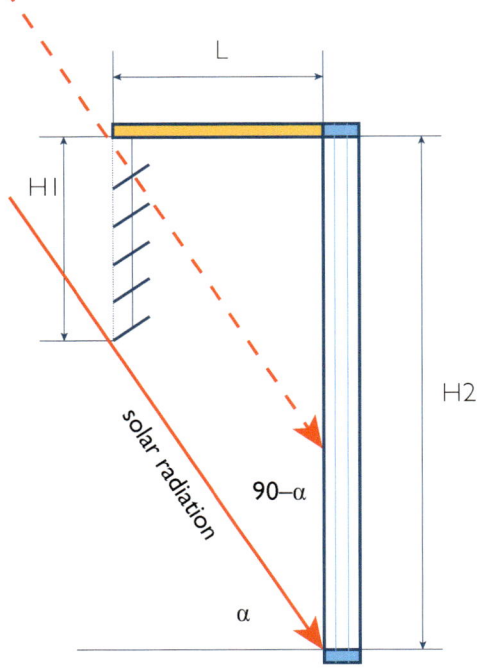

Figure 113   Combined overhang and louvre

10.4 DESIGN OF OVERHANG AND LOUVRE GEOMETRY FOR A SPECIFIED TIME PERIOD

A common design problem occurs when the solar shading needs to be designed for a range of dates and times, so that direct solar gain is completely eliminated in that period. We will demonstrate this using the calculations for single solar altitude angles already explained in this chapter.

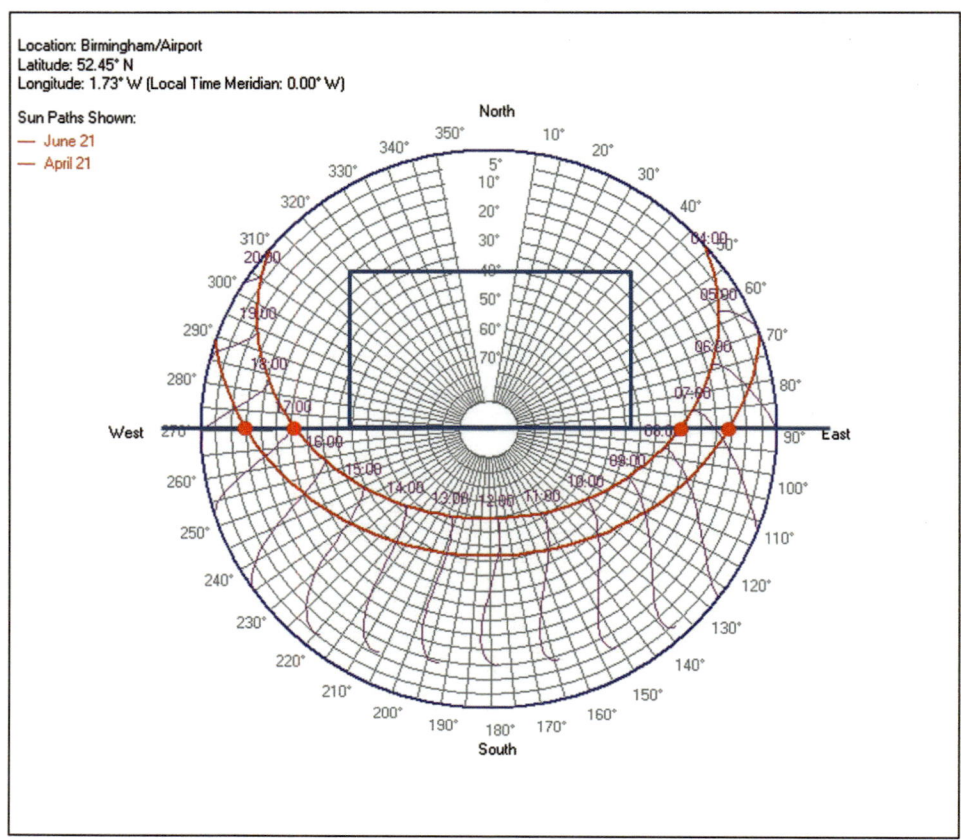

Figure 114   Solar angles April to June to August – south-facing building

Let us assume that the solar gain needs to be eliminated from 21st April to 21st August. As these two dates are symmetric with respect to the summer solstice on 21st June, and therefore the respective date curves in the sun path diagram are identical, we only need to find solar altitude angles for 21st April.

Figure 114 shows the case of a south-facing building, from which we find that both the morning and evening solar altitude angles are 15°. We can now use this information in Equation (25) to calculate the overhang length as follows, assuming the height of glazing of 2.5 m:

$$L = 2.5 \times \tan(90 - 15) = 9.33 \, \text{m} \tag{29}$$

The result is a rather large and disproportionate overhang, and it is very unlikely that it will be acceptable to the design team and the client. We will therefore investigate the combination of the overhang and external louvre as an alternative solution. We first calculate the height between the louvre blades as per Equation (27) and Table 30 as shown in Table 31..

TABLE 31	CALCULATION OF LOUVRE BLADE HEIGHT	
	Parameter (units)	Value
1	α (°)	15.00
2	β (°)	30.00
3	$\sin((90-\alpha) \cdot \pi/180)$	0.97
4	$\sin((\alpha+\beta) \cdot \pi/180)$	0.71
5	a (m)	0.50
6	b (m)	0.37

From Table 31, we then obtain the height between the louvre blades as 0.37 m. This now enables us to reduce the overhang length, by adding the louvre blades to a desired overall louvre height. Using Equation (28) and assuming the same glazing height of 2.5 m, the relationship between the overhang length L and the louvre height $H1$ (Figure 113) becomes

$$L = \left(2.5 - 0.37 \times (n-1)\right) \times \tan(90 - 15) \tag{30}$$

where n is the number of louvre blades. This now enables us to vary the louvre heights and obtain overhang lengths as follows:

TABLE 32	CALCULATION OF OVERHANG LENGTH AND LOUVRE HEIGHT		
n	$H1 = (n-1) \cdot 0.37$ (m)	$2.5-H1$ (m)	L (m)
2	0.37	2.13	7.95
3	0.74	1.76	6.57
4	1.11	1.39	5.19
5	1.48	1.02	3.81
6	1.85	0.65	2.43
7	2.22	0.28	1.04

The values from Table 32 now enable the designer to choose suitable dimensions for shading geometry.

Analysis for a different orientation is similar, except that morning and afternoon angles will not be symmetrical, as shown in Figure 115. For instance, for orientation 30° west of south the exposed facade will receive solar radiation on 21st April from 9 am until sunset. Due to varying angles and solar intensities during these hours, the most practical solution would be to specify adjustable louvre shading. If fixed louvre shading is preferable, it will be necessary to select the shading end time, e.g. 5 pm in an office, and calculate the shading geometry accordingly.

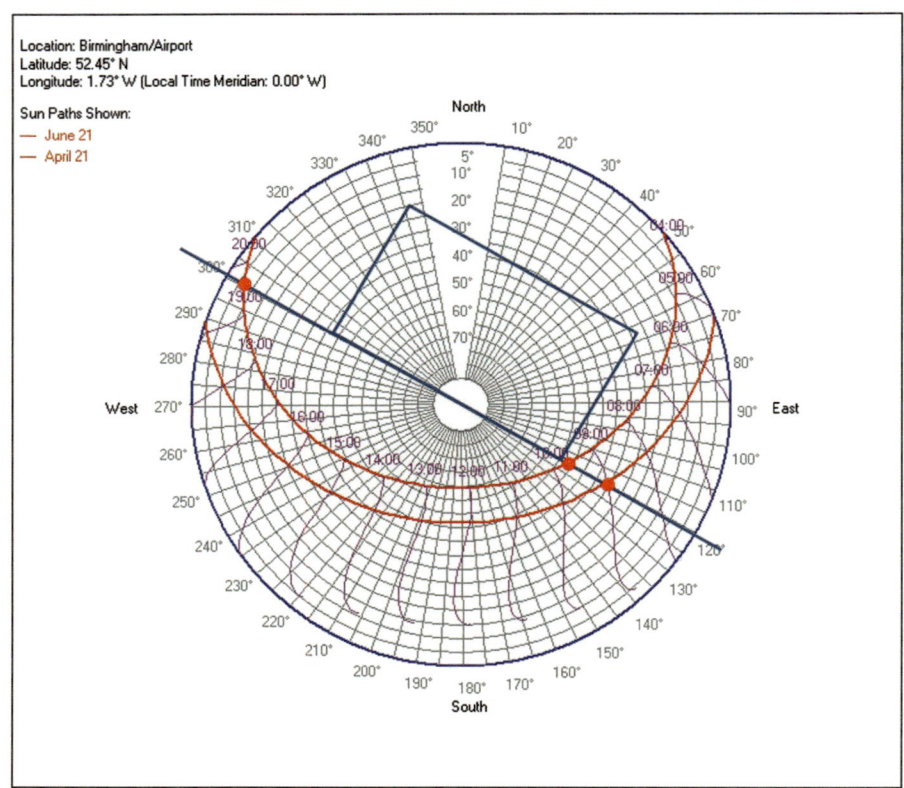

Location: Birmingham/Airport
Latitude: 52.45° N
Longitude: 1.73° W (Local Time Meridian: 0.00° W)

Sun Paths Shown:
— June 21
— April 21

Figure 115   Solar angles April to June to August −30° west of south-facing

10.5 TESTING OF OVERHANG AND LOUVRE GEOMETRY USING SIMULATION EXPERIMENTS

In our test case model (Figure 43), the glazing height is 4.47 m and the horizontal length of the overhang is 2.76 m. If we would like to shade the south facade between 09:00 and 15:00 on 21st April in Portsmouth, we find from Figure 116 that the critical solar altitude angle is 38°. Using Equation (30) and substituting corresponding values for overhang length, glazing height and solar altitude angle as follows

$$2.76 = \left(4.47 - 0.37 \times (n-1)\right) \times \tan(90 - 38)$$

we find that

$$n = \frac{1}{0.37} \times \left(4.47 - \frac{2.76}{\tan(90-38)}\right) + 1$$

and from there we find that $n = 7.25$ m. As the overhang doubles up as the top louvre blade, we need 6 louvre blades in total.

We now need to test this calculation using the sun-casting module in the DSM. Whilst running the SunCast module for 21st April with our eye position aligned with the sun, we need to check whether we can see any glazing between 09:00 and 15:00 hours. The results, generated using IES VE, are shown in Figure 117. The bottom edge of the south-facing glazing is hidden behind the shading as expected, confirming that our calculations were correct. However, the left and right edges of the glazing at the beginning and the end of this time interval are exposed to solar radiation, and further design is needed to take care of these areas. We can also notice that east-facing glazing is completely exposed to radiation until 11:00 and additional design is needed for this area as well.

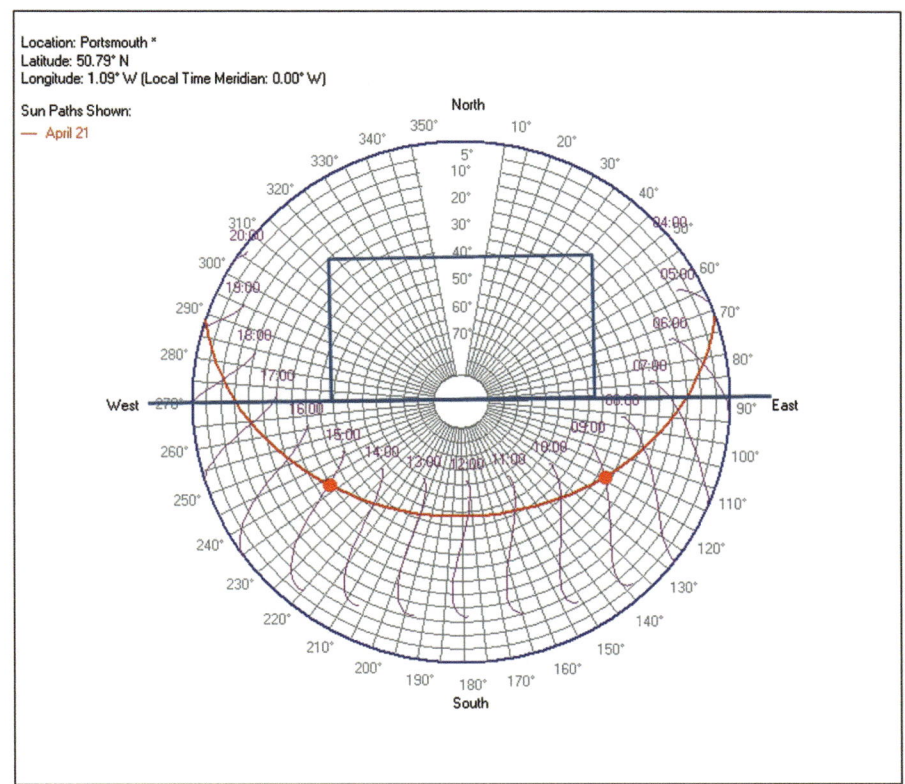

Location: Portsmouth *
Latitude: 50.79° N
Longitude: 1.09° W (Local Time Meridian: 0.00° W)

Sun Paths Shown:
— April 21

Figure 116   Sun path diagram for Portsmouth

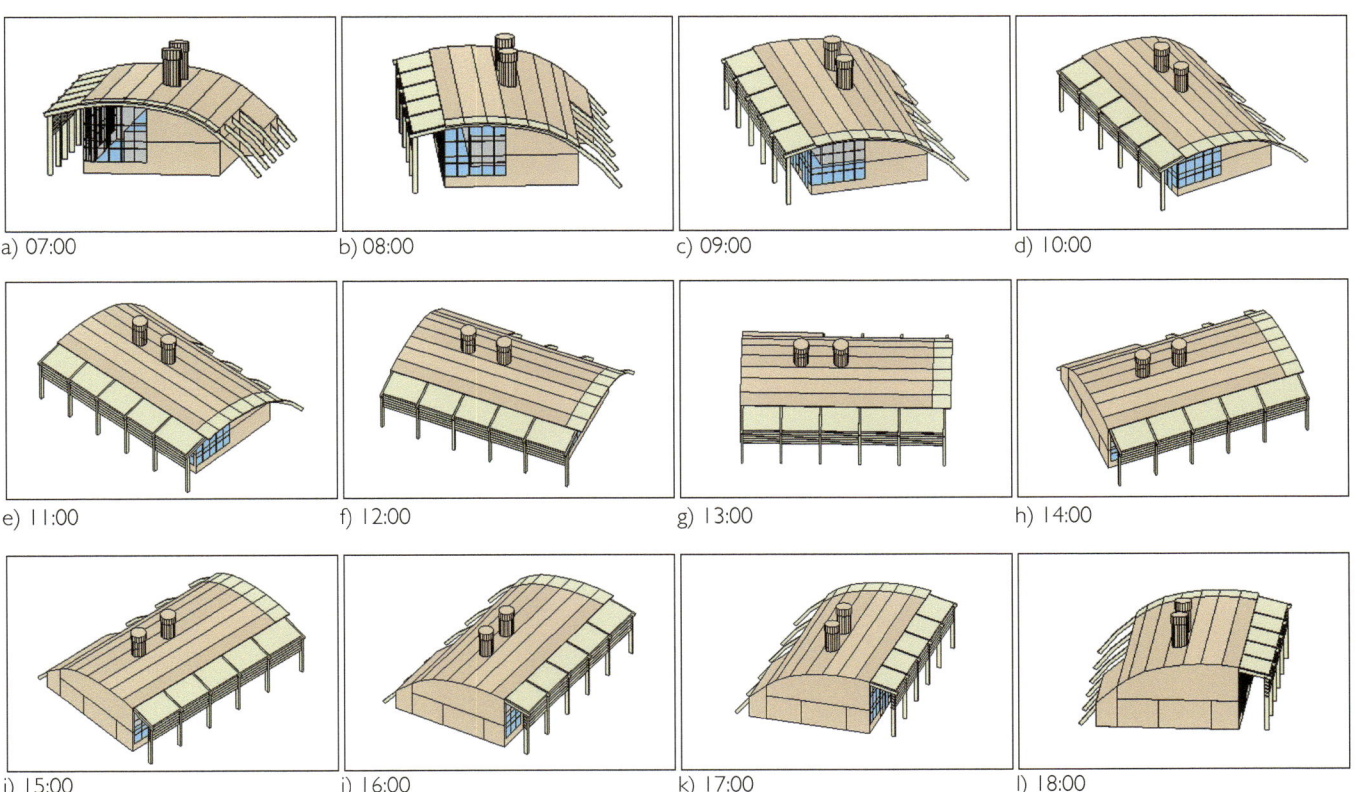

a) 07:00   b) 08:00   c) 09:00   d) 10:00

e) 11:00   f) 12:00   g) 13:00   h) 14:00

i) 15:00   j) 16:00   k) 17:00   l) 18:00

Figure 117   SunCast images from IES VE analysis for Portsmouth on 21st April

It is important to note that in this type of analysis, sun shadow casting observed from a fixed eye position would not be of much use. Aligning the eye position with the sun enables us to see what the sun 'sees', and will highlight any problem areas.

The working model and the spreadsheets from this chapter are available for download from the supplementary material website www.ljankovic.com.

10.6 TASKS FOR SIMPLE SIMULATION EXPERIMENTS

1 A glazed south-facing facade, 6 m tall and 10 m wide, with 40% of glazing on the sun-facing side, needs to be shaded between 21st May and 21st July using an overhang. Calculate shading geometry using the methods explained in this chapter and the weather data location of your choice. Validate the calculations using DSM solar simulation with eye position aligned with the sun. Reiterate this process until satisfactory shading is achieved.
2 Use the same model as above in free-running mode, with no heating and cooling, and compare hourly internal air temperatures before and after the external shading is applied. Use cumulative frequency of occurrence analysis of hourly air temperatures for numerical comparison. Discuss the results with colleagues/classmates.
3 Set heating to 19°C and cooling to 23°C in the model above and compare hourly and annual heating loads before and after the external shading is applied. Use cumulative frequency of occurrence analysis of hourly heating and cooling loads for numerical comparison. Discuss the results with colleagues.
4 Repeat the above analysis, replacing the overhang with louvres; and with a combination of overhang and louvres.

10.7 SUMMARY OF DESIGN PRINCIPLES

Solar shading:

• Use passive shading devices to protect the building from unwanted solar gain.
• Calculate shading geometry using the methods explained in this chapter.
• Validate the calculations using DSM solar simulation with eye position aligned with the sun.
• Reiterate this process until satisfactory shading is achieved.

CHAPTER 11

THERMAL MASS

Thermal mass is associated with heat absorption and retention in a building. In this chapter, it will be explained what thermal mass is. This will be followed by simulation experiments to show how thermal mass influences building behaviour.

11.1 WHAT IS AND WHAT IS NOT THERMAL MASS?

Thermal capacitance of a single type of material is defined as

$$C = m \times c \tag{31}$$

where m is the mass of the material and c is the specific heat. Thermal capacitance alone only gives information on how much heat is needed to change the temperature of the material by a unit temperature, but it does not give any indication as to how much time it will take.

The parameter that gives the time aspect to thermal mass is called a time constant. In general, the time constant is the time required for an observed system parameter to reach 0.632 of the change between its initial and final value after the system receives a step input that causes that change. It is widely used in the field of electronics as a measure of the speed of response of voltage or current in a circuit receiving an impulse or a step input. In the context of this text, the time constant represents the speed of response of a building air temperature to a heat input, and it is obtained by dividing the thermal capacitance by the overall transmittance-area product.

Thermal mass is therefore encapsulated in the relationship for the building time constant $t_c = m \times c /(U \times A)$. If we express mass m as a product of density and volume ($m = \rho \times V$), the time constant summed over all building components from 1 to n becomes

$$t_c = \frac{\sum_1^n (\rho V c)_i}{\sum_1^n (UA)_i} \tag{32}$$

This means that the higher the density and specific heat of the materials used and the lower the overall transmittance-area product, the higher the time constant, the longer (slower) the building response, and the higher the thermal mass. This also indicates that thermal mass only exists within an insulated envelope of a building.

Definition: *Thermal mass is a relationship between the building thermal capacitance C and the overall transmittance-area product UA of the building, which through the time constant t_c gives information on the speed of response of a building to a heat input.*

Having established what thermal mass is, it is worth noting what thermal mass is not. Thermal mass is not the total mass of all building materials, as this may or may not contribute to a desired dynamic thermal behaviour of the building. Thermal mass is not merely the thermal capacitance C of all materials in the building, as this does not tell us how much time it takes to change the building temperature under a certain heat input. For instance, for $C = 40$ MJ/K, a building with the overall transmittance-area product of UA = 200 W/K will have the time constant $t_c = (4 \times 10^6$ J/K)/(200 J/(s·K)) = 200,000 s = 55.6 hours, and a building with UA = 400 W/K will have the time constant two times shorter, namely $t_c = 27.8$ hours.

Please see Section 20.3.1 in Chapter 20 for a detailed explanation of how thermal mass can be measured experimentally.

11.2 POSITIONING THE THERMAL MASS

Not only that thermal mass must be positioned within the insulated building envelope, but it also must be easily exposed to solar heat input. Primary thermal mass is defined as thermal mass directly exposed to solar gain and secondary thermal mass is defined as thermal mass exposed to internal heat gains and heat gains from the space heating/cooling system. Instead of using rules of thumb on how much high-density material to use and where, optimisation with DSM will always achieve better results.

11.3 EFFECTS OF THERMAL MASS ON TEMPERATURES IN BUILDINGS

In this section, the effects of thermal mass on summer time and winter time temperatures in buildings will be investigated, and in the next section, the consequences of thermal mass on building energy efficiency will be discussed.

Figure 118 shows a conceptual comparison of summer time temperatures in a building with high and low thermal mass over a period of a few days. This analysis applies to temperate climates, where summer night-time temperatures are significantly lower than daytime temperatures. The building with low thermal mass will experience high temperature fluctuations and the building with high thermal mass will experience low temperature fluctuations. After a sequence of a few clear days with high solar radiation, thermal mass will accumulate heat, and there will be a temperature build-up in the building (Figure 118a). To prevent the temperature build-up, night-time ventilation must be deployed to discharge thermal mass and make it ready to absorb more heat the following day (Figure 118b).

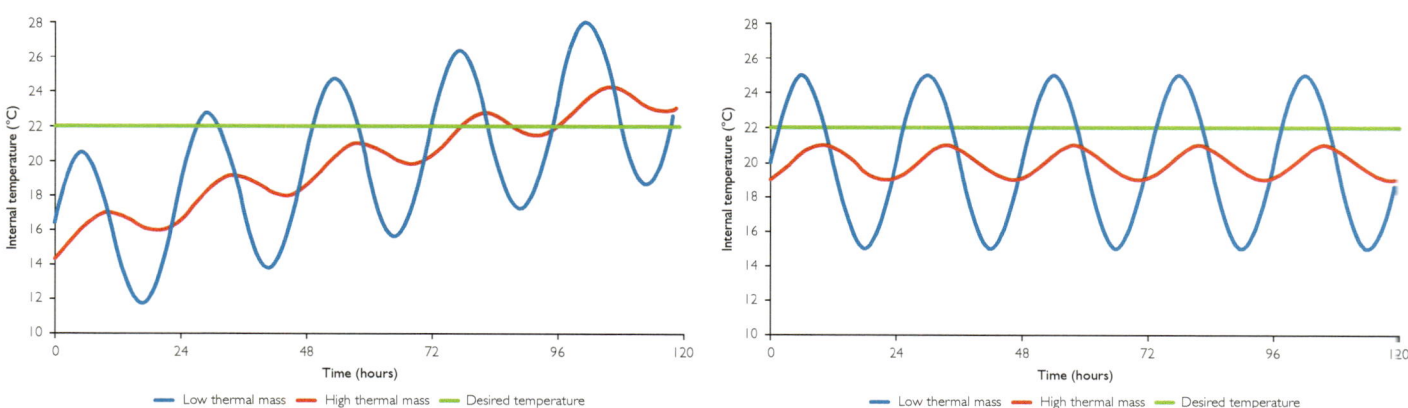

Figure 118   The effect of thermal mass on internal building temperature

We will now analyse internal building temperatures in winter time, with auxiliary heating controlled by a room thermostat (Figure 119). Room thermostats have a characteristic called a 'dead band' temperature difference, which prevents the heating system from switching on and off too frequently. The upper and lower limits of the dead band temperature difference are shown as horizontal lines in Figure 119. When the heating system starts, the building temperature will start increasing until the upper temperature limit is reached. At this point, the thermostat will switch the heating off. The building will then start cooling down, until the lower temperature limit is reached and at that point the thermostat will start the heating again.

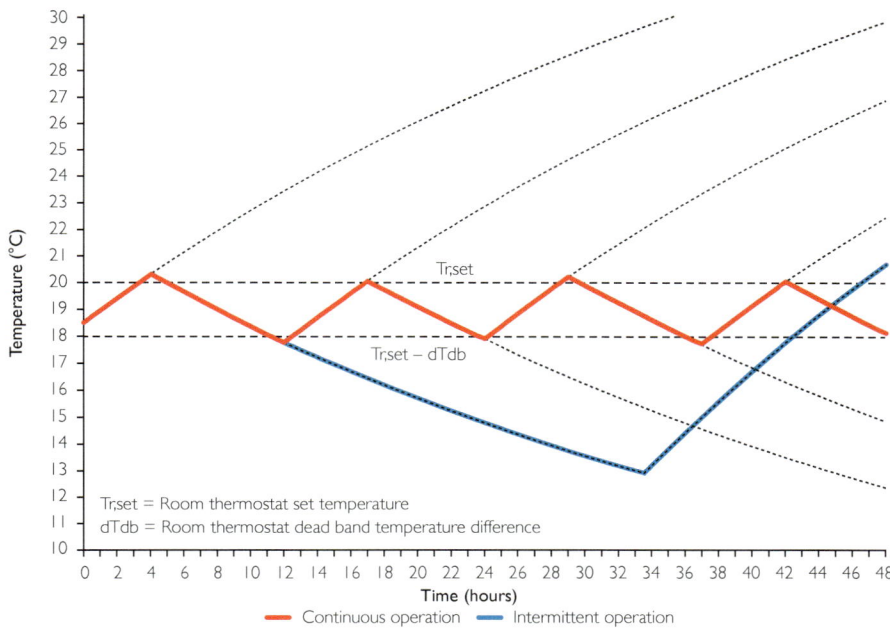

Figure 119   Conceptual operation of a building with thermostat-controlled heating

11.4 CONSEQUENCES ON BUILDING ENERGY EFFICIENCY

11.4.1 Summer time operation

The horizontal line in Figure 118 indicates the desired temperature in the building, and it can be seen that this temperature will be exceeded in the building with low thermal mass, whilst the building with high thermal mass will never reach this temperature with night-time ventilation in operation (Figure 118b). This means that the building with low thermal mass will require mechanical cooling/air-conditioning in order to maintain the internal temperature at the desired level, and that the building with high thermal mass will not. High thermal mass, if operated in conjunction with night-time cooling, will therefore act as a natural comfort cooling system, resulting in energy-efficient summer time operation. If however no night-time ventilation is deployed or if night-time temperatures are not much lower than daytime temperatures, it will be difficult to control the temperature build-up in the building with high thermal mass (Figure 118a).

11.4.2 Winter time operation

High thermal mass will make the heating and cooling process slower, and therefore a good quality thermostat with a narrow dead band will be required for continuous operation. If the heating is run intermittently so that it is switched off for longer periods of time, prolonged heating-up periods will occur after the heating is switched on (Figure 119). During these heating-up periods, the desired temperature will not be reached for a while. Intermittent heating in a building with high thermal mass is therefore likely to cause higher use of energy and lower levels of comfort. Continuous heating controlled by a narrow dead band thermostat will achieve lower energy consumption and a higher level of thermal comfort.

11.5 MATERIALS USED FOR THERMAL MASS

Sensible heat transfer occurs when addition or deduction of heat results in a change of temperature of a material. Conventional thermal mass materials use sensible heat transfer to store and release energy, on the basis of their thermal capacitance as per Equation (31). Concrete, stone, brick and other high-density and high-specific-heat materials are generally suitable for this purpose, bearing in mind the relationship in Equation (32).

Other types of materials used for thermal mass are based on latent heat transfer, which results in the change of phase from solid to liquid when heat is added to the material and from liquid to solid when heat is deducted from the material (Figure 120). These materials are commonly referred to as PCM – Phase Change Material. They can be chemically programmed during the manufacturing process to change phase (melt) at near room temperatures, and can therefore be used in buildings. As the latent heat of melting is generally considerably greater than the heat required for the same material to change temperature by a few degrees, layers of phase change materials can be about 10 times thinner than layers of conventional thermal mass materials, and can still provide comparative heat absorption.

Sensible heat transfer materials can be used in any building component, such as floors, walls, and ceilings. As phase change materials are generally soft and come in thin metal or plastic 'sandwiches', they can only be used on walls and ceilings.

Having introduced thermal mass material into the building, it is very important not to subsequently insulate it from the internal environment. Floor finishes on top of thermal mass materials need to either match the properties of the chosen material (ceramic or stone tiles), or to have high conductivity (thin layer of plastic covering or paint) so that the material is thermally coupled to the internal space. Similar requirements apply to wall and ceiling coverings on top of either sensible or phase change thermal mass materials.

Simulation experiments with the PCM in this chapter will be based on a BioPCM® material, for which the simulation capability was readily available in DesignBuilder simulation software at the time of writing this text.

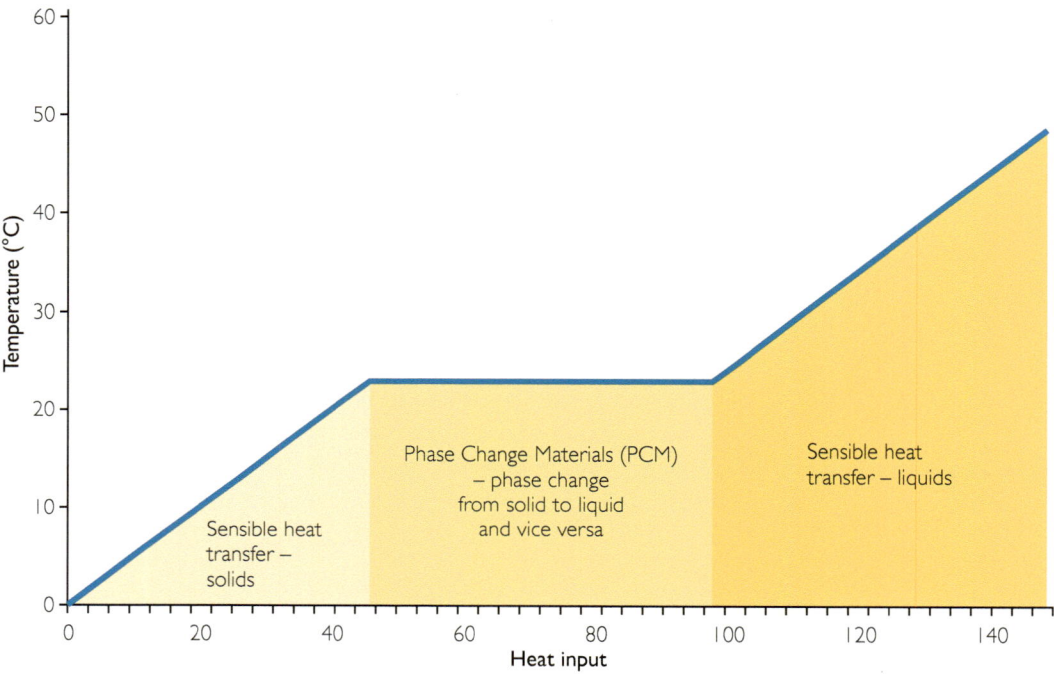

Figure 120   Characteristics of sensible and phase change thermal mass materials

11.6 SCALE MODEL EXPERIMENTS

The brief for thermal mass experiment was as follows:

Thermal mass: Create a cardboard model with dimensions 20 cm × 12 cm × 6 cm, and inner box with dimensions 18 cm × 10 cm × 6 cm. Experiment 1: fill the gap between the outer and inner box with thermal insulation. Place a container with boiling water into the inner box, cover it with a lid and record internal temperature and time in 1 minute intervals until the box air temperature has reached its maximum and started decaying. Experiment 2: repeat the above procedure with sand instead of thermal insulation between the inner and outer box. Compare the results from the two experiments and comment on the findings.

The model is shown in Figure 121. First, the cavity between the inner and outer boxes was filled with recycled newspaper (Figure 121a). A temperature sensor was suspended using a plastic cable tie and connected to a data logger. A container was filled with boiling water, placed in the box, and a transparent lid was placed on top of the model and secured with adhesive tape. Internal air temperature was recorded in one-minute intervals over 60 minutes. This process was repeated with sand instead of thermal insulation in the cavity between the inner and outer boxes (Figure 121b), and the temperature was recorded for another 60 minutes.

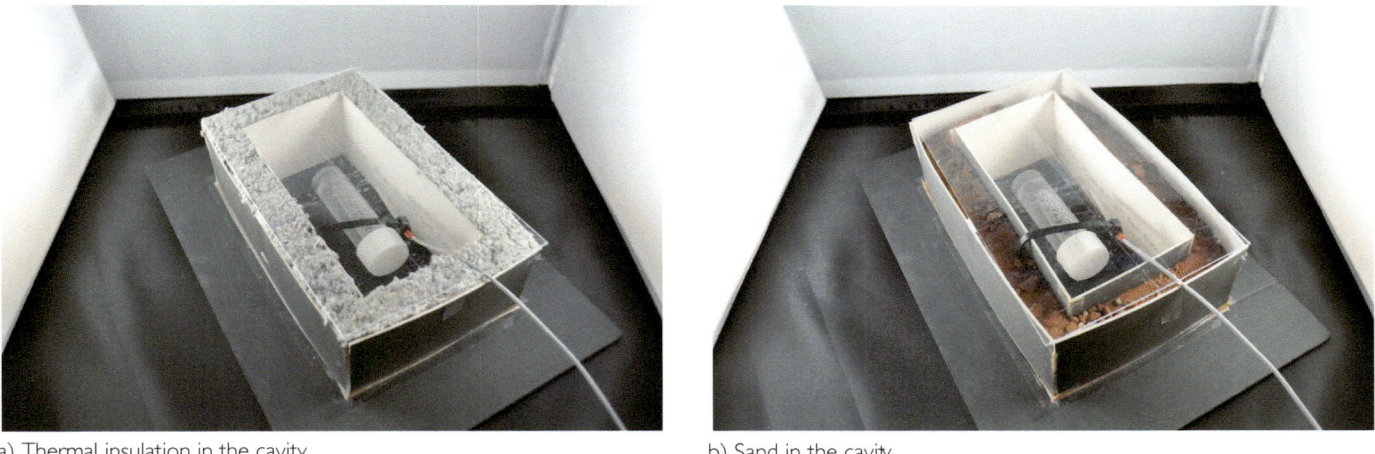

a) Thermal insulation in the cavity   b) Sand in the cavity

Figure 121   *Thermal mass scale model experiment (model by students Simon Pope, Parminder Dhillon, Etienne Amion, Antonios Papanastasiou and Pierre Arnou).*

The results of monitoring are shown in Figure 122. It is clear that the box with insulation in the cavity, representing low thermal mass, heats up much quicker than the box with sand in the cavity, representing high thermal mass, as a result of higher absorption of heat in the sand. The latter cools down slower than the former, as a result of heat stored in the thermal mass of sand. During both experiments, the outside air temperature, representing the air temperature of the room in which the experiments were conducted, remained fairly stable, so that no adjustments had to be made to make the results of the two experiments comparable.

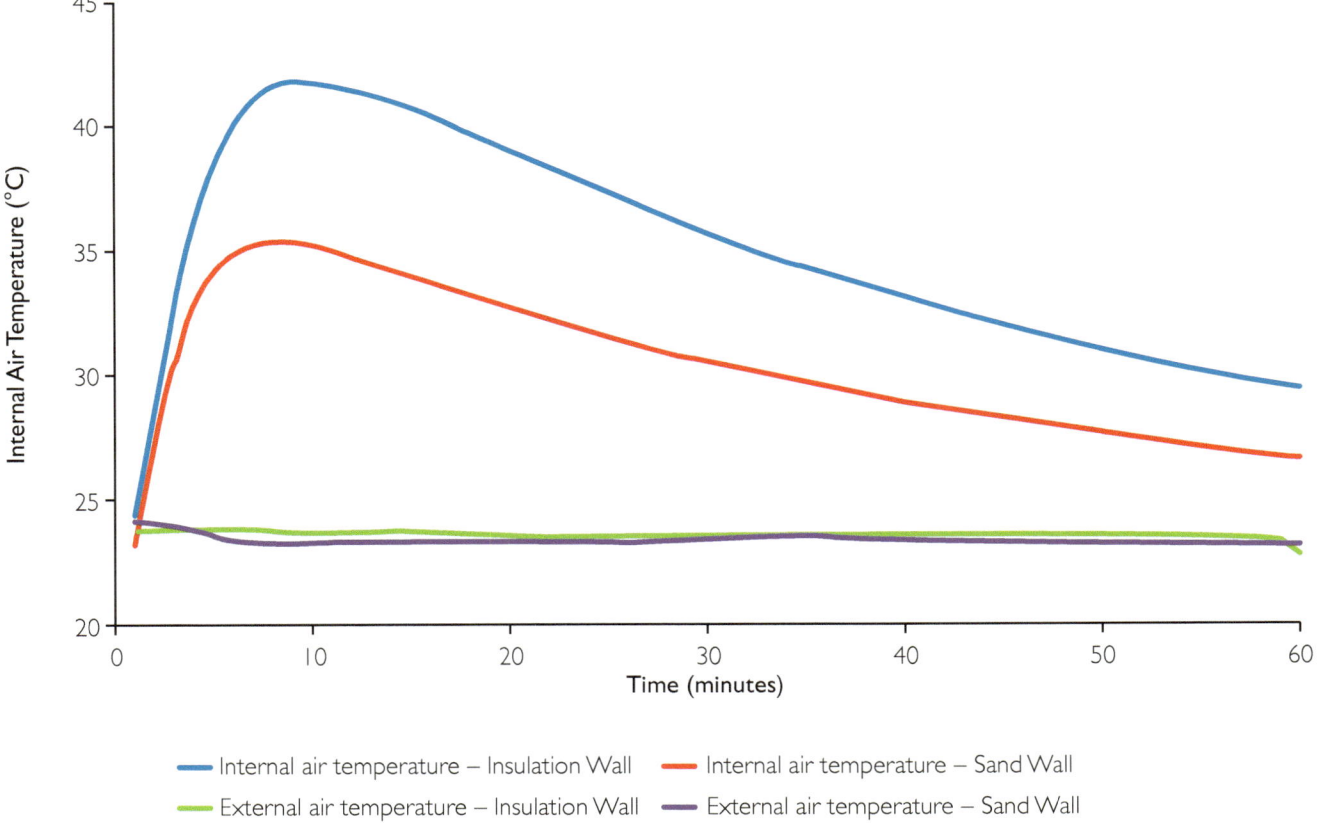

Figure 122   *Results of monitoring of two scale models in the thermal mass experiment.*

As we will see later in this chapter, the behaviour of the models in this scale model experiment is consistent with the dynamic simulation experiments in the next section.

11.7 DYNAMIC SIMULATION EXPERIMENTS

On the basis of Equation (32), we have established that increasing density, volume, and specific heat, and reducing overall transmittance-area product will increase thermal mass. We will now discuss how to put this into practice in a dynamic simulation model.

A simple box model as in Figure 65, created in DesignBuilder, will be used to investigate how different materials influence building thermal behaviour. In order to compare like with like, the overall conductance-area product UA will be kept the same between different comparison cases. As all these cases use the same geometry and therefore all areas are the same, ensuring that all U values are the same will achieve that the overall conductance-area product is also the same between the cases. This can be achieved by varying thermal insulation in the construction types specified in Table 33. Working models of the simulation cases are available for download from the supplementary material website www.ljankovic.com.

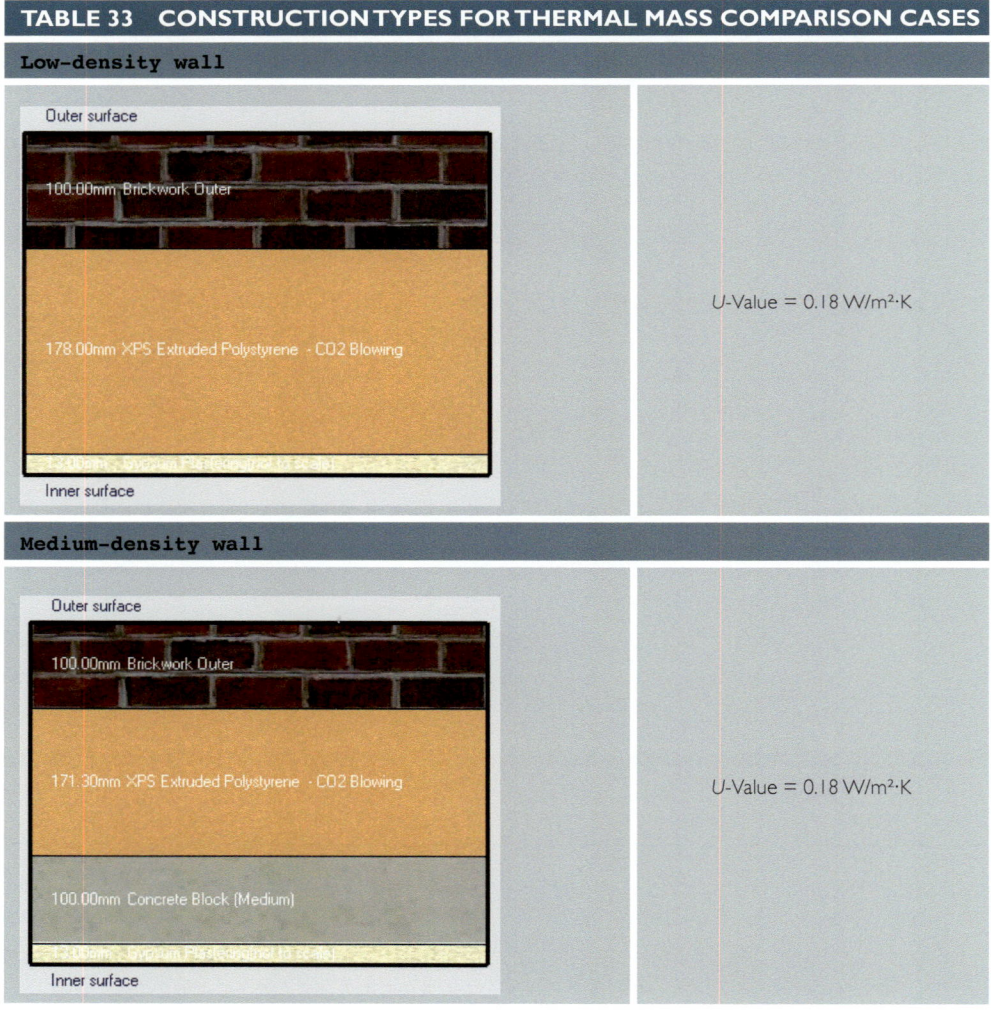

TABLE 33   CONSTRUCTION TYPES FOR THERMAL MASS COMPARISON CASES

Low-density wall

Outer surface

100.00mm Brickwork Outer

178.00mm XPS Extruded Polystyrene - CO2 Blowing

Inner surface

U-Value = 0.18 W/m²·K

Medium-density wall

Outer surface

100.00mm Brickwork Outer

171.30mm XPS Extruded Polystyrene - CO2 Blowing

100.00mm Concrete Block (Medium)

Inner surface

U-Value = 0.18 W/m²·K

TABLE 33 CONTINUED

High-density wall

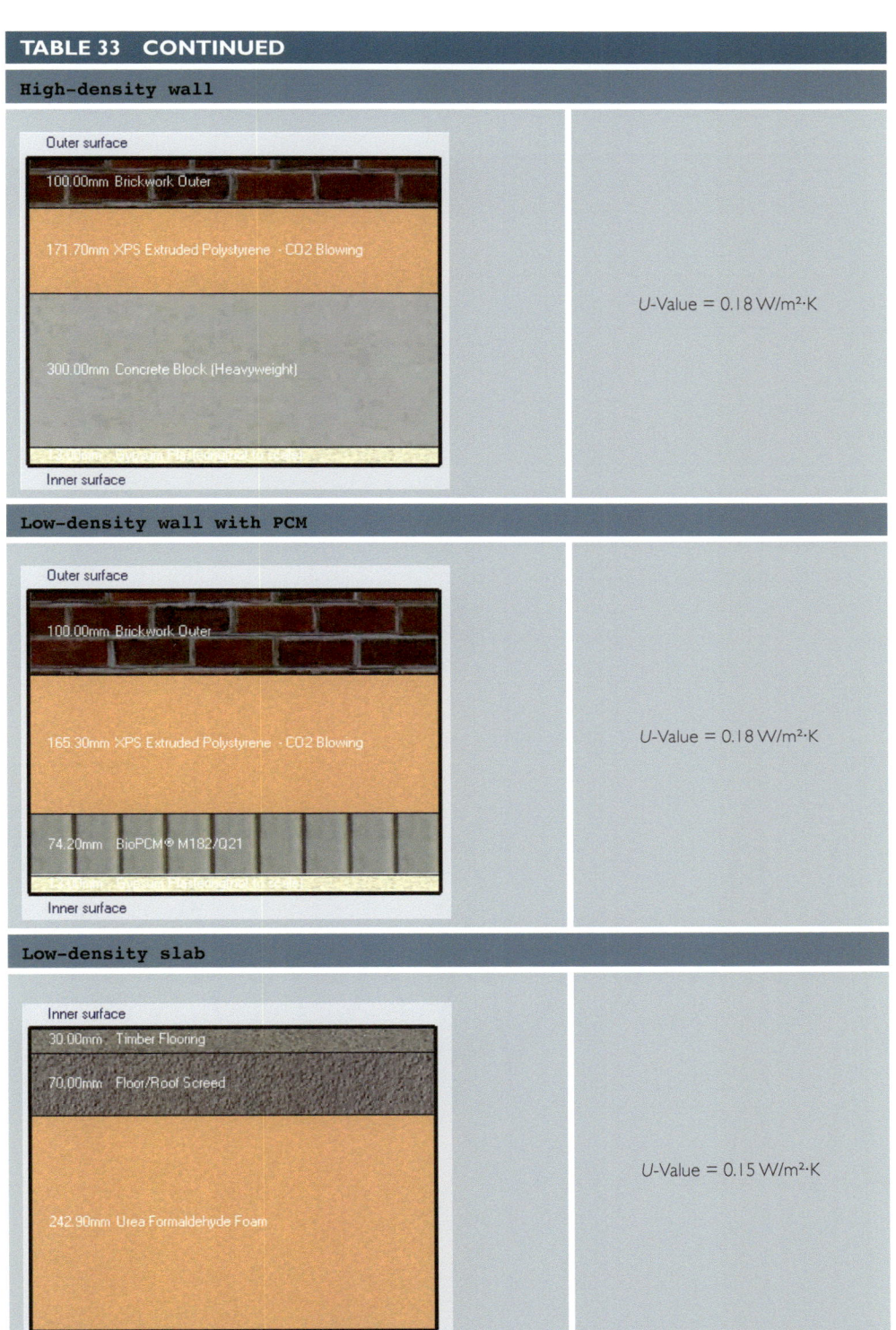

Outer surface

100.00mm Brickwork Outer

171.70mm XPS Extruded Polystyrene - CO2 Blowing

300.00mm Concrete Block (Heavyweight)

13.00mm Gypsum Plasterboard

Inner surface

U-Value = 0.18 W/m²·K

Low-density wall with PCM

Outer surface

100.00mm Brickwork Outer

165.30mm XPS Extruded Polystyrene - CO2 Blowing

74.20mm BioPCM® M182/Q21

13.00mm Gypsum Plasterboard

Inner surface

U-Value = 0.18 W/m²·K

Low-density slab

Inner surface

30.00mm Timber Flooring

70.00mm Floor/Roof Screed

242.90mm Urea Formaldehyde Foam

Outer surface

U-Value = 0.15 W/m²·K

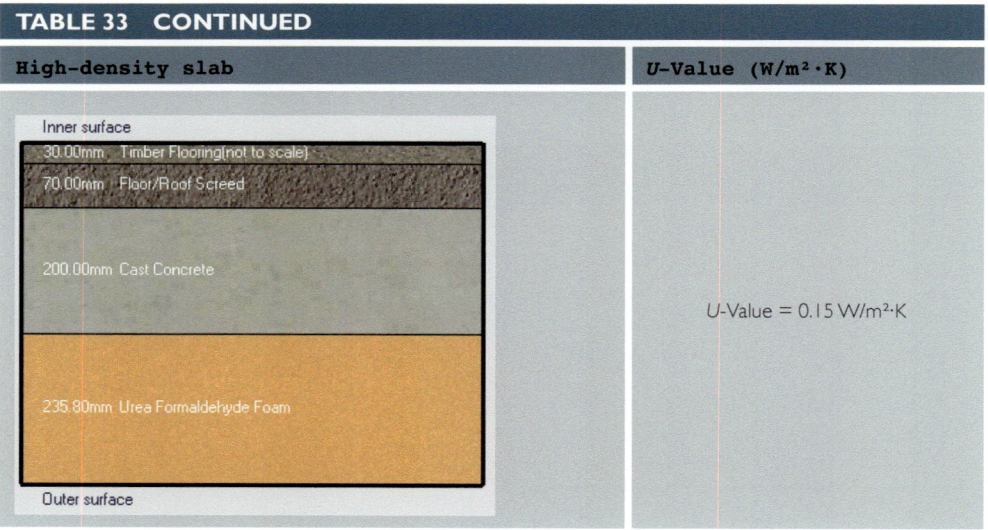

TABLE 33 CONTINUED	
High-density slab	U-Value (W/m²·K)
Inner surface 30.00mm Timber Flooring(not to scale) 70.00mm Floor/Roof Screed 200.00mm Cast Concrete 235.80mm Urea Formaldehyde Foam Outer surface	U-Value = 0.15 W/m²·K

We will first compare two floor constructions: 70 mm screed versus 70 mm screed plus 200 mm concrete, in combination with a low-density wall construction for all external walls (Table 33). Figure 123 shows that the heavier floor responds relatively slowly and consistently with the theoretical case in Figure 118 during the days with higher solar radiation, and that it keeps the heat longer during the days with lower solar radiation.

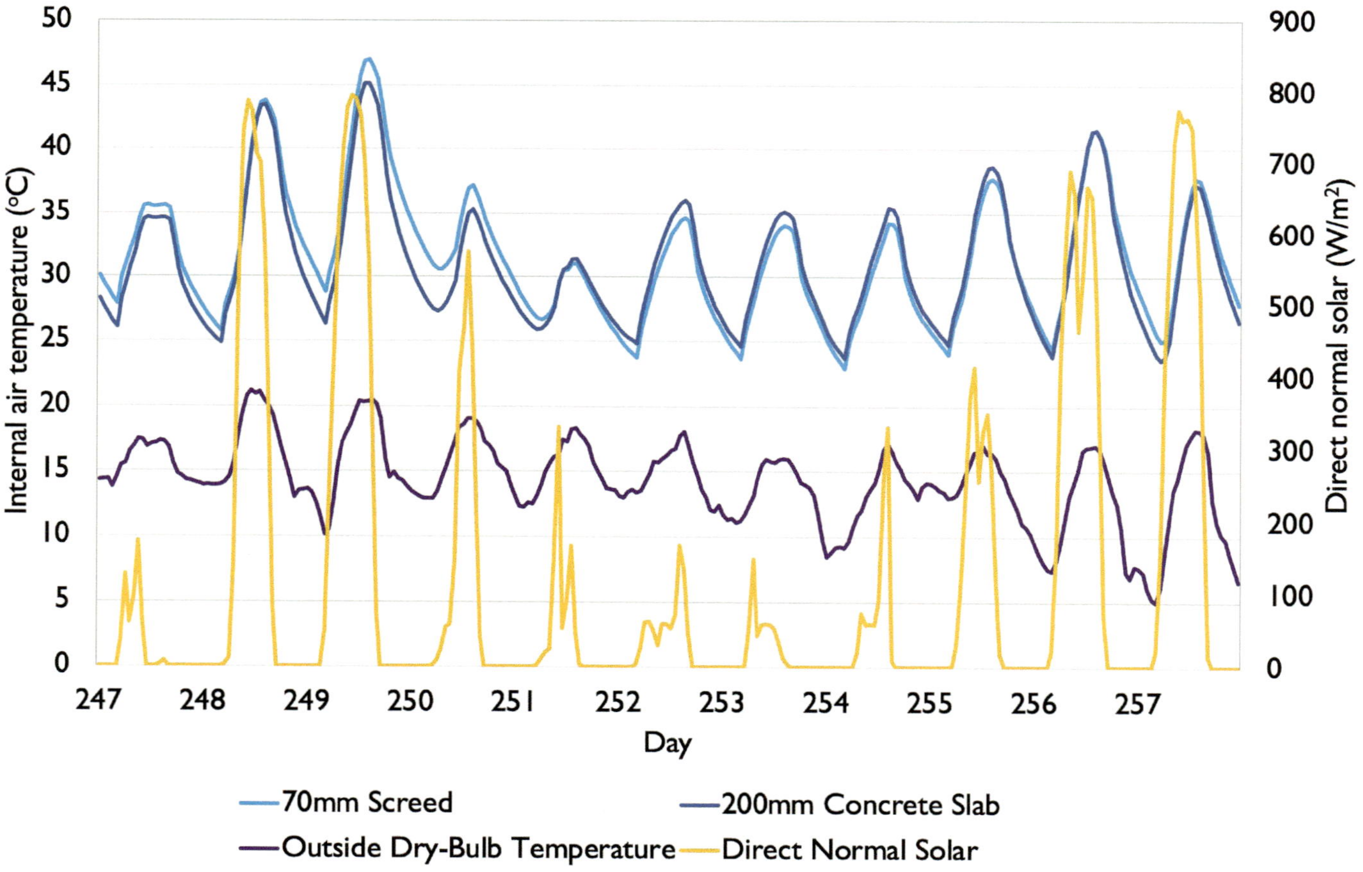

Figure 123   Comparison between low and high thermal mass floor.

Next, we will compare three different wall construction types (Table 33), using 70 mm screed and 100mm concrete floor in all three cases. Figure 124 shows that the low-density wall has the fastest response, the medium-density wall a medium response, and the high-density wall the slowest response, and this is also consistent with the theoretical case in Figure 118. We will also compare the low-density wall to which a layer of phase change material (PCM) is added. Figure 124 shows the effect of this material in the context of the three densities of external walls. It appears that the wall with PCM performs slightly better than the low-density wall, but slightly worse than the medium-density wall. This example demonstrates how thermal response can be modified with PCM without the use of high-density materials, such as concrete.

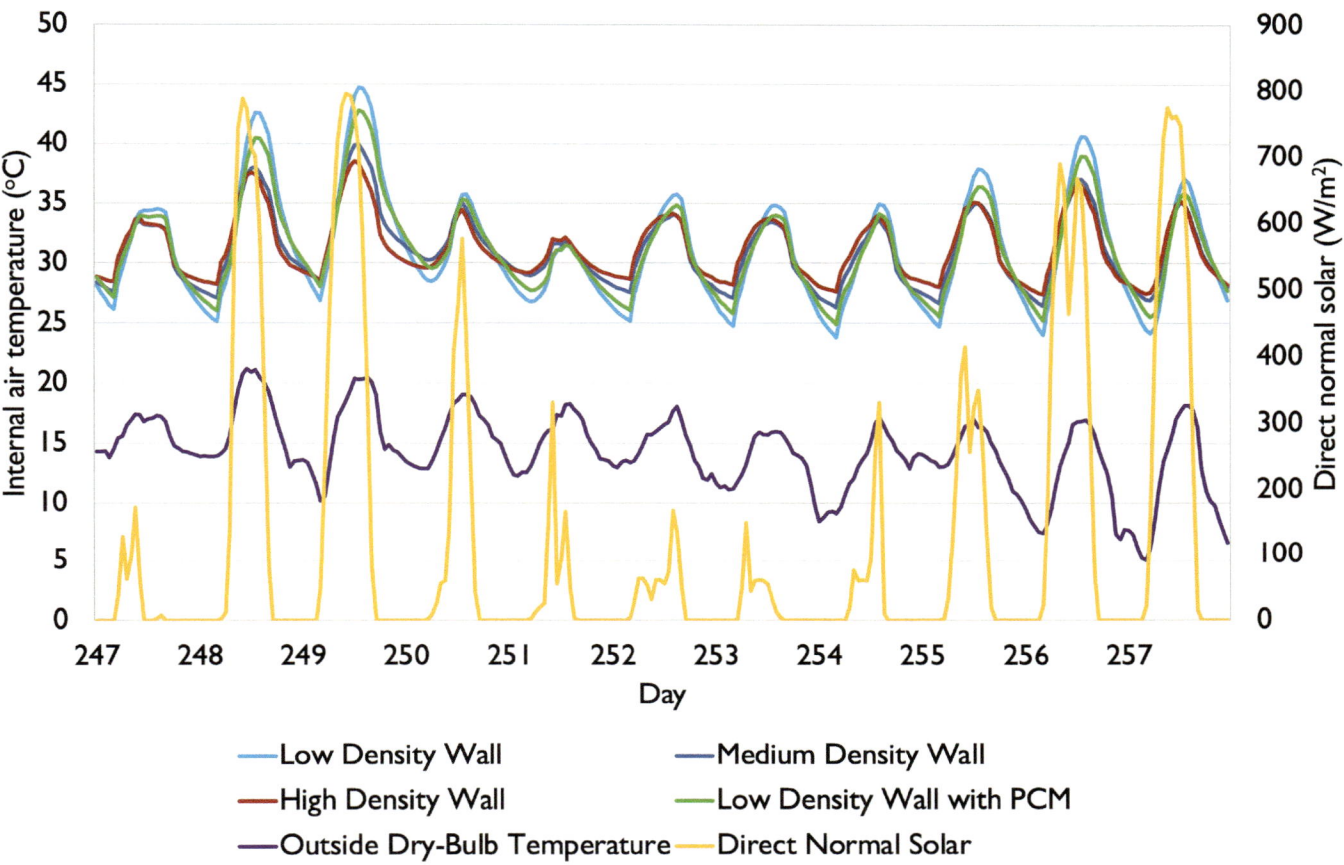

Figure 124   Comparison between three different wall densities and phase change material.

11.8 TASKS FOR SIMPLE SIMULATION EXPERIMENTS

1   Simulate a dynamic heating test and establish the time constant of a building by going through the following steps:
   a   Create a model of a box with dimensions of 10 m × 6 m × 3 m. Point the larger wall towards the south and install 40% glazing on it.
   b   Set the floor construction type to include at least 100 mm of thermal insulation and 100 mm of cast concrete.
   c   Set a daily temperature profile for heating control to increase from an initial unheated temperature to 80°C in a single step at 8 am and stay on the rest of the day.
   d   Set another daily temperature profile for heating control at 80°C throughout the day.
   e   Set a weekly heating control profile with the first day controlled by the temperature profile from c and the rest of the week days controlled by the temperature profile from d.
   f   Set another weekly temperature profile with all days controlled by the temperature profile from d.
   g   Set an annual temperature profile with the first seven days controlled by the weekly temperature profile from e and the rest of the year controlled by the temperature profile from f.
   h   Set electricity heating with 100% efficiency and limit it to 6 kW. Switch off cooling and domestic hot water.

i   Exclude all internal heat gains from people and appliances, and set infiltration air exchange to 0.167 ach.

j   Install internal blinds in the glazing that are lowered at zero W/m² and risen at 1300 W/m², therefore practically always on, thus excluding solar gain from this experiment.

k   Set the initial internal air temperature to the initial external air temperature and the simulation preconditioning period to zero days.

l   Carry out dynamic simulation with an hourly time step over a period of 15 days starting from 1st January.

m   Export the hourly internal room air temperature into a spreadsheet.

n   Number the lines of data in the spreadsheet, starting from the first rise of the internal air temperature until the end of the data set.

o   Calculate the difference between the final and the initial temperature, and multiply that difference by 0.632.

p   Add the calculated value to the initial temperature.

q   Find the closest value in the simulation results to the calculated temperature in the previous step.

r   Read the corresponding line number, using the numbering from step m. This number corresponds to the time constant in hours.

s   Think how you would apply this test in practice in a real building.

t   Discuss the process and the outcome with your colleagues.

This example has similarities with the co-heating test example in Chapter 8, but it also has important differences. It was developed for and tested in IES VE, but it can be customised for other simulation tools.

11.9 SUMMARY OF DESIGN PRINCIPLES

- High thermal mass reduces fluctuation of temperatures in buildings.
- Heat build-up in thermal mass must be mitigated with night time ventilation.
- Lower temperature peaks result in lower or no requirements for comfort cooling.
- Primary thermal mass needs to be exposed to solar gains.
- Secondary thermal mass helps to absorb internal heat gains and auxiliary heating output.
- Thermal mass is increased by increasing the density, volume, and specific heat of building materials, and by reducing the overall conductance-area product.
- Different design options should be investigated using DSM in order to achieve optimum results.
- For meaningful comparisons of how different materials effect temperature fluctuations in a building, the corresponding construction types must have identical U values, achieved by varying thermal insulation thickness.
- Phase change materials
  - Contribute to increasing thermal mass in buildings by reducing temperature fluctuations in a narrow range around the phase change temperature.
  - They achieve this with a fraction of thickness of the conventional materials.
  - Outside the phase change temperature range these materials do not contribute to the thermal mass.

CHAPTER 12

NATURAL AND MECHANICAL VENTILATION

When attending a simulation conference in Porto, Portugal, I was most surprised not to feel hot inside a glazed walkway connecting a local shopping centre with the university hospital on a very sunny day. I had to take a closer look at how the walkway was designed.

Figure 125a shows considerable amount of south-facing external glazing, resulting in internal solar gain in Figure 125b. A closer inspection of the walkway detail in Figure 125c showed a gap between the glazing and the floor deck, with an equivalent gap between the glazing and the roof. This meant that the walkway was fully naturally ventilated, resulting in comfortable internal conditions.

a) External view

b) Internal view

c) Ventilation opening detail

Figure 125   Naturally ventilated glazed walkway.

This chapter will give an overview of the principles and their implementation in dynamic simulation. For background reading on natural ventilation principles and experimental case studies, the reader is advised to consult a book by Ghiaus and Allard (2016).

12.1 NATURAL VENTILATION MECHANISMS

There are two mechanisms that generate natural ventilation. They are called

• wind effect
• stack effect

12.1.1 Wind effect

This phenomenon occurs as a result of a wind blowing across the building. This creates a positive pressure (compressed air) on the windward side of the building, and at the same time it causes a negative pressure (partial vacuum) on the leeward side of the building. This difference in pressure will drive the air through the building, either through openings or through cracks in the building envelope. If the openings are designed for ventilation then the building will have natural ventilation as a result of the wind effect. If the openings (cracks) are not designed for natural ventilation then the building will have unwanted ventilation called infiltration (Figure 126).

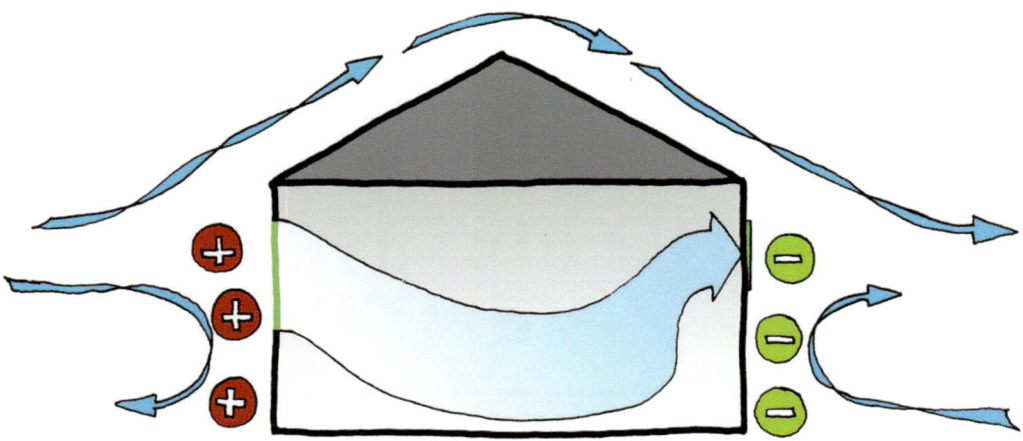

Figure 126   Wind effect (illustration by Holly Doron).

The pressure that drives wind effect can be calculated using the following equation from Germano et al. (2016):

$$\Delta p = 0.5 \times C \times \rho \times v^2 \tag{33}$$

where

Δp – pressure difference through the building ($N/m^2 = Pa$)
C – pressure coefficient resulting from building characteristics and wind direction
ρ – density of air (kg/m^3)
v – wind velocity (m/s)

Example. Calculate pressure difference through the building using an internal air density of 1.16 kg/m³, pressure coefficient of 0.7, and air velocity of 3 m/s. Substituting these parameters in the above equation, we obtain the pressure difference that drives wind effect $\Delta p = 0.5 \times 0.7 \times 1.16 \times 3^2 = 3.65$ Pa.

12.1.2 Stack effect

Another natural ventilation mechanism is called the stack effect. The stack effect occurs as a result of buoyancy of air. Warm air has lower density than cold air and, as a result, the volume of warm air has a tendency to rise above the volume of cold air. Cold air effectively pushes warm air upwards, as it has a tendency to drop down towards the floor. If we provide openings at the base and top of a tall space (which we will call a stack or chimney), we will find that warm air will escape from the top openings, creating reduced pressure in the building, drawing colder air to come through from the outside through the opening at the base (Figure 127a). The higher the temperature difference between the top of the stack and the outside air, the more intensive the stack effect will be.

The openings at the top and bottom need to be roughly the same size. If there are larger openings at the base and smaller openings at the top, this will create a bottleneck effect and natural ventilation will not work well. If there are larger openings at the top of the chimney and smaller openings at the bottom, there will be an inverse bottleneck effect, and again natural ventilation will not work well. Equal openings at the top and bottom will provide balanced air flow through the chimney.

The pressure that drives the stack effect and the resultant air flow can be calculated using simple formulae, as reported by Germano et al. (2016):

$$\Delta p = \rho_i \times g \times h \times \frac{T_i - T_e}{T_e} \tag{34}$$

and

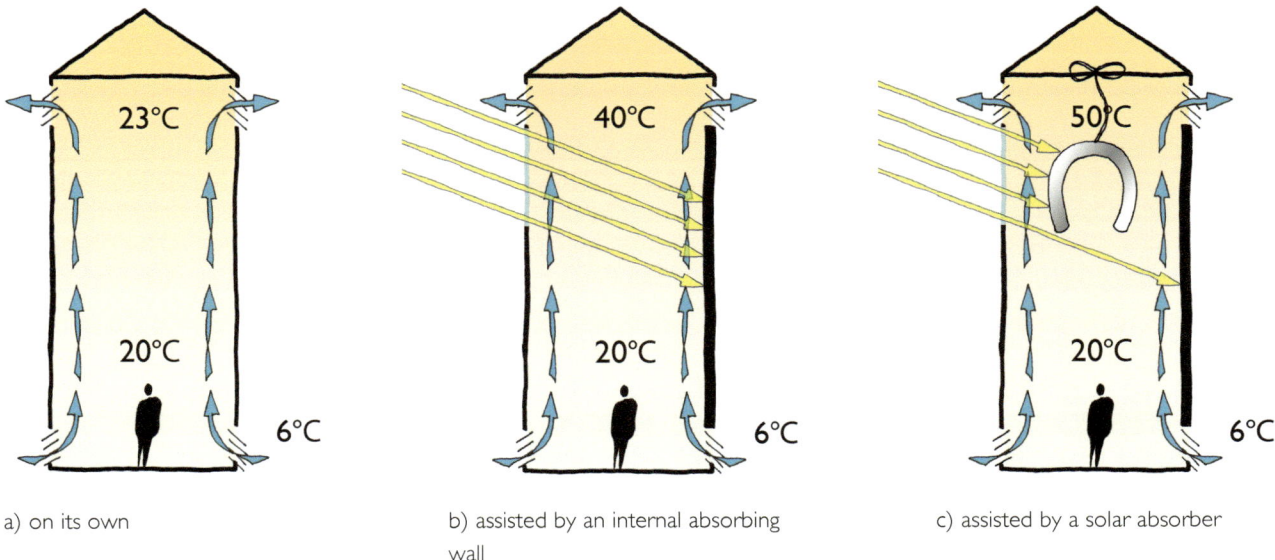

a) on its own

b) assisted by an internal absorbing wall

c) assisted by a solar absorber

Figure 127   Stack effect (illustrations by Holly Doron)

$$m = \rho_i \times A \times C \sqrt{2 \times g \times h \times \frac{T_i - T_e}{T_i + T_e}} \qquad (35)$$

where

Δp – stack pressure – pressure difference between inside and outside air ($N/m^2 = Pa$)
ρ_i – density of internal air (kg/m^3)
g – gravitational acceleration ($9.81\ m/s^2$)
h – height of the stack (m)
T_i – internal air temperature ($K = °C + 273$)
T_e – external air temperature ($K = °C + 273$)
m – air flow (kg/s)
A – area of the smaller opening (m^2)
C – discharge coefficient for openings (-)

Example. The internal air temperature at the top of the stack is 30°C, and the external air temperature is 10°C. Calculate stack pressure and air flow, using an internal air density of 1.16 kg/m³, a stack height of 6 m, an opening area of 2 m², and a discharge coefficient of 0.7. Recalculate both parameters for an internal temperature of 40°C. Increase the opening to 3 m² and recalculate both parameters again.

The results of these calculations are shown in Table 34.

TABLE 34	CALCULATIONS OF STACK PRESSURE AND AIR FLOW		
		Δp (Pa)	m (kg/s)
$T_i = 30°C$	$A = 2\ m^2$	4.8	4.2
	$A = 3\ m^2$	4.8	6.3
$T_i = 40°C$	$A = 2\ m^2$	7.2	5.1
	$A = 3\ m^2$	7.2	7.7

We can see from this calculation that increasing internal temperature by 10°C the stack pressure increased by 2.4 Pa. By increasing the opening area from 2 to 3 m² the air flow increased by 2.1 kg/s at 30°C and by 2.6 at 40°C internal temperature.

It is worth mentioning that stack effect will not work if the openings are only at the bottom or only at the top of the chimney. Although this seems like stating the obvious, it is surprising how many of us forget.

12.1.3 Using solar radiation and orientation to enhance natural ventilation

The stack effect can be enhanced by making the top of the chimney transparent and south-facing and allowing solar radiation to come through and heat up the inside of the back of the chimney. If the back part of the top of the chimney is made of a high-density material, it will absorb heat from the sun, increasing the local temperature and the intensity of the stack effect (Figure 127b). This effect can be enhanced even further by installing a simple solar absorber inside the top of the chimney – a piece of metal coated with a dark or selective surface (Figure 127c). This collector can be standalone. It does not need to be connected to any other system but it just needs to be suspended from the top of the stack to absorb solar energy. The heat losses from the collector will heat the air at the top of the stack and increase the intensity of the stack effect. As this region of warm air is well above the occupancy level, it will not affect thermal comfort of people at the occupancy level. It therefore does not matter that the air at the top of the stack will be hot as it will still be possible to achieve comfortable conditions at the floor level. In this way natural ventilation can be enhanced by making the solar absorber and solar energy work almost like an 'engine' of the stack effect.

12.1.4 Using thermal mass to enhance natural ventilation

High thermal mass in the building is a prerequisite for a successful operation of natural ventilation. If the conditioned space gets too hot, then natural ventilation will not be sufficient to maintain comfort conditions, and comfort cooling may be required instead. If, however, temperature fluctuations in the space are stabilised by having enough thermal mass, then natural ventilation will have a good chance to provide the required internal conditions.

When used in conjunction with thermal mass, natural ventilation needs to work at night to discharge thermal mass and precondition it for heat absorption on the following day. This was discussed in Chapter 11 with reference to Figure 118, showing the effect of night-time ventilation on reducing temperature build-up in the building. If there is a sequence of sunny days when thermal mass gets charged with heat (Figure 118a), then natural ventilation will not be able to provide comfort temperatures if it starts at the end of that sequence, during the day when it is very hot in the building. As night-time air in temperate climates is much colder than day-time air during most of the year, we can use night-time ventilation to discharge thermal mass at night, precondition it for the next day to absorb more heat, and prevent temperature build-up, thus achieving cyclic operation without temperature build-up as shown in Figure 118b. We need to design controlled openings to enable this type of operation at night and we therefore need to have control scenarios for day and night.

12.1.5 Using orientation, chimneys, spoilers, and louvres to enhance wind effect

Configuration of buildings will affect the success of natural ventilation through the wind effect. If there are other buildings obstructing the building that we would like to ventilate using the wind effect, then the benefit of prevailing wind will be reduced and wind effect may not work. The building orientation needs to be towards the prevailing wind, so that the pressure differences between windward and leeward side are maximised. Tall chimneys will help catch the cross wind, generating a suction effect at the top of the chimney and will help with the extraction of air. A spoiler on top of the chimney, working like an inverse airplane wing, will enhance the suction effect of the chimney as result of the wind effect. Louvres can be used to target the air circulation arising from the wind effect towards certain areas in internal space.

12.1.6 Building geometry

Natural ventilation will work to about six metres of the plan depth from the window in a single aspect building with windows only on one side or up to 12 metres in double aspect buildings with windows on two opposite sides, with typical ceiling heights of 3 m. In a much deeper plan building with the same ceiling height, natural ventilation will not work, and mechanical ventilation or even comfort cooling will need to be introduced, resulting in much higher capital and running costs, and in higher energy consumption and carbon emissions. See also 'Shallow plan versus deep plan' in Chapter 7. CIBSE Application Manual AM10 (Butcher, 2011) gives the rule of thumb for limiting the depth for natural ventilation of twice floor to ceiling height in single aspect buildings (windows on one side only) and 2.5 times floor to ceiling height in double aspect buildings (windows on two opposite sides).

If we are ducting air in natural ventilated systems, the ducts need to be short and wide, with similar restrictions as in the plan depth, so that the resultant pressure drop in the ducts does not slow down or stop the air movement.

12.1.7 Computer control

In larger buildings we need to have computer controlled openings, as it would be very tedious for a person to go between different openings at high and low levels and operate them in response to temperature or CO_2 readings. Therefore, even though natural ventilation does not use fans, it needs to use computer control in larger buildings in order to synchronise the opening of all vents in response to temperature and air quality conditions in the building. In smaller buildings, and especially in smaller houses, this can be done manually.

12.1.8 Air pollution and filtering

Normally natural ventilation cannot be used in heavily polluted areas without special filters. Conventional filters, such as dry and wet filters will introduce too high a pressure drop into the natural ventilation system and will effectively stop its operation. Electrostatic filters, which rely on making dust particles in air electro-statically charged and attracting them towards electrodes of the filter, can be used to reduce particle count in the incoming air in naturally ventilated buildings, without a detrimental effect on the air flow.

12.1.9 Day, night, summer and winter scenarios

From the above analysis a suitable set of scenarios of operation in summer and winter at day and night begins to emerge. In summer, the building needs to be protected from excessive solar radiation during the day (Figure 128a) and thermal mass needs to be discharged and preconditioned for the following day using night-time ventilation (Figure 128b). In winter, the

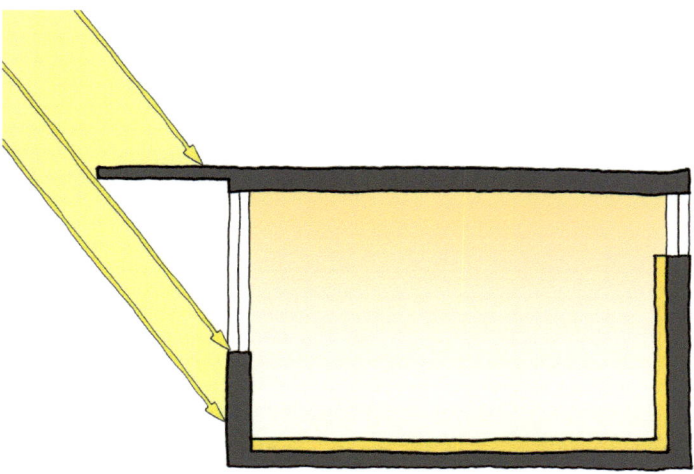

a) Summer day: protect the building from solar gain

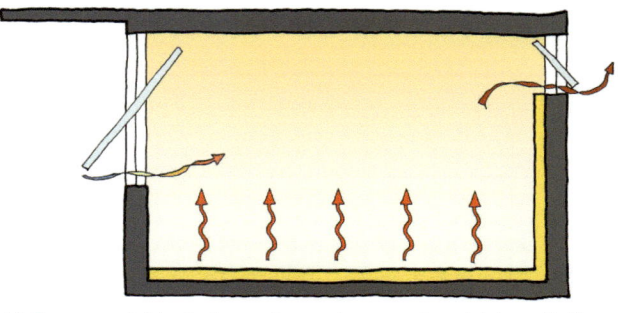

b) Summer night: discharge thermal mass using night ventilation

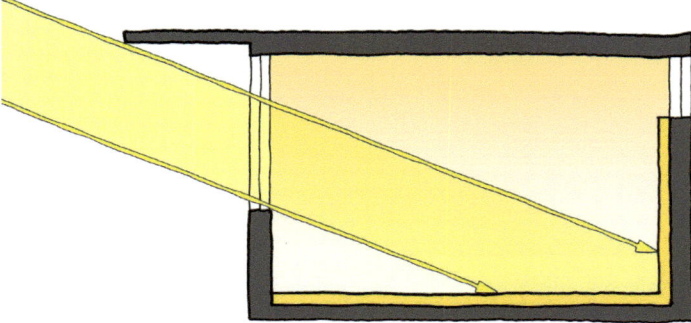

c) Winter day: maximise heat gain from solar and charge thermal mass

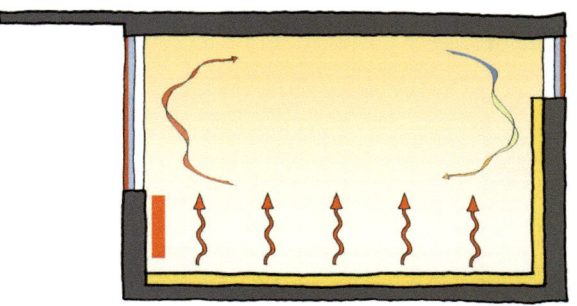

d) Winter night: protect the building from heat loss using thermal blinds top up heat input from thermal mass with conventional heating

Figure 128   Scenarios of operation in summer and winter, day and night (illustrations by Holly Doron)

building needs to receive as much solar gain as possible to charge thermal mass during the day (Figure 128c), and it needs to be protected from excessive heat losses to utilise the heat from the thermal mass during the night (Figure 128d).

12.2 MECHANICAL VENTILATION WITH HEAT RECOVERY

Mechanical ventilation is used where natural ventilation cannot provide one or more of the following:

• consistency of supply and control
• quantity and quality of required air
• isolation from external environment to prevent pollution, noise, or increased security.

Mechanical ventilation will use electricity for the operation of fans and will extract some of the useful heat from the building, resulting in increased energy use and increased carbon emissions.

We can reduce fan electricity consumption by controlling mechanical ventilation on the basis of CO_2 concentration in parts per million (ppm), so that the fans are operated only if the concentration setting is exceeded.

Mechanical ventilation effectively throws warm air out. We can reduce the amount of useful heat rejected from the building by using a heat recovery system. Typically there are two types of heat recovery system: a thermal wheel system (Figure 129a) and a cross-flow system (Figure 129b).

The thermal wheel heat recovery system is based on a metal wheel that rotates between the cold incoming air and warm exhaust air ducts (Figure 129a). The wheel consists of thin metal tubes in line with the air stream. As the wheel rotates, it carries the tubes between the warm and cold ducts, collecting heat from the former and depositing the heat in the latter. This is an active system with moving parts and an electric motor that runs the wheel. Consequently a certain amount of energy will be required to run the system.

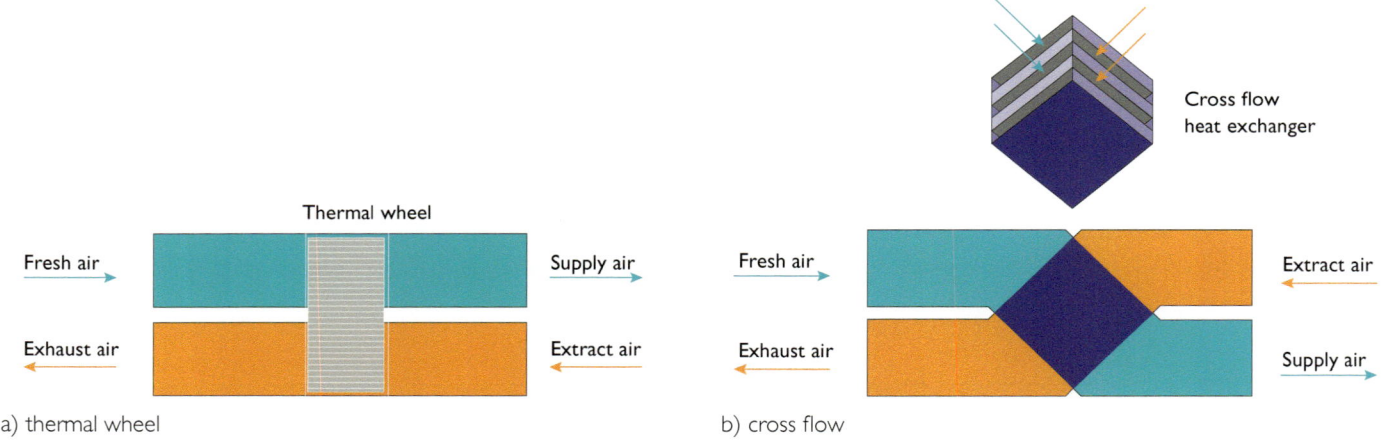

Figure 129   Mechanical ventilation heat recovery (MVHR) systems

The cross-flow system is based on a very simple idea of parallel thin sheets of metal, with edges sealed on alternating sides, thus achieving cross-flow between two ducts without the two streams of air mixing (Figure 129b). Heat between the warm and cold air ducts is exchanged by conduction through the thin metal sheets of the cross-flow heat exchanger. One can easily replicate the geometry of this heat exchanger using sheets of paper positioned in parallel and sealing alternate edges accordingly. This type of heat exchanger has no moving parts and can recover over 90% of heat, making it more efficient than the thermal wheel system. The system can be very compact, and in domestic applications, it can fit into a kitchen wall element (Figure 130a), or in larger cases into a utility cupboard (Figure 130b).

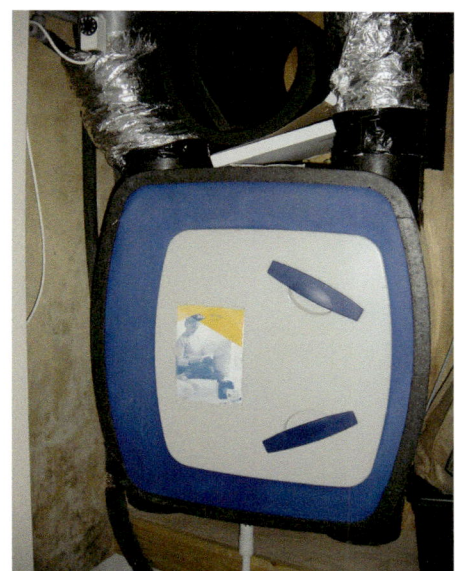

| a) In a kitchen wall cupboard | b) In a utility cupboard |

Figure 130   Mechanical ventilation heat recovery units

If mechanical ventilation is specified in a building, heat recovery systems are absolutely necessary in order to reduce energy consumption and carbon emissions. Efficiency from manufacturers' data sheets should be used when specifying either of the two heat recovery systems in a simulation model.

12.3 SCALE MODEL EXPERIMENTS

A scale model experiment of natural ventilation is shown in Figure 131. The apparatus (Figure 131a) consists of a vertical rainwater pipe with a side opening near the top, covered with a transparent sheet. This transparent sheet in front of the dark back wall of the rainwater pipe represents a solar collector when exposed to radiation heat input. A dark piece of steel is used to enhance the absorption capacity of the solar collector. A base-level opening enables air to enter the pipe. A 400-watt halogen lamp provides radiation heat input. The rainwater pipe configured in this way represents a solar chimney. A smoke source is used to trace air flow through different stages of the experiment.

At the start of the experiment the lamp was switched on, the top of the pipe was covered with a lid to allow internal surfaces in the pipe to heat up, and the smoke source was placed near the base-level opening just outside of the vertical line of the pipe, and the room air was kept still to allow the smoke flow to stabilise. Figure 131b shows the smoke flow rising vertically outside of the pipe in this first step of the experiment whilst the lid was kept in place. Temperatures measured at the back of the base and top of the pipe reached 28°C and 125°C, respectively, at this stage of the experiment. In the next step, the lid from the top of the pipe was removed, and the higher buoyancy of air at the top of the pipe, caused by the hot internal surfaces, attracted the smoke to start flowing into the base of the pipe (Figure 131c). In order to find out whether air buoyancy would be enhanced by a solar absorber, a dark steel blade was inserted into the top of the pipe behind the transparent sheet and enough time was allowed for it to heat up. This resulted in a sharper angle of smoke flow entering the base of the pipe, indicating stronger buoyancy of air inside the pipe (Figure 131c). These qualitative results correspond well to the stack configurations introduced in Figure 127.

We can see from this example how scale model experiments are a very simple and effective way of increasing our understanding of how things work in buildings.

a) The test apparatus for stack ventilation

b) The smoke flow rising vertically outside of the pipe (bottom) while the lid is in place (top)

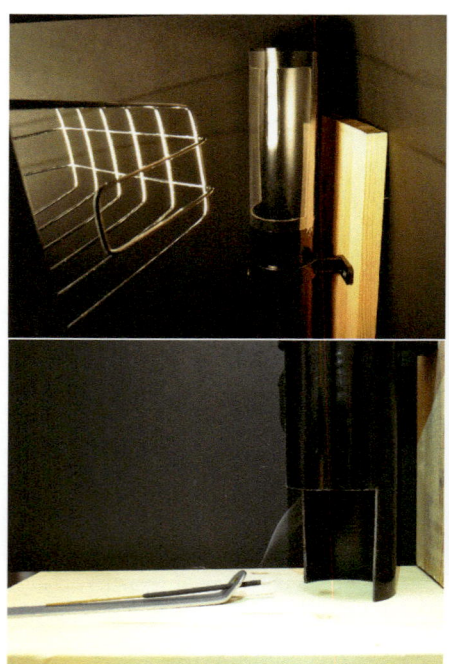

c) The smoke is attracted into the base of the pipe (bottom) when the lid is removed (top)

d) Sharper angle of smoke flow entering the base of the pipe (bottom) resulting from higher heat absorption by the dark steel blade (top)

Figure 131   Scale model experiment of natural ventilation

12.4 DYNAMIC SIMULATION EXPERIMENTS

12.4.1 A simple office with and without ventilation

This section explains simple simulation experiments looking into the reduction of air-conditioning requirements in a south-facing cellular office in London, New York, and Miami. The results are expressed in terms of a fraction of the total time when air conditioning is required. In the first case without mechanical ventilation, in London and New York there is a similar yearly pattern where air conditioning is needed from April till October, whilst in the Miami office, air conditioning appears to be needed all the time (Figure 132a).

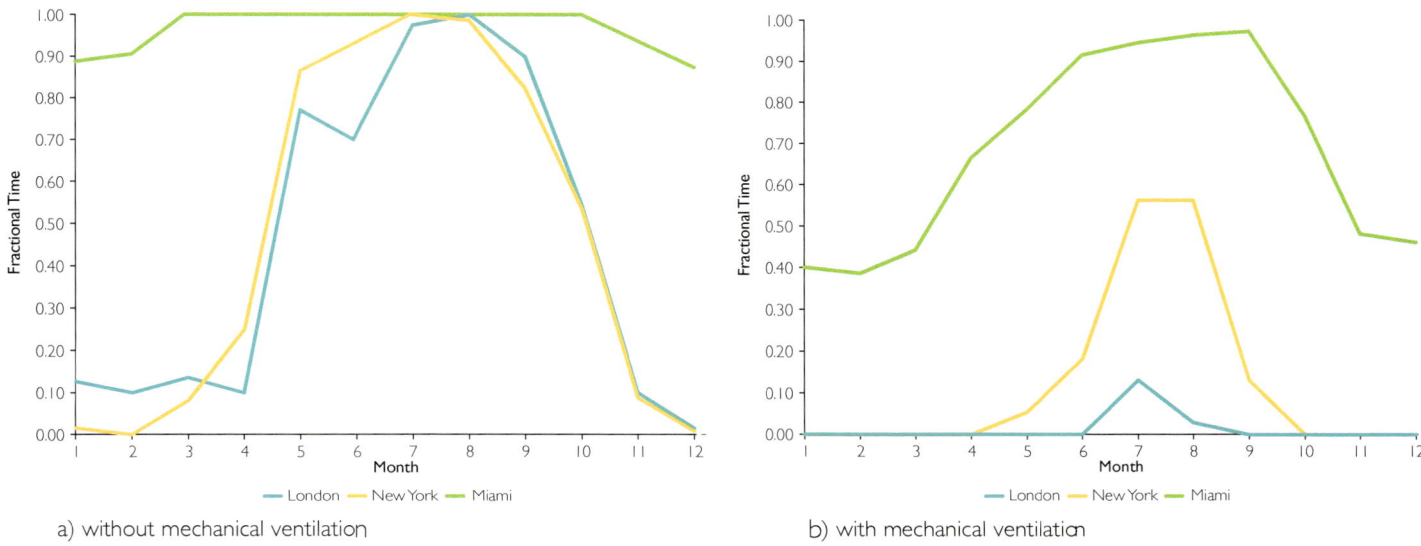

a) without mechanical ventilation b) with mechanical ventilation

Figure 132   The effect of mechanical ventilation on comfort cooling requirement

After the introduction of mechanical ventilation with one air change per hour, there is a drastic reduction in the air-conditioning requirement (Figure 132b). In the Miami office, the reduction is to about 40% in January and December with the rest of the year air-conditioning requirement being reduced significantly. In the New York office, there is some requirement for air conditioning from May till September, and in the London office, the requirement for air conditioning is reduced to two days in July and one day in August. In the London case, it therefore may be more cost effective to send the workforce to work from home during the three days in question and save on the capital and running costs of air conditioning, replacing it with mechanical or even natural ventilation.

12.4.2 Greenpower office with stack ventilation

The model used to investigate natural ventilation in this chapter is described in more detail as Model 'a' in supplementary material on www.ljankovic.com. In this section, the main characteristics of this model will be summarised and the effects of bulk air flow modelling will be investigated.

This is a south-facing building with a shallow plan depth that enables utilisation of daylight and natural ventilation. Thermal insulation is based on UK Part L 2010 building regulations with *U*-values of thermal elements as follows:

ELEMENT	W/(M²K)
Exposed floor	0.2499
Exterior walls	0.3495
Roof	0.2497
Doors	2.1944
Glazing	1.9773

High level of air tightness with a permeability-based infiltration of 0.167 air changes per hour is specified. The building is based on the direct gain system, with a large amount of south glazing, and high-density concrete ground floor, mezzanine floor, and internal partitions comprising thermal mass that absorbs solar gain and releases it slowly into the building.

Natural daylight is achieved using a high ceiling (6.15 m), and high south glazing. The building plan, relative to the ceiling height, is shallow (11.33 m) and provides good conditions for natural daylight.

Electrical lighting is based on 3.75 W/m²/(100 lux) in office and lecture space and 4.46 W/m²/(100 lux) in the toilets. The lighting level is set to 500 lux in the open plan office and the lecture space, 200 lux in the toilets, and 100 lux in circulation spaces, store, and the plant room.

Conventional heating is based on a condensing gas boiler with a seasonal efficiency of 93%, and with underfloor heating system.

The natural ventilation system consists of two high-level passive stacks and four low-level vents, controlled on the basis of temperature and CO_2 concentration as shown in Table 35.

TABLE 35 NATURAL VENTILATION CONTROL RULES IN THE SIMULATION MODEL	
Description	**Control formula**
Low-level glazing	$t_a > 23°C$ & $t_a > t_o$
Doors	$t_a > 23°C$ & $t_a > t_o$
Natural ventilation vents	$t_a > 23°C$ & $t_a > t_o$ OR $CO_2 > 500$ ppm

t_a – internal air temperature; t_o – external air temperature

Two simulation runs were made: once without natural ventilation and once with natural ventilation. The comparison of hourly temperatures between the two runs is shown in Figure 133.

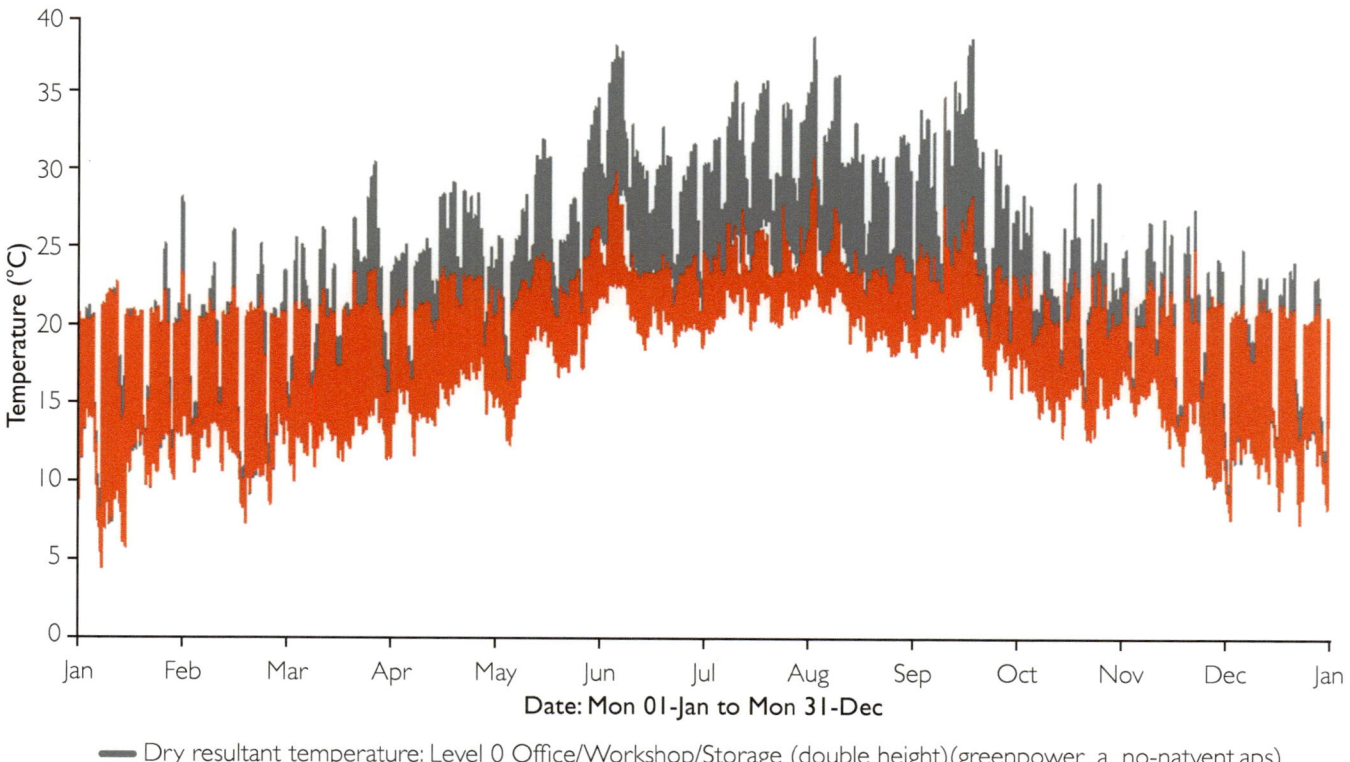

— Dry resultant temperature: Level 0 Office/Workshop/Storage (double height)(greenpower_a_no-natvent.aps)
— Dry resultant temperature: Level 0 Office/Workshop/Storage (double height)(greenpower_a_natvent.aps)

Figure 133   Comparison of hourly internal temperatures with and without natural ventilation

We can see from this figure that considerable overheating occurs in the model without natural ventilation between March and October, and that overheating appears to be much more under control in the case with natural ventilation.

We can evaluate the magnitude of overheating by analysing the cumulative frequency of the occurrence of hourly temperatures, as shown in Figure 134. The analysis shows that in the case with natural ventilation internal temperatures were below 24°C 95% of the time, whereas in the case without natural ventilation temperatures were below 32°C 95% of the time.

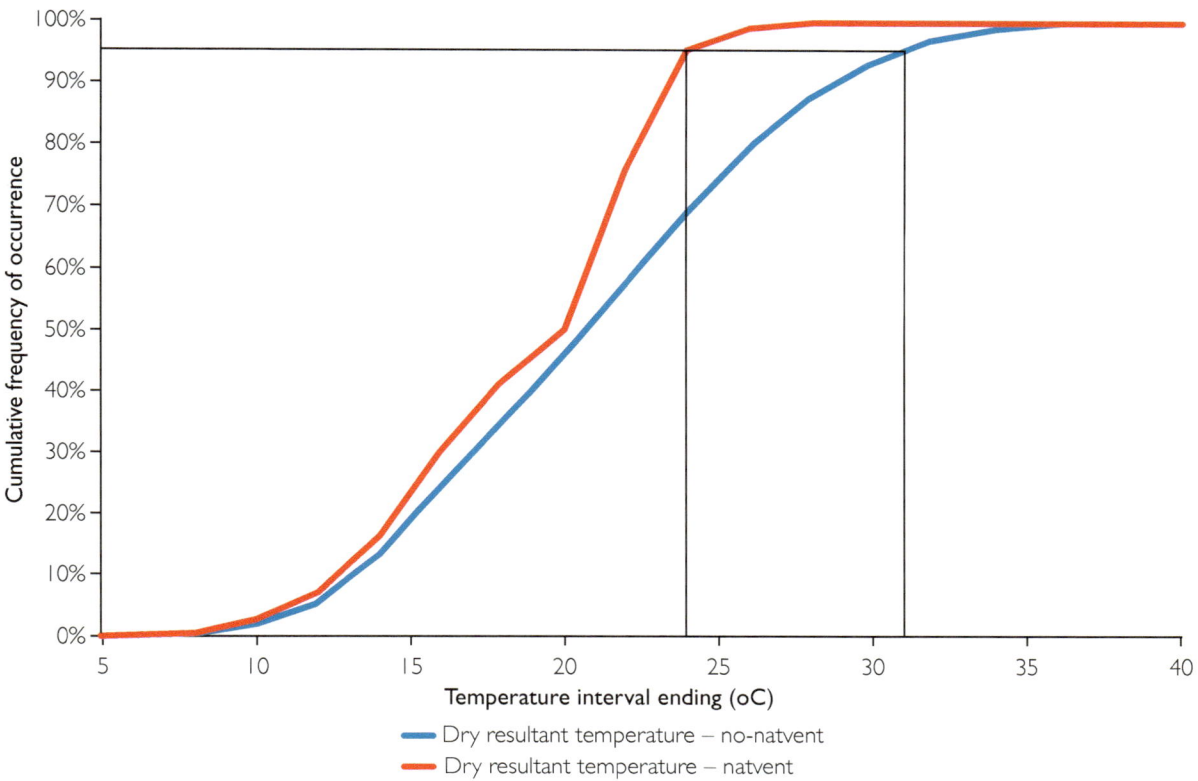

Figure 134   Cumulative frequency of occurrence of internal temperatures with and without natural ventilation

Natural ventilation therefore helps considerably with controlling excessive internal temperatures, but what does it do to energy consumption and carbon emissions? The comparison is shown in Table 36, from where we can see in the naturally ventilated case an increase of 6.2 MWh per annum in energy consumption and an increase of 1.233 tonnes of CO_2 emissions, resulting from increased energy requirements in the heating season.

TABLE 36 COMPARISON OF ANNUAL ENERGY CONSUMPTION AND CARBON EMISSIONS WITH AND WITHOUT NATURAL VENTILATION		
Model description	Total yearly energy consumption (MWh)	Total carbon dioxide emissions (kg CO_2)
Not naturally ventilated	24.1	10420.0
Naturally ventilated	30.3	11653.3

We are therefore dealing with conflicting design constraints: on the one hand, we can save energy and reduce emissions by not naturally ventilating the building, but on the other hand, we cannot use the building if it is overheated in the absence of natural ventilation, as well as due to poor air quality. It is therefore advisable to optimise the design by adjusting the control parameters in Table 35 and re-running the model until a satisfactory trade-off between thermal comfort, energy consumption, and carbon emissions is reached.

During the design process, it is useful to visualise that natural ventilation air flow is working as expected. We can do this in IES MacroFlo with arrows that represent the air flow intensity and direction (Figure 135). This however gives us only qualitative information.

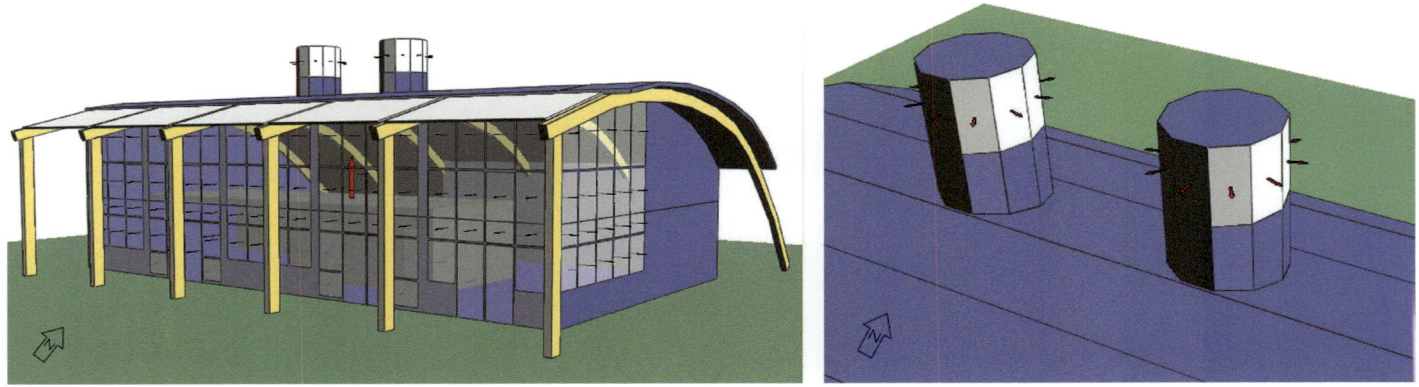

Figure 135   Visualisation of bulk air flow in our IES VE model. Left: overall circulation diagram; Right: close-up of circulation from the ventilation stacks

For more in-depth quantitative and qualitative analysis, we can use computation fluid dynamics (CFD). In IES, we first need to run a thermal simulation in Apache, generate the required output for CFD analysis, and export boundary conditions for a specified day and hour into a file. We then go into the CFD module called MicroFlo, import the boundary conditions, and run the CFD simulation for the specified day and hour. We carried out this analysis on 21st June at 12 noon, and the results are shown in Figure 136 through to Figure 141.

The velocity surfaces diagram in Figure 136 shows that natural ventilation is working, with air flow going from the ground floor and mezzanine into the stacks. The temperature surfaces diagram in Figure 137 shows temperature stratification along the

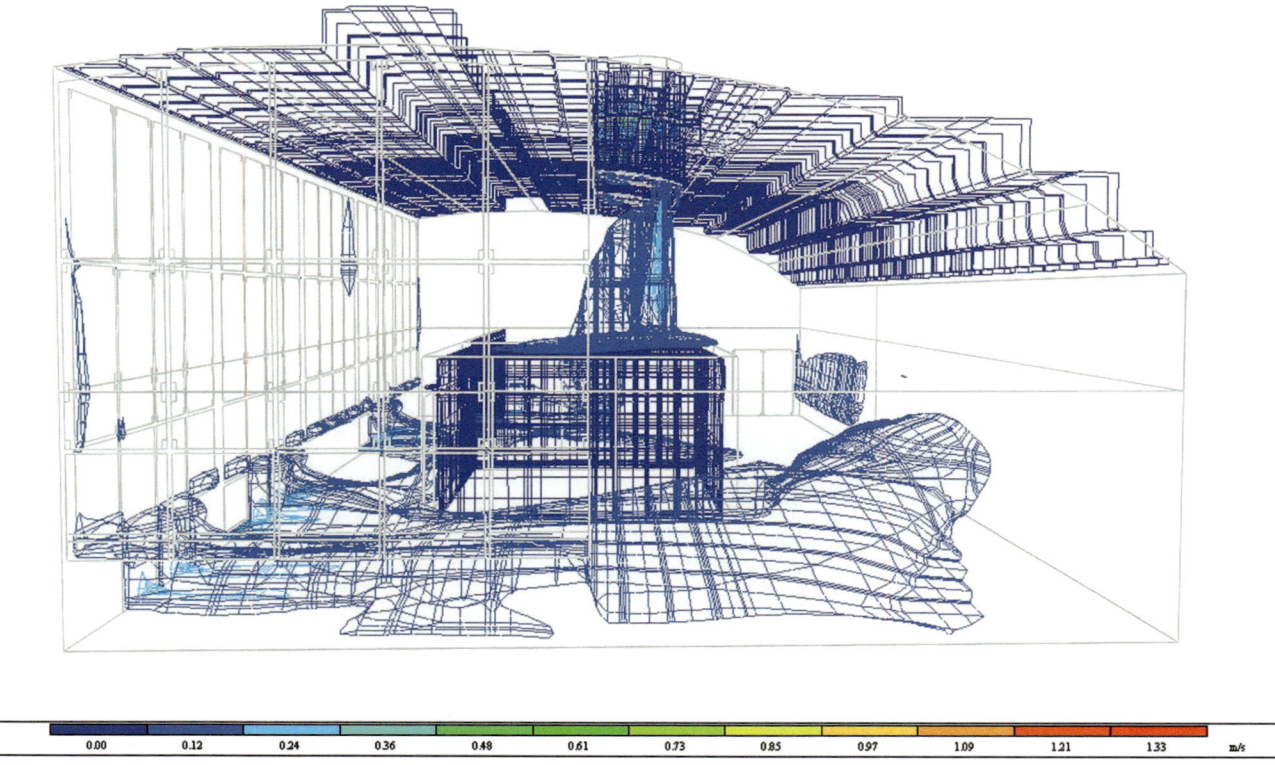

| Velocity | 0.00 | 0.12 | 0.24 | 0.36 | 0.48 | 0.61 | 0.73 | 0.85 | 0.97 | 1.09 | 1.21 | 1.33 | m/s |

Figure 136   CFD velocity surfaces

height of the internal space. The CO_2 surfaces diagram in Figure 138 shows that the CO_2 concentration is around the control threshold of 500 ppm and that CO_2 is being removed through the stack chimney.

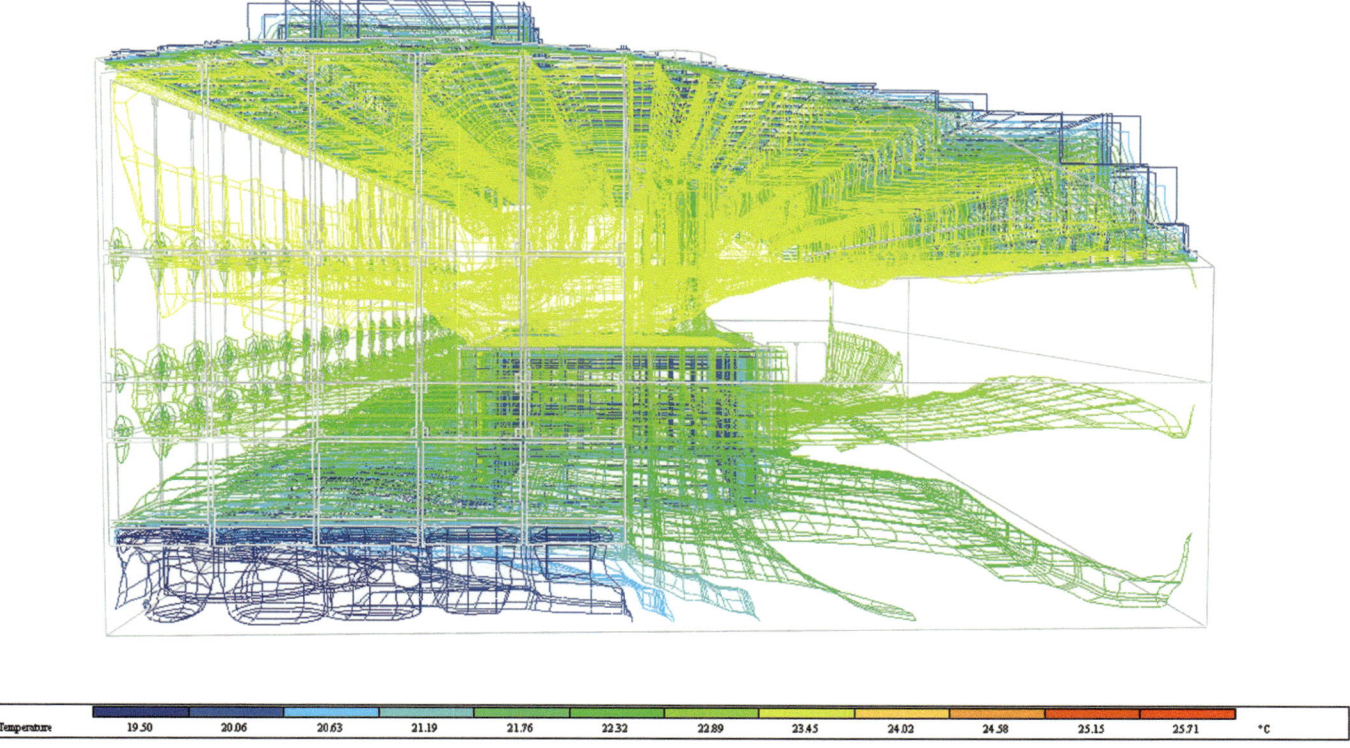

| Temperature | 19.50 | 20.06 | 20.63 | 21.19 | 21.76 | 22.32 | 22.89 | 23.45 | 24.02 | 24.58 | 25.15 | 25.71 | °C |

Figure 137   CFD temperature surfaces

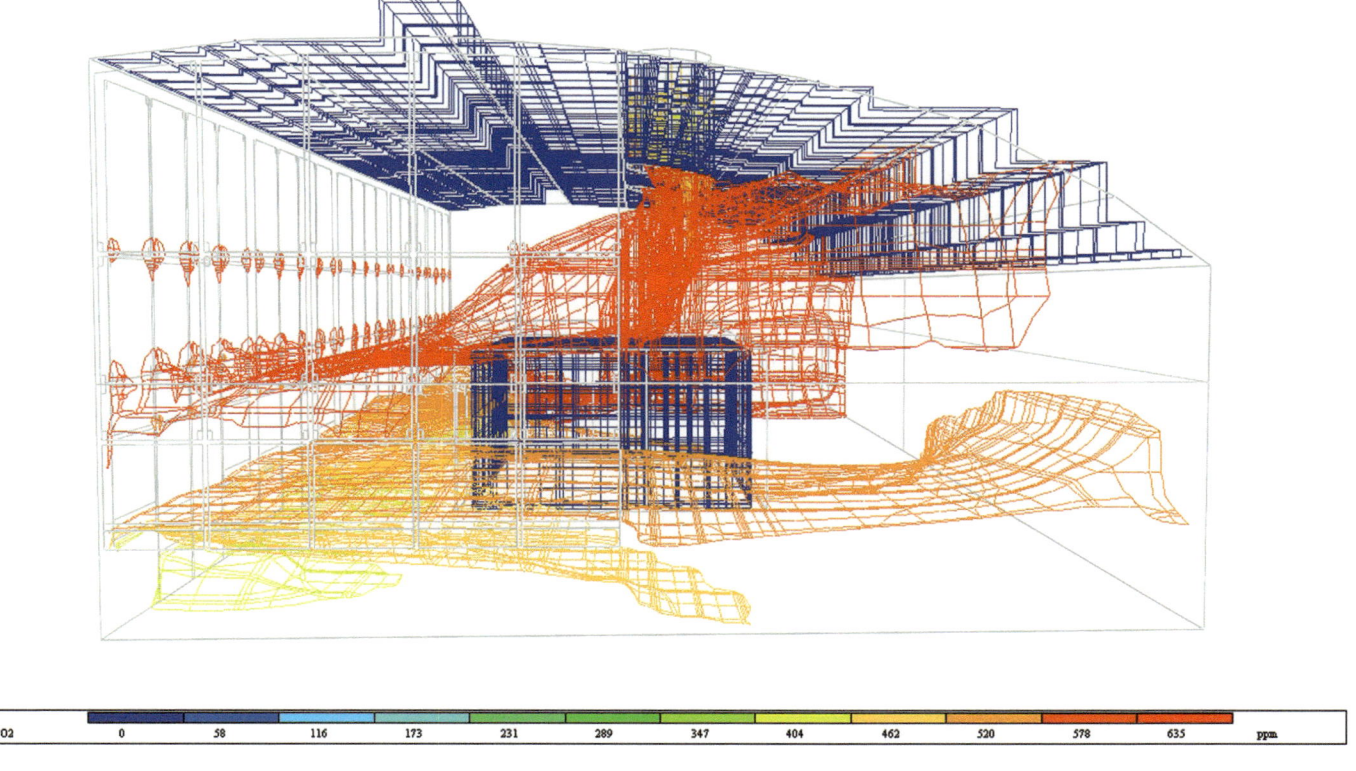

| CO2 | 0 | 58 | 116 | 173 | 231 | 289 | 347 | 404 | 462 | 520 | 578 | 635 | ppm |

Figure 138   CFD CO_2 surfaces

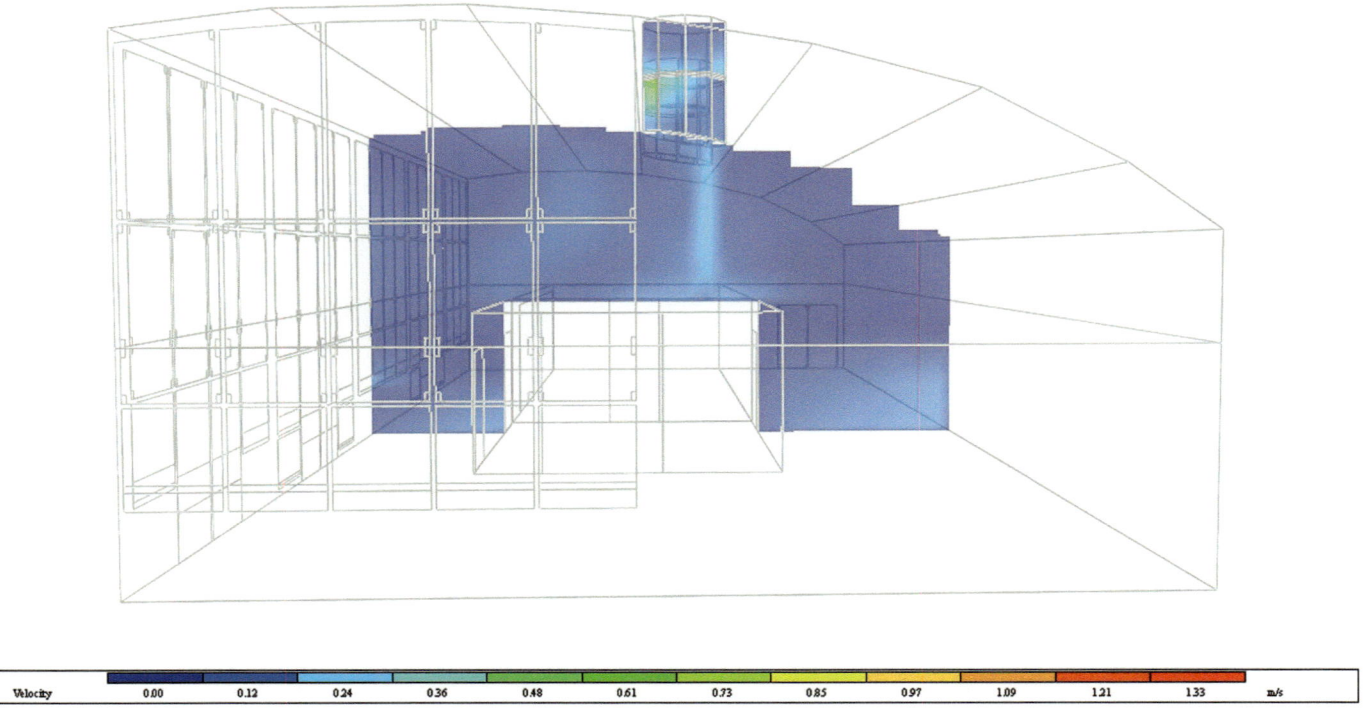

| Velocity | 0.00 | 0.12 | 0.24 | 0.36 | 0.48 | 0.61 | 0.73 | 0.85 | 0.97 | 1.09 | 1.21 | 1.33 | m/s |

Figure 139   Filled velocity contour slice

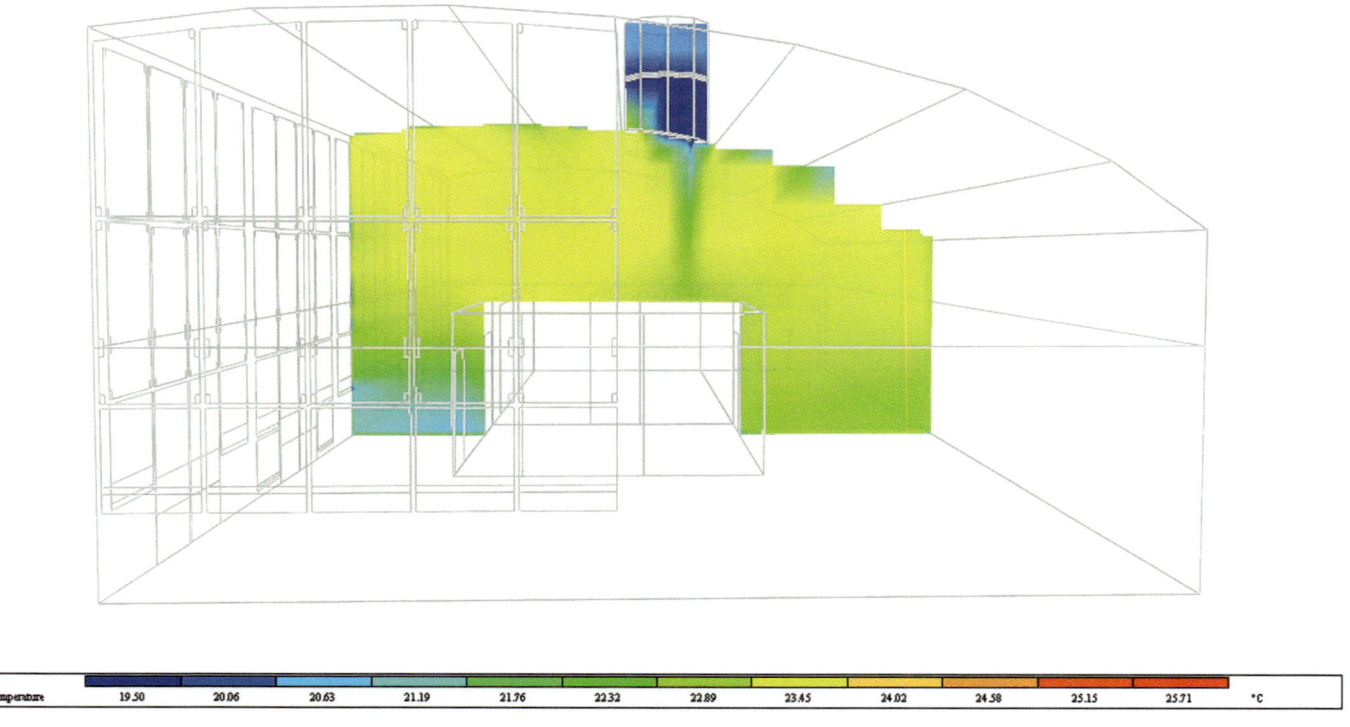

| Temperature | 19.50 | 20.06 | 20.63 | 21.19 | 21.76 | 22.32 | 22.89 | 23.45 | 24.02 | 24.58 | 25.15 | 25.71 | °C |

Figure 140   Filled temperature contour slice

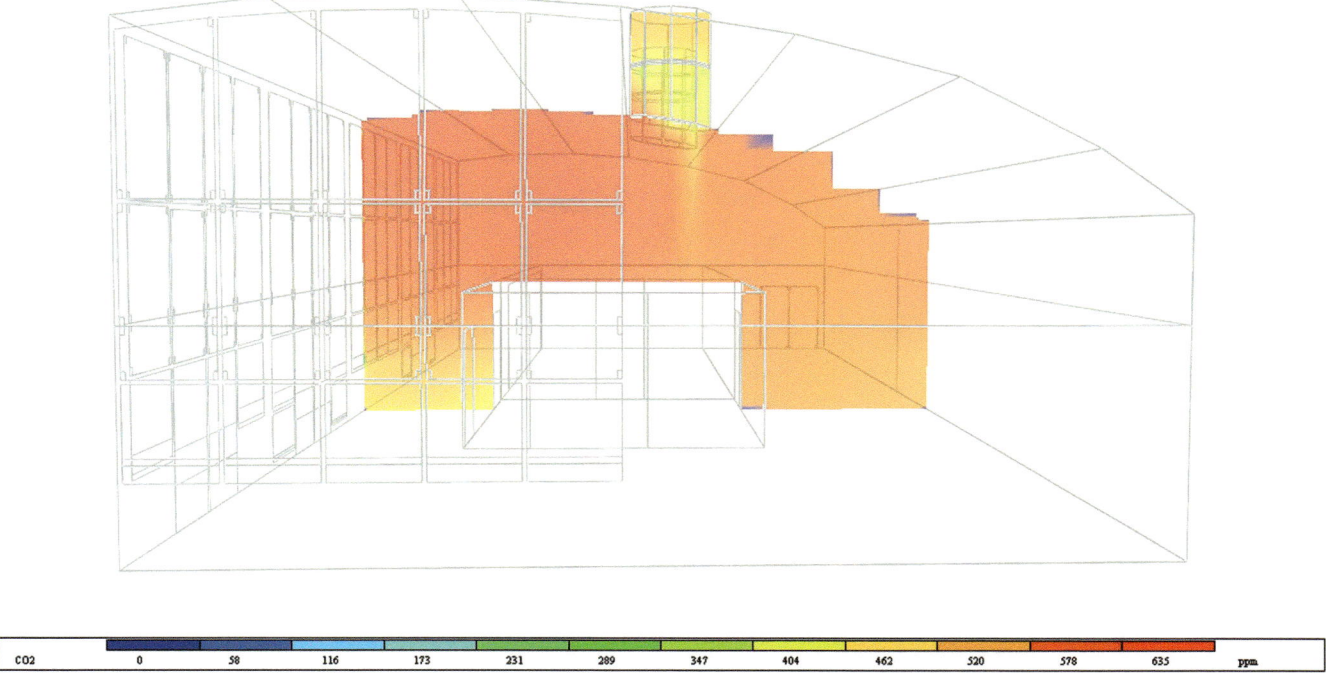

| CO2 | 0 | 58 | 116 | 173 | 231 | 289 | 347 | 404 | 462 | 520 | 578 | 635 | ppm |

Figure 141   Filled CO$_2$ contour slice

We can also generate slices of filled contours in x, y, and z directions, giving us continuous fields of velocities, temperatures, and CO$_2$ (Figures 139 through 141). We can use this information to fine-tune window openings, introduce solar absorbers to boost air buoyancy, and thus optimise air flow and internal conditions.

12.5 TASKS FOR SIMPLE SIMULATION EXPERIMENTS

1 Create a simulation model of a box 10 m × 6 m × 3 m with two north-facing windows 3 m wide and 1 m high. Duplicate this model so that two exactly the same boxes are side by side, not overshadowing each other. Insert a round chimney 3 m high on the south end of the flat roof of one of the boxes. Insert glazing in the upper half of the south end of the chimney Make all windows 30% openable, above 25°C internal temperature whilst the external air temperature is lower than the internal air temperature. Compare hourly room temperatures in the box with the chimney and in the box without the chimney using the weather data of your choice.

2 Swap windows location from north to south and chimney location from south to north and re-run the above task.

3 Increase internal material density and solar absorptance in the chimney and re-run the above task.

4 Use cumulative frequency of occurrence analysis of hourly room air temperatures for quantitative comparison of all three cases above.

5 Run a CFD analysis of the model above for a selected summer day and hour and investigate temperature and velocity distribution in x, y, and z dimensions through the box and the chimney.

6 Discuss results with colleagues.

12.6 SUMMARY OF DESIGN PRINCIPLES

Natural ventilation – general

• Use a shallow building plan.
• Use large thermal mass to reduce temperature fluctuations.
• Use night ventilation to discharge thermal mass.

- Use short and wide ducts.
- Use computer control of grilles and openings.
- Develop control scenarios for day and night.

Natural ventilation – wind effect

- Take into consideration the configuration of surrounding buildings, trees, and terrain.
- Take into account the availability of wind and pollution on site.
- Choose orientation towards the prevailing wind.
- Provide controllable openings.
- Configure the openings so as to provide the desired ventilation path.
- Do not exceed 12 metres window to window plan depth.
- Use louvres to direct internal air flow.
- Use tall chimneys with cross wind to enhance ventilation.

Natural ventilation – stack effect

- High air temperature differences drive the stack effect.
- Use tall chimneys to achieve sufficient temperature difference and sufficient buoyancy of air.
- Make top of the chimney transparent.
- Use a solar absorber at the top of the chimney to increase temperature difference that drives the stack effect.
- Provide openings of equal size at the top and bottom of the chimney: larger openings at the bottom or at the top of the chimney will create a bottleneck or inverse bottleneck effect respectively, and in both cases the stack ventilation will not work well.
- Computer control is essential for co-ordinated openings in different parts of the building.

Mechanical ventilation

- Use CO_2 sensors to operate mechanical ventilation on demand so as to minimise running time of the ventilation fans.
- Use heat recovery systems to minimise heat losses from the exhaust air.

Simulation

- Control window, door, and vent openings using temperature and CO_2 concentration formulae.
- Analyse hourly temperatures and their frequency of occurrence to establish the effectiveness of natural ventilation.
- Visualise the air movement using bulk air flow diagrams.
- Conduct CFD analysis and obtain surface representations and filled contours of velocity, temperature, CO_2 and other parameters for detailed qualitative and quantitative analysis.

Repeat the process until design objectives are achieved.

CHAPTER 13

NATURAL DAYLIGHT

This chapter explores major characteristics of daylight and explains simulation-based daylight design. For background reading on fundamental principles and case studies, the reader is advised to consult the book by Nick Baker and Koen Steemers (Baker & Steemers, 2002).

13.1 DEFINITION AND PROPERTIES OF DAYLIGHT

13.1.1 What is daylight?

Solar radiation is distributed across a range of wavelengths as shown in Figure 87, comprising the solar spectrum. Only 47% of this spectrum, between wavelengths of approximately 380 and 780 nm, is visible to the human eye. This is what we call daylight.

Colour is the sensation produced when light of different wavelengths falls on the human eye. Although the spectrum in Figure 87 is continuous, the human eye groups regions of the visible spectrum into distinct colours (Figure 142), so that violet colour starts at 380 nm, and red colour ends around 780 nm. In between there are indigo, blue, green and yellow colours and each colour has its distinct wavelength range.

Violet Indigo Blue Green Yellow Orange Red

380 nm 780 nm

Figure 142   Colours of the visible spectrum

Sunlight is defined as direct unobstructed light coming from the sun. Daylight (Figure 143) is a combination of direct sunlight (blue sky), sunlight diffused through clouds (white and grey sky), and sunlight reflected from the water, ground, plants, and other objects.

13.1.2 Colour temperature

Colour temperature is related to emissive energy coming from a body. The higher the energy in the body of the same size, the higher the colour temperature, and the body will emit more blue light. The lower the energy in the body of the same size, the lower the temperature, and the body will emit more red light. Colour temperature of daylight at noon under the clear sky is approximately 6,000 K, and is approximately equal to the temperature of the source – the sun. The temperature of the clear blue sky is 10,000 K and of candlelight is 1,000 K.

Human perception of colour differs from the colour temperature scale. In other words, colour temperature is counterintuitive to our perception of red as a warm colour and blue as a cold colour. We perceive colours below 3,300 K as warm even though they come from lower energy sources, and colours above 5,300 K as cold, even though they come from higher energy sources. Colours in between are intermediate – neither warm nor cold (Parry, 2022).

Figure 143 *Daylight at San Sebastián/Donostia beach, in the Basque Autonomous Region of Spain – a combination of direct sunlight (blue sky), sunlight diffused through clouds (white and grey sky), and sunlight reflected from external objects (water, ground, plants)*

13.1.3 Human response to daylight and its colour temperature

We might think that light only affects us in an aesthetic or in an emotional way, but it in fact affects us in a much more profound physiological way by influencing the production of a hormone called melatonin in the human body. In the absence of light or at low colour temperatures melatonin is produced in larger quantities and it is suppressed at high colour temperatures, such as by bright white or blue light. As the regulator of the body's biological clock, melatonin, which is produced by the pineal gland, will make us sleepy. It therefore follows that at higher colour temperatures, we will not feel sleepy and at lower colour temperatures we will feel sleepy.

13.1.4 Advantages of daylight

Natural daylight is a renewable resource and its use can considerably reduce energy consumption in buildings and resultant carbon emissions. It also affects human wellbeing in buildings by making the indoor environment healthy and enjoyable, influencing us not only aesthetically and emotionally through our perception colours as warm or cold, but also physiologically by influencing the production of the hormone melatonin. It is therefore important to design buildings that maximise the use of daylight.

13.2 DAYLIGHT DESIGN

The main design objectives are to maximise visual comfort and minimise energy use. To achieve these objectives, measurable parameters are needed to determine the degree of visual comfort. This chapter will first define units of lighting, parameters of daylight, and key aspects of visual comfort, followed by an iterative and experimental approach to daylight design.

Starting with the initial room and building geometry, an initial design will be validated through simulation. Incremental changes will then be made in order to improve the design until design objectives are achieved.

13.2.1 Units of measure

In order to define units of lighting, the first step is to define the unit of a three-dimensional angle, measured in steradians. One steradian, symbol 'sr', corresponds to a cone within a sphere, starting from the centre of the sphere and projecting to its surface, so that the area of the base of the cone on the sphere's surface is equal to the square of the sphere's radius (Figure 144).

This now helps to define the units of measure of light. The luminous intensity of light, called candela, symbol 'cd' is one of the seven fundamental units in the SI system. In the SI system *'one candela is the luminous intensity, in a given direction, of a source that emits monochromatic radiation of frequency 540 x 10^{12} Hz and has a radiant intensity in that direction of (1/683) W/sr'* (BIPM, 2023). The reason for defining this unit of luminous intensity in this way is that it corresponds to the luminous intensity of a common candle.

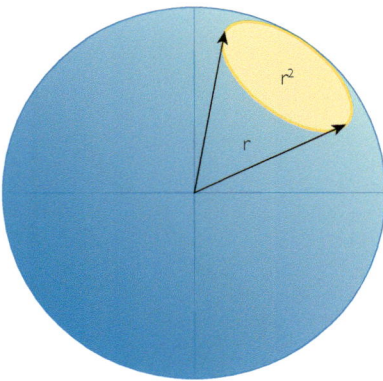

Figure 144   Definition of steradian

Luminance (emitted light) is defined as luminous flux – a rate of energy flow per unit of time per unit of area across a surface. The unit of luminance is lumen (lm), defined as one candela steradian: 1 lm = 1 cd.sr.

Illuminance (received light) is then defined as luminance per unit area and is measured in units of lux (lx). Therefore, 1 lx = 1 lm/m². As 1 lm = 1 cd.sr, one lux can also be defined as 1 cd.sr/m². Hence, shining a source of light equivalent to a typical candle (one candela) through one steradian (Figure 144) on one square metre of surface area will get illuminance of one lux onto that surface.

13.2.2 Standard skies

Standard skies have been defined by the International Commission on Illumination (CIE, 2003) to provide illuminance values for design purposes.

A standard CIE overcast sky is a sky model that assumes that illuminance does not vary with azimuth (horizontal displacement from the direction of north) but it varies with altitude so that illuminance in the zenith is three times the illuminance on the horizon (CIE, 2003).

A standard CIE clear sky is a sky model in which illuminance varies both with azimuth and altitude, with the highest illuminance near the sun (CIE, 2003).

Simulation software tools may contain a number of different standard sky models and their variations, and it is essential to check sky model properties and choose an appropriate sky model accordingly before commencing any simulations, as results can be radically different for different skies.

13.2.3 Daylight factor

The daylight factor is a measure of internal room daylight characteristics. It defines how room geometry modulates external light into internal light. The daylight factor is defined as the ratio between the average internal illuminance and the outside unobstructed horizontal illuminance, as shown in Figure 145.

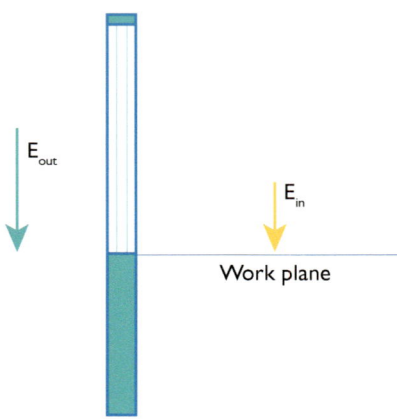

Figure 145   Definition of the daylight factor

Using the notation from Figure 145, the daylight factor DF is defined as

$$DF = \frac{E_{in}}{E_{out}} \times 100 \ (\%)$$ (36)

where

E_{in} – average internal illuminance on a horizontal work plane

E_{out} – external illuminance on a horizontal surface

Rooms with daylight factors of 5% or above are considered to be predominantly daylit, and rooms with daylight factors of 2% or less are considered to require additional lighting. If we apply a standard sky with an average illuminance of 5,000 lx, then the room with DF = 5% will have an average illuminance of 250 lx. If we apply a standard sky with an average illuminance of 10,000 lx, then the room with DF = 5% will have an average illuminance of 500 lx.

The SLL Code for Lighting from CIBSE (Parry, 2022) gives guidelines for internal illuminance levels for different applications and different tasks. We can compare illuminance values obtained from daylight simulation with the guidelines and establish whether the daylighting provided in the space we are designing is sufficient for the intended use of that space.

In a dynamic simulation, illuminance in the room will change as the sky illuminance changes. This will be used in Chapter 14 to control electrical light in a thermal energy simulation, on the basis of daylight sensor readings and daylight-sensitive controls. This approach can lead to substantial energy savings in buildings.

An example of how a good daylight design results in energy savings can be found in the passenger terminal at Porto Airport in Portugal (Figure 146). Internal ambient electrical lights are controlled with daylight sensors, and on many occasions, one cannot see a single electrical light switched on. In contrast, in most airport buildings around the world ambient electrical lighting seems to be permanently on.

13.2.4 Colour rendering

Colour rendering is a property of light that enables the human eye to differentiate between eight different reference colours. A colour rendering index has been defined to indicate colour rendering properties of a light source. The way the colour rendering index works is to use eight sample colours, illuminate them with the light source being tested, and also illuminate the same set of colours with a reference lamp. If there is perfect agreement between the differentiation of colours illuminated by the reference lamp and the light source being tested, then the colour rendering index is equal to 100. A lower agreement will result in a lower colour rendering index.

This is easy to understand through the following practical example. If we take a paper chart of eight different colours and illuminate it with a halogen lamp, we will easily differentiate between the eight sample colours. However, if we go

Figure 146   Passenger terminal at Francisco SA Carneiro (Oporto) Airport, Porto, Portugal: with sufficient daylight electrical lights are switched off

out at night under a street light, it might be that we will not be able to differentiate between some colours on the chart. The colour rendering index of the street light will therefore be lower than the colour rendering index of an internal halogen light.

Different applications will have different colour rendering requirements. For accurate colour matching, the colour rendering index needs to be greater than 90. Where colour differentiation is of no importance the colour rendering index can be less than 40. Details of colour rendering index values and corresponding applications can be found in *The SLL Code for Lighting* from CIBSE (Parry, 2022). This is why we cannot choose just any kind of artificial light that has the lowest energy consumption, without considering the colour rendering requirements.

Good colour rendering will help with the perception of different objects and will result in reduced illuminance levels required for specific tasks. Excellent colour rendering is another advantage of daylight.

13.2.5 Visual comfort

The human eye is adaptable to illuminances ranging between 100,000 and 0.1 lx, but it needs some time to adjust. Simultaneous occurrence of radically different illuminances will cause discomfort. Poor lighting conditions for specific tasks will also cause discomfort and result in eyestrain, headache, irritation and fatigue, and can also lead to accidents when operating machinery.

In order to provide visual comfort in internal environments, we must provide suitable illumination for tasks, avoiding radically different illuminances (glare), whilst achieving appropriate colour rendering and variety of lighting. Daylighting has all of these qualities, and is the easiest way to achieve good visual comfort.

13.2.6 Building geometry

In Chapter 7, Figures 72–74 demonstrate advantages of the shallow building plan and disadvantages of the deep plan, as the former is conducive to natural daylight and natural ventilation and the latter excludes both of the two natural phenomena. This applies to standard floor-to-ceiling heights of about 3 metres. The room proportions can now be looked at in more detail, to give the flexibility of increasing the plan depth as the width, the height, and the reflectance of the room increase.

Room proportions for a naturally daylit depth L are shown in the following relationship as a function of room height H, room width W, and reflectance R_b at the back of the room:

$$\frac{L}{W} + \frac{L}{H} \leq \frac{2}{(1 - R_b)} \tag{37}$$

In order to investigate variations of the room dimensions and their effect on daylit depth, the above equation needs to be re-arranged into a more useful form. This is done by extracting L as a common factor on the left-hand side

$$L \times \left(\frac{1}{W} + \frac{1}{H} \right) \leq \frac{2}{(1 - R_b)} \tag{38}$$

and solving it for L as follows:

$$L \leq \frac{\dfrac{2}{(1 - R_b)}}{\left(\dfrac{1}{W} + \dfrac{1}{H} \right)} \tag{39}$$

The above equation enables us to investigate daylit depth as a function of height H, width W, and reflectance R_b at the back of the room. However, this equation gives the room proportions for average conditions. Once the room dimensions have been established on this basis, they need to be optimised through simulation.

A spreadsheet that calculates L on the basis of the above equation is available on the supplementary material website, and an example calculation is shown in Table 37.

TABLE 37 SAMPLE ROOM GEOMETRY CALCULATION	
Description	**Value**
L	6.00
W	2.00
H	3.00
R ceiling	0.90
R floor	0.30
R left wall	0.70
R right wall	0.70
R back wall	0.30
R front wall	0.70
R_b	0.60

13.3 SCALE MODEL EXPERIMENTS

Scale model experiments of daylight were conducted by my students. The brief for the natural daylight experiment was as follows:

Natural daylight: Create two cardboard models: Model 1–20 cm × 12 cm × 6 cm, with 2 × 6 cm × 4 cm openings on the larger wall (wall with dimensions of 20 cm × 6 cm). Model 2 – the same as model 1, but two times deeper plan, with dimensions of 20 cm × 24 cm × 6 cm, and two window openings of 6 cm × 4 cm on the wall with dimensions of 20 cm × 6 cm. Experiment with a lamp that shines light onto the openings, simulating summer and winter sun angles. Take photographs of the interior of the model from east side (if we call the 'south side' the side where the openings are). Introduce louvre blinds in front of the window openings. Repeat the illumination of the model and observe the effect on the interior illumination. Take photographs of the interior. Compare the results from the two experiments and comment on the findings.

Results of experiments conducted by a group of students, using solar geometry for 52.4° latitude, are shown in Figure 147.

a) Shallow plan, noon on 21st June

b) Deep plan, noon on 21st June

c) Shallow plan, noon on 21st December

d) Deep plan, noon on 21st December

e) With louvre blinds added to case a)

f) With louvre blinds added to case b)

g) With louvre blinds added to case c)

h) With louvre blinds added to case d)

Figure 147   Scale model experiment of daylight illumination (models and photos by students Lik Hung Tang, Zhen Xi Piao, and Najwa Majar)

Figures 147 (a) and (b) simulate daylight illumination of a shallow plan and deep plan room on the summer solstice at noon on 21st June. Winter solstice illumination at noon on 21st December is shown in Figures 147 (c) and (d). Figures 147 (e–f) are equivalent to Figures 147 (a–d), but with horizontal blinds on.

As expected, the shallow plan room is well illuminated both in summer and winter (Figures 147a and c). The back of the deep plan room is not well illuminated at summer solstice (Figure 147b), but it is much better illuminated at winter solstice (Figure 147d). Horizontal louvre blinds prevent illumination by daylight at summer solstice in both shallow and deep plan rooms (Figures 147e and f). At winter solstice the horizontal blind allows some daylight to get to the back of the shallow plan room (Figure 147g), whilst the deep plan room remains predominantly dark (Figure 147h). This is perhaps even more obvious from a photo of both models exposed to solar radiation (Figure 148), from where we can see that the back end of the deep plan model has a much darker appearance.

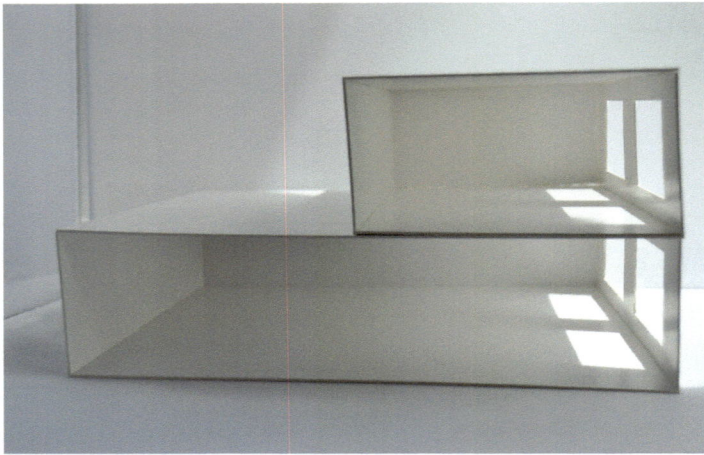

Figure 148   *Shallow and deep plan models from the previous figure exposed to solar radiation*

We can see once again from the above how a simple scale model experiment can help with understanding of how things work in buildings.

13.4 DYNAMIC SIMULATION EXPERIMENTS

In this section, simulation experiments with different room geometries and different standard skies will be conducted in order to increase our understanding of their effects on design.

A simple box model as shown in Figure 149 will be used, with dimensions of 10 m × 6 m × 3 m. Two variations of this model will also be used: a model with double depth, and a model with double depth and double height (Figure 149). The simulations will be conducted in RadianceIES, for the Birmingham location at 52.45° northern latitude, at noon on 21st September, with all three models facing south. Each model will be simulated under two different skies: the standard CIE clear sky, and the 10 k lux standard CIE overcast sky. A comparison between simulation results will help with the understanding of how the changes in room dimensions influence the internal daylit appearance under different sky conditions.

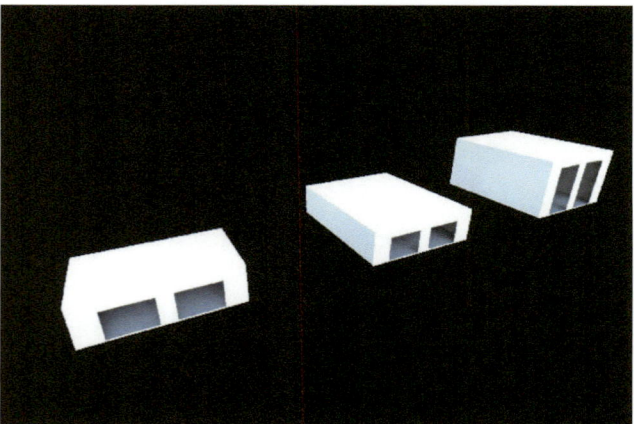

Figure 149   *Variations of the simple box model: shallow plan model (10 m × 6 m × 3 m); double depth model; and double depth double height model*

The simulation results for the internal axonometric views for the three models are shown in Figure 150, and the results for the working plane simulations are shown in Figure 151. Both sets of results are for the standard CIE clear sky.

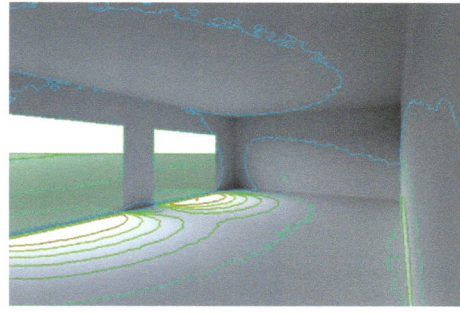
a) in the shallow plan model

b) in the double depth model

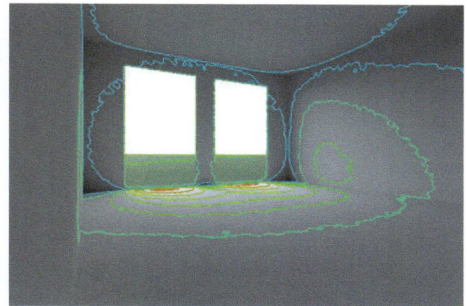
c) in the double depth double height model

Figure 150   Internal illumination under standard CIE clear sky

We can see from Figure 150 that the initial shallow plan model (a) is well daylit, that the double-depth model (b) is poorly day-lit, and that the double-depth double-height model (c) appears to be as well lit as the initial shallow plan model. As this is only a quantitative result, we will now analyse the working plane images (the working plane is a horizontal plane at 0.85 m height from the floor) with a superimposed grid of illuminance values in Figure 151.

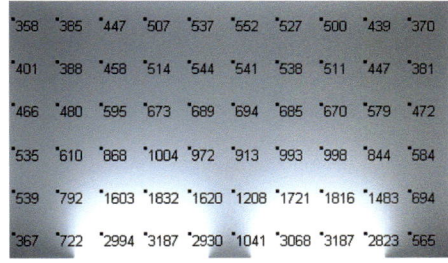

98	93	97	99	113	116	104	102	97	92
100	88	96	96	105	104	103	97	89	90
112	103	103	110	122	125	122	114	105	94
133	130	127	148	150	157	153	154	120	134
163	164	181	205	211	205	198	200	184	158
225	217	238	235	281	276	278	263	231	220
299	277	321	354	377	383	378	354	322	275
375	368	434	484	505	514	505	480	434	368
468	474	594	676	689	687	695	671	590	478
543	610	894	1040	1020	942	1022	1055	895	615
527	744	1645	1962	1812	1250	1778	1960	1660	759
330	581	3072	3314	3160	954	3124	3331	3076	600

535	581	615	647	667	680	671	642	620	582
526	539	567	591	614	615	619	590	563	533
557	557	586	610	637	646	628	605	591	551
601	614	643	680	688	707	688	665	642	601
686	704	745	793	816	826	806	780	739	684
780	812	884	942	958	959	973	936	867	811
884	955	1099	1174	1214	1202	1211	1164	1079	930
994	1119	1343	1488	1524	1477	1495	1485	1323	1086
1073	1306	1665	1874	1827	1761	1847	1854	1610	1220
1073	1409	2067	2336	2218	1974	2258	2316	1988	1306
920	1401	2501	2863	2582	2026	2692	2844	2351	1248
543	1023	3160	3398	3030	1399	3220	3378	2983	816

358	385	447	507	537	552	527	500	439	370
401	388	458	514	544	541	538	511	447	381
466	480	595	673	689	694	685	670	579	472
535	610	868	1004	972	913	993	998	844	584
539	792	1603	1832	1620	1208	1721	1816	1483	694
367	722	2994	3187	2930	1041	3068	3187	2823	565

a) in the shallow plan model

b) in the double depth model

c) in the double depth double height model

Figure 151   Internal working plane illumination in the shallow plan model (a), double depth model (b), and double depth double height model (c) under standard CIE clear sky

This figure shows that the double-depth double-height model (Figure 151c) is better daylit than the other two models, with illuminance values in the far end of the room in the range of 535–680 lux. The double-depth model (Figure 151b) is poorly daylit in the far end of the room, with illuminance values in the range of 92–116 lux. The shallow plan model (Figure 151a) is reasonably well daylit, with illuminance values in the range of 358–552 in the back of the room.

Using the recommendations for internal illuminances from the *CIBSE: SLL Code for Lighting* (Parry, 2022), we can assess whether daylight performance in the depth of the room is sufficient, or whether additional electrical light will be needed. For instance, the recommended illuminance for classrooms and tutorial rooms is 300 lux. If the above models were classrooms, then models (a) and (c) would be predominantly daylit and model (b) would be predominantly artificially lit.

The importance of the zoning of artificial lighting proportional to the distance from the window becomes apparent in all three models when the change of the illuminance values is taken into account. The zoned electrical lighting should be operated using daylight-sensitive control to respond to variations in external lighting and its effect on the internal daylight.

Simulation results for the 10 k lux standard CIE overcast sky are shown in the next set of images. The axonometric views of interiors of the three models are shown in Figure 152, and the working plane views are shown in Figure 153.

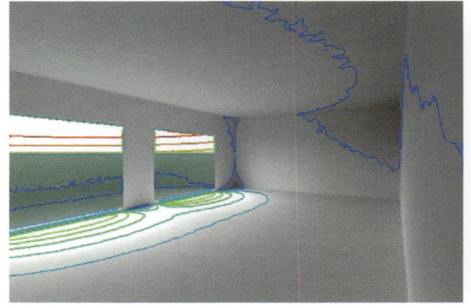 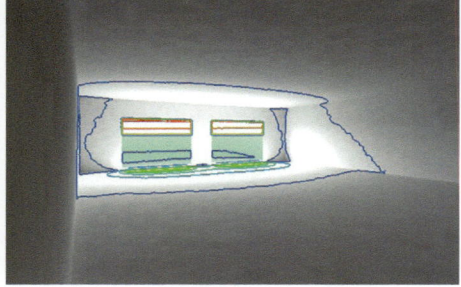 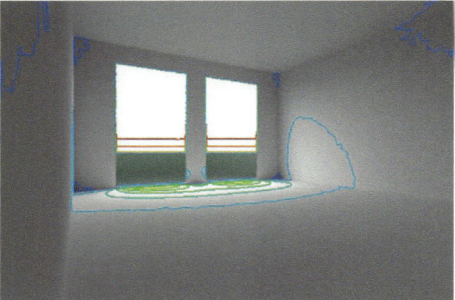

a) in the shallow plan model   b) in the double depth model   c) in the double depth double height model

Figure 152   *Internal illumination under 10 k lux standard CIE overcast sky*

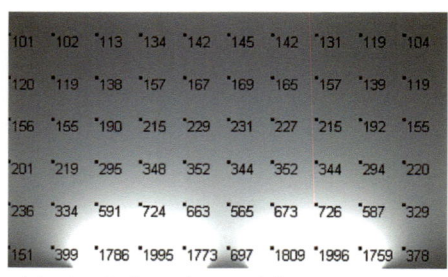 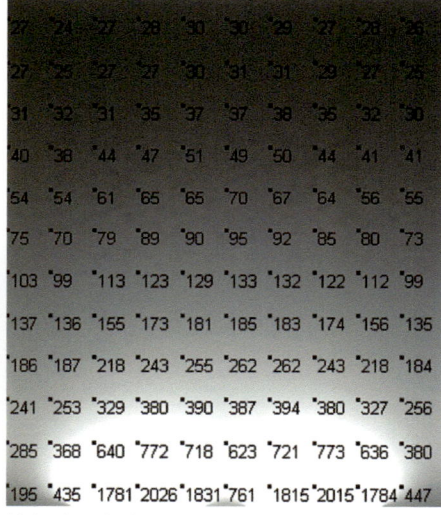 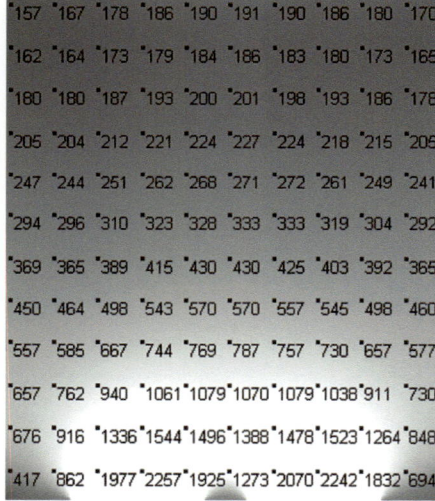

a) in the shallow plan model   b) in the double depth model   c) in the double depth double height model

Figure 153   *Internal working plane illumination under 10 k lux standard CIE overcast sky*

We can see from these two figures that the same pattern of daylighting repeats, so that model (c) is the best and model (b) is the worst, however the illuminance values on the working plane are now lower across all three models. This is clearly the consequence of the reduced illuminance in the overcast sky model.

The shallow plan model now shows the back-end illumination on the working plane between 101 and 145 lux; the double-depth model between 24 and 30 lux; and double-depth double-height model between 157 and 191 lux. The difference in results between the clear sky and overcast sky models on the centre of the back of the room is significant: 407 lux in model (a); 86 lux in model (b); and 489 lux in model (c). According to this second set of results, and assuming that internal illuminance of 300 lux needs to be achieved, all three models will be predominantly artificially lit.

Figure 154   Three comparison models shown in plan with external illuminance values on the horizontal roof

We can also use the values of internal illuminance from Figure 153 and external illuminance from Figure 154 to calculate the daylight factor in different parts of the room. For instance, if we want to know what the daylight factor is in model (c) in the location which shows illuminance of 191 lux (Figure 153c), we need to divide that illuminance by the external illuminance of 3358 lux from Figure 154. We therefore obtain the daylight factor as 191/3358 = 0.056 (or 5.6%). Similarly, the daylight factor in model (a) (Figure 153a) in the location which shows illuminance of 145 lux is 145/3358 = 0.043 (or 4.3%). Daylight factor at the back of model (b) is 30/3358 = 0.009 (or 0.9%).

The first two daylight factors that we have calculated are greater or close to 5%, which is recommended by *The SLL Code for Lighting* from CIBSE (Parry, 2022) for predominantly daylit performance during daytime. The third daylight factor is 0.9%, considerably less than 5%, indicating predominantly artificially lit space. We must remember that this calculation is for a single point and single hour of the year, whilst CIBSE recommendations are for the average daylight factor.

Unlike the absolute illuminance values, daylight factor is a relative measure of daylight performance. This means that absolute illuminance values alone are not the best indicators of daylit performance as they are influenced by the choice of the sky used in the simulation. We therefore always need to cross-check the absolute values of internal illuminances with calculated daylight factors in order to assess daylight performance of the space we are designing.

We can therefore conclude that the choice of the sky model will significantly influence simulation results and that we need to understand differences between different sky models in order to interpret results correctly. We can also conclude that the six metre depth rule for daylit spaces applies to ceiling heights of three metres. Deep plan spaces with ceiling heights of three metres will be poorly daylit, but increasing the ceiling and window height will significantly improve daylight performance.

We need to be conscious that the above analysis corresponds to a single hour in the year, and that it needs to be repeated at various times of the year in order to achieve more general results. We will show in Chapter 14 how this can be achieved using daylight sensor readings from a lighting simulation to operate electrical lights in an energy simulation.

Once again we can see that simulation is the only tool that can provide us with sufficient qualitative and quantitative information in the design process.

13.5 TASKS FOR SIMPLE SIMULATION EXPERIMENTS

• Create simulation models of three simple buildings for a location of your choice:
  • 10 m × 6 m × 3.5 m
  • 10 m × 12 m × 3.5 m
  • 10 m × 12 m × 7 m
• Place 8 m wide × 1 m high window on the south-facing side.

- Investigate daylight illuminance levels on the 0.8 m high work plane for window sill heights of 0, 1, and 2 m on the first two models, and continuing to 4 and 5 m on the third model.
- Repeat this process for a few different orientations of the models, from south to north.
- What can you conclude about different building depths, different ceiling heights, different sill heights, and different orientations? Discuss results with your colleagues.

13.6 SUMMARY OF DESIGN PRINCIPLES

Daylight d esign
- Good daylight design will result in
  - visual comfort and wellbeing of building users
  - energy savings from reduced need for electrical lighting
- Room proportions for good daylight can be initially set using a simple equation.
- Detailed simulation analysis needs to be carried and room proportions adjusted in order to achieve good daylight performance.
- Rooms with ceiling height of about 3 metres will be daylit up to 6 metres of plan depth.
- Rooms with ceiling height of 3 metres and deeper plan than 6 metres will be poorly daylit.
- Increasing room height and window height will improve daylight performance in deeper plan rooms.
- Use of different sky models will influence simulation results.
- Good understanding of the sky models used is essential for correct interpretation of results.
- Simulation results will enable us to obtain illuminance values in different parts of the room.
- We can calculate daylight factor by dividing internal horizontal illuminance obtained from simulation by the simulated external illuminance on a horizontal surface.
- We can use internal illuminance recommendations and absolute illuminance values to assess the quality of daylight design.
- We should cross-check the absolute internal illuminance values with the calculated daylight factors, as the former depends on the choice of sky used in the simulation, in order to assess the quality of daylight design.

CHAPTER 14

ELECTRICAL LIGHTING AND ITS INTEGRATION WITH NATURAL DAYLIGHT

Electrical lighting accounts for a considerable proportion of energy consumption and consequent carbon emissions where electricity is primarily generated using fossil fuels. In this context, it is essential to make electrical lighting as efficient as possible in order to reduce energy consumption and carbon emissions.

We can increase the efficiency of electrical lighting in two ways:

1 By selecting energy-efficient lamps
2 By increasing the control efficiency through
   • Zoning
   • Occupancy control
   • Daylight-sensitive switching
   • Daylight-sensitive dimming
   • Low ambient lighting and task lighting
   • High-frequency control gear

In this chapter, I will explain the above measures and the way they are used in a simulation model.

14.1 ENERGY-EFFICIENT LAMPS

A general overview of different lamp types, efficiencies, and applications can be found in the *CIBSE SLL Code for Lighting* (Parry, 2022). A detailed explanation of various lamp types is beyond the scope of this text, as technical details can be found in relevant manufacturers' catalogues. However, several lamp types are compatible with zero carbon design of buildings due to their energy efficiency.

Lamp efficiency, called luminous efficacy, is expressed in Lumens per Watt or lm/W. Compact fluorescent lamps, which are widely used in homes, are characterised by efficacies of around 55 lm/W. Fluorescent-triphosphor lamps, such as T5, are high-efficacy lamps, with up to 100 lm/W). LED lamps are based on light-emitting diodes. Their efficacy can be greater than 125 lm/W. This means that they are more efficient than the most efficient fluorescent lamps and far more efficient than the most efficient compact fluorescent lamps.

This information needs to be converted into the lighting power density, expressed in $W/m^2/(100\ lux)$, in order to be taken into account as a contribution towards heat gains in a simulation model. A simple method to estimate the lighting power density is explained in Chapter 15.

14.2 LIGHTING CONTROL

14.2.1 Zoning

Appropriate zoning that takes daylight into account connects all lamps that are at daylight distance from windows to one electrical circuit controlled by a single switch (Figure 155). We can set this in the simulation model by dividing larger spaces into sub-spaces and assigning individual lighting to these spaces. The space division in the simulation model needs to be made without partitions so that the zoning does not prevent heat and air movement between zones.

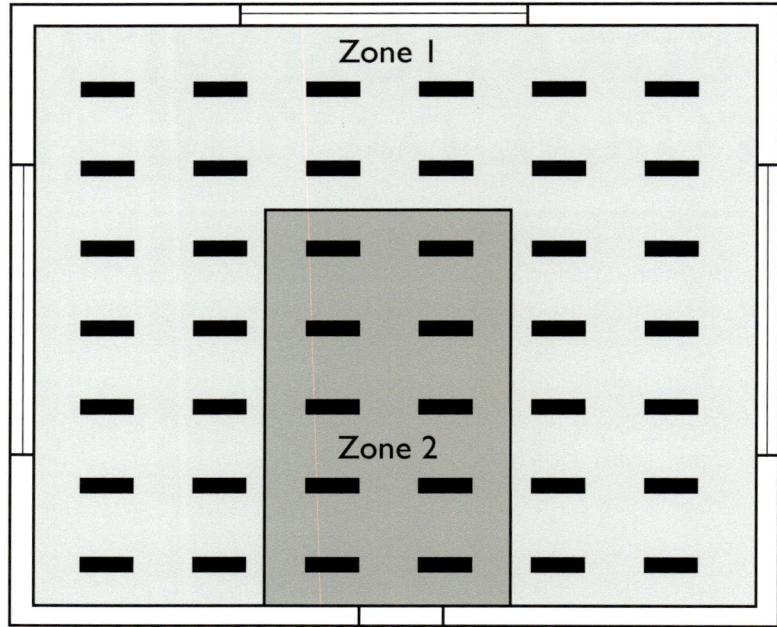

Figure 155   Zoning of electrical lights that takes into account distance from windows

14.2.2 Occupancy control

This is set by controlling the lights using the corresponding occupancy profile in the heat gains from lighting, replacing daylight controls in the variation profile and dimming profile in Figure 161 with a suitable occupancy profile, such as the example profile in Figure 156.

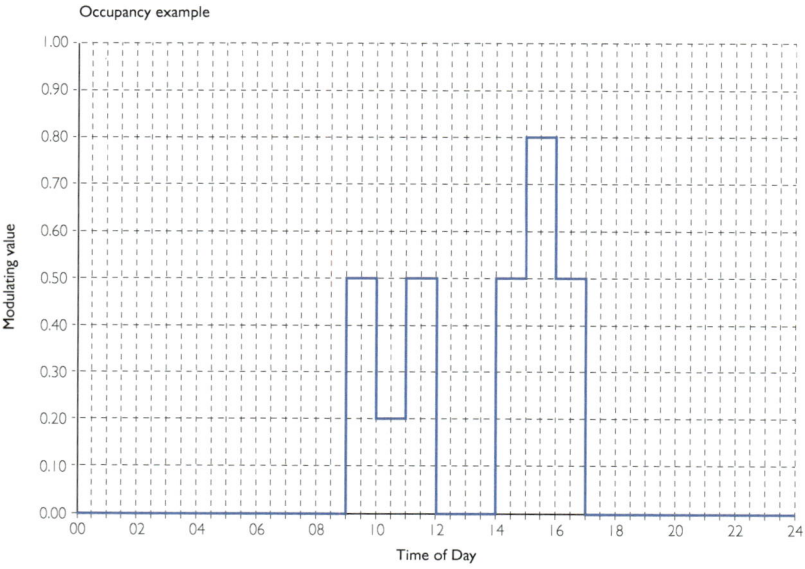

Figure 156   Example profile for occupancy-controlled lighting

14.2.3 Daylight-sensitive switching

Daylight-sensitive switching assumes that each lamp is controlled independently. For this type of switching of the electrical lights, we need to obtain internal daylight levels and switch on electrical lights below a certain level of internal daylight. In IES Virtual Environment, we can do that through cross-referencing between the daylight simulation module and the thermal simulation module. In the daylight simulation module, called RadianceIES, we first set daylight sensors in individual spaces we wish to control (Figure 157).

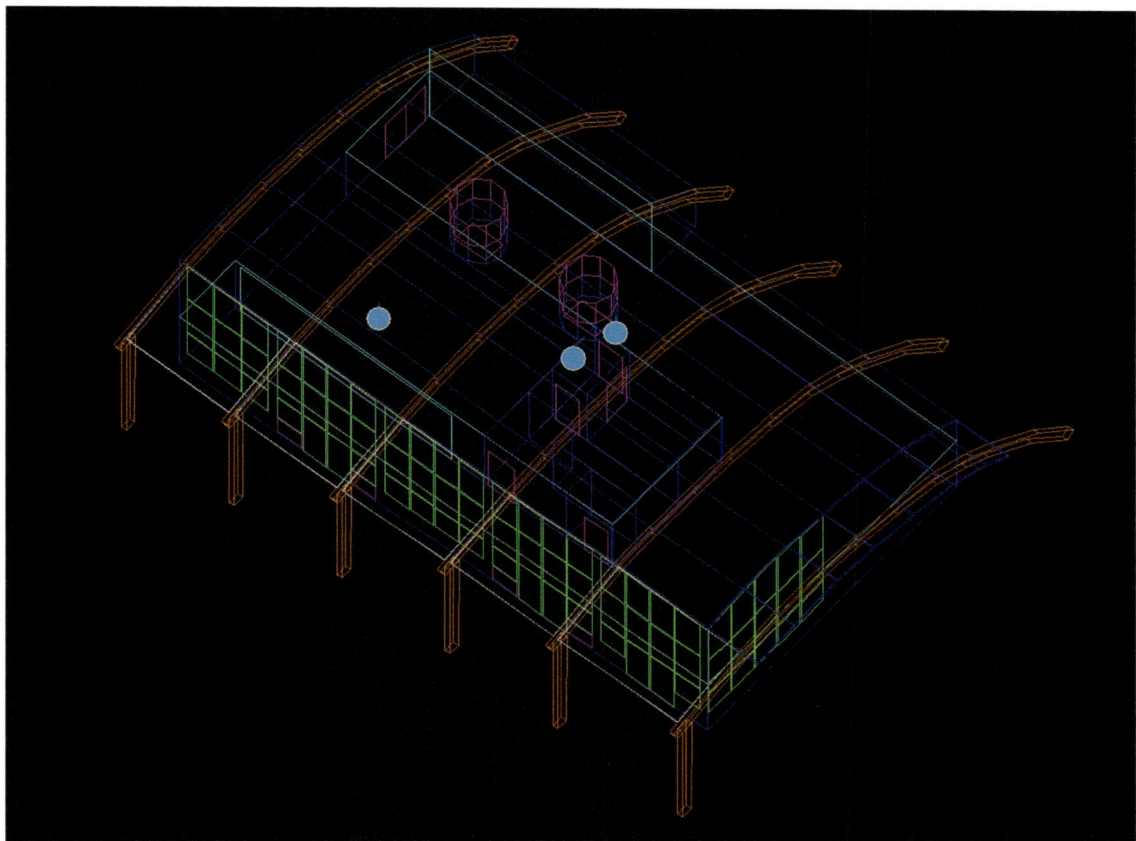

Figure 157   Daylight sensors shown as blue circles in an IES VE model

We then run an annual daylighting simulation that creates sensor readings for the thermal simulation (Figure 158).

Having done this, we then go into the thermal simulation module called Apache and create an operating profile that uses information from the lighting sensors to switch the lights on below a certain level of daylight. Figure 159 shows an example of a modulating profile that operates electrical lights to the maximum illuminance and corresponding power density when the daylight sensor reading is less than 300 lux.

14.2.4 Daylight-sensitive dimming

A daylight-sensitive dimming is effectively 'daylight following'. Electrical lights are modulated from 0 to 1 of the maximum illuminance in a way that complements the level of daylight read by a daylight sensor. For instance, if the maximum illuminance of electrical lights is ELmax = 300 lux, then electrical lights will be off when the daylight sensor gives a reading of above DL = 300 lux. Below the daylight level of 300 lux, the dimming profile sets the electrical lights so that the total of the daylight illuminance and the electrical light illuminance gives the required illuminance, e.g. DL + EL = 300 lux or EL = 300 − DL. The installed power density of electrical lights is modulated according to this dimming profile. A pre-requisite for this type of operation is the same as in the daylight-sensitive switching above: we need to set daylight sensors, run daylight simulation, and use the

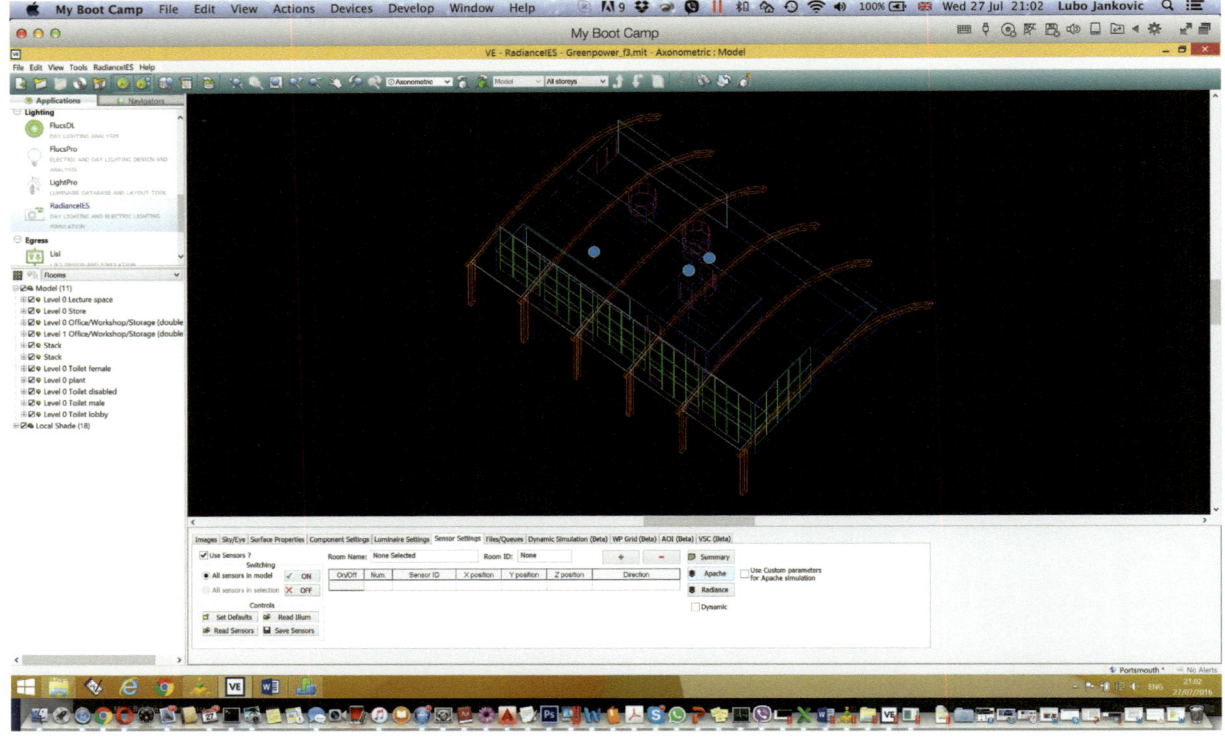

Figure 158   Apache button generates daylight sensor readings for use in thermal simulation

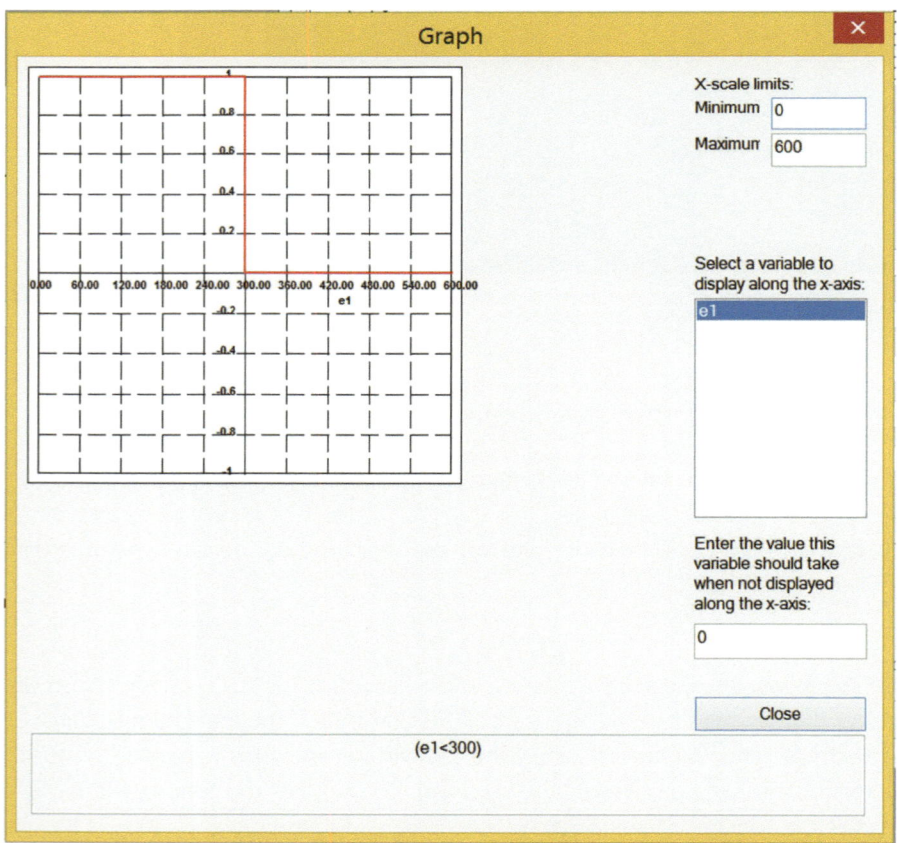

Figure 159   Daylight-sensitive switching profile

results in the thermal simulation. An example of this type of dimming profile is shown in Figure 160 where electrical lights are dimmed gradually from full intensity to zero intensity between 0 and 300 lux of the daylight sensor reading.

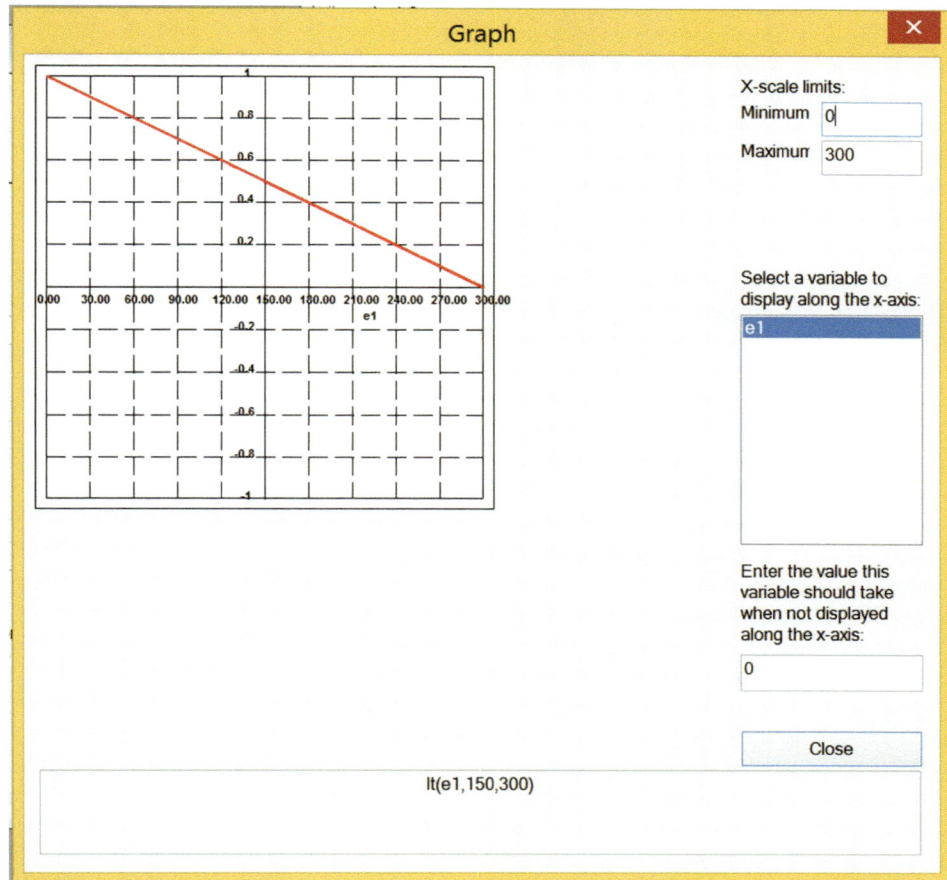

Figure 160   Daylight-sensitive dimming profile

In order to instruct the simulation software to conduct daylight-sensitive switching and dimming, we need to implement these control profiles in the section that sets internal gains from lighting for individual spaces (Figure 161).

14.2.5 Low ambient lighting and task lighting
In large internal spaces, such as open-plan offices, ambient lighting is often fixed at the illuminance level required by this type of space, delivering between 300 and 500 lux at the working plane. If we however set ambient lighting to a much lower level, suitable for circulation only, and provide desk lamps that top up this level to between 300 and 500 lux, this can lead to significant savings.

Ambient lighting in a thermal simulation model is set as part of internal heat gains from lighting. We first select the illuminance level and power density in the internal gains from the lighting section, and set daylight-sensitive controls in the variation and dimming profile (Figure 161). Task lighting is set as additional heat gain from lighting, but this time it is controlled using a suitable occupancy profile, for instance as the profile shown in Figure 156.

To ensure that daylight sensors are taken into account and that electrical lighting controls are operating in response to daylight in the thermal simulation, we must tick the box next to 'Radiance link' in the simulation setting (Figure 162) before we run the thermal simulation.

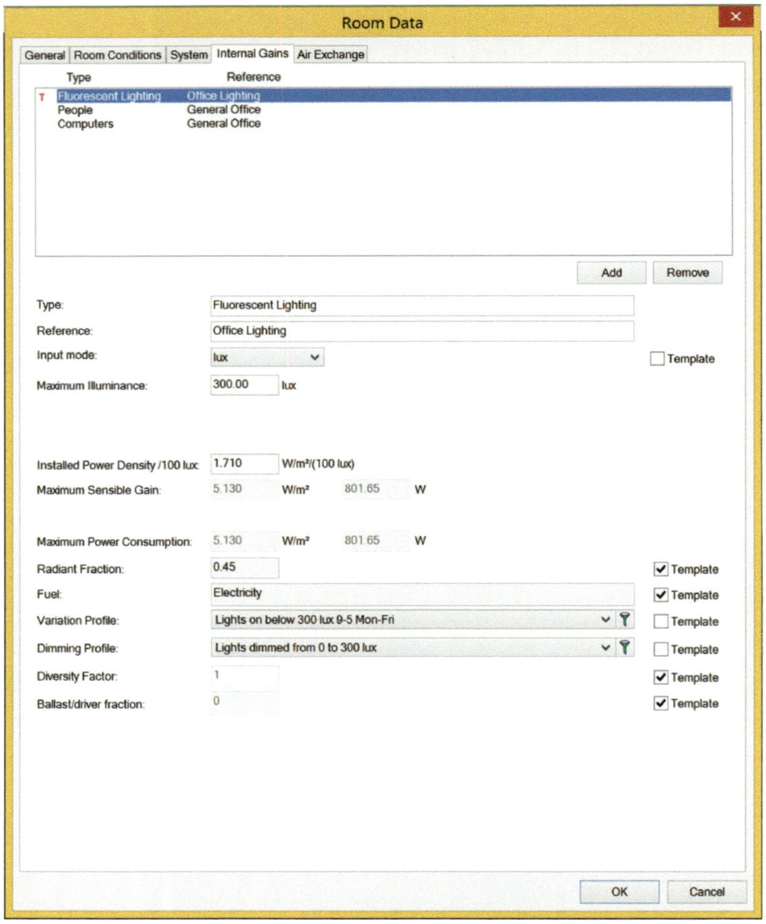

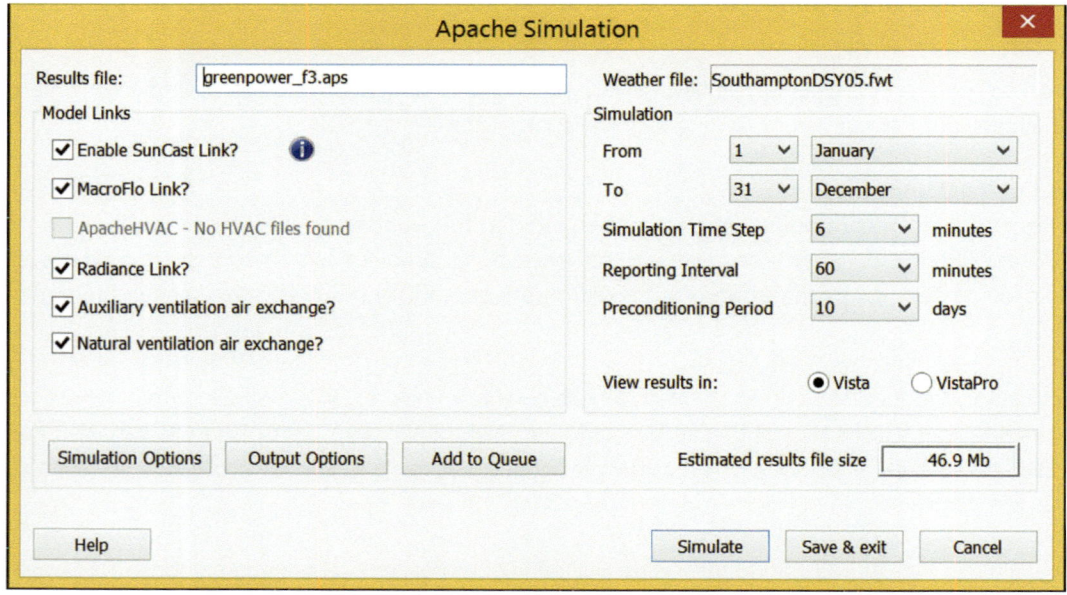

Figure 161   Heat gains from lighting

Figure 162   Simulation setting with Radiance link checked

14.2.6 Power factor and high-frequency control gear

In an alternating current (AC) system, the current lags behind the voltage by a fraction of the full cycle. As the full cycle is expressed as an angle of 360°, the lag between the voltage and the current is also an angle. The greater the lag angle, the more energy needs to be supplied to an appliance, a lamp, a system, or an entire grid in order to deliver the required output. The cosine of this lag angle, $\cos(\Phi)$, is called the power factor.

The power factor can be expressed as

$$\cos(\Phi) = \frac{P}{V \times A}$$

(40)

where

P – power (W)
V – voltage (V)
A – current (A)
Φ – lag angle

If, for instance, the lag angle is 60°, and considering that $\cos(60°) = 0.5$, the amount of energy and carbon emissions will need to be doubled in order to provide the required output, in comparison with a situation in which the power factor is equal to 1.

The lag between the voltage and the current will reduce as the AC frequency increases. In the example in Figure 163b, the frequency is three times higher than the frequency in Figure 163a, and the lag between voltage and current is three times shorter. A considerable increase in frequency will therefore make the power factor approach its maximum value of 1, resulting in increased energy efficiency.

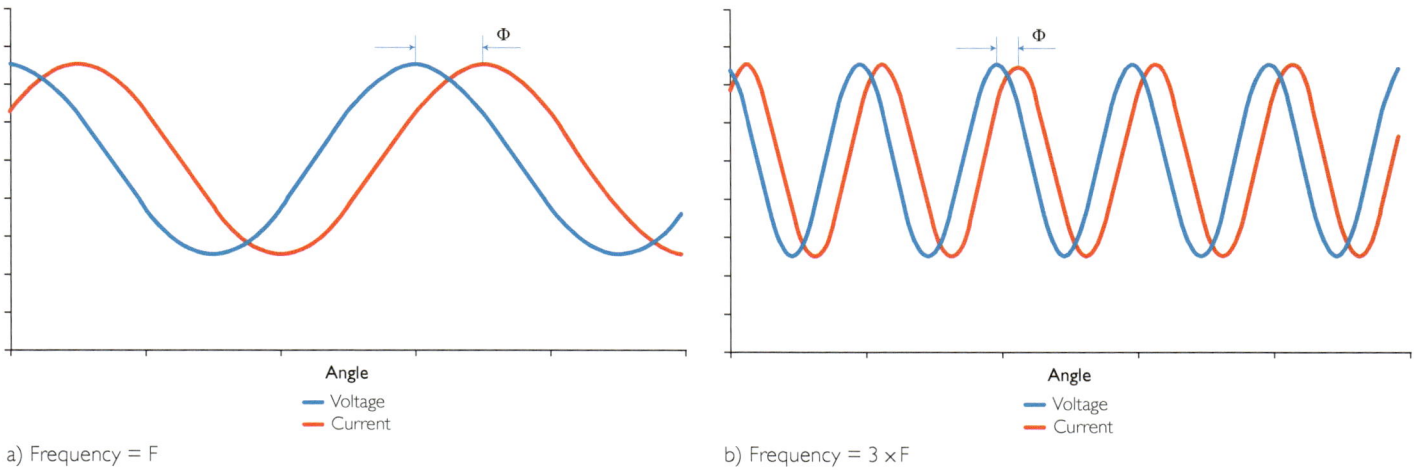

a) Frequency = F

b) Frequency = 3 × F

Figure 163   Power factor as a function of frequency

This is the basis of the high-frequency control gear in fluorescent lighting systems. The frequency of alternating current is 50 Hz, but high-frequency control electronics increases the frequency up to 50 kHz, or up to 1000 times, resulting in the power factor approaching unity. The benefits of high-frequency control are: up to 20% energy savings; up to 15% higher luminous efficacy resulting from higher ionisation of the gas in the lamp; longer lamp life; and virtually eliminated flicker and humming noise of fluorescent lamps, resulting in better internal conditions for building users and lower incidence of sick building syndrome.

Electricity suppliers charge customers for delivered Watts, but they actually supply Volts and Amperes; hence, they aim to maintain the power factor at a minimum of 0.9. However, higher power factors resulting from better controls can be specified in some simulation models. It is advisable to check whether the simulation software is capable of dealing with different values of power factors, and use the power factor that corresponds to a specified control system accordingly.

14.3 TASKS FOR SIMPLE SIMULATION EXPERIMENTS

1   Create a simple box model 10 m × 6 m × 3 m in IES VE. Insert two windows 3 m × 2 m on the south-facing side. Use default construction types and thermal templates. Insert a sensor in Radiance. Run Radiance to provide sensor readings for Apache. Create a daylight-sensitive switching and a daylight-sensitive dimming profiles. Investigate the reduction of CO_2 emissions resulting from the application of the first profile and of both profiles together. Use a weather data file for a location of your choice.

2   Export the above model from IES VE using File > Export > Green Building XML (gbXML). Import this exported model into DesignBuilder using File > Import > Import BIM Model. Check for model consistency between the two simulation tools and make any adjustments as necessary. Apply a lighting template of the same lighting power density in W/m²/100 lux as in the IES VE model. Run the annual simulation using uncontrolled and controlled lighting template (Linear/off lighting control for daylight following) and compare the resultant CO_2 emissions.

14.4 SUMMARY OF DESIGN PRINCIPLES

Design principles

• Divide internal spaces in a simulation model into lighting zones, depending on the distance from windows, so that the zones close to the windows are controlled separately from the zones further away (6 metres or more) from the windows.
• Ensure that the lighting zones are not divided by solid partitions so that heat and air can travel between them in a simulation model.
• Run daylight simulation in Radiance or other suitable software, creating daylight sensors in spaces where lighting is to be controlled in response to daylight.
• Make daylight sensor readings available to the thermal simulation module.
• Create occupancy profiles for lighting control.
• Create daylight-sensitive switching profiles.
• Create daylight-sensitive dimming profiles.
• Select energy-efficient lamps and use their technical data to set heat gains from lighting using recommended illuminance and resultant power density.
• Apply occupancy control, daylight-sensitive switching and daylight-sensitive dimming in the simulation model in order to reduce energy consumption and carbon emissions from lighting.
• Apply high-frequency control gear for further reductions of energy consumption and carbon emissions from lighting.

CHAPTER 15

INTERNAL HEAT GAINS, HEATING, AND COOLING

15.1 BUILDING HEAT BALANCE

Internal temperature in a building is a consequence of a natural tendency to establish equilibrium or balance between heat gains and heat losses. Heat gains and heat losses will occur simoultaneously as a result of dynamic changes in external and internal influences and a time-delayed response resulting from the building's thermal mass. These discrepancies will be eliminated and the equilibrium re-established through this natural tendency, resulting in consequent temperature changes in the building.

This is a continuous process that occurs as a result of the dynamic thermal behaviour of buildings, and results in continuous changes in internal temperatures. If heat gains are temporarily higher than losses, internal temperature will rise, causing higher heat losses and re-establishing the heat balance. Inversely, if heat gains are temporarily lower than heat losses, internal temperature will fall, causing lower heat losses and re-establishing the heat balance.

Building heat balance is influenced by various heat gains and kisses as illustrated in Figure 164.

We can represent the building heat balance in a very simplified form as

$$Q_{loss} = Q_{solar} + Q_{internal} \pm Q_{auxiliary} \qquad (41)$$

where

Q_{loss} – heat losses
Q_{solar} – external heat gains from solar energy
$Q_{internal}$ – internal heat gains
$Q_{auxiliary}$ – auxiliary heating (positive) or cooling (negative).

External heat gains from solar energy, entering the building by means of direct transmission through glazing or indirect transmission through walls and roofs, will make a contribution towards changing the heat balance in a building and towards elevating internal operative temperature[1] above the external air temperature.

Internal heat gains from lighting, appliances, and people will further contribute towards changing the heat balance and the resultant operative temperature will be elevated further.

If the operative temperature, achieved through these two sources of heating (solar and internal gains) and the consequent change in the heat balance, is still below the required temperature, additional auxiliary heating either from conventional or renewable sources will be needed to elevate the operative temperature further. If however, the operative temperature is above the required comfort conditions, additional auxiliary cooling will be needed to reduce the operative temperature to a required level.

Figure 164   Heat balance diagram (illustration by Holly Doron)

Heat losses were dealt with in Chapter 8 and solar gains in Chapter 9. In this chapter, internal heat gains and how to specify them in a dynamic simulation will be explained.

A split between external heat gains, internal heat gains and auxiliary heating varies between different building types and corresponding energy efficiencies. An example of this split in six passive solar houses of similar type is shown in Figure 165. The variations between different houses, which were of similar type were due to different occupancy patterns and use. There will be a higher proportion of auxiliary heating in conventional buildings and a lower proportion of auxiliary heating in low or zero carbon buildings.

15.2 INTERNAL HEAT GAINS

Internal heat gains can be divided into three categories:

• Heat gains from electrical lighting
• Heat gains from people
• Miscellaneous heat gains from appliances

15.2.1 Heat gains from electrical lighting
Different levels of internal lighting are recommended for different space uses. A list of detailed recommendations can be found in the The SLL Code for Lighting from CIBSE (Parry, 2022), denoted as 'maintained illuminance' and expressed in units of lux. Lamp output on energy labels is normally specified as luminance in units of lumen, together with power rating in units of Watt.

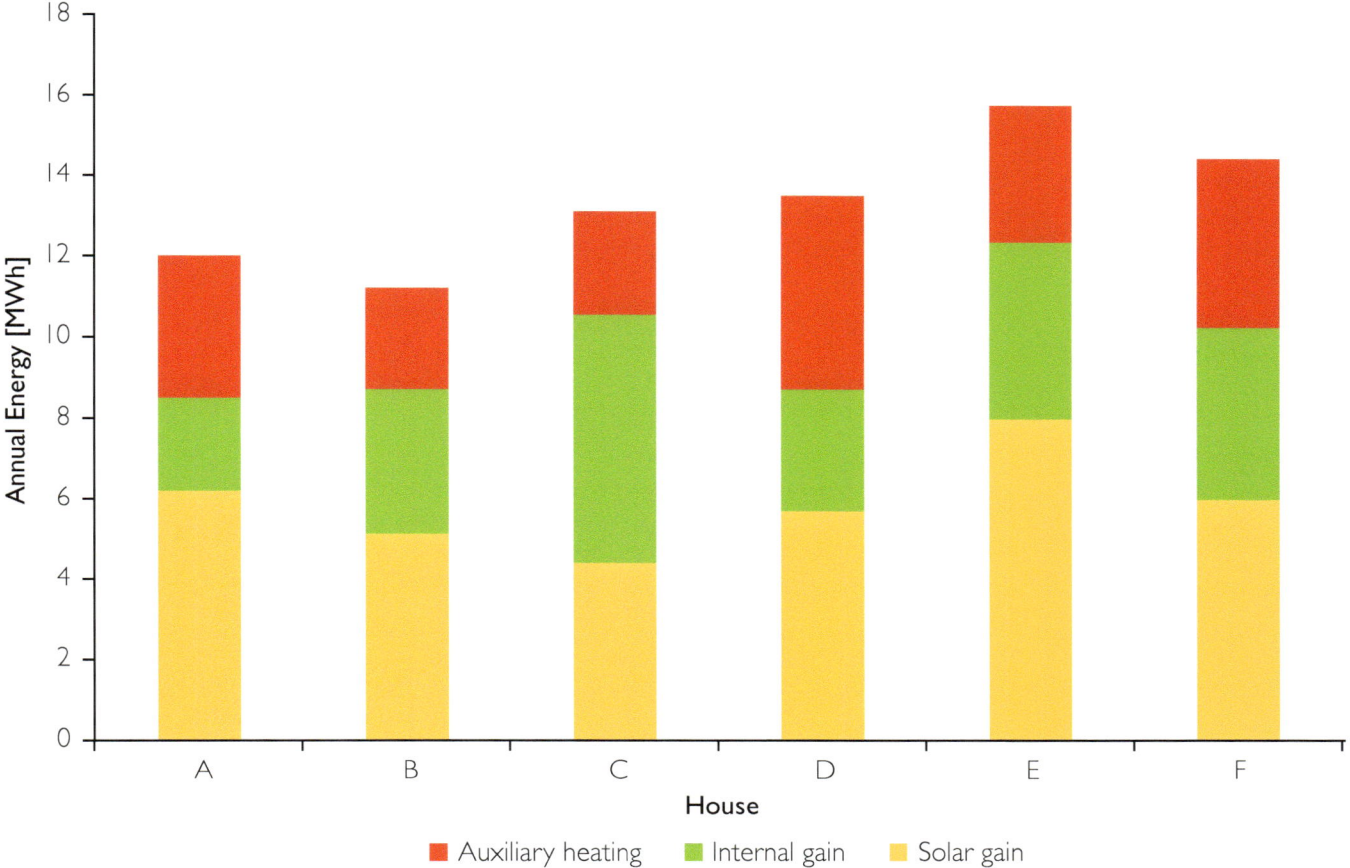

Figure 165 *Variation of solar gains, internal gains, and auxiliary heating in six passive solar houses of similar type*

As already defined in Chapter 13, illuminance can be expressed as

$$\text{illuminance} = \frac{\text{luminance}}{\text{surface area}} \tag{42}$$

As a very rough approximation, which does not take room geometry into account, the total room illuminance can be expressed as

$$\text{illuminance} = \frac{\text{number of lamps} \times \text{lamp luminance}}{\text{surface area}} \tag{43}$$

This will now help with deriving the lighting power density from illuminance in order to enable the simulation model to calculate heat gains from lighting.

Example. The lighting recommendation for tutorial rooms in educational buildings is 300 lux. Let us assume that the room size is 20 m² of floor area and that a compact fluorescent lamp rated at 600 Lumen and 11 Watts is used for electrical lighting. From Equation (43), we can derive

$$\text{number of lamps} = \frac{\text{illuminance} \times \text{surface area}}{\text{lamp luminance}} \tag{44}$$

and by substituting the corresponding values, we get

$$\text{number of lamps} = \frac{300 \times 20}{600} = 10$$

As each lamp is rated at 11 Watts, we get the total energy used for lighting as $10 \times 11 = 110$ Watts.

Installed power density is then obtained by dividing the total power by the floor area. In this particular example, the power density is $110W/20m^2 = 5.5$ W/m². Power density is often expressed in terms of W/m²/100 lux. As illuminance in our example is 300 lux, power density expressed in this way is $110W/20m^2/(3 \times 100lux) = 1.83$ W/m²/(100 lux). The simulation model will use this information on the installed power density to calculate the internal heat gain from electrical lighting.

Improving the energy efficiency of lighting will result in lower installed power density and lower internal heat gains from electrical lighting. Although there will be a resultant increase in annual heating energy requirement due to more auxiliary heating that will be needed to compensate for the reduced heat gains from lighting, this will be by far outweighed by savings in electricity energy and consequent carbon emission reductions.

15.2.2 Heat gains from people
As a result of the internal metabolic rate in the human body, building occupants will be the source of sensible and latent heat gains. The metabolic rate will depend on the activity (see Table 41 in Chapter 17). For a mixture of sedentary, standing, and walking activities in an office building, for example, sensible heat gain is typically 90 W/person and latent heat gain 60 W/person. After the occupancy density is specified in terms of the number of people per square metre of the floor area, the heat gain from people for a particular space will be the product of the occupancy density and the heat gain from each person arising from the metabolic rate, as follows:

$$Q_s = O_d \times M_{s,p} \tag{45}$$

$$Q_l = O_d \times M_{l,p} \tag{46}$$

where

Q_s – Power density of sensible heat gain from people (W/m²)
$M_{s,p}$ – Sensible heat gain per person arising from the metabolic rate (W/person)
Q_l – Power density of latent heat gain from people (W/m²)
$M_{l,p}$ – Latent heat gain per person arising from the metabolic rate (W/person)
O_d – Occupancy density (people/m²)

15.2.3 Miscellaneous heat gains
This category of heat gains can be further sub-divided into heat gains from machines, computers, electrical equipment, cooking, and other sources. We need to clearly differentiate between heat gains that only have sensible heat components (such as gains from computers), and heat gains with both sensible and latent components (such as gains from food preparation).

Subsequently, we need to quantify the heat gains and specify them either as absolute gains in terms of power (W), or as relative gains in terms of power per square metre of floor area, also called power density (W/m²). We can do this by obtaining the equipment specification, its rating and number of pieces, and expressing it as the total power, or dividing it by the floor area to obtain power density.

15.3 HEATING

Auxiliary heating is calculated as a residual from the heat balance equation as follows:

$$Q_{h, \text{auxiliary}} = Q_{\text{loss}} - Q_{\text{solar}} - Q_{\text{internal}} \tag{47}$$

It is advisable to give unlimited capacity to the heating system in the simulation model and subsequently derive heating plant size from analysis of hourly boiler loads generated by the model.

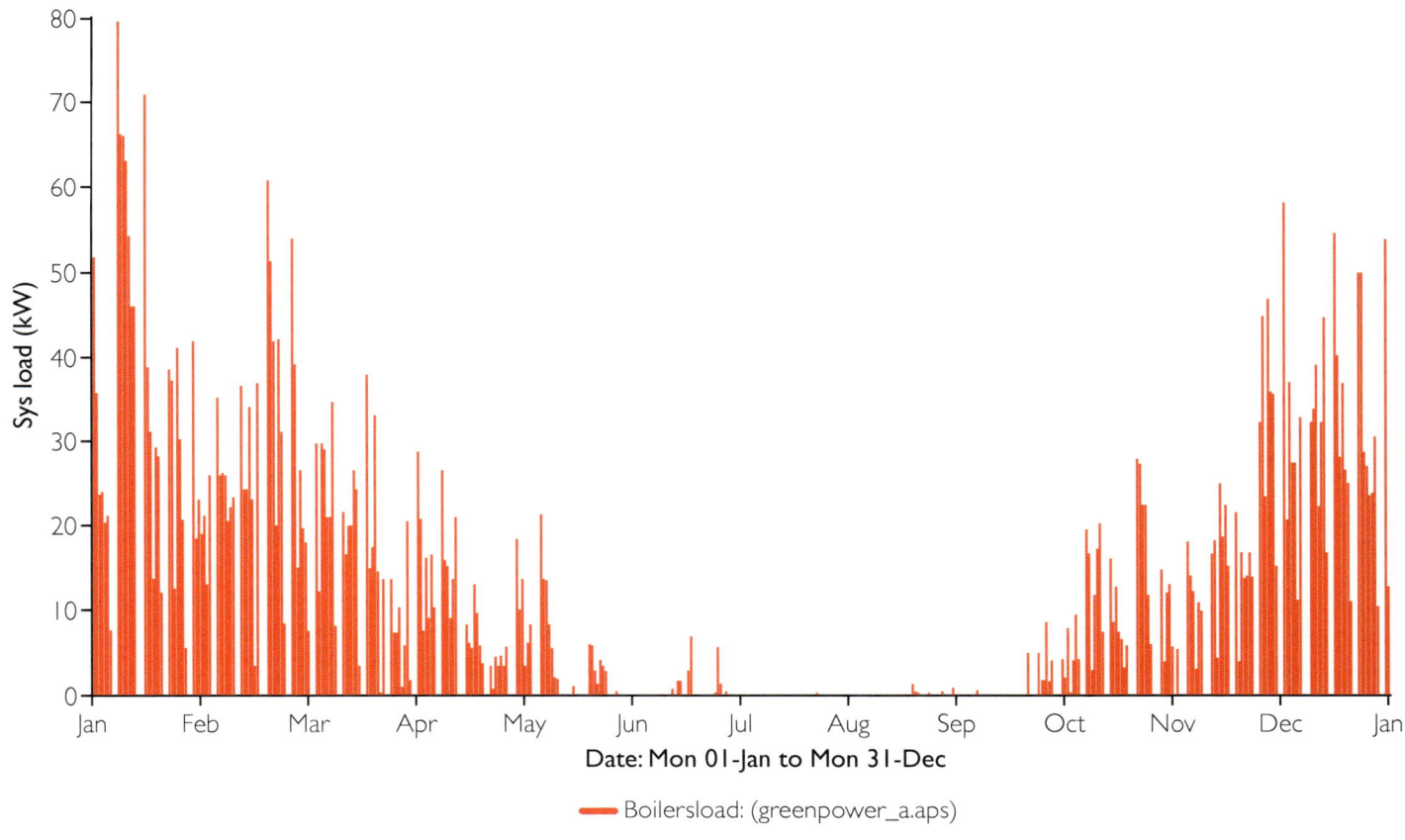

Figure 166   Boilers load in Model 'a', Figure 197

This can be explained using the results of simulations of Model 'a' from Figure 197 in Chapter 18. Figure 166 shows hourly boiler load and a peak load of 80 kW. As the peak load occurs infrequently, specifying the heating system on this basis may lead to considerable over-sizing. To get a better understanding of the distribution of boiler loads, a frequency of occurrence analysis of hourly values of the load was carried out. The results are shown in Table 38, from where we can see that the peak load occurs only during one hour of the year, representing 0.01% of the total time. It would therefore be more meaningful to choose the plant size on the basis of the majority of hourly heating loads rather than on the basis of the peak load.

TABLE 38	FREQUENCY OF OCCURRENCE ANALYSIS OF BOILER LOADS		
Heating plant size (kW)	**Absolute frequency**	**Percentage frequency**	**Cumulative percentage frequency**
5	8,065	92.07%	92.07%
10	252	2.88%	94.94%
15	183	2.09%	97.03%
20	101	1.15%	98.18%
25	52	0.59%	98.78%
30	42	0.48%	99.26%
35	21	0.24%	99.50%
40	17	0.19%	99.69%

TABLE 38 (CONTINUED)

45	11	0.13%	99.82%
50	4	0.05%	99.86%
55	6	0.07%	99.93%
60	1	0.01%	99.94%
65	2	0.02%	99.97%
70	1	0.01%	99.98%
75	1	0.01%	99.99%
80	1	0.01%	100.00%
Total	8,760	100.00%	

If we for instance decide to cover 97% of all hourly boiler loads, the corresponding heating plant size obtained from Table 38 will be 15 kW rather than 80 kW. This size then needs to be adjusted further to take into account heating up times, as recommended by the CIBSE Guide B1 (Palmer, 2016). The boiler choice made on this basis will result in considerably lower capital cost and higher operational efficiency.

The same method can be used to determine the cooling plant size in cases where cooling is applicable.

15.4 COOLING

Auxiliary cooling is calculated in a similar way as auxiliary heating, as a residual between heat gains and losses, but with a different sign:

$$Q_{c, \text{auxiliary}} = Q_{\text{solar}} + Q_{\text{internal}} - Q_{\text{loss}} \tag{48}$$

Methods for comfort cooling will now be described in the next three sections.

15.4.1 Compression refrigeration cooling

There are various ways in which auxiliary cooling can be provided in a building. The highest carbon emissions will be generated from a compression refrigeration cycle shown in Figure 167. The compressor, driven by electricity, drives the refrigeration cycle. A low-pressure liquid in the heat exchanger called the evaporator changes phase from liquid to gas as it draws latent heat of evaporation from the surroundings. The resultant low-pressure gas is drawn from the evaporator into the compressor where its pressure is increased. As this compressed high-pressure gas comes into a heat rejection heat exchanger called the condenser, it is cooled down and it becomes high-pressure liquid. The liquid then travels towards the expansion valve, where its pressure is considerably reduced due to the small size of the orifice in the valve. Past this valve, the liquid is at low pressure as it enters the evaporator, where it evaporates into gas and the process repeats.

The compression refrigeration cycle is commonly used in domestic fridges. The evaporator is the cold surface inside the fridge, and the condenser is the warm radiator at the back of the fridge.

If we place the equivalent of the 'inside of the fridge' in the building air and the equivalent of the 'back of the fridge' outside, then we have a comfort cooling system. If we, however, bury the equivalent of the 'inside of the fridge' in the ground and place the equivalent of the 'back of the fridge' in the building, then we have a heat pump.

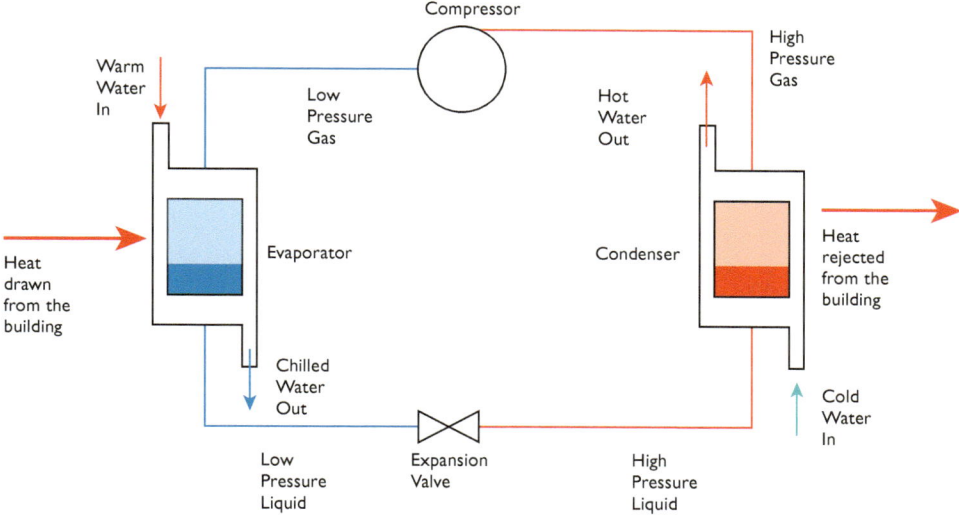

Figure 167   Compression refrigeration cycle

15.4.2 Absorption cooling

There are also so-called 'absorption cooling machines' that have much lower carbon emissions than the compression refrigeration machines. Their refrigeration cycle is driven by heat. A gas fridge is a consumer application of this machine. A diagram of the absorption refrigeration cycle is shown in Figure 168. Machines based on the absorption refrigeration cycle are virtually maintenance-free with an operating life of more than 20 years. They have very low electricity consumption as the refrigeration cycle is driven by heat. Although there are several types of absorption cooling cycles, the cycle based on water and ammonia will be described here for illustration.

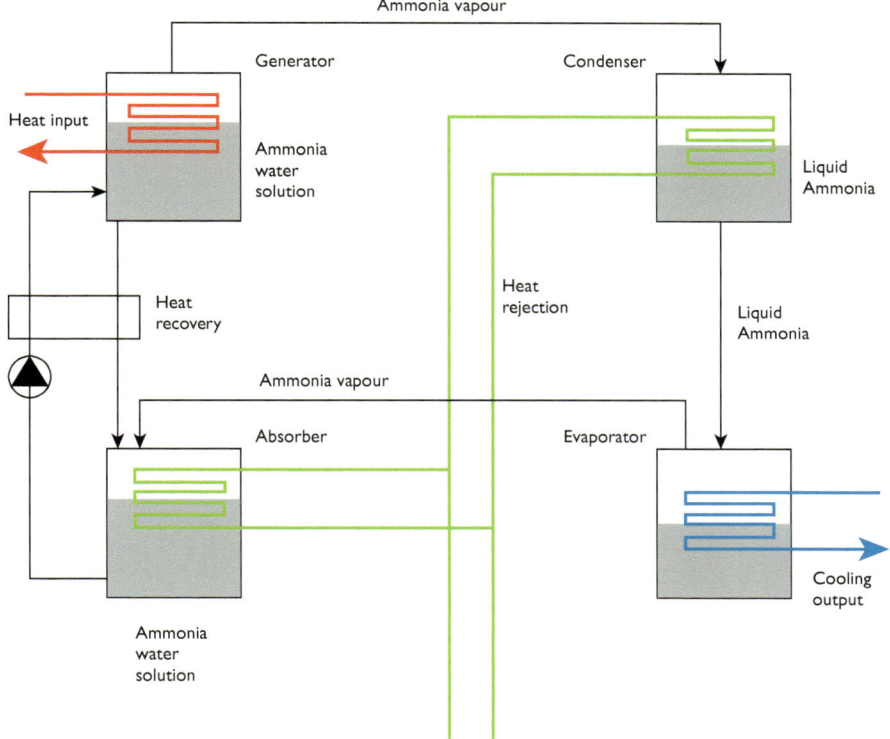

Figure 168   Absorption refrigeration cycle

In this cycle, heat input is supplied into the generator, causing ammonia to evaporate from the ammonia-water solution. Subsequently, ammonia vapour loses its heat in the condenser. Thus liquid ammonia is gathered inside the condenser and from there it goes into the evaporator. In the evaporator, liquid ammonia absorbs heat from the surroundings and changes into vapour. This enables cooling output to be drawn from the evaporator. The ammonia vapour then goes into the absorber where it is absorbed by the ammonia-water solution. In this process, the condenser and the absorber produce surplus heat which is removed by the heat rejection loop. In this way, the cooling output from the evaporator is produced as a result of heat input into the generator.

The absorption refrigeration machines are used where there is surplus heat and a simultaneous requirement for cooling. They are often used together with Combined Heat and Power systems, as explained in Chapter 16, to form a tri-generation system that provides electricity, heating, and cooling.

15.4.3 Evaporative cooling

Evaporative cooling works on the basis of latent heat of water evaporation. There are two types of evaporative coolers: direct (Figure 169a) and indirect (Figure 169b). As part of the evaporative cooling process, water is circulated across evaporative pads which magnify its surface area. As air is blown from outside to inside through the evaporative zone across the evaporative pads, the water draws its latent heat of evaporation from the air. In the case of the direct system, the air is mixed with evaporating water and its humidity increases as it enters the building. In the case of the indirect system, the air is ducted through a heat exchanger within the evaporative zone and its humidity does not change.

Evaporative cooling works well in hot and dry climates. Its efficiency ratio increases with temperature and decreases with relative humidity.

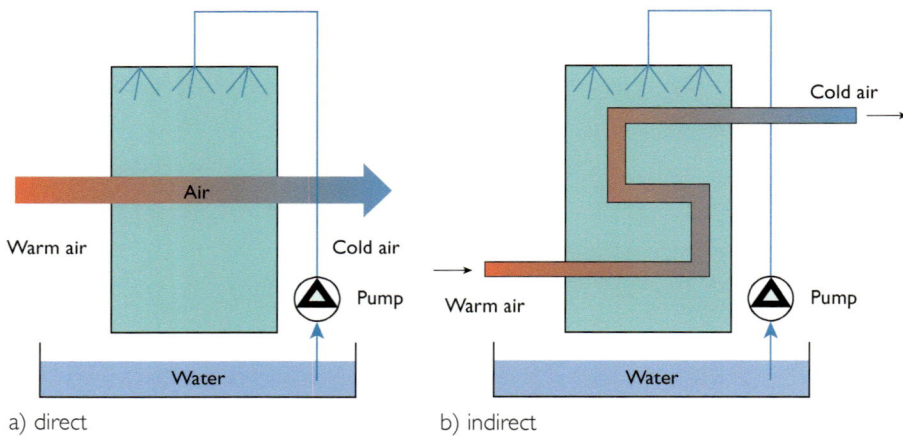

a) direct   b) indirect

Figure 169   Evaporative cooling systems

15.4.4 COP, SCOP, EER, and SEER

When specifying heating or cooling systems in a dynamic simulation model, information on their efficiency needs to be supplied. Efficiencies of heat pumps and chillers will however differ from efficiencies of combustion heating devices. A combustion heating device, such as a biomass boiler, will have the efficiency of less than one expressed as follows:

$$\text{Efficiency} = Q_{\text{heat output}} / Q_{\text{fuel input}} \qquad (49)$$

The concept of efficiency in the conventional sense does not apply in the case of the compression refrigeration cycle described above. In the case of a heat pump, the heat output will be greater than fuel input, as the output corresponds to the amount of heat elevated from a heat source at a lower temperature, such as from deep ground, to a higher temperature, usable for heating in a building. This process of 'pumping' heat from a lower to a higher temperature is driven by external electrical energy input, however, the output is typically several times greater than the input. As efficiency cannot be greater than one by definition, a different concept of a Coefficient Of Performance or COP has been defined for heat pumps as

$$COP = Q_{heat\ output} / Q_{electrical\ input} \qquad (50)$$

In the case of a chiller, the performance is expressed as Energy Efficiency Ratio or EER

$$EER = Q_{heat\ output} / Q_{electrical\ input} \qquad (51)$$

On the basis of the above equations, one can argue that the COP and the EER are identical, however it is not as simple as that. Firstly, the name difference helps differentiate between heating and cooling. Secondly, heat output in EER is sometimes expressed in Btu/h and the electrical input in Watts. In those cases EER is not dimensionless and a conversion between COP and EER must take into account the unit conversion between Btu/h and Watt, as follows

$$EER = 3.413 \times COP \qquad (52)$$

The output of the compression refrigeration cycle, whether it is used for heating or cooling, will change with temperature difference between the evaporator and the condenser. The lower that temperature difference, the lower the output. For this reason a Seasonal EER or SEER has been introduced, to take into account part-load performance throughout the year.

COP, EER, and SEER will be provided in technical specifications of heating and cooling generators. The reader is advised to check how the EER and SEER have been expressed in relation to the cooling equipment they are intending to use in a simulation and how these parameters are defined in the simulation software they are using, as a mismatch between the two can lead to inaccuracies in simulation results.

There are other parameters that are used for defining system efficiencies. A SCOP is a Seasonal System Coefficient of Performance (UK DLUHC, 2021) which is inclusive of the heating generator COP and heat losses from ducts and pipes. A SSEER is Seasonal System Energy Efficiency Ratio (UK DLUHC, 2021) and it includes the cooling generator SEER as well as sensible and latent heat gains in the system. Needless to say that correct assumptions about the system need to be made in order to minimise inaccuracies in the simulation results.

15.5 HEATING AND COOLING WITH RENEWABLE ENERGY

Before even considering renewable energy for heating or cooling we need to ensure that we have exhausted all passive means, such as solar gain, stabilising temperature fluctuations with thermal mass, natural ventilation and others, as discussed in the preceding chapters of this book.

As renewable energy systems are discussed in detail inChapter 16, for now it is enough to say that all of these systems, including biomass boilers, heat pumps, and solar thermal systems can easily be incorporated in most simulation models using manufacturers' specifications, although each simulation tool may have different ways of specifying these systems.

Renewable cooling systems, such as absorption cooling or evaporative cooling can also be modelled in most simulation tools using information on their EER and SEER. However this will lead to inaccuracies in some cases. In the case of evaporative cooling for instance, EER will be very dependent on external air temperature and relative humidity. A more accurate way of modelling evaporative cooling would be to choose a simulation tool that has a built in capability for modelling it, which will take external parameters directly from the weather data file, calculate the cooling output and apply it to building heat balance in every simulation time step.

15.6 TASKS FOR SIMPLE SIMULATION EXPERIMENTS

1 Create a well insulated box, 10 m × 6 m × 3 m, with 40% of glazing on the larger south facing side. Set all heating, cooling, and domestic hot water to off, as well as internal heat gains. Set walls, ground floor slab, and roof U-values to 0.1 W/m²K, triple glazed windows to 0.65 W/m²K and insulated personnel door on the north side to 0.8 W/m²K. Set infiltration air exchange to 0.167 ach. Set the location to a cool climate, such as Birmingham. Set the initial internal room temperature to the outside air temperature, and the simulation preconditioning period to 10 days.

a) Run the first annual simulation.

b) Set internal gains from people to 90 W/person sensible gain and 60 W/person latent gain, and with density of 10 m²/person throughout the 24-hour period every day. Run the second annual simulation, giving the results file a different name to enable the comparison with the first simulation.

c) Set fluorescent lighting to operate from 8 am till 6 pm, with illuminance of 300 lux and lighting power density of 1.8 W/m²/100 lux. Run the third simulation, giving a separate name to the results file.

d) Add miscellaneous gain of 2.5 W/m² over 24 hours per day. Run the fourth simulation with a separate name for the results file.

e) Compare room temperatures of the four simulations in a single graph. At what point the building becomes heated in winter with internal heat gains only?

f) At what point the building becomes overheated in summer as result of internal heat gains?

g) How would you balance the conflicting influence of heat gains in summer and winter, so that the building can run on internal gains only throughout the year, without any additional heating or cooling?

h) Discuss the findings with your colleagues.

15.7 SUMMARY OF DESIGN PRINCIPLES

Internal gains:

- from lighting
  - obtain internal illuminance recommendations and lamp efficacy
  - derive the number of lamps and total power
  - derive power density by dividing total power by corresponding floor area
- from people
  - determine activity type
  - determine occupancy density
  - derive power density of latent heat gain and sensible heat gain from people
- miscellaneous
  - use equipment rating and number of pieces of equipment to determine total heat gain
  - derive power density of miscellaneous heat gains by dividing the total power by the corresponding floor area
- Auxiliary heating:
  - obtain hourly heat generator load from dynamic simulation
  - conduct frequency of occurrence analysis of loads and decide on the percentage of hourly loads to be covered (e.g. 95%–97%)
  - adjust the generator size to take into account heating up times
- Auxiliary cooling
  - apply the same principles to sizing of the cooling generator as described above for the heating generator
  - investigate the use of passive cooling first, based on a combination of thermal mass and natural ventilation
  - if mechanical cooling is needed consider using absorption cooling or evaporative cooling

NOTE

1 See Dry Resultant Temperature or Operative Temperature in Chapter 17.

CHAPTER 16

RENEWABLE ENERGY

This chapter will first discuss what is and what is not renewable energy. It will then give a brief overview of renewable energy systems, followed by an analysis of renewable energy options using simple spreadsheets and dynamic simulation methods.

16.1 WHAT IS AND WHAT IS NOT CONSIDERED TO BE RENEWABLE ENERGY

Whilst attending a conference on renewable energy at the Royal Institution of British Architects in London a few years ago, I was most surprised when a speaker told the audience that passive solar energy was not considered to be renewable. In the same context, natural daylight and natural ventilation were also not considered to be renewable. The argument put forward was that passive solar systems, natural daylight, and natural ventilation contribute to improvements in building energy efficiency and prepare the building for implementation of other systems that are considered to be renewable. As this definition is widely adopted, the methods in this book will be structured accordingly.

16.2 OVERVIEW OF RENEWABLE ENERGY SYSTEMS

The following energy systems are considered to be renewable:

• Combined heat and power (CHP)
• Biomass heating
• Solar water heating
• Ground coupled heating and cooling
• Photovoltaic systems
• Wind energy

16.2.1 Combined heat and power (CHP)

CHP is effectively an internal combustion engine, just like a lorry engine. If we instead of running a lorry attach this engine to run an electricity generator, the work of this engine will be converted into electricity energy (Figure 170). Just like a lorry engine needs to be cooled in order to maintain its operating temperature, a CHP engine will also need to be cooled, and the heat removed in this way is used for heating buildings. For a 100% of fuel input into the CHP engine, there will be 50% of heat output and 30% of electricity output. A CHP system will therefore give energy output in the ratio of 5/3 between heat and electricity, and the overall efficiency of the system will be around 80%. CHP engines can run on natural gas or on diesel. If bio-fuel is used to run this engine, the system becomes a biomass CHP. Bio-fuel can take the form of vegetable oils, landfill gas, or other renewable materials of organic origin.

This is a very effective way of generating energy. Heat is a by-product of the electricity generation process and as it is recycled, so that the system generates both electrical and thermal output.

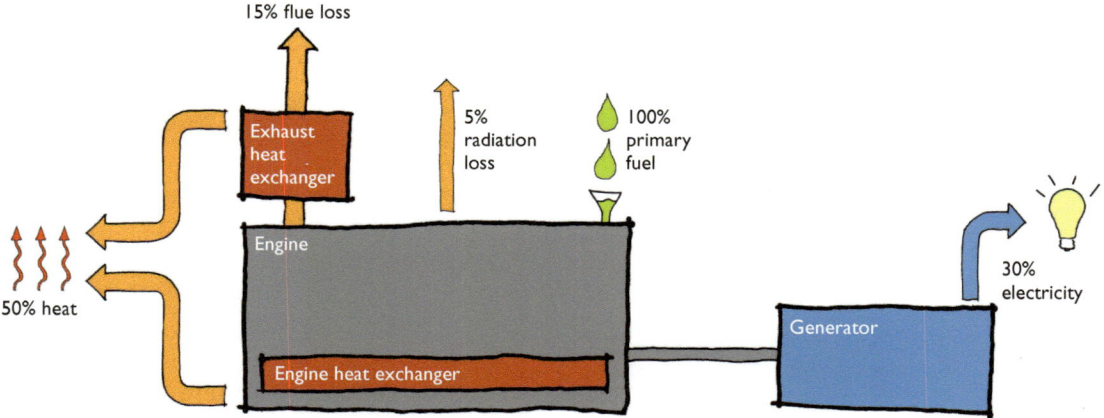

Figure 170   Conceptual diagram of a CHP system (illustration by Holly Doron)

In summer, running a CHP system in order to fulfil electricity demand will result in surplus heat when heat may not be required. CHP systems are therefore not suitable for applications where there is a considerable difference in heat demand between summer and winter. What can be done to run CHP in summer without wasting heat?

We have seen in Chapter 15 that absorption refrigeration machines are driven by heat. If such a machine is connected to the heat output of the CHP system, that will provide cooling by drawing heat out from the CHP. A system consisting of a CHP and an absorption refrigeration plant is called a tri-generation system. It provides heating, cooling, and electricity.

Considering that CHP is an internal combustion engine, its operation will be noisy. It will therefore not be advisable to have it running very near the building. Enough space is required to place the plant away from the building. Although these come packaged in acoustic boxes, the noise is one of the considerations that need to be considered when designing a CHP system for a building.

16.2.2 Biomass heating

The key idea behind biomass heating is that the material needs to have grown and captured carbon recently, within the past few years. Returning this recently captured carbon back into the atmosphere as a result of combustion of biomass does not increase the overall carbon emissions but effectively keeps the carbon dioxide in balance.

Biomass comes in various forms as follows:

• Plants grown purposely for bio-fuel on a short rotation cycle
• Plant waste, such as forest residues
• Food crop waste, such as straw or rejected vegetables
• Food crops converted to bio-fuel
• Agricultural waste
• Municipal waste
• Landfill gas

The above forms of biomass come with different degrees of usability. Different fuels need to be cleaned from impurities before they can be used in machinery that converts them into heat. This is especially the case with liquid or gas biomass used in CHP engines.

Wood chips and wood pellets (Figure 171) are used as bio-fuel for biomass boilers. Wood chips can be up to 30 mm in length and are a direct product of wood shredding. Wood chips are less dense than wood pellets, have lower calorific value, and are less expensive. Wood pellets are much more compact and are specially manufactured to achieve uniform shape and size. They are of similar length as wood ships and are typically 7 mm in diameter. In comparison with wood chips they have higher density and higher calorific value, but for this reason, they are more expensive. They generate higher carbon emissions per kilogram, but fewer amounts are needed in comparison with wood chips.

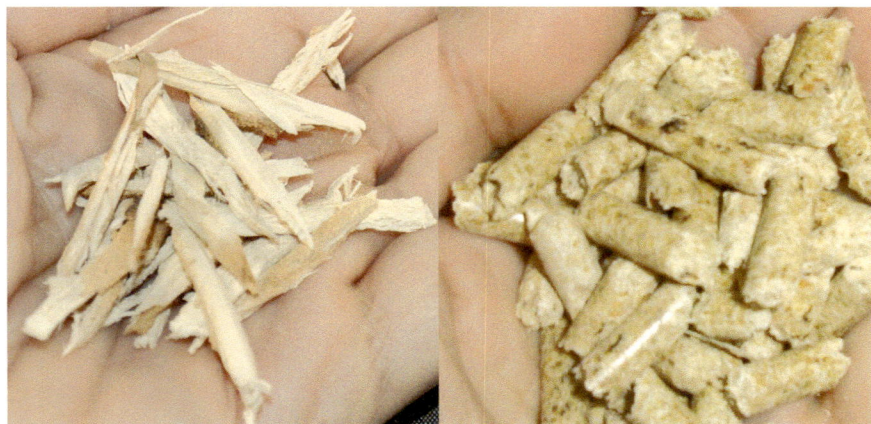

Figure 171   Wood chips (left) and wood pellets (right)

Figure 172 shows a biomass-heating application in a retail building near Milton Keynes, UK. The biomass boiler is in the plant room, which is separate from the main building. Hot water from the biomass boiler is pumped into the building through underground pipes, and then into an underfloor heating system.

Figure 172   Biomass boiler (left) and the auger (right) that carries wood pellets from the store into the boiler

The storage for biomass was designed as an extension to a standalone plant room building with V-shaped internal sides so that biomass is drawn by gravity towards the bottom of the V shape. In the base of this V shape is the auger, a screw mechanism, which draws wood pellets from the pellet store and feeds it into the boiler. The boiler and the pellet store were sized using a DSM, and the final size of the store was determined on the basis of energy demand and the maximum number of deliveries that were acceptable during the year. The two pipes on the outside are used for injecting the wood pellets into the store and for letting the air out that is displaced by the pellets filling the store (Figure 173). When a delivery tanker full of wood pellets comes, the driver connects a 100 mm diameter hose to the inlet of the biomass store, and the wood pellets are pumped into the store. As they fly like bullets horizontally into the store, they hit a rubber mat that is suspended vertically from the ceiling, and drop down to the pile of pellets already in the store. As the store fills up, the air goes out the other way.

Figure 173   Wood pellet store with longer supply pipe and shorter air exit pipe

16.2.3 Solar water heating

There are two types of commonly used collectors: flat plate and evacuated tube collectors.

Flat plate collectors are heat exchangers in a box insulated from five sides and protected by glazing from the sixth side. The absorber is coated with a selective surface to absorb energy from the part of the solar spectrum with high-energy content, thus reducing energy losses. Collectors connected into arrays are exposed to solar energy at an optimum angle and supply hot water into storage tanks used primarily for domestic hot water. Storage tanks are topped up with heat from conventional sources if heat from solar collectors is not sufficient to meet demand. An array of flat plate collectors is shown in Figure 174. The collectors are installed on the roof and sloped at an appropriate angle. The difference between roof angles and optimum angles for flat plate collectors can be a major factor that prevents seamless integration into buildings.

Figure 174   Flat plate solar collector array on the roof of the Scandinavian Green Roof Institute in Malmö, Sweden

Another type of solar collector is the evacuated tube collector. Evacuated glass tubes were the basis of all electronics until the 1960s and, with transition to solid state electronics, an entire industry was going to become obsolete. A new application was found some years later in the field of solar energy, when the evacuated tube collector was invented. A copper tube with two fins is placed into the evacuated glass tube, with a tip of the copper tube protruding out of the evacuated tube ('hot tip'). The fins are coated with a selective surface to absorb solar energy. The copper tube is also evacuated and it contains a small amount of liquid, such as alcohol, which does not fill it completely, comprising a 'heat pipe'. When the heat pipe receives heat from the sun, the liquid inside it boils and the resultant vapour rises to the hot tip. The hot tip is placed into a manifold in which a stream of water takes the heat out of the hot tip and into a storage tank. The gas in the hot tip condenses in the process and returns back to the copper pipe where it receives solar heat again and the process repeats.

Evacuated tube collectors are more efficient than flat plate collectors and can be integrated into a building much more easily. The entire evacuated tube, together with the fixed fins inside, can be rotated in most models into the optimum position towards the sun. An example of an evacuated tube array is shown in Figure 175-left and its operation principle is shown in Figure 175-right. The array is vertical but the fins inside the evacuated tubes are rotated to the optimum angle. As a result of this adjustable feature, evacuated tube collectors can be vertical, horizontal, or at any angle that is independent from the optimum angle, yet the collector fins can be set to the optimum angle. This gives designers a high degree of flexibility for integration into buildings.

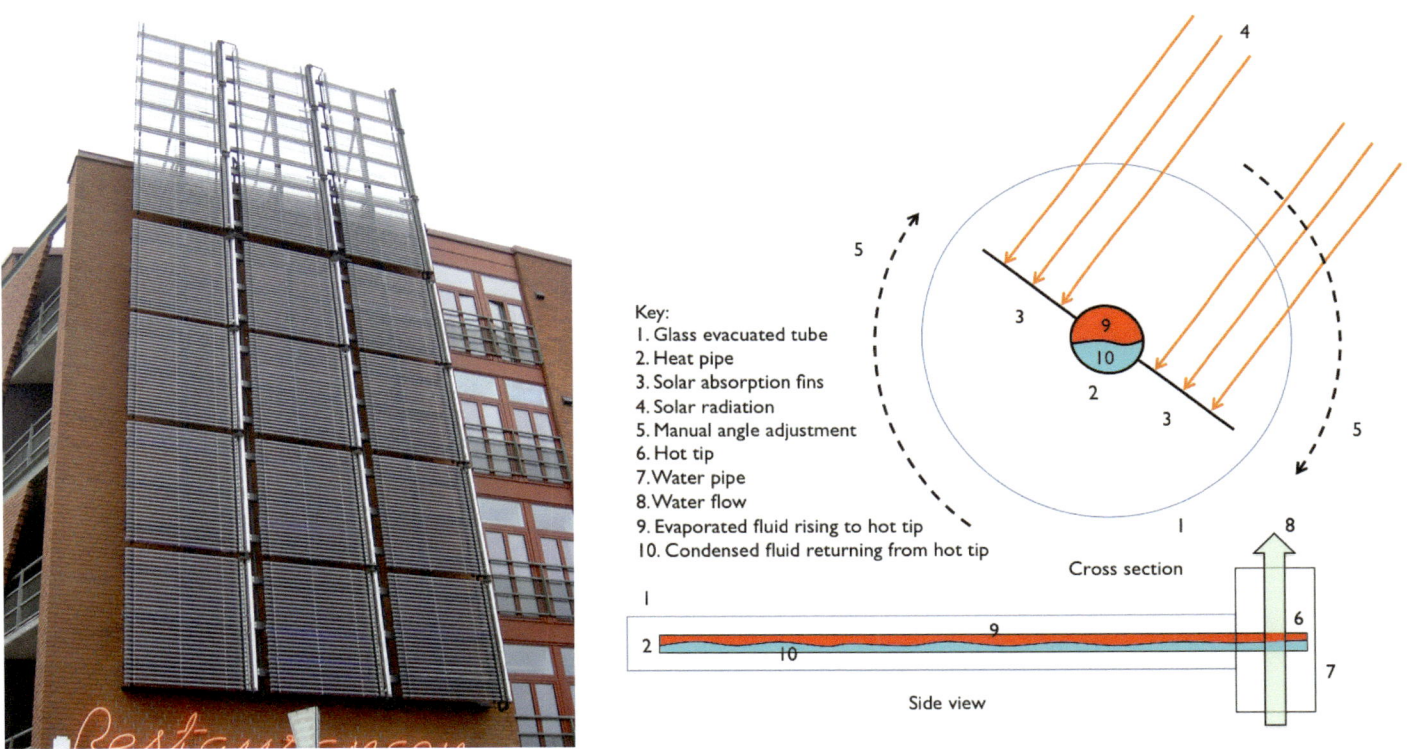

Figure 175   *Evacuated tube collector on a building in Västra Hamnen development in Malmö (left) and its operation principle (right)*

The optimum angle. The specification of the optimum angle is deliberately left open in the text above. There are rules of thumb that suggest what this angle is, but the reader is advised to find the optimum angle for each individual application using the dynamic simulation of several different angles and analysing the respective collector output. This will ensure the correct positioning of solar collectors in each individual project.

16.2.4 Ground coupled heating and cooling

This type of renewable energy system is based on a refrigeration machine similar to domestic fridges. We know well that it is cold inside the fridge and warm at the back of the fridge. The refrigeration machine therefore takes the heat out of the inside of the fridge and transfers it to a higher temperature at the back of the fridge. If we imagine taking the fridge apart and placing the heat exchanger from inside of the fridge into the ground and connecting the heat exchanger at the back of the fridge to pipework under the building floor, then we have created a conceptual ground source heat pump.

These machines will then take heat from the ground and supply it to the underfloor heating in the building. Heat pumps will deliver about 1 kW of output per 10 linear metres of the ground loop, depending on the model. Vertical systems have their ground loop in a bore hole. Horizontal systems have the ground loop in trenches about 1.5 metres deep. The vertical loop is more expensive but requires less surface area of the ground. The horizontal loop is less expensive but requires more surface area of the ground.

The performance of ground source heat pumps is characterised with the Coefficient of Performance or COP, which is the ratio between the energy delivered for end use and energy supplied to run it. As there is no energy production involved in this process but only energy transfer from lower to higher temperature (hence the name 'heat pump') the COP is always greater than unity, and typically greater than three in some models.

How the heat pump works (Figure 176). The operation of the heat pump is very similar to the operation of the compression refrigeration machine explained in Chapter 15. The main difference is in the location of the evaporator and the condenser. If the evaporator is in the ground and the condenser in the building then it is a heat pump. If however the evaporator is in the building and the condenser is outside the building, then this is a comfort cooling device.

Why is the heat pump considered to be a renewable energy system? Where does the energy in the ground come from? What happens if we use up all of the energy from the ground? Energy in the ground comes from solar radiation in the summer. Due to the large mass of the ground, solar energy is stored in it seasonally. In other words, it is absorbed in summer and it stays in the ground over winter until it is again renewed the following summer. It is unlikely that a heat pump will deplete all energy from the surrounding ground if bore holes or trenches are kept sufficiently apart as per manufacturers' recommendations. However, if the ground loop is too dense, the ground will freeze.

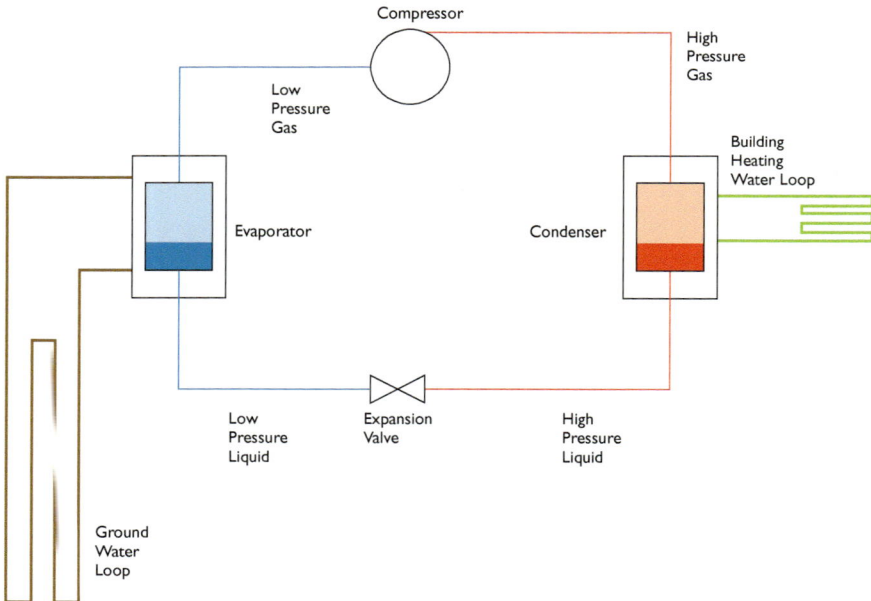

Figure 176   Operation of a ground source heat pump

Some types of heat pumps can be operated in reverse. This means that they will be able to take heat out of the building in summer and transfer it into the ground for inter-seasonal storage. This heat will be reused in winter by operating the heat pump in its normal mode.

16.2.5 Photovoltaic systems

Photovoltaic systems are another type of renewable energy systems. The technology is based on silicon cells which convert solar energy into DC electricity. There are several types of PV cells:

• amorphous thin film;
• polycrystalline;
• monocrystalline;
• hybrid.

The list above is ordered in an increasing level of efficiency, which changes with technological development and performance characteristics should be used as provided by technical data sheets for specific systems. Amorphous cells can be placed on curved surfaces whilst the other types of cells are rigid. Figure 177 shows examples of installations of PV systems which are also used as overhangs to provide solar shading.

Figure 177   Examples of PV installations in Västra Hamnen development (left) and Scandinavian Green Roof Institute (right) in Malmö

16.2.6 Wind energy

Wind energy systems convert the kinetic energy of wind into the rotary movement of a turbine attached to an electricity generator. There are two main types of wind turbine: horizontal axis and vertical axis.

Horizontal axis wind turbines are suitable for large-scale applications such as wind farms (Figure 178), where tens or hundreds of turbines can be installed in a group, either off or on shore. The operation of horizontal axis wind turbines depends on wind direction, and for that reason their design includes a swivel mechanism for tracking the wind.

An example of a vertical-axis wind turbine is shown in Figure 179. This type of wind turbine has a lower rated output than the horizontal axis turbine, but because of its scale it is suitable for applications on buildings. Vertical-axis wind turbines are independent of wind direction because of their vertical design, and they are particularly suitable for urban areas.

Figure 178   Horizontal axis wind turbine Enercon E-40 rated at 500 kW (left) and a wind farm in South Tenerife with total nominal power of 18.375 MW (https://www.thewindpower.net/windfarm_en_31893_chimiche-ii.php)

Figure 179   Vertical-axis wind turbine on the roof of the Aston University Student Village building in Birmingham

16.2.7 Feed in tariff

There are various names in different countries for what is called Feed in Tariff (FIT) in the UK but the principle is the same (Figure 180). In the early days of the FIT, the owner of a PV system effectively became an energy generator, and as a result, they got paid by the utility company a standard rate per kWh generated. This payment was receivable regardless of whether the owner used all the electricity they had generated, or whether they sold it by exporting it to the grid. Surplus electricity was sold to the grid either at a fixed price per kWh or at a market price. In 2019, the FIT scheme was closed to new applicants in the UK, and it was replaced by Smart Export Guarantee (SEG) scheme. The SEG enables small producers of renewable electricity to get paid for electricity they export to the grid. Under this scheme, licensed electricity companies with more than 150,000 customers are obliged to provide at least one SEG tariff.

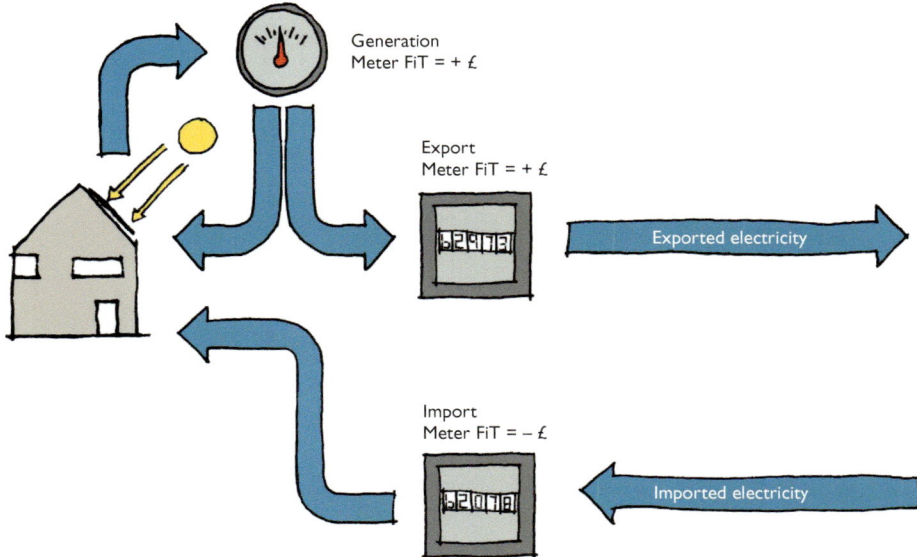

Figure 180   Energy metering in a PV system (illustration by Holly Doron)

16.3 EVALUATION OF RENEWABLE ENERGY OPTIONS USING DYNAMIC SIMULATION

In this section, each individual type of renewable energy discussed in the previous section will be evaluated using dynamic simulation. These will be applied on the Greenpower Centre model in IES Virtual Environment (Figure 181) based on model 'a' described in Chapter 18. This is a standard model that meets UK building regulations and for the purpose of this analysis it is heated by a natural gas boiler combined with an underfloor heating system. This section explains what to specify in order to carry out an analysis of renewable energy systems and how to interpret results.

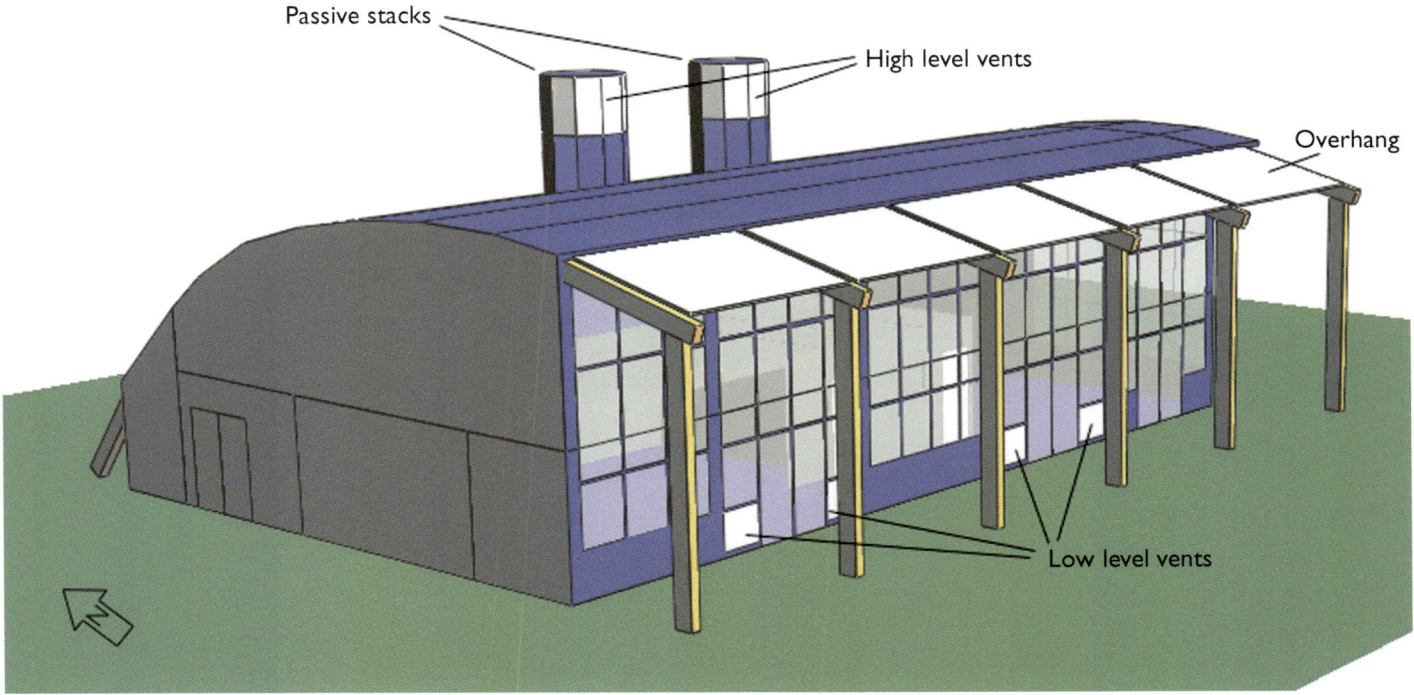

Figure 181   Greenpower Centre model in IES Virtual Environment

Biomass heating. In the systems section of the simulation model, the fuel type is chosen as biomass, with appropriate system efficiency (93%).

Combined heat and power (CHP). In the systems section, the gas heating plant is specified as a CHP plant, and then in the renewables section specifications are entered for the heat output (12.48 kW), thermal efficiency (50%), and power efficiency (28%) for a CHP (Figure 182).

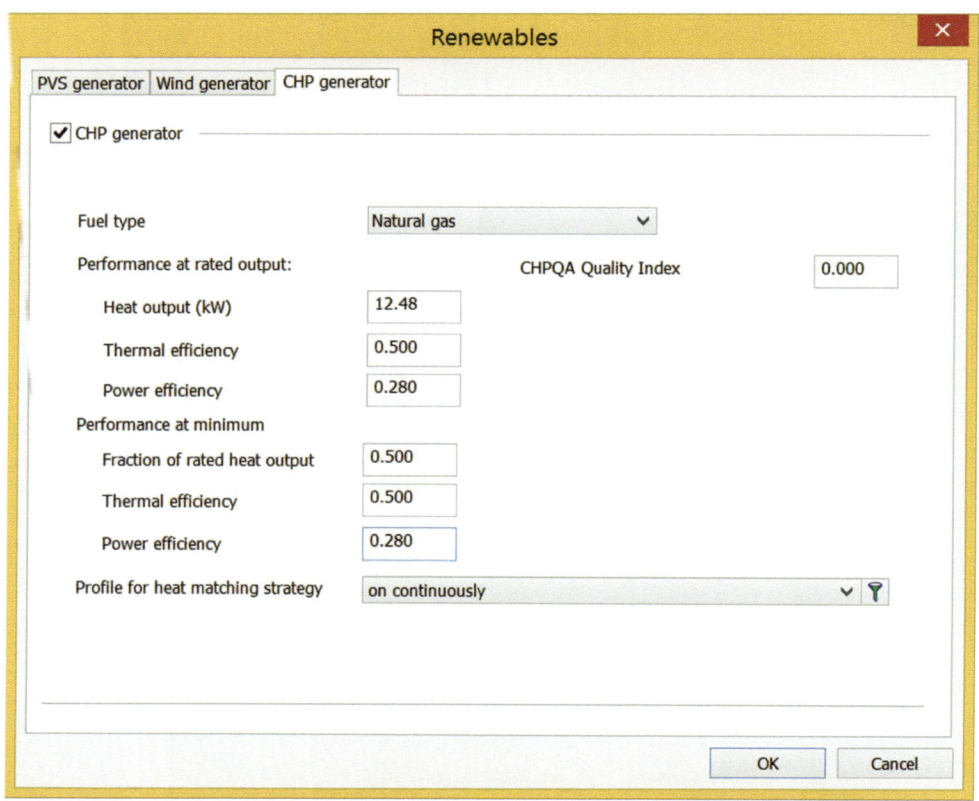

Figure 182   Specifying CHP in the model

Ground source heat pump. This is specified in the systems section from a pull-down menu, with a corresponding seasonal coefficient of performance (3.45).

Air source heat pump. As above.

Solar water heating. This is specified in Systems -> Domestic Hot Water -> Solar Water Heating (Figure 183).

Photovoltaic system. This is specified in the renewables section under the PVS generator (Figure 184).

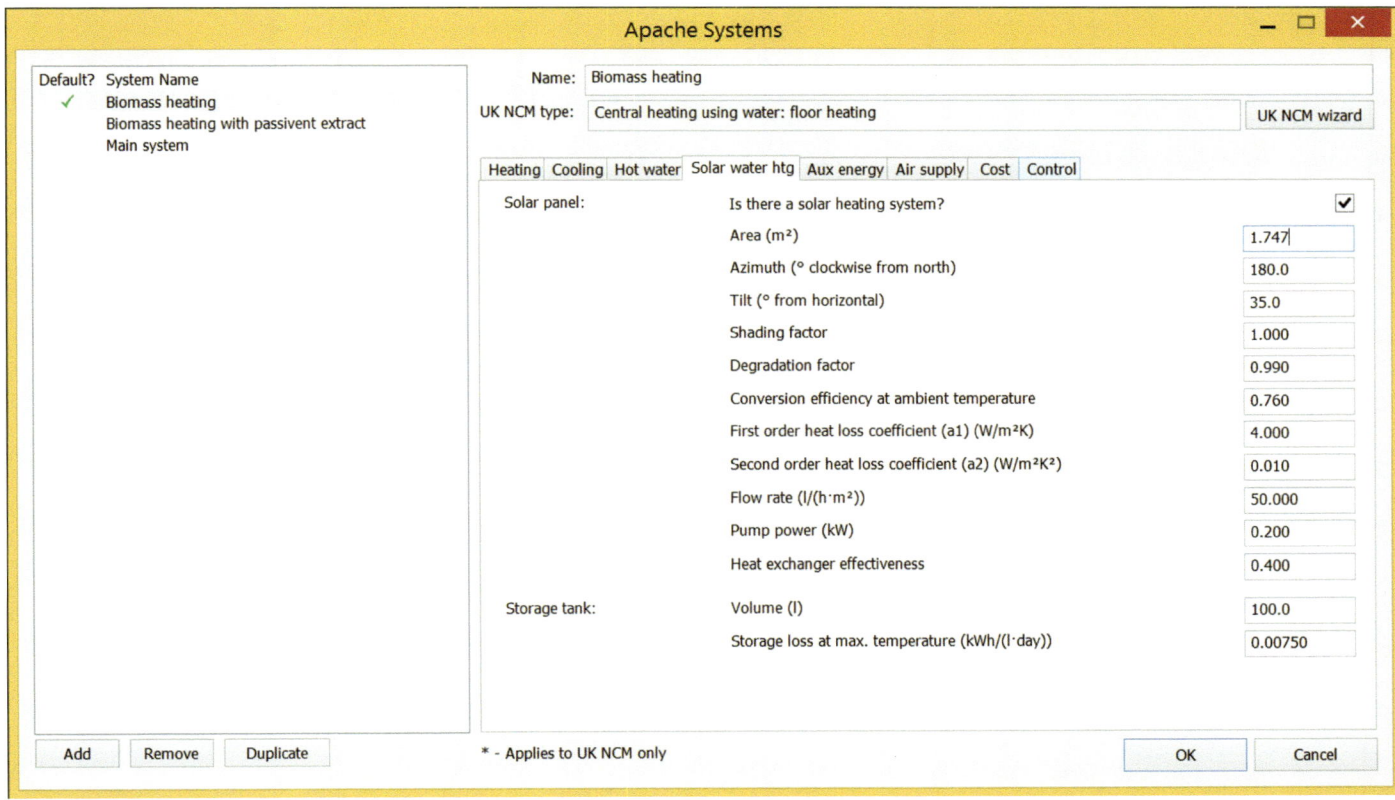

Figure 183   Specifying solar water heating in the model

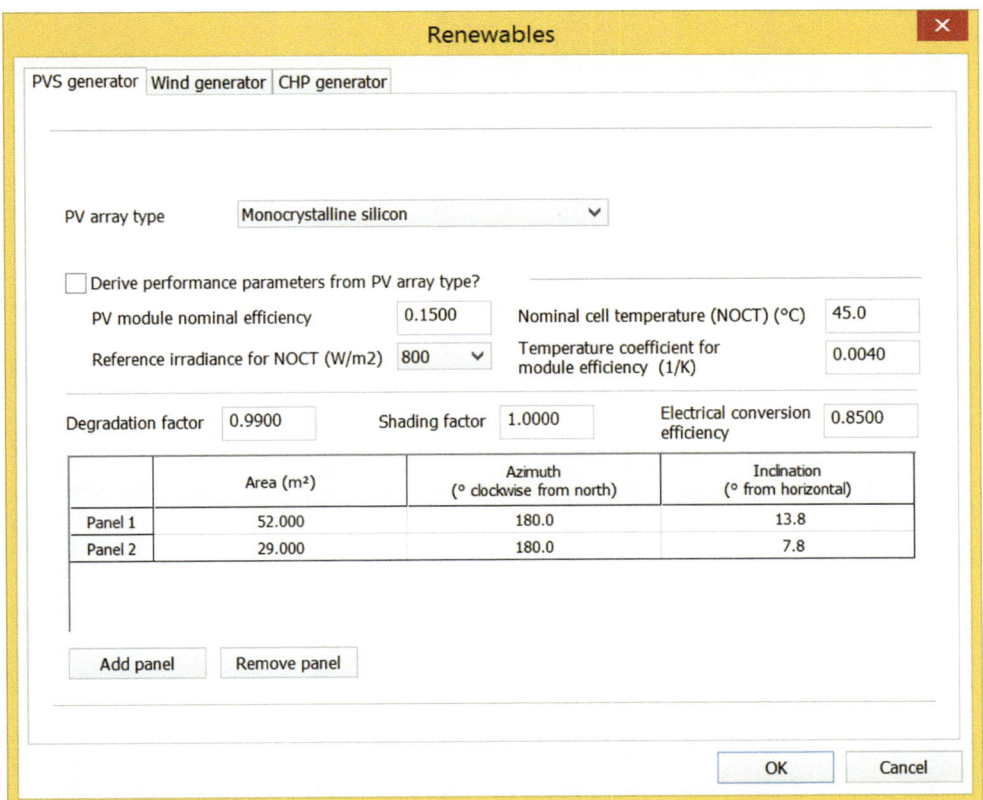

Figure 184   Specifying PV system in the model

Wind power. The wind power system is specified in the renewables section under the Wind generator (Figure 185). It is also necessary to specify how the wind turbine converts wind velocity into power. This is specified as a 'Power curve' in Figure 186.

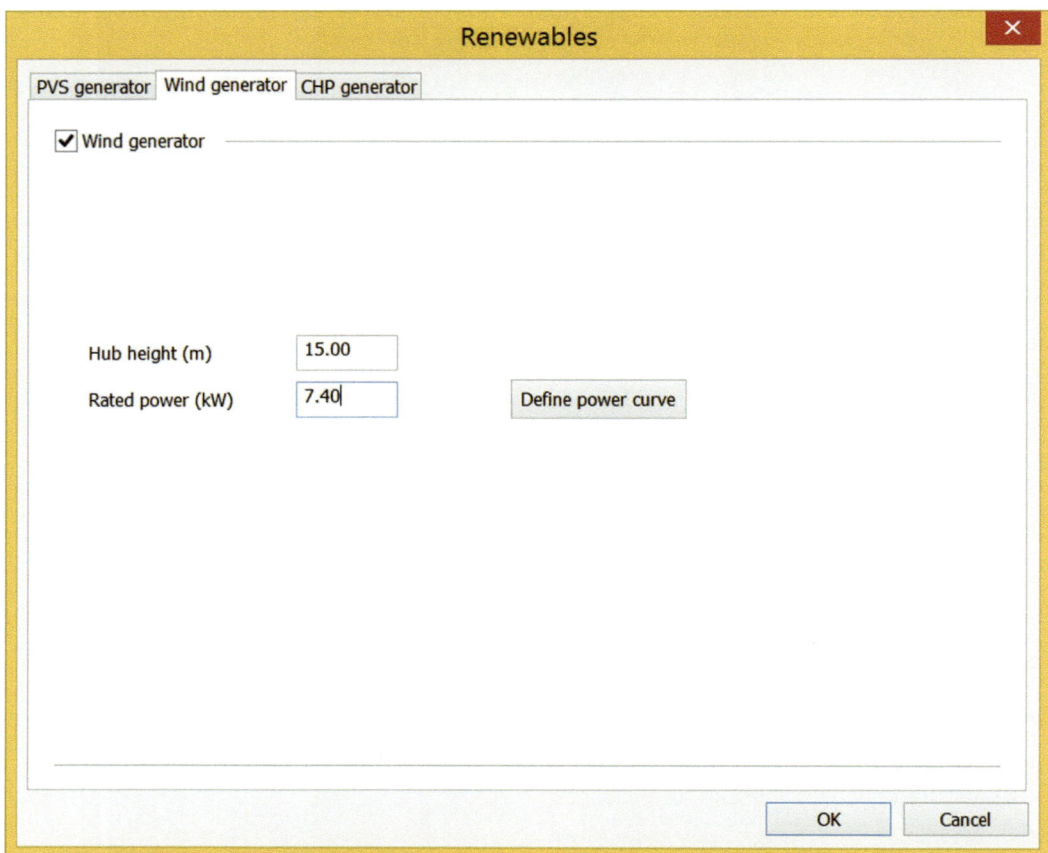

Figure 185   *Specifying wind generator in the model*

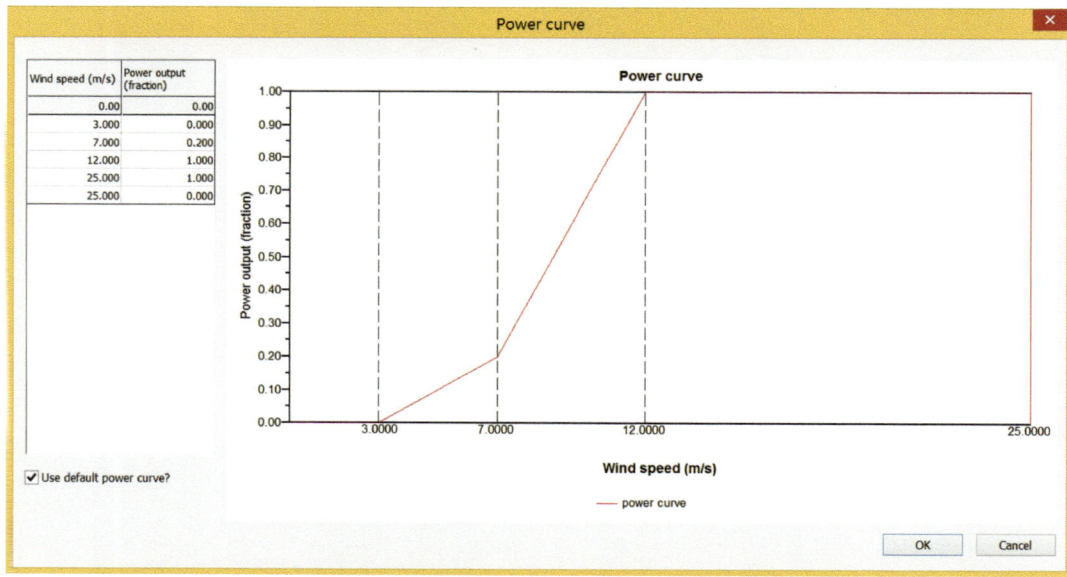

Figure 186   *Specifying wind velocity to power conversion function (the power curve) in the model*

Simulations for all of the above renewable options cases were run one at a time, and the carbon emission results were retrieved from the simulation summary report (Heating and cooling report). Carbon emissions for the baseline model without any renewable options were 11653.3 kg CO_2. The carbon emissions corresponding to the performance of the model with one of the renewable options were then subtracted from this number. The difference between these two sets of emissions was divided by the floor area of the building and is shown in column 'kgCO$_2$/m² floor area' in Table 39.

TABLE 39	RENEWABLE ENERGY SYSTEMS COMPARISON BASED ON DYNAMIC SIMULATION					
		System size	CO$_2$ savings			Demand met
	Renewable option	kW/m² floor area	kgCO$_2$/m² floor area	kgCO$_2$/m² panel	kgCO$_2$/kW	(%)
Sized to meet proportion of specified end use demand	Biomass heating		6.68		287.55	100% SH+DHW
	CHP	0.0390	2.15		55.06	100% SH+DHW
	Ground source heat pump		2.30		99.18	100% SH+DHW
	Air source heat pump		2.30		99.18	100% SH+DHW
	Solar water heating		-0.31		−575.89	50% DHW
Discretionary sizing	PV		25.87	73.00	663.63	100% e CO$_2$
	Wind	0.0231	0.38		1320.70	100% e CO$_2$

As it can be seen from this table, the renewable systems provide a range of carbon savings, except the solar water heating system. As the demand for domestic hot water in this particular example is very low, the emisisons from running this system outweigh the emissions savings from the solar hot water, with the overall result shown with a negative sign. The reader is encouraged to do their own analysis of the carbon emisisons savings in their design projects, as the results shown here are for the particular example only and cannot be generalised.

16.4 TASKS FOR SIMPLE SIMULATION EXPERIMENTS

1 Create a model of a simple box 10 m × 6 m × 3 m, with 40% glazing on the south-facing side and with a 30° pitched roof. Run an annual simulation for a location of your choice, using a heating set temperature of 19°C and a cooling set temperature of 23°C. Compare the total external incident solar flux on the roof surface with the total annual heating and cooling load.

2 Add a PV array to the pitched roof with 80% surface utilisation and 15% efficiency. Re-run the simulation and compare the annual PV generated output with the incident solar flux and total annual heating/cooling load from the above task.

3 Repeat these tasks for several different locations, from low to high latitudes.

4 What can you conclude about energy availability and PV efficiency? If the number of storeys is proportional to the amount of solar energy received on the roof, what would be the height distribution of buildings between 0° and 60° latitude?

16.5 SUMMARY OF DESIGN PRINCIPLES

Design principles

• Establish a baseline simulation model of the building and note CO_2 emissions resulting from this model.
• Implement renewable energy options one at a time in the baseline model and calculate the difference in emissions between the baseline model and the same model with the selected renewable option.
• Final assessment of the feasibility of renewable options should be made only after a full life cycle analysis.

Practical design issues.

- Biomass CHP
  - High investment cost
  - Fuel storage requirement
  - Fuel delivery and access for delivery requirements
- Biomass heating
  - High investment cost
  - Fuel storage requirement
  - Fuel delivery and access for delivery requirements
  - Land required to grow biomass
  - Potential jamming of the auger mechanism
  - Maintenance contract is highly advisable
- Solar water heating
  - Geometry of the building needs to be designed so that solar collectors are well integrated
  - Flat plate collectors can be integrated well into pitched roofs
  - Evacuated tube collectors offer much more flexibility for integration on surfaces of different orientations
- Ground source heating and cooling
  - High installation cost
  - Land is required for horizontal ground loop
  - Nature of the land is important (soft, hard, rocks, water table depth, etc.)
- Photovoltaic systems
  - Installation cost
  - Space requirements
  - Efficiency
  - Architectural integration into the building
  - Energy generation/export incentives exist in different local settings
- Wind energy
  - Land required for horizontal axis systems
  - Vertical-axis systems installed on buildings will cause vibrations and will need to be structurally decoupled from the building

CHAPTER 17

DESIGNING FOR THERMAL COMFORT

Although good design needs to save energy and reduce carbon emissions, buildings are designed for people, and thus thermal comfort is of paramount importance. Therefore, the first step towards designing true zero will be designing for thermal comfort, as discussed in this chapter. This will be followed by designing for negative operational emissions in Chapter 18, and designing for a combination of embodied and operational emissions that targets a zero emissions year in Chapter 19.

In this chapter, factors that influence thermal comfort will be explained, and assessment of thermal comfort will be discussed. Adaptation of people to internal conditions will also be discussed and how to carry out adaptation analysis with simulation tools will be explained. It will also be explained how and why thermal comfort results in simulation outputs may be inaccurate, and how to achieve more accurate assessment of comfort through post-processing of simulation outputs. It will then be demonstrated in later chapters, how dynamic simulation can be used to maximise thermal comfort whilst maintaining zero carbon performance.

17.1 DEFINITION OF THERMAL COMFORT

Thermal comfort is defined as a condition of mind that expresses satisfaction with the thermal environment. Although this suggests that thermal comfort is subjective, there are certain measurements that can be used to assess what the majority of people will feel like in an internal environment.

Thermal comfort is based on the equilibrium between heat gains inside the human body and heat losses from the body to the environment. When the heat gain from metabolism is equal to heat loss, then we are in balance with the environment and that makes us feel comfortable.

We can represent this balance conceptually as weighing scales. On one side of the scales in Figure 187 is metabolism, the internal chemical 'engine' that generates heat, and on the other side of the scales is the heat exchange through conduction, convection, radiation, and evaporation. If metabolism on the left is in balance with heat transfer mechanisms on the right, we feel comfortable. If metabolism is low or the heat transfer mechanisms are high, we feel cold, and if metabolism is high and heat transfer mechanisms are low, we feel hot.

17.1.1 Measurement of comfort

When I ask my students in the class to let me know by show of hands how they feel: comfortable, slightly warm, hot, slightly cold, or very cold, there is always non-uniform distribution of responses. In other words, there are always more hands up in one category and less in other categories. In order to assess thermal comfort of a large number of individuals, Povl Ole Fanger, who was an early pioneer in thermal comfort research, introduced a seven-point scale (Fanger, 1972) as shown in Table 40.

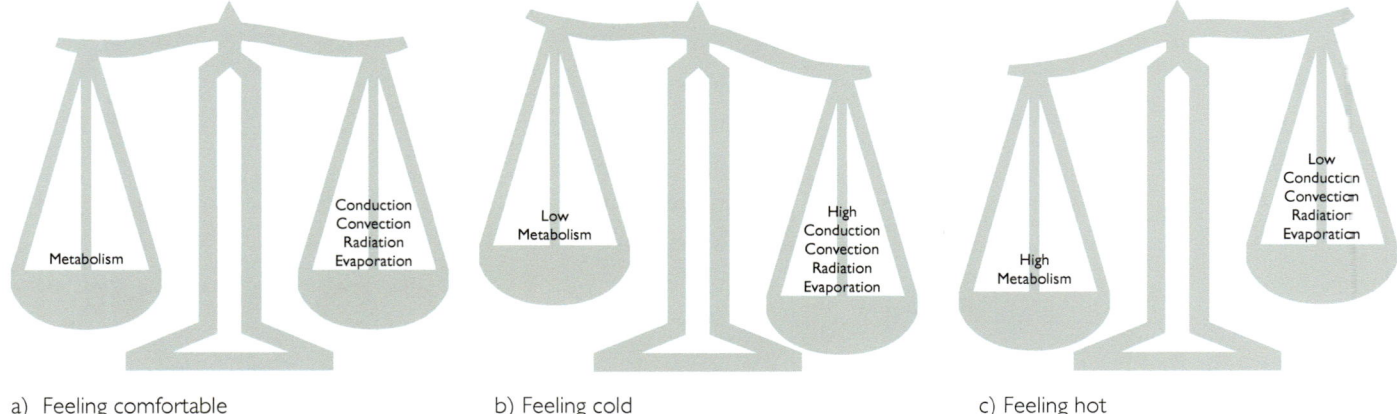

a) Feeling comfortable   b) Feeling cold   c) Feeling hot

Figure 187   Relationship between heat gains and heat losses in human body

TABLE 40 SEVEN POINT SCALE	
Predicted Mean Vote (PMV)	**Meaning**
−3	Cold
−2	Cool
−1	Slightly cool
0	Neutral
+1	Slightly warm
+2	Warm
+3	Hot

This table was used as the basis of voting by a large number of volunteers who were asked to express how they felt under different activity levels, clothing levels and internal environmental conditions. Through analysis of experimental results, it was found that these votes can be predicted. A concept of predicted mean vote (PMV) (Fanger, 1972) was established as a function of six parameters that influence thermal comfort as follows:

$$PMV = f(M, Rc, Ta, Tmrt, v, rh)$$ (53)

where

M – metabolic rate
Rc – resistance of clothing
Ta – air temperature
$Tmrt$ – mean radiant temperature
v – air velocity
rh – air relative humidity

Details of PMV calculations in Equation (53) consist of four extensive equations; however, there are several external sources that automate these equations within standard spreadsheet software.

Furthermore, the predicted percentage of dissatisfied (PPD) people was found to correlate to PMV as follows:

$$PPD = 100 - 95 \times e^{-\left(0.03353\ PMV^4 + 0.2179\ PMV^2\right)}$$

(54)

Graphical representation of this equation is shown in Figure 188. A striking feature of this equation is that it has a minimum of PPD = 5% for thermally neutral vote. This means that there will always be at least 5% of dissatisfied people in every building. If we get our design slightly wrong resulting in PMV deviating further from the neutral vote, the PPD will increase exponentially.

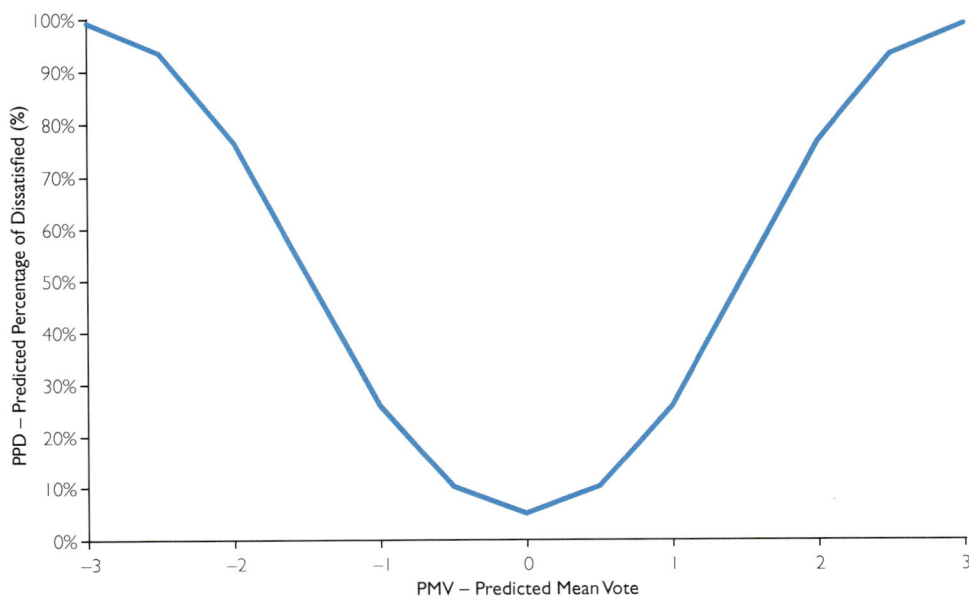

Figure 188   *Predicted percentage of dissatisfied as function of predicted mean vote*

In the second half of this chapter, I will show how PMV and PPD are used in dynamic simulations, and how thermal comfort can be improved by making changes in the design and using these parameters to assess alternative designs. But first the parameters that influence thermal comfort, as shown in the PMV equation above, will be discussed.

17.1.2 Comfort parameters and how to influence them

Metabolic rate is internal heat production in the human body. Fanger (1988) introduced a unit of metabolic rate called 'met' and established a simplified list of activities and corresponding metabolic rates as shown in Table 41.

| TABLE 41   ACTIVITIES AND METABOLIC RATES ||
Activity	Metabolic rate (1 met = 58 W/m²)
Laying at rest	0.8
Sitting	1.0
Standing	1.4
Walking 3 km/h	2.0
Walking 5 km/h	3.0
Running 10 km/h	8.0

Higher metabolic rates require a higher intensity of heat transfer with the environment in order to maintain constant body temperature and feeling of thermal comfort (Figure 187). Designers can influence metabolic rates of building occupants only to some extent by stimulating physical activity. For instance, it is possible to make the use of stairs more inviting than the use of lifts by making flights of stairs short. However, metabolic rates in buildings are primarily influenced by the building use type, such as office, school, home, etc. A more detailed list of activities and corresponding metabolic rates is available in CBE Thermal Comfort Tool (Tartarini et al., 2020).

Resistance of clothing is categorised in a similar way as metabolic rates, by introducing the unit of thermal resistance named 'clo'. Fanger (1988) defines the thermal resistance of typical clothing arrangements as shown in Table 42.

TABLE 42 RESISTANCE OF CLOTHING	
Description of clothing	**Resistance of clothing (1 clo = 0.155 m²K/W)**
Underpants	0.1
Shorts, short sleeve shirt, sandals	0.5
Lightweight trousers, short sleeve shirt, shoes	0.5
Boiler suit	0.8
Suit and tie	1.0
Padded overall	1.5
Winter coat, boots, gloves, scarf, fur hat	3.0

Resistance of clothing can be influenced by designers only to an extent. For instance, well insulated buildings will make people more comfortable in less clothing. A more detailed list of clothing resistances is available in CBE Thermal Comfort Tool (Tartarini et al., 2020).

Air temperature. This temperature controls the conduction and convection heat transfer mechanisms between the human body and its environment (Figure 187). If it reaches the body temperature of 37°C, the corresponding heat transfer mechanisms will stop functioning and thermal comfort will be affected considerably. The maintenance of thermal comfort will in that case need to be taken over by the remaining two heat transfer mechanisms: radiation and evaporation.

In summer months, we can influence internal air temperature by providing sufficient thermal mass to stabilise temperature fluctuations and by providing adequate natural ventilation to prevent overheating, combined with adequate solar shading. In winter months we can influence internal air temperature by providing adequate space heating, thermal mass and thermal insulation.

Mean radiant temperature. This temperature controls the radiation heat transfer mechanism between the human body and the environment (Figure 187). It is an area-weighted average temperature of all internal surfaces in an internal space. To get a mean radiant temperature, the surface temperatures of all internal surfaces are multiplied with corresponding surface areas, added all together, and divided by the sum of all internal surfaces:

$$T_{mrt} = \frac{\sum_i^N T_i \times A_i}{\sum_i^N A_i}$$

(55)

where

T_i – temperature of i-th surface
A_i – area of i-th surface
N – total number of surfaces in internal space

If the mean radiant temperature approaches 37°C, the radiation heat transfer mechanism for maintaining heat balance in the human body (Figure 187) will stop functioning and will need to be taken over by the other three mechanisms. If the other two temperature-based mechanisms (conduction and convection) have also stopped functioning because of high air temperature, the only remaining mechanism will be evaporation.

A large amount of standard glazing will have a low surface temperature in cold weather, and this will reduce the mean radiant temperature. We can influence this temperature by reducing the amount of glazing or improving the specification of glazing to a higher **U**-value. We can also influence it by increasing the amount of thermal insulation in opaque surfaces.

Dry resultant or operative temperature. This is a combination of air temperature and mean radiant temperature. For air velocities below 0.1 m/s, this temperature is defined as

$$T_r = \frac{(T_a + T_{mrt})}{2} \tag{56}$$

Dry resultant temperature is measured at the centre of a blackened globe of a diameter of 40 mm. The globe exchanges heat with air and with all surfaces in an internal space and is a simplified representation of the human body. This is why this temperature is referred to as comfort temperature, and it is the basis of temperature recommendations for internal spaces.

We can see from the above equation that we can increase dry resultant temperature T_r by either increasing air temperature or mean radiant temperature. We can also keep dry resultant temperature unchanged by reducing air temperature and increasing mean radiant temperature by the same amount, or the other way round. Most of these changes can be achieved by making changes in the building fabric or in the ventilation strategy, and all of that can be tested using dynamic simulation.

Air velocity influences thermal comfort in combination with air temperature and turbulence intensity. The influence on PPD can be expressed as

$$PPD = f\left(t_a, v_a^c, tu\right) \tag{57}$$

where

t_a – air temperature
v_a – air velocity
tu – air turbulence
c – constant

This means that PPD changes linearly with air temperature and air turbulence, and exponentially with air velocity. If turbulence is kept constant, then an increase in air temperature will require an exponential increase of air velocity to maintain the PPD unchanged. Further details of how air velocity influences thermal comfort can be found in Fanger (1988) and *CBE Thermal Comfort Tool* (Tartarini et al., 2020). Designers can respond to the need to increase air velocity exponentially by making adequate provisions for natural ventilation, which can be tested using dynamic simulation.

Air relative humidity. A high level of relative air humidity will progressively reduce and ultimately eliminate the evaporation heat transfer mechanism from the human body (Figure 187), and can cause serious problems to individuals if the other three temperature-based mechanisms (conduction, convection, and radiation) have already been eliminated due to high temperatures. We can control this parameter by adequate ventilation in summer months, assuming that outside air humidity is not very high, and in winter months by adequate heating and ventilation.

17.2 ADAPTIVE THERMAL COMFORT DUE TO PHYSIOLOGICAL ADAPTATION

The pioneering work by Fanger (1972, 1988) and subsequent related work provided input into several standards and other technical documents that are used by simulation tools as the basis for thermal comfort assessment. The importance of physiological adaptation of people to internal conditions has also been recognised in recent years. Based on relevant standards and subsequent work, CIBSE TM52 (Nicol, 2013) established three criteria for defining overheating in free-running buildings. Using a temperature difference between the operative temperature T_{op} and a maximum acceptable temperature T_{max}, rounded to the nearest integer and referred to as $\Delta T = T_{op-}T_{max}$ below, the CIBSE TM52 criteria are as follows:

1 Hours of exceedance – the number of hours during which ΔT is greater than or equal to one Kelvin shall not exceed more than 3% of occupied hours; the maximum acceptable temperature T_{max} is defined using the running mean of the outdoor temperature T_{rm} as $T_{max} = 0.33\,T_{rm} + 21.8$;
2 Daily weighted exceedance – a daily sum of a product of ΔT and the corresponding hours of occurrence shall not exceed 6;
3 Upper limit temperature – the absolute maximum of the operative temperature for which ΔT shall not exceed 4 Kelvin.

In addition to detailed explanation and guidance on these criteria for adaptive thermal comfort in CIBSE TM52 (Nicol, 2013), a comprehensive overview of the subject can be found in Nicol et al. (2012).

Simulation tools now provide options for physiological adaptive thermal comfort analysis. In DesignBuilder, ASHRAE 55, EN 15251, CIBSE TM52 and others can be included in the analysis (Figure 189), and in IES VE, CIBSE TM52 adaptive thermal comfort can be specified (Figure 190). As it will be seen from the remainder of this chapter, this analysis of physiological adaptation is not matched by the analysis of behavioural adaptation and can lead to an erroneous assessment of thermal comfort.

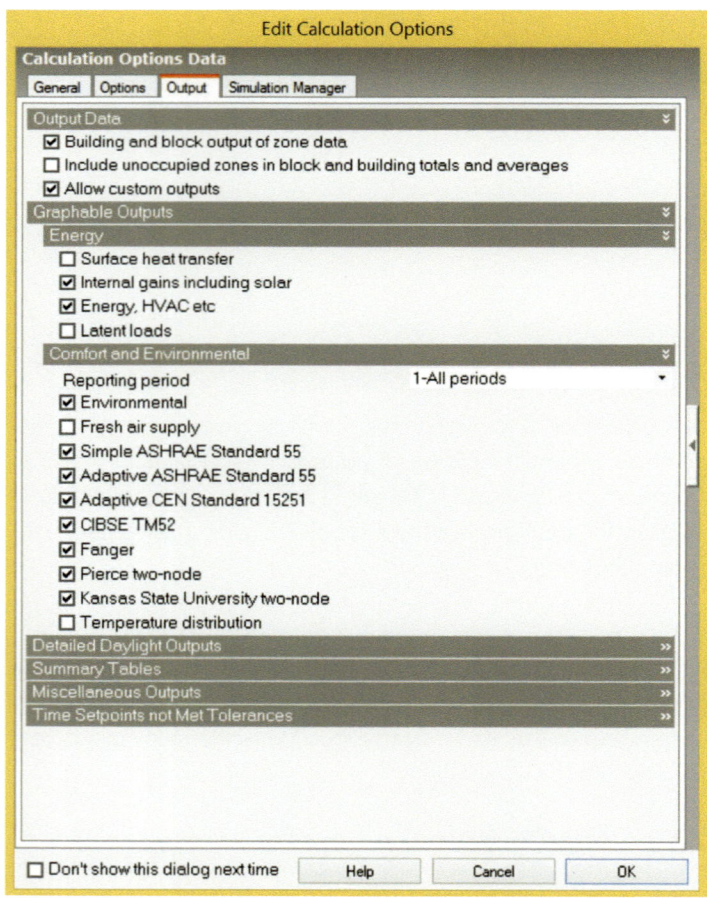

Figure 189   Calculation options dialogue box in DesignBuilder showing adaptive thermal comfort output options

17.3 INTERPRETATION OF THERMAL COMFORT RESULTS IN SIMULATION OUTPUTS

In order to demonstrate the issues associated with the interpretation of thermal comfort results in simulation outputs, a simulation of the simple box as in Figure 104 was carried out using IES Virtual Environment.

In IES Virtual Environment thermal comfort is evaluated as a post-simulation process, after the dynamic simulation is completed. This post-processing occurs within Vista/VistaPro of the IES VE suite, based on the settings in the dialogue box shown in Figure 190. As it can be seen from this figure, the activity and clothing levels are adjustable, but once they are set they will remain constant over the entire simulation period. Although the same figure shows that CIBSE TM52 Adaptive Comfort is evaluated, keeping the activity level and clothing level constant throughout the simulation year will inevitably produce a high percentage of dissatisfaction at some point during the year. Other simulation tools, such as DesignBuilder for instance, specify winter and summer clothing and a single activity level, leading to similar thermal comfort assessment issues.

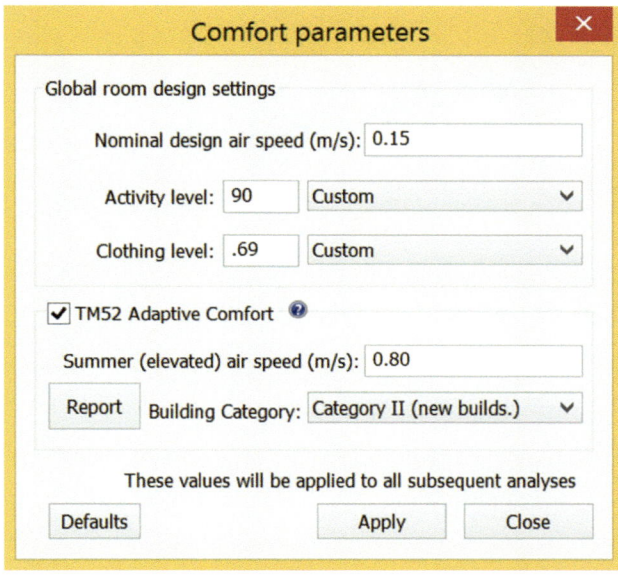

Figure 190   Comfort parameters dialogue box in IES Virtual Environment

This is illustrated in IES VE simulation outputs in Figure 191. The model from which these results are obtained is heated to 19°C and is free-running in summer, with free cooling operating at 5 ach from 25°C and above.

Whilst there are reasonable levels of satisfaction with thermal comfort in winter based on an activity level of 90 and clothing level of 0.69 clo (light office wear), the results are almost completely reversed with the clo level of 0.1 (underwear), where there are high levels of dissatisfaction in winter and much less so in summer, except during the highest temperatures (Figure 191). Although this method of personal adaptation to internal conditions would not be acceptable in an office environment, it illustrates the point that keeping the comfort parameters set to the same clothing throughout the year will produce high levels of dissatisfaction during certain periods. This makes a case for the analysis of adaptive clothing behaviour, which will be introduced in the next section.

17.4 IMPROVING THERMAL COMFORT ASSESSMENT THROUGH SIMULATED ADAPTIVE CLOTHING BEHAVIOUR

In real life, people can adjust their activity and clothing levels according to internal conditions and can also move between rooms thus following the most appropriate conditions. This 'comfort tracking' is especially practised in hot and humid climates, and it should not be confused with adaptive thermal comfort due to physiological adaptation discussed earlier in this chapter.

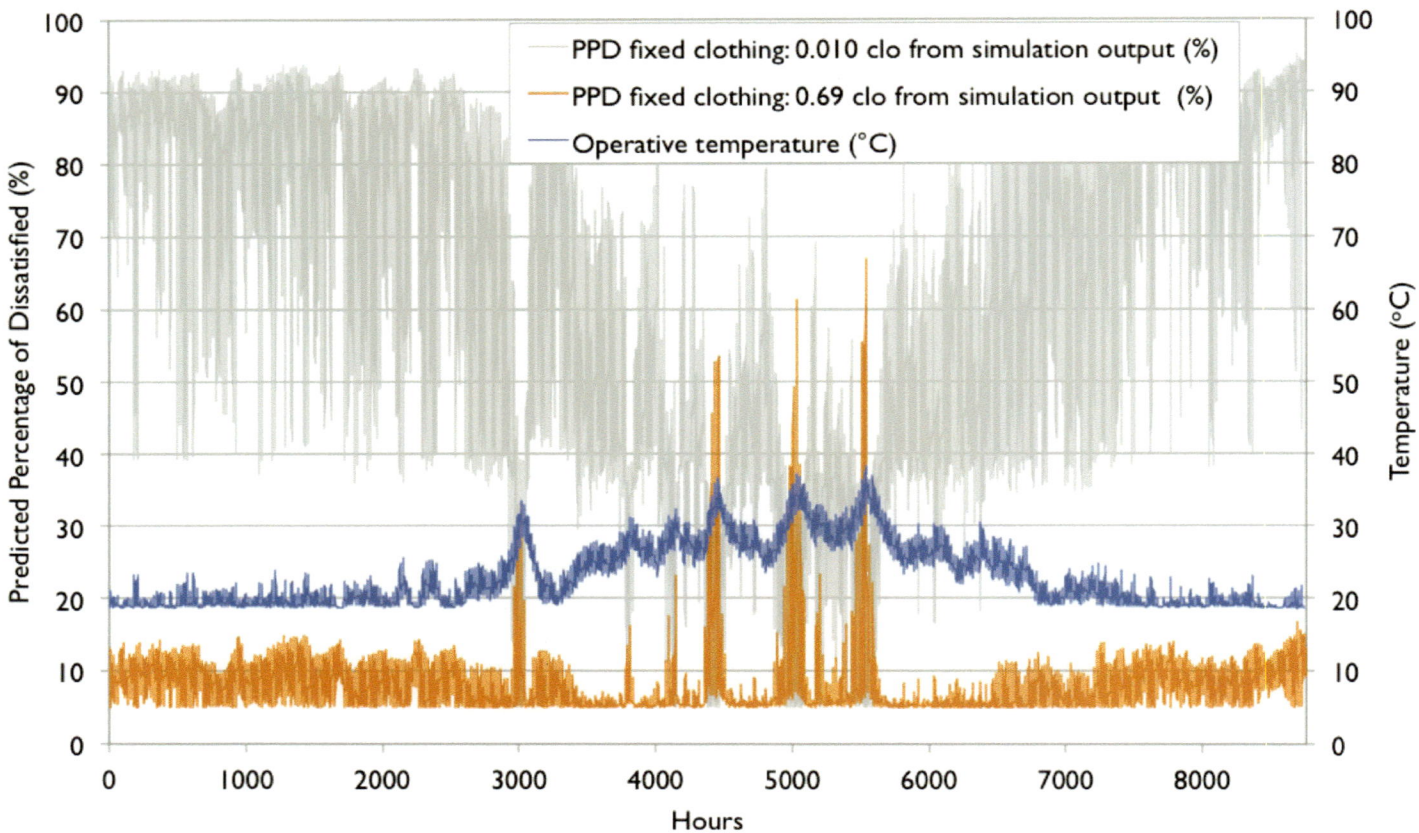

Figure 191   Assessment of thermal comfort based on two different levels of fixed clothing

Therefore, the interpretation of thermal comfort results in simulation outputs requires special attention and further post-processing, in order to consider expected human behaviour, rather than to rely on constant activity and one or two constant clothing levels throughout the year.

In order to simulate the behavioural adaptation, the following steps were taken:

1   Hourly air temperature, mean radiant temperature, and relative humidity from the simulation were exported into a spreadsheet, adding fixed air velocity and activity level to this dataset;
2   An initial clothing level was used, as set in the simulation model;
3   The clothing level was subsequently modified on an hourly basis using the formulae in Table 43; essentially the clothing level was increasing as PPD was reducing, and it was reducing as PPD was increasing;
4   The PMV and PPD were recalculated using Equations (53) and (54).

This is a further development of a method initially reported by Huws and Jankovic (2014), where fixed clothing levels were used in the interval boundaries corresponding to the rows of Table 43. That method however resulted in significant step changes in PPD when PMV crossed the interval boundaries and did not provide sufficient adaptation. A further development of the method reported here results in smoother changes of the PPD due to the cumulative and therefore progressive changes of clothing levels.

The resultant PPD taking into account adaptive clothing behaviour represents a dynamic and more accurate assessment of thermal comfort. The comparison between fixed clothing and adaptive clothing behaviour from Table 43, and based on the IES model from the previous section, is shown in Figure 192.

TABLE 43 CLOTHING ADJUSTMENT IN RESPONSE TO PMV	
Predicted Mean Vote (PMV)	**Adjusted clothing clo**
PMV < −1 (Cold)	$clo_{(t)} = clo_{(t-1)} + 0.25$
−1 <= PMV < 0 (slightly cold)	$clo_{(t)} = clo_{(t-1)} + 0.15$
PMV = 0 (neutral)	$clo_{(t)} = clo_{(t-1)}$
1 >= PMV > 0 (slightly warm)	$clo_{(t)} = clo_{(t-1)} - 0.15$
PMV > 1 (warm)	$clo_{(t)} = clo_{(t-1)} - 0.25$

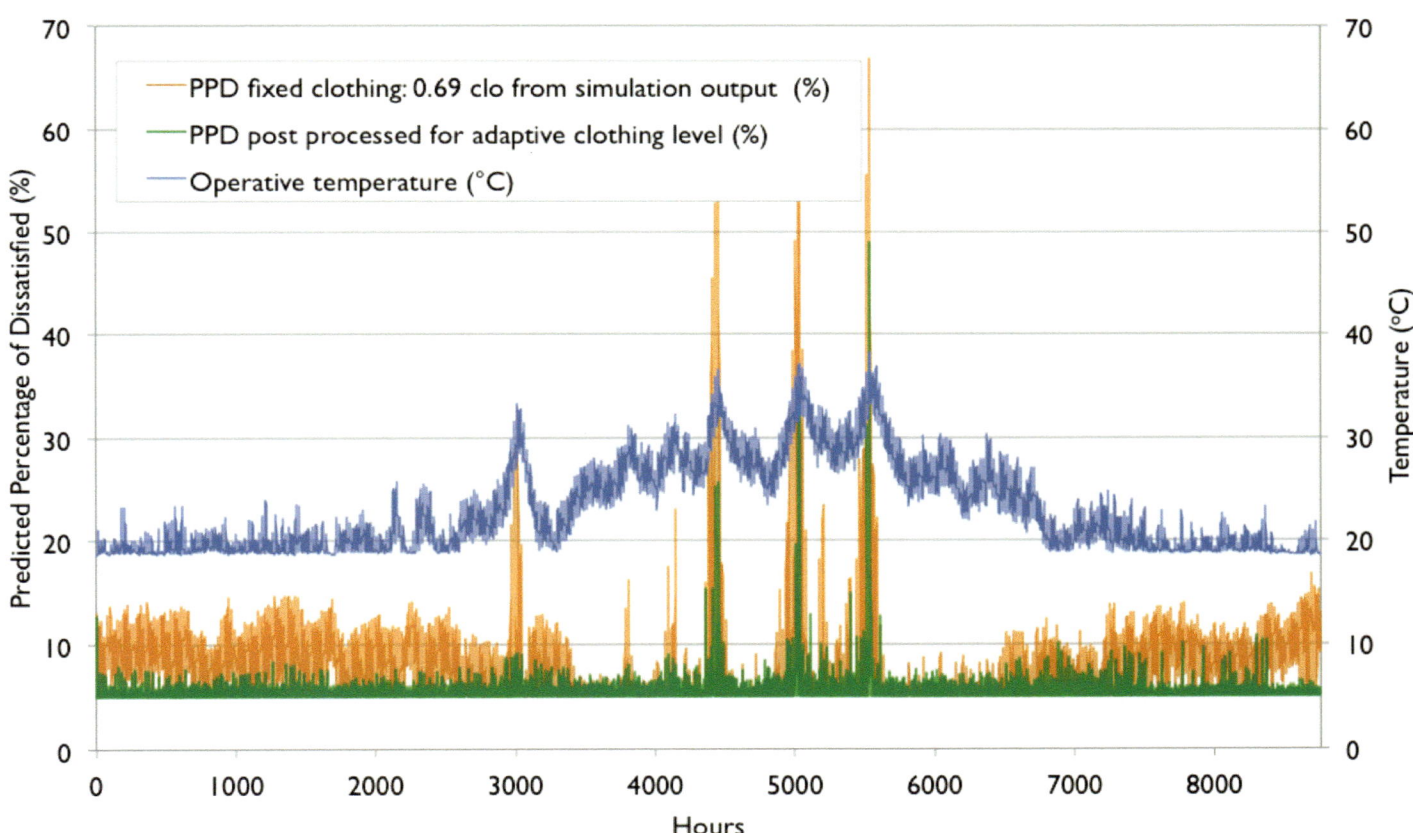

Figure 192   Assessment of thermal comfort based on adaptive clothing behaviour

As it can be seen from this figure, adaptive clothing behaviour corresponds to a considerably reduced PPD (reduced number of dissatisfied people), and therefore this kind of analysis should be conducted routinely through post-processing of simulation results.

A similar method can be used for simulation of adaptive activity behaviour through post-processing of simulation results, by increasing activity level when PMV is decreasing and decreasing activity level when PMV is increasing.

Whilst preparing the material for this chapter, I attempted to apply this method of adaptive activity behavour to the above example before the introduction of free cooling and with much higher overheating than in Figure 192. However, in that instance, it did not make much difference. It only started making a noticeable difference after summer temperatures were brought down by free cooling. We can therefore say that the method of adaptive activity behaviour can be useful after the threshold for its application has been lowered by design interventions.

17.5 TASKS FOR SIMPLE SIMULATION EXPERIMENTS

1 Carry out the analysis of thermal comfort using the adaptive clothing method introduced in this chapter. First, conduct an annual simulation of a building and export hourly values of air temperature, mean radiant temperature, relative humidity, and predicted percentage of dissatisfied for a selected room into a spreadsheet. Implement the clothing adaptation in the spreadsheet on the basis of the example in Table 43, and calculate PPD resulting from this adaptation. Compare the calculated PPD with the PPD exported from the simulation model. Experiment by changing the constants from Table 43 that adjust the clothing level and introduce upper and lower limits to clothing levels. Investigate the effect of these changes on the calculated PPD.

2 As above, but instead of adjusting the clothing level, use the same method to adjust activity level/metabolic rate as a post-simulation process in a spreadsheet. Compare the calculated PPD with the PPD obtained from the simulation model and reflect on the accuracy of this approach in comparison with the fixed activity level PPD.

3 Develop a post-processing spreadsheet that combines the clothing level adjustment and activity level adjustment simultaneously. Compare the calculated PPD with the PPD obtained from the simulation model and compare the two.

4 Discuss the results with colleagues.

17.6 SUMMARY OF DESIGN PRINCIPLES

In summary, we looked at what thermal comfort is, what influences it, and what designers can do about it. We also looked at how thermal comfort can be assessed and how simulation outputs can be interpreted and post-processed to overcome limitations of simulation models, thus achieving higher accuracy. The summary of design principles is as follows:

Simulation software:

• Select activity level and clothing level in the simulation tool settings, having checked the units used for these two parameters (they are not always expressed in terms of 'met' and 'clo' in different simulation tools).
• If the values for activity level and clothing level are fixed for the majority of (or for the entire) simulation period, carry out post-processing of simulation outputs using spreadsheet software.

Post-processing using spreadsheet software:

• Calculate PMV on an hourly basis using Equation (53) or an equivalent readymade macro script.
• Using spreadsheet formulae, increase clothing level when PMV is decreasing and decrease clothing level when PMV is increasing.
• Following the same approach, increase activity level when PMV is decreasing and decrease activity level when PMV is increasing.
• Balance the activity and clothing levels as may be appropriate for the corresponding climate and type of space.
• Calculate PPD on an hourly basis using Equation (54) and PMV from the previous three steps.
• Use the result of this calculation for dynamic and more accurate assessment of thermal comfort.

CHAPTER 18

DESIGNING FOR NEGATIVE OPERATIONAL EMISSIONS

This chapter describes a method for operational zero carbon design using multi-objective optimisation, in which technical, financial, and thermal comfort analyses are conducted in parallel. The results present the user with a range of trade-off solutions for operational zero carbon design.

In addition to investigating a large number of candidate solutions, the multi-objective optimisation addresses the complex non-linear interdependence between design variables in a much more systematic way in comparison with a series of single simulations run by the user one at a time.

18.1 WHY WE NEED NEGATIVE OPERATIONAL EMISSIONS AND HOW ARE THEY ACHIEVED?

As we saw in in Chapter 1, emissions embodied in building materials represent a high starting point when the construction of a new building or a retrofit of an existing building is completed. In order to get this high starting point of emissions down to zero, we need negative operational emissions to gradually reduce the total emissions. The reduction to zero is not straightforward, as there will also be positive operational emissions that will be adding to this starting point on an ongoing basis. The overall sum between positive and negative operational emissions needs to be negative in order to get from the high starting point down to zero. This is the essence of the Zero Equation expressed either as a jam jar with a smaller and larger spoon (Figures 4 and 5), or as a mathematical equation – Equation (1), and its derivatives in the form of Equations (2), (3) and (4) in Chapter 1.

Until such time when energy grids are completely decarbonised, and unless a building is completely off the grid, there will always be some positive carbon emissions arising from using a certain amount of fossil fuels in the building on a day-to-day basis, for heating, cooling, lighting, etc. These positive emissions need to be exceeded sufficiently by negative operational emissions from renewable energy systems associated with the building, so that zero emissions are achieved by a certain year, as per Zero Equation.

The negative operational emissions, which we need in order to reduce the starting embodied emissions and the ongoing positive operational emissions to zero, arise from renewable energy systems such as photovoltaic solar systems or wind generators. These systems not only suppress the use of fossil fuels but also supply new energy to the electricity grid, thus reversing overall carbon dioxide emissions.

18.2 CASE STUDY BUILDING FOR APPLICATION OF THE OPERATIONAL ZERO CARBON DESIGN METHOD

In this analysis, we will take an example of a building near Portsmouth, named Greenpower building (Figure 193). The original design was carried out using dynamic simulation methods to evaluate a range of different design options and achieve excellent environmental performance. In this and the next chapter, we will make several assumptions that will enhance the existing design beyond the actual building, effectively using it as a vehicle for operational zero carbon design.

The building site sits between a busy road and private residential properties and has a south orientation. The road is behind a row of trees at the north end of the site. Noise from the road is deflected by the curved green roof, whilst trees are retained to blend the building into the site and protect the neighbours' privacy.

a)   b)

Figure 193   (a) Greenpower building on the opening day on 16 July 2010. The rear of the building (b) consists of a curved green roof that reduces heat losses and deflects noise from the road

The walls, floor, roof, and glazing have insulation properties that are up to 40% better than the UK building regulations at the time of construction, reducing heat losses and the demand on the heating system. Heat losses are further reduced by the green roof and its curved geometry with no windows on the north-facing side.

The external skin is fabricated using lightweight materials to maximise insulation, whilst the core walls and ground and mezzanine slabs are made of dense concrete providing the building with a high degree of thermal mass to attenuate internal temperature fluctuations and reduce heating and cooling demand.

The green roof has several roles: it reduces heat gains in summer, helping natural ventilation to maintain comfortable internal conditions; it reduces the harmful effect of UV radiation, extending the lifetime of the building materials; it attenuates noise from the busy road at the rear; and it makes the roof more robust in windy conditions due to its porous surface, thus preventing the aerofoil effect and the risk of wind damage.

A large amount of south-facing glazing allows the sun to heat the building in winter from low-incidence angles, whilst the front canopy, louvre blinds, and a deciduous tree in the front shade the building to prevent overheating in summer.

The building is naturally ventilated using passive ventilation stacks on the roof, in combination with low-level front vents (Figure 194). The stack and vents are computer controlled.

a) Low-level vents   b) High-level vents leading to passive stacks

Figure 194   Natural ventilation

As a result of high levels of natural daylight from the south-facing glazing the need for electrical light is reduced during most days. Daylight-sensitive control of electrical lighting has been specified as a result of dynamic simulation analysis, considerably reducing the use of electricity.

Space heating is designed on the basis of an air source heat pump, which delivers hot water to the underfloor heating system (Figure 195). The building has rainwater collection system based on gutters that run on top of the structural arches and deliver rainwater to collection funnels on the ground, feeding the water into an underground tank (Figure 196).

a)   b)

Figure 195   (a) Air source heat pump that feeds hot water into (b) the underfloor heating system

Figure 196   Rainwater collection system: a gutter running on top of the structural arch (top) and collection funnel on the ground (bottom) that collects water into underground tank

Although the original building was not designed to have negative carbon emissions, it was energy A-rated, with plans for upgrades for the future achievement of zero carbon performance, and the building owners were happy to make the building available for experimentation towards that goal. The originally intended solar photovoltaic system was omitted from the initial phase of development, but it is used in this analysis as one of the assumptions that enhances the design beyond the actual building.

18.3 SETTING UP THE INITIAL MODEL AND RUNNING SINGLE SIMULATIONS

I developed the initial model in IES Virtual Environment (Figure 197) whilst working on the design team for the actual building. In this chapter, I demonstrate a further evolution of that model in DesignBuilder. To get from IES Virtual Environment to

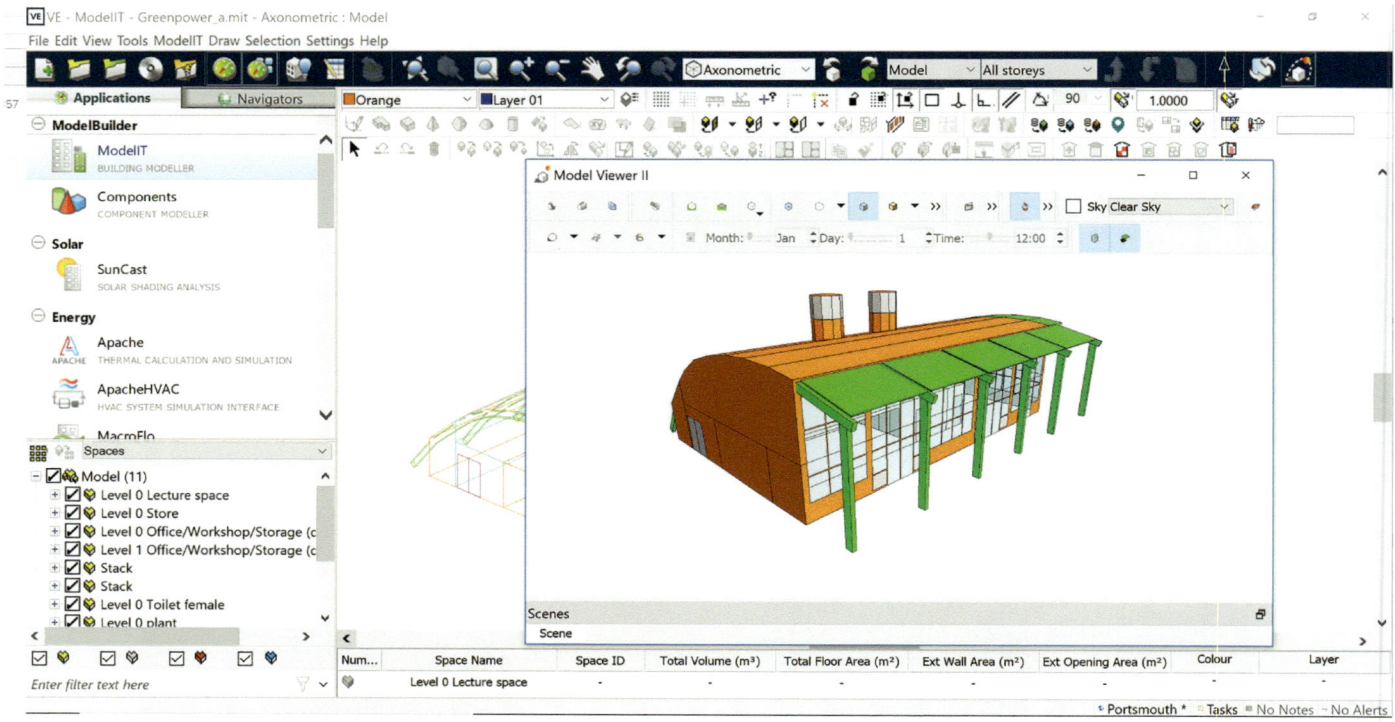

Figure 197   The model initially developed in IES Virtual Environment (Model 'a')

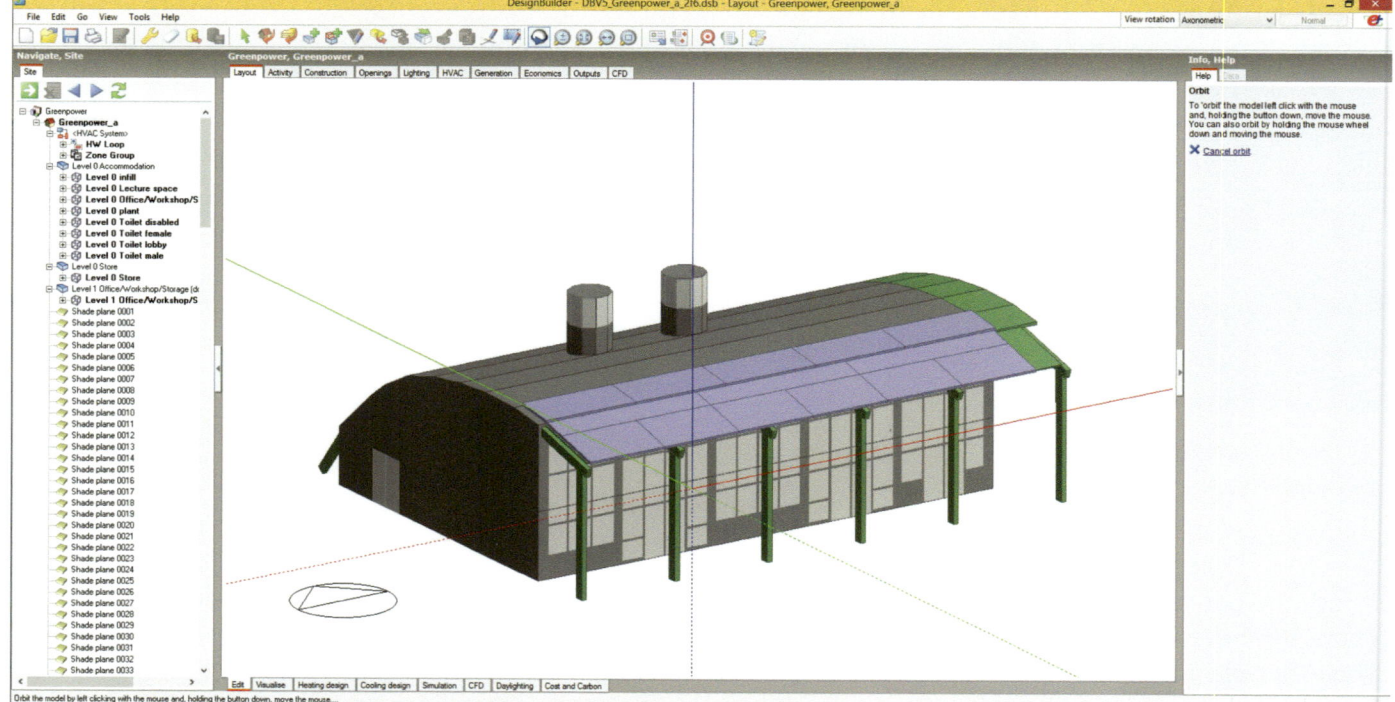

Figure 198   The model from IES Virtual Environment imported and amended in DesignBuilder

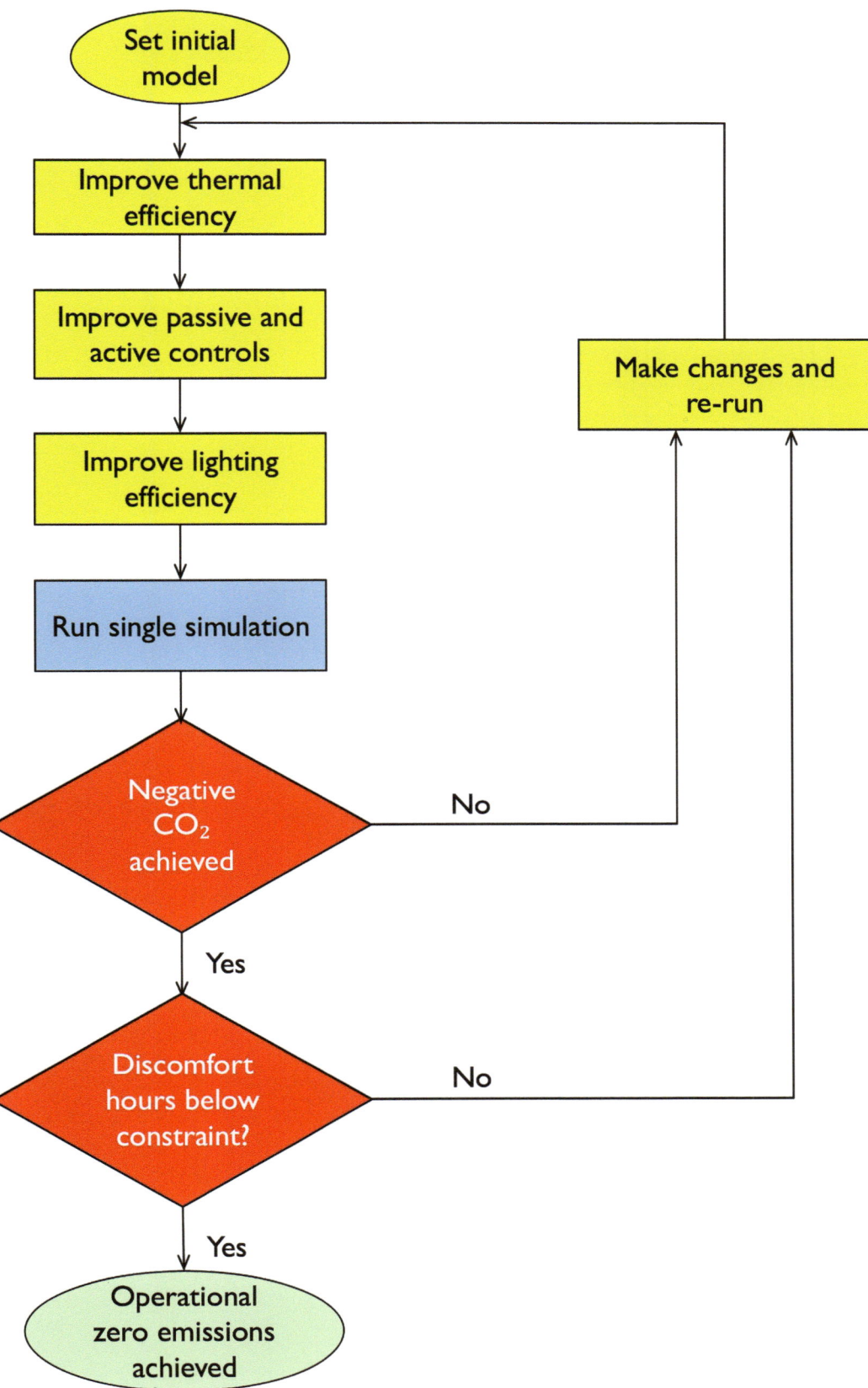

Figure 199   Single simulations design workflow

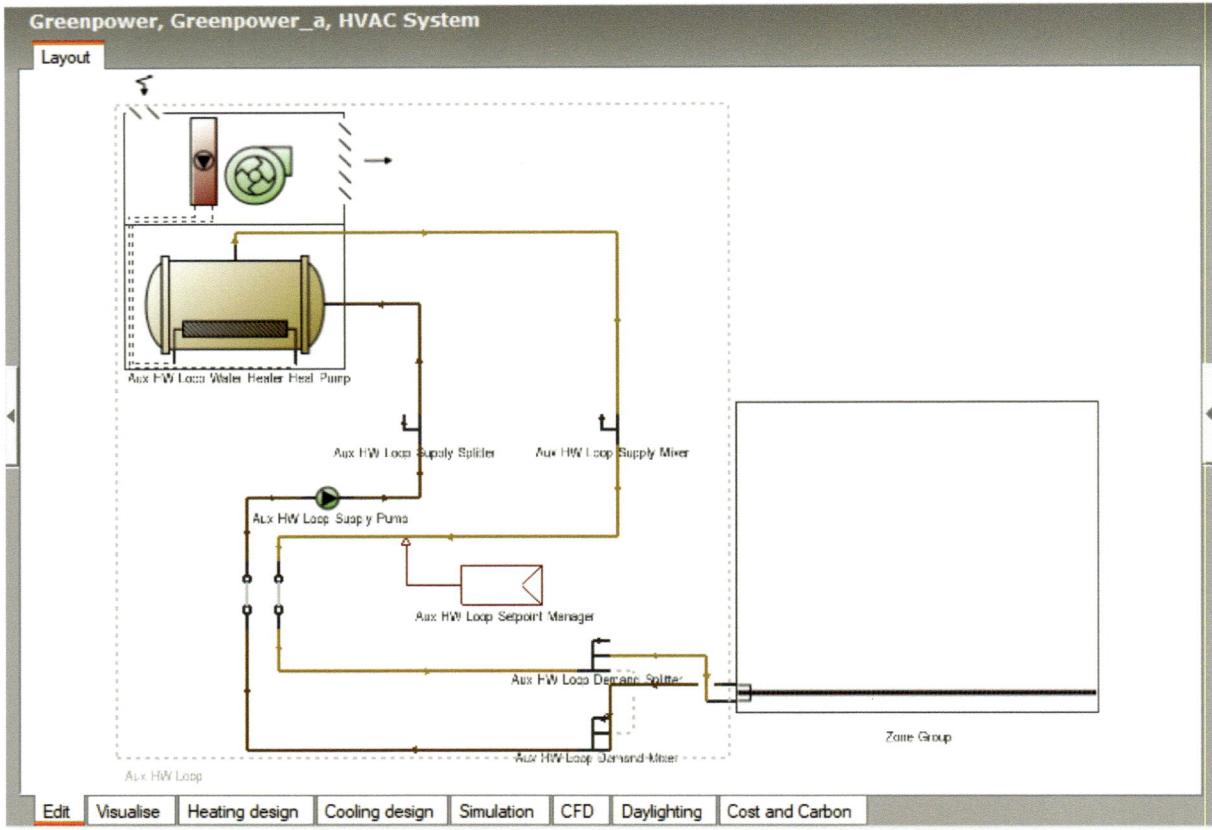

Figure 200   The model HVAC system

DesignBuilder, I exported the model from the former simulation tool and imported it into the latter using gbXML format (Figure 198). That model was subjected to a single simulation design workflow in Figure 199.

The model was set to use an air-to-water heat pump with underfloor heating as in the actual building (Figure 200). Additionally, the model included PV panels arranged in 14 sets of 6.494 m² clusters each (Figure 198).

After following the single simulation design workflow in Figure 199, the model achieved negative operational emissions of −87.97 kg CO_2/year and 186.62 discomfort hours per year, using the values of design variables as in Table 44.

No.	Description	Value	
1	Air tightness (m³/h/m²)	3.6	
2	Wall *U*-value (W/m²·K)	0.10	
3	Lighting power density (W/m²/100 lux)	1 W/m²/100 lux (Controlled)	
4	PV Cluster/Area (m²)	PV14/90.916	
5	Ground floor slab *U*-value (W/m²·K)	0.10	
6	External window opening %	30	
7	Green roof *U*-value (W/m²·K)	0.10	
8	Glazing	Triple LoE Clr 3mm/13mm Argon	
9	Thermal mass internal	100 mm concrete	

TABLE 44   VALUES OF DESIGN VARIABLES OF THE FIRST NEGATIVE OPERATIONAL EMISSIONS MODEL

18.4 HOW CAN WE BE SURE THAT THIS IS THE BEST SOLUTION?

The simulation model developed until now has gone through several improvements, shaping it into good performance. It has achieved negative carbon emissions and discomfort hours lower than the minimum specified. But how can we be sure that this is the best solution?

This question arises because of the many variables used in the model as shown in Table 44. There are nine variables in this table, and each one can take multiple values. This is effectively a problem in nine dimensions, which is incomprehensible for us who have worked mostly in three (x, y, z) or four (x, y, z, time) dimensions.

If we represent the multi-dimensional space with the more familiar three-dimensional space (Figure 201), this can help to explain how single simulations work. In the rugged terrain of valleys and hills illustrated in this figure, the individual valleys represent local optimum results, and the lowest valley represents a so-called global optimum result.

What we have effectively done when using single simulations is to start a serial search from a single point and settle on the first and most likely a local optimum (Figure 201a), without checking whether there are any better optima. For that reason, we cannot be sure that what we have found with a single simulation, or a small number of single simulations, is the best solution.

Fortunately, there are ways of checking if there are better solutions and finding such solutions. Computational methods called genetic algorithms can perform a parallel search of the solutions space, starting simultaneously from multiple points of that solution space, as illustrated in Figure 201b, as part of a design optimisation process.

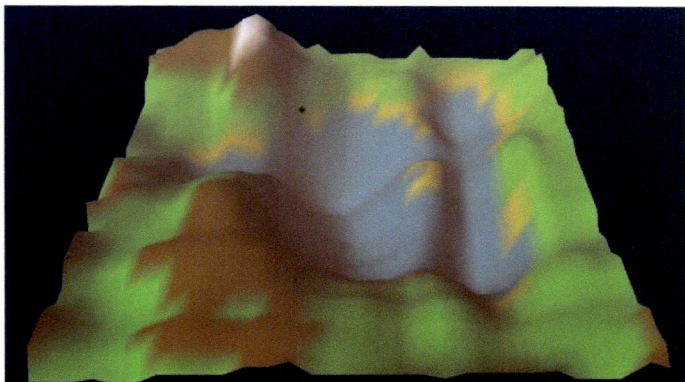
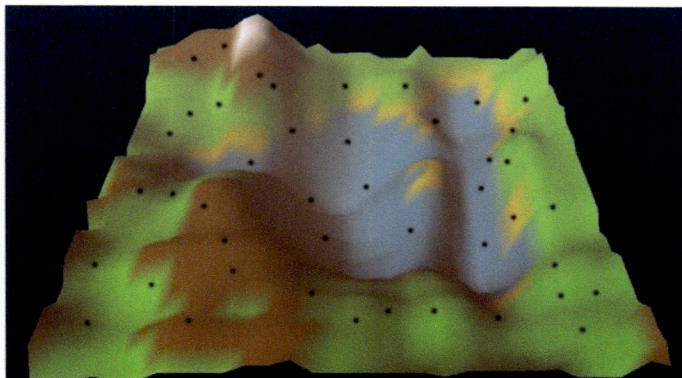

a) Serial search starting from a single point
b) Parallel search starting from multiple points

Figure 201   A conceptual representation of a multi-dimensional solution space and different numbers of starting points for searching the optimum

18.4.1 A brief overview of genetic algorithms

A Genetic Algorithm (GA) that operates in the background of the optimisation process requires the population size and the number of generations to be set. What is the meaning of these parameters and how do they relate to building simulation?

Although a detailed discussion of GAs is outside of the scope of this text, it is important to give here a few details that are relevant to building simulation, especially in terms of helping with the choice of the GA parameters. Readers interested in further details of the GAs are encouraged to consult an excellent introduction to the field by Goldberg (1989).

GAs are search methods based on the simulation of natural genetics. In natural genetics, genes from parents are combined by a crossover to create an offspring. The crossover occurs at random positions of the gene strings. Genes also undergo random mutations in parts of the individual strings. Natural evolution favours the survival of the fittest, so that each new offspring will succeed if their fitness, achieved by their genetic makeup, is sufficient to keep them alive and make them prosper. Hence, not all parents (solutions) are selected to create the next generation. Instead, there is a 'tournament' of fit parents (good solutions) that decides which of the best solutions will be selected for creating the next generation.

Unlike other search methods which start from a single point, GAs start from a population of points. The advantage of this approach can be easily seen in the context of a conceptual representation of a multi-dimensional solution space (Figure 201). If we start searching a complicated solution landscape of valleys and hills from a single point in series (Figure 201a), the horizon of this landscape will look similar in all directions. Choosing to go in a particular direction, we are much less likely to find the global minimum and much more likely to get locked into a local minimum. In comparison, the GA searches the solution landscape in parallel from a population of points spread across the entire landscape. Thus sequential generations of artificial genes 'crawl' up and down the solution space starting from numerous points in parallel (Figure 201b), and that makes it much more likely not only to find the global minimum but also to find it quickly.

18.4.2 Genetic algorithms in the context of building simulation

This now enables us to take the discussion into the domain of building simulation, to define the population, the generations, and the fitness, and enable us to make informed decisions about setting up a multi-objective optimisation.

GAs operate as black box models and are only concerned with the relationship between the gene properties and the fitness, and not with the internal workings of the system that is made from these genes. This is fortunate, as it does not require the GA to 'know' all internal workings of the building simulation model.

What is then a gene in the context of building simulation optimisation? When setting up a multi-objective optimisation, we set design variables together with the respective ranges of values they will take any steps within these ranges. A gene in this context represents the encoding of each value that each design variable can take. A collection of genes corresponding to a specific combination of values of design variables, called a 'chromosome', is a recipe for a design solution and is represented as a single point in the solution space (Figure 201b). A population is a group of chromosomes representing different design solutions and thus occurring in different parts of the solution space. The larger the initial population, the denser the coverage of the solution space represented in Figure 201b, and the more efficient search for the optimum solution. The fitness of each chromosome is the distance from the origin of the coordinate system defined by the chosen objectives of the multi-objective optimisation. As we will see later in this chapter, these objectives are Comfort, CO_2, and cost.

Chromosomes with high fitness are put into a pool to create a new generation either as a crossover recombination between pairs of parents or on the basis of mutation. The crossover between two parents will create a solution that lies on the line determined by the parents' characteristics. The mutation will create a solution that lies outside of the line determined by the characteristics of the parents. The optimisation will run for a specified number of generations, and it will then stop, giving the results in the form of a Pareto chart.

Therefore, increasing the number of design variables will increase the number of genes and will make the solution space larger and more challenging, and the chromosomes more comprehensive. Increasing the initial population size will increase the coverage of the solution space by the chromosomes. In other words, it will increase the number of points within the solution space from which a parallel search for the optimum solution will be carried out. However, if the search is predominantly based on the crossover, the initial randomly distributed population will limit the search for solutions between the pairs of the initial population members. Increasing the mutation rate will extend the search outside the initial population spread. Increasing the number of generations will make the search more exhaustive, but it will also make the search process longer. Specifying a small number of generations may result in an incomplete search for the optimum solution, and therefore give sub-optimum results. Specifying a large number of generations may result in a lot of unnecessary computation after the optimum solution has been reached.

In order to make the most of multicore simulation servers, a population size equivalent to the number of parallel processes that can be run on the server can be specified. For instance, the simulation server used for the work described here is capable of 32 simultaneous energy simulations, and therefore the simulations that generated the results for this chapter used a population size of 32.

18.5 OPERATIONAL ZERO CARBON DESIGN AS A MULTI-OBJECTIVE OPTIMISATION PROCESS

As the objectives of operational zero carbon design are to achieve negative operational emissions, minimum cost, and maximum comfort, this can be considered to be a multi-objective optimisation problem, with three objective functions and a number of independent variables associated with these functions. The method explained in this chapter will therefore align operational zero carbon design with multi-objective optimisation. The solutions obtained by the method provide the design team with choices that enable trade-offs to be made between conflicting design constraints.

18.5.1 Software and hardware tools

At the time of writing this text, there were several simulation tools capable of multi-objective optimisation analysis. My familiarity with DesignBuilder as well as its multi-objective optimisation capability determined the choice of simulation tool used in this chapter.

Multi-objective optimisation can run on a local machine, such as a personal computer or a laptop, but that mode of operation could be very computationally intensive for a small machine. Whilst running on the local machine, multi-objective optimisation could easily use up most of the processor resources and can take many hours to complete, thus making it hard to run any other less intensive tasks at the same time. This problem can be overcome by using a dedicated simulation server, a remote multicore machine where simulation jobs can be sent to over the network for faster execution.

In DesignBuilder, this remote simulation facility is provided via JESS, a third-party JEPlus Cloud Simulation Server, where JEPlus is an EnergyPlus simulation manager for parametric simulations (Zhang, 2022).

18.5.2 Planning the multi-objective optimisation

DesignBuilder, the simulation tool chosen for this analysis, uses two objectives at a time to carry out multi-objective optimisation. However, operational zero carbon design requires simultaneous achievement of three objectives: optimisation of comfort, cost and CO_2 emissions. One of the ways to map the three objectives required for operational zero carbon design to the software capability of two objectives is to formulate the problem as a pair of two multi-objective optimisations with two objectives each: Comfort versus CO_2 emissions optimisation and Cost versus CO_2 emissions optimisation. This is illustrated in Figure 202.

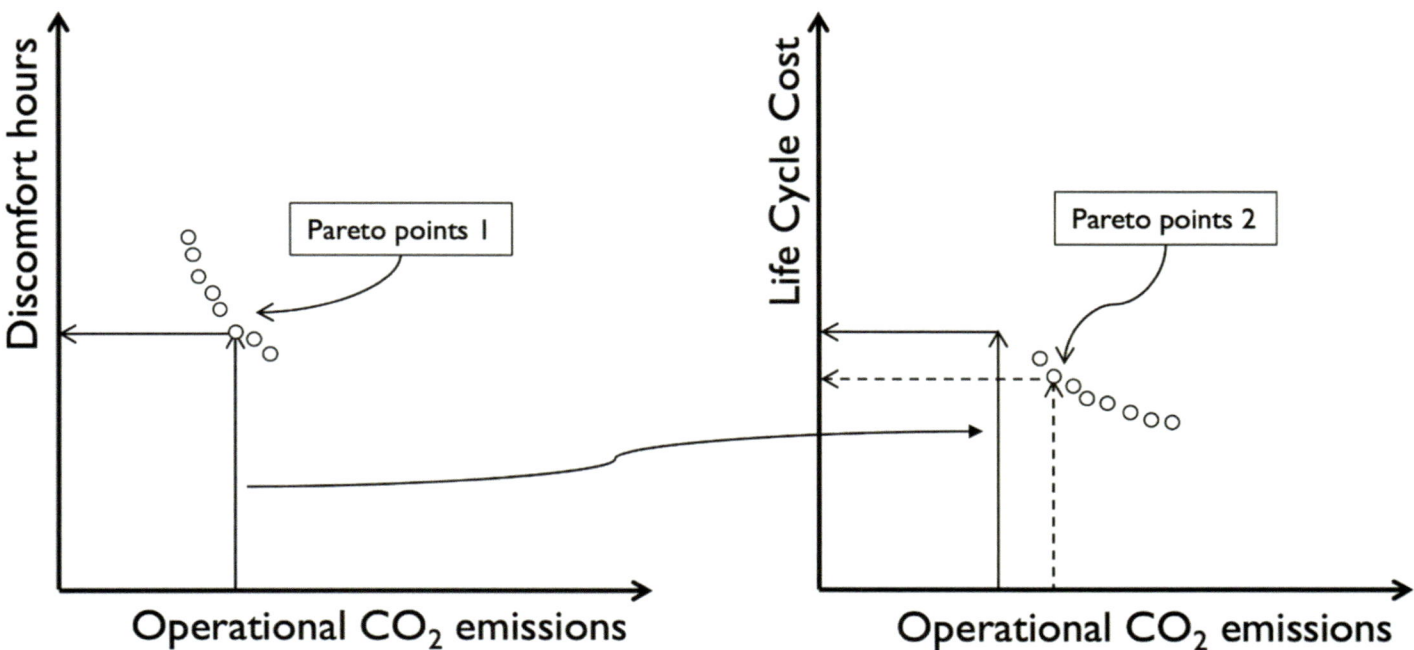

Figure 202   An illustration of a double two-objective optimisation scenario

For this analysis to work, both sets of the multi-objective optimisation results need to converge to the same values of CO_2 emissions. In other words, when the straight solid lines with arrows from the left coordinate system in Figure 202 are transferred to the right coordinate system, they need to coincide with the dashed lines with arrows in relation to at least one of the points in the right coordinate system. In practice, I found that the two sets lines are unlikely to coincide.

Therefore, an alternative approach is adopted: a single two-objective Cost versus CO_2 emissions optimisation, restricting the results to a minimum level of discomfort hours (in other words, setting thermal comfort as a constraint), as illustrated in Figure 203. The rationale for this approach is that we cannot compromise on thermal comfort. Therefore, it will be sufficient to decide on an acceptable thermal comfort range and use it as a constraint in this optimisation.

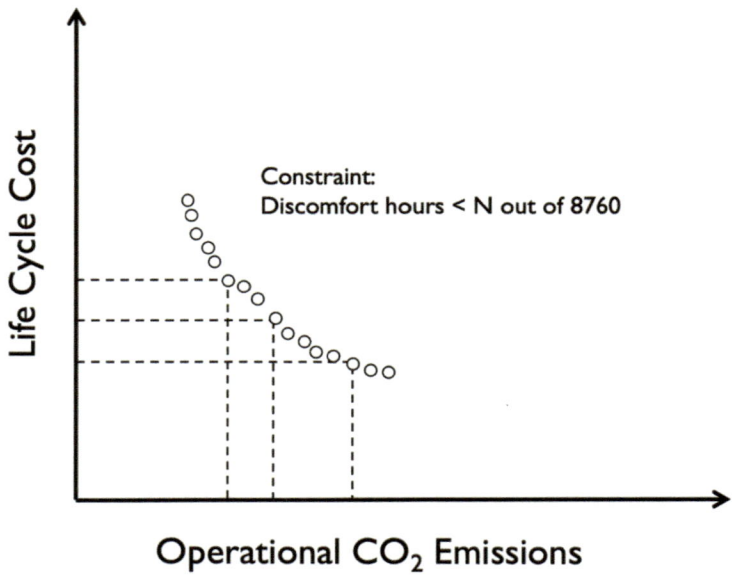

Figure 203   *A single two-objective optimisation method for operational zero carbon design with constraints*

The multi-objective optimisation with multiple independent variables will not provide a single optimum as a result. Instead, a range of optima will be available, represented by the discrete points in Figures 202 and 203. These points cluster on a so-called Pareto front, named after an Italian scientist Vilfredo Pareto. They give a distribution of values of the objective functions that are the closest to the origin of the coordinate system, and thus represent the lowest possible values of the objective functions.

Different combinations of values of design variables will result in different optimum points on the Pareto front, and hence there is no single optimum produced as result of the multi-objective optimisation. This helps design decision-making. The dashed lines in Figure 203 indicate the alternative choices that the designer can make from a range of options, as trade-offs available from the Pareto front.

From this point on, we will adopt the workflow as shown in Figure 204 in order to find out if there are any better solutions than the one achieved using a single simulation earlier in this chapter.

These trade-offs will be used in combination with embodied emissions in Chapter 19, in order to achieve cumulative zero emissions, embodied and operational, with certainty and by a specified year.

18.5.3 Choosing design variables and their variation range

The independent variables in a multi-objective optimisation process are in fact design variables for the building. These design variables need to be taken through a wide enough range in order to ensure that operational zero carbon design can be found in that range. In other words, we need to cast our net wide in order to ensure that operational zero carbon design will occur in it. The design variables also need to be taken in sufficiently small steps through their respective ranges, in order to ensure that we hit rather than miss the optimum value.

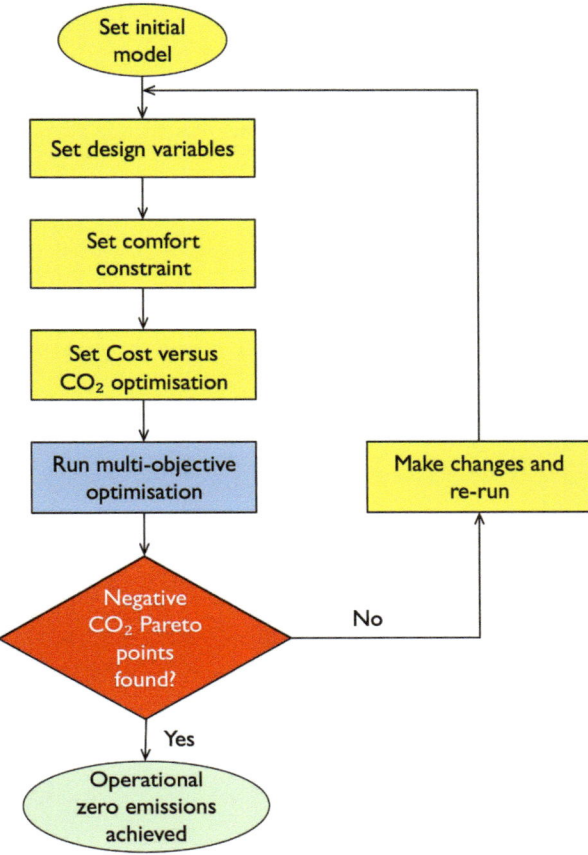

Figure 204   Multi-objective optimisation design workflow

Hence, the following design variables are chosen, as well as their ranges and steps, as shown in Table 45. As it can be seen from this table, the total number of possibilities for a single design is over 3.3 million! This is a very large number of possibilities which we refer to as a solution space. That solution space will be searched through using an evolutionary algorithm. The evolutionary algorithm will not simulate all parameter combinations, but it will streamline the search down to hundreds of the most promising solutions, and ultimately produce a collection of optimum solutions to choose from. This is fortunate, as it will reduce the waiting time for simulation results to a reasonable time scale. As it takes just over 11 minutes for each annual simulation from Table 45, the total time that the user would need to wait for the results would be 69 years on a single processor machine or over two years on a 32-processor machine! Instead, with help from the evolutionary algorithm, where not all but only the most promising simulations are carried out, the results will be returned from the simulation server in a matter of hours.

TABLE 45 DESIGN VARIABLES AND VALUES CHOSEN FOR INITIAL MULTI-OBJECTIVE OPTIMISATION				
Variable	**Minimum value**	**Maximum value**	**Step**	**Number of cases**
	or List			
Air tightness ($m^3/h/m^2$)	0.6	4	0.2	18
Wall construction U-value ($W/m^2 \cdot K$)	0.10, 0.18, 0.26			3
Roof construction U-value:	0.10, 0.16, 0.25			3
Thermal mass – Ground floor U-Value ($W/m^2 \cdot K$)	Floors (ground): U-Value=0.10, 0.16, 0.22			3
Thermal mass – Internal floors: material, thickness	Floors (internal): 25 mm timber, 100 mm concrete, 150 mm concrete			3

TABLE 45   CONTINUED

Variable	Minimum value	Maximum value	Step	Number of cases
	or List			
Glazing	Low emissivity double, triple, quadruple			3
External window opening	0%	80%	10%	9
Lighting power density	1 W/m²/100 lux Uncontrolled 1 W/m²/100 lux Controlled 2 W/m²/100 lux Uncontrolled 2 W/m²/100 lux Controlled 3 W/m²/100 lux Uncontrolled 3 W/m²/100 lux Controlled			6
PV (m²) – using all 14 clusters	6.494	90.916	6.494	14
		The size of solution space: product of the above		3,306,744

18.5.4 Setting up life cycle cost analysis

Life cycle cost analysis takes into account construction and operation costs. Construction costs in DesignBuilder are entered in the construction tab for each individual construction type. Operational costs are entered in the economics tab, via tariff settings (Figure 205).

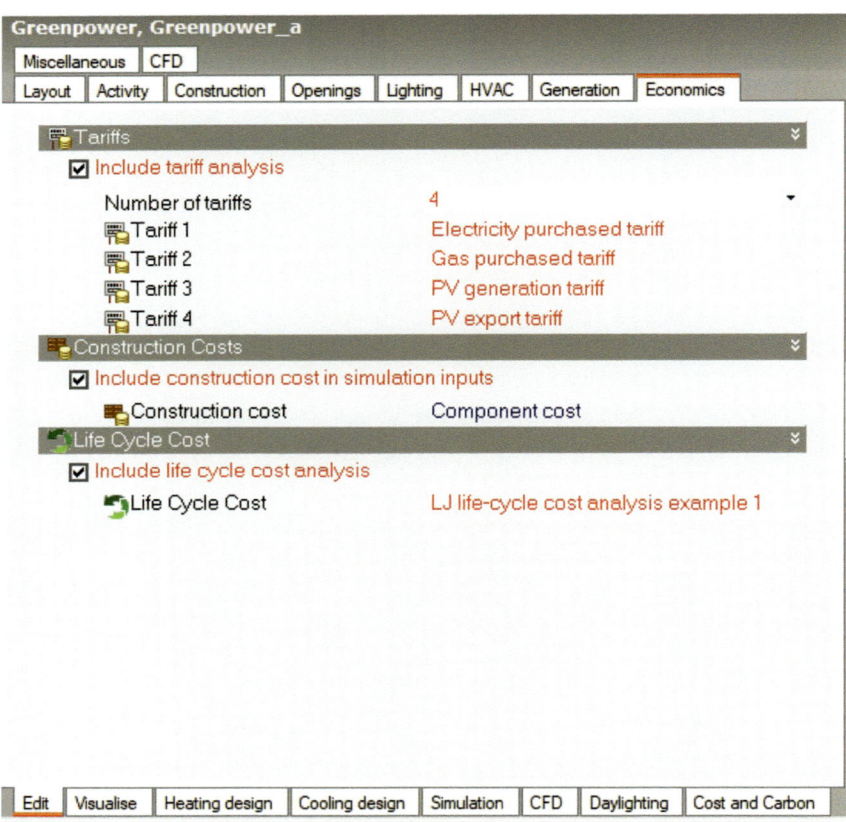

Figure 205   DesignBuilder Economics tab for setting life cycle cost analysis

The tariff setting values are shown in the dialogues in Figure 206. As energy systems in this model include heat and electricity supplied by the utility provider, as well as a PV system, there are four tariffs that need to be set in general analysis: purchased heat, purchased electricity, generated electricity, and exported electricity. In the particular case, heating is by a heat pump, and therefore purchased heat will be part of purchased electricity. The purchased heat tariff is kept here as a placeholder to make this example more general. Although the generated electricity tariff has been discontinued in the UK in 2019, it is also kept here as a placeholder for regions where it may be applicable, and has been set to zero in this particular example. These tariff settings need to be associated with individual energy meters in the model. It is therefore advisable that the names of available meters be checked in the 'Summary' tab in the results of a single simulation prior to setting up the optimisation, and to then use these meter names in the dialogue boxes shown on the left hand side of Figure 206 (a–d), in the field for 'Output meter name'. This is where the units of energy that correspond to the costs are also chosen, and where the service charges are entered, as well as where we set whether energy is bought from or sold to utility.

Energy costs are then entered in the dialogue boxes on the right-hand side of Figure 206 (a–c) and (d). Note that purchased energy unit costs are positive and sold energy unit costs, including generated (where applicable) and exported electricity, are negative.

In addition to the correct setting of the above tariffs, it is essential to set 'Source variable' to 'totalEnergy' in the dialogue boxes in the right-hand side of Figure 206. Otherwise, all energy costs will be zeroed out, making the life cycle cost analysis invalid.

It is therefore advisable to run a single test simulation first and check in the 'Summary' tab that all energy costs are calculated as non-zero values and with the correct sign: purchased as positive, sold as negative.

The reason why the tariff setting is given with this level of detail here is to ensure the reproducibility of this analysis by the reader. It is also recommended that the reader goes through the other model settings from the economics tab (Figure 205), available from the supplementary content web site www.ljankovic.com before embarking on their own analysis.

18.5.5 Running the multi-objective optimisation with constraints and interpreting results
The design variables for multi-objective optimisation are set as shown in Figure 207.

The next step is to set objective functions for multi-objective optimisation, as shown in Figure 208.

The final step before running the optimisation is to set discomfort hours as the optimisation constraint, shown in Figure 209.

After running the optimisation with the comfort constraint for a while, no negative CO_2 Pareto points could be found in the results (Figure 210). For that reason, the optimisation was stopped in order to analyse the values of design variables and make adjustments.

As can be seen from Figure 210(a), the best Pareto solution (-387.54 kg CO_2/year and 121.95 discomfort hours per year) is better than the one obtained by single simulations earlier in this chapter. However, there is a hint of an even better solution that achieves less than -500 kg CO_2/year, however, that solution fails the comfort constraint, seemingly because the external window opening is set to 0% (Figure 210b). For that reason, we will now adjust the design variables as shown in Table 46 and re-run the multi-objective optimisation. The PV clusters are now reduced from a total of 14 to the largest two, and the external window opening range is adjusted from 0%–80% to 20%–100%. A comparison between the sizes of solution spaces between Tables 45 and 46 shows a reduced solution space from over 3.3 million to just under 0.5 million.

After running the optimisation with amended values of design variables, a single Pareto point has emerged with negative CO_2, shown as a red point in Figure 211.

This solution extracted from the numerical results file is shown in Table 47. The model has achieved a performance of -532.57 kg CO_2/year and 107.15 discomfort hours per year. In this way, operational zero carbon design has been achieved and

(a)

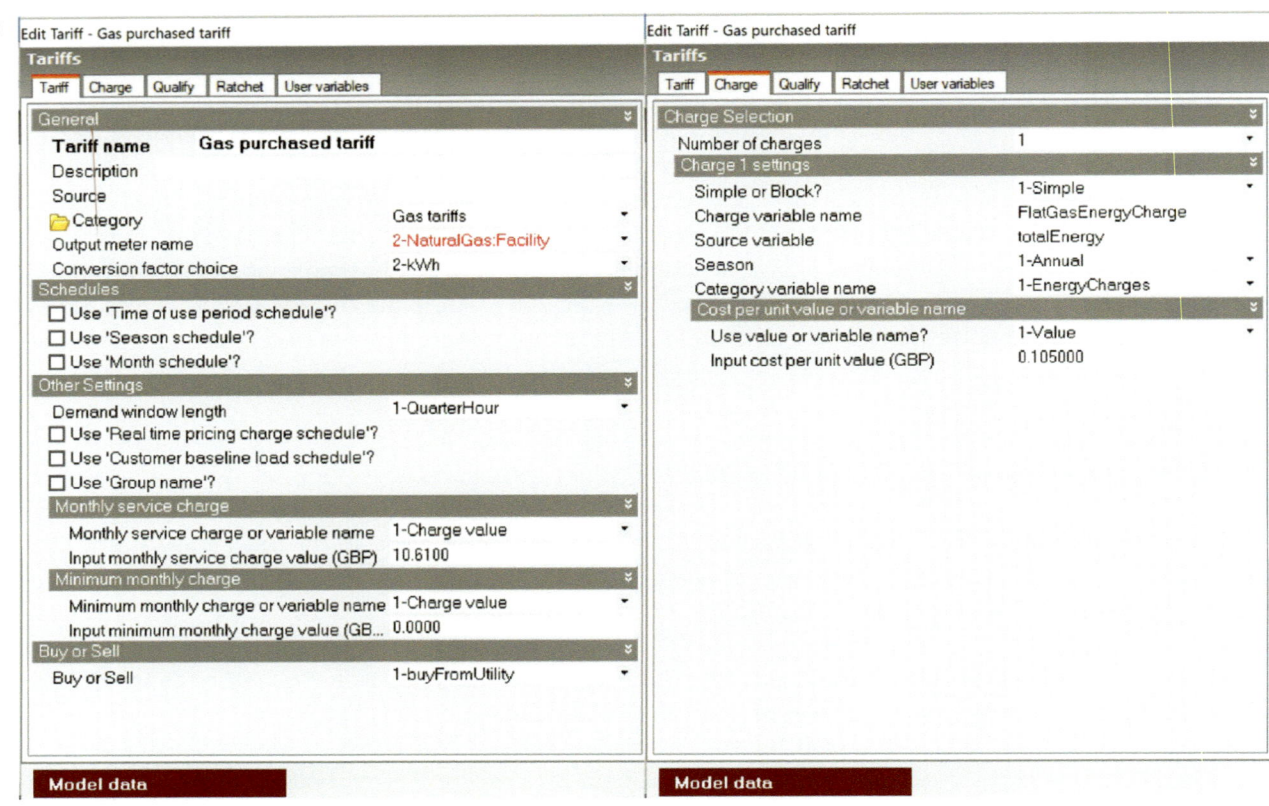

(b)

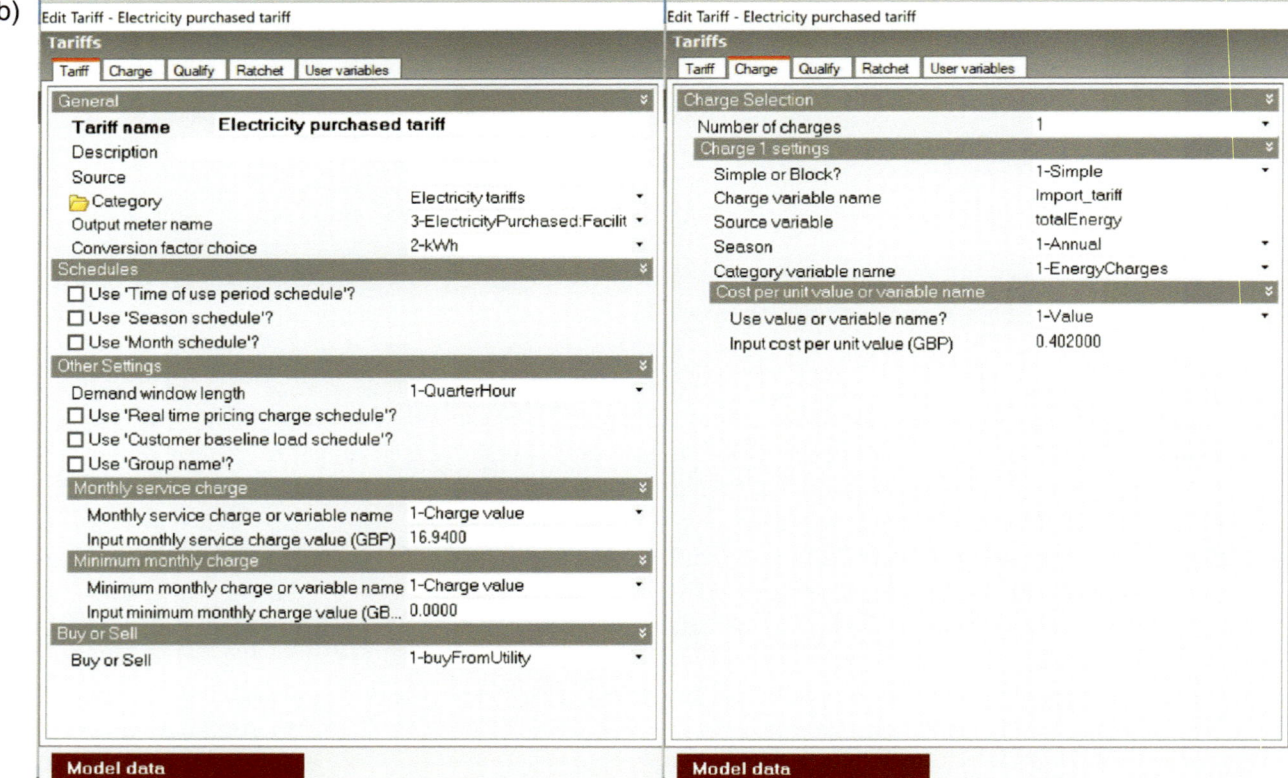

Figure 206 Energy tariff settings for life cycle cost analysis in DesignBuilder

(Continued)

(c)

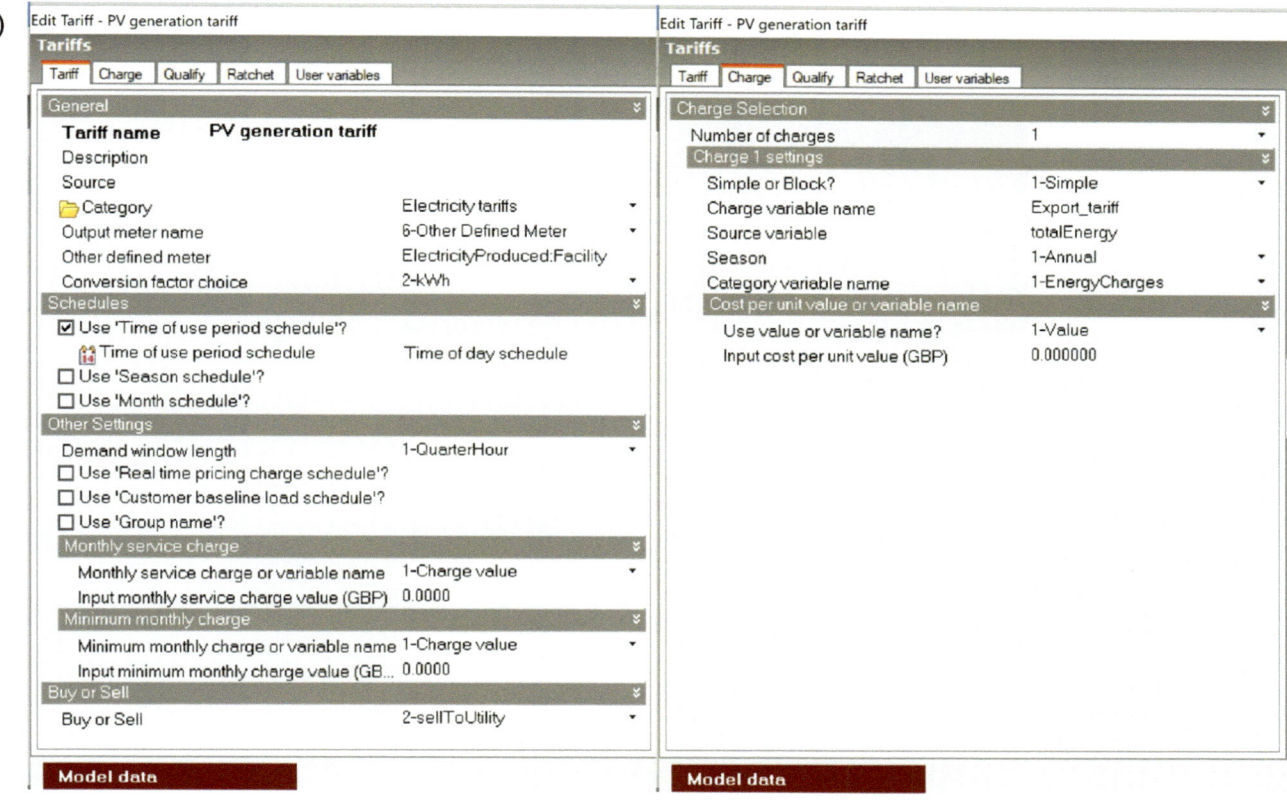

(d)

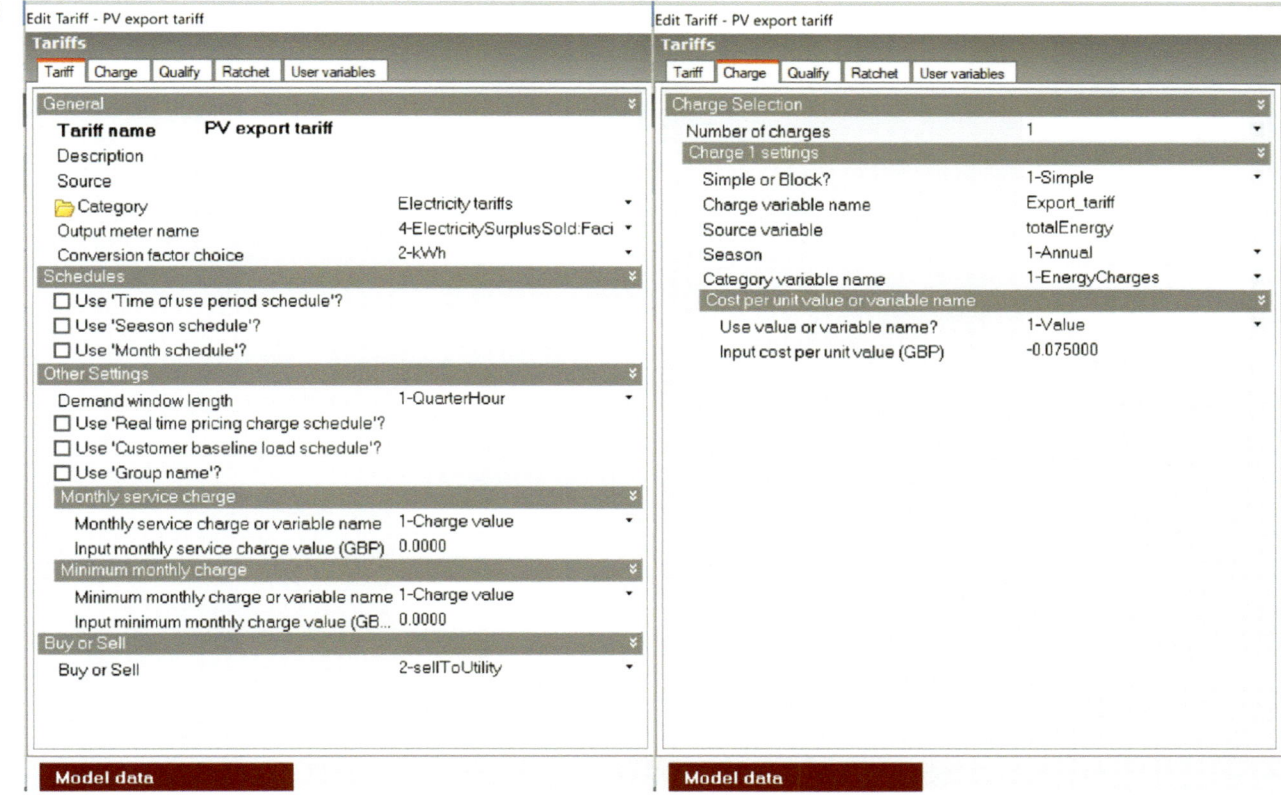

Figure 206   (Continued)

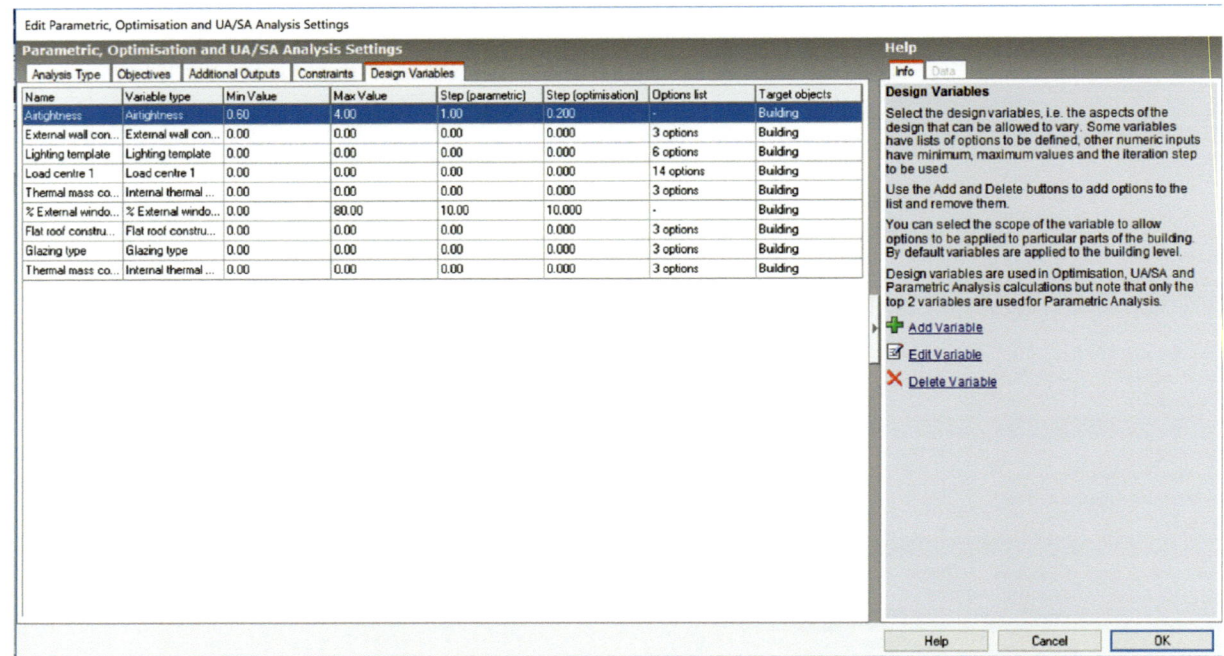

Figure 207   Setting up design variables and corresponding parameter ranges in DesignBuilder

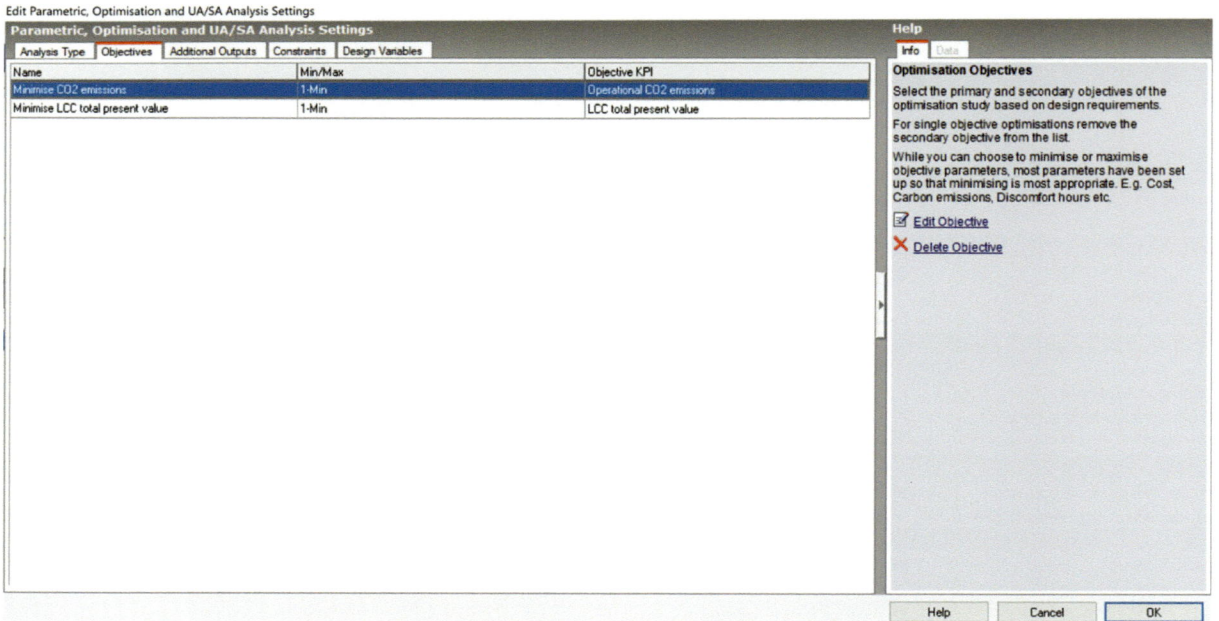

Figure 208   Setting up Cost versus CO_2 optimisation in DesignBuilder

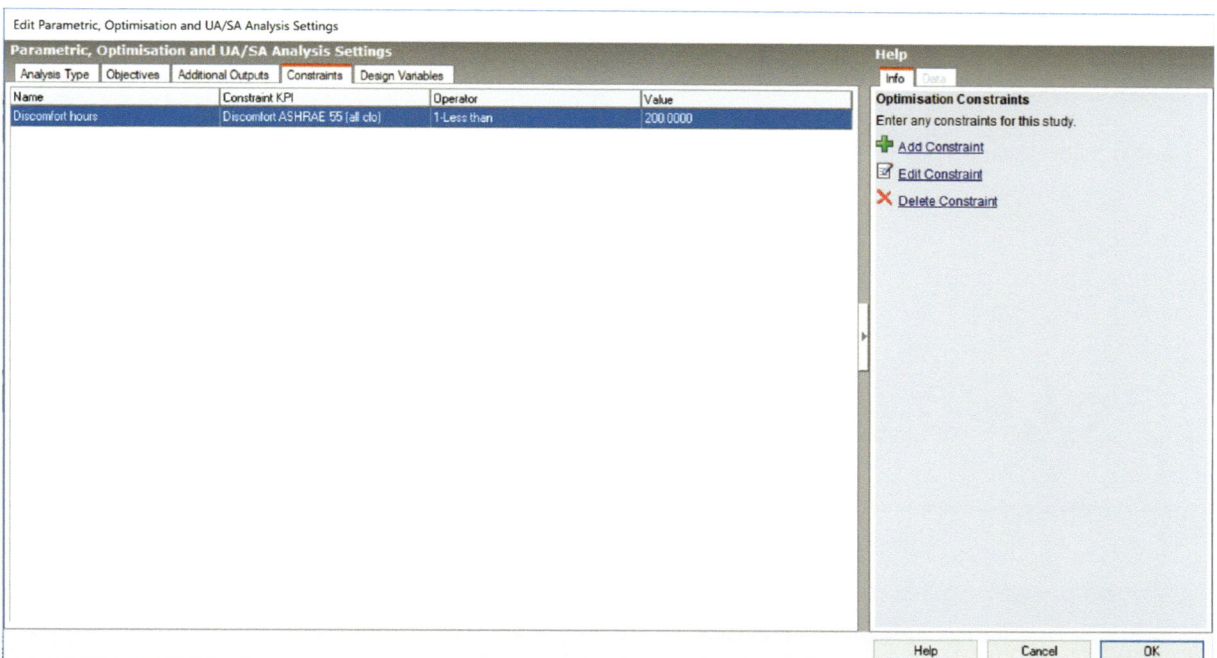

Figure 209   Setting the optimisation constraint in DesignBuilder

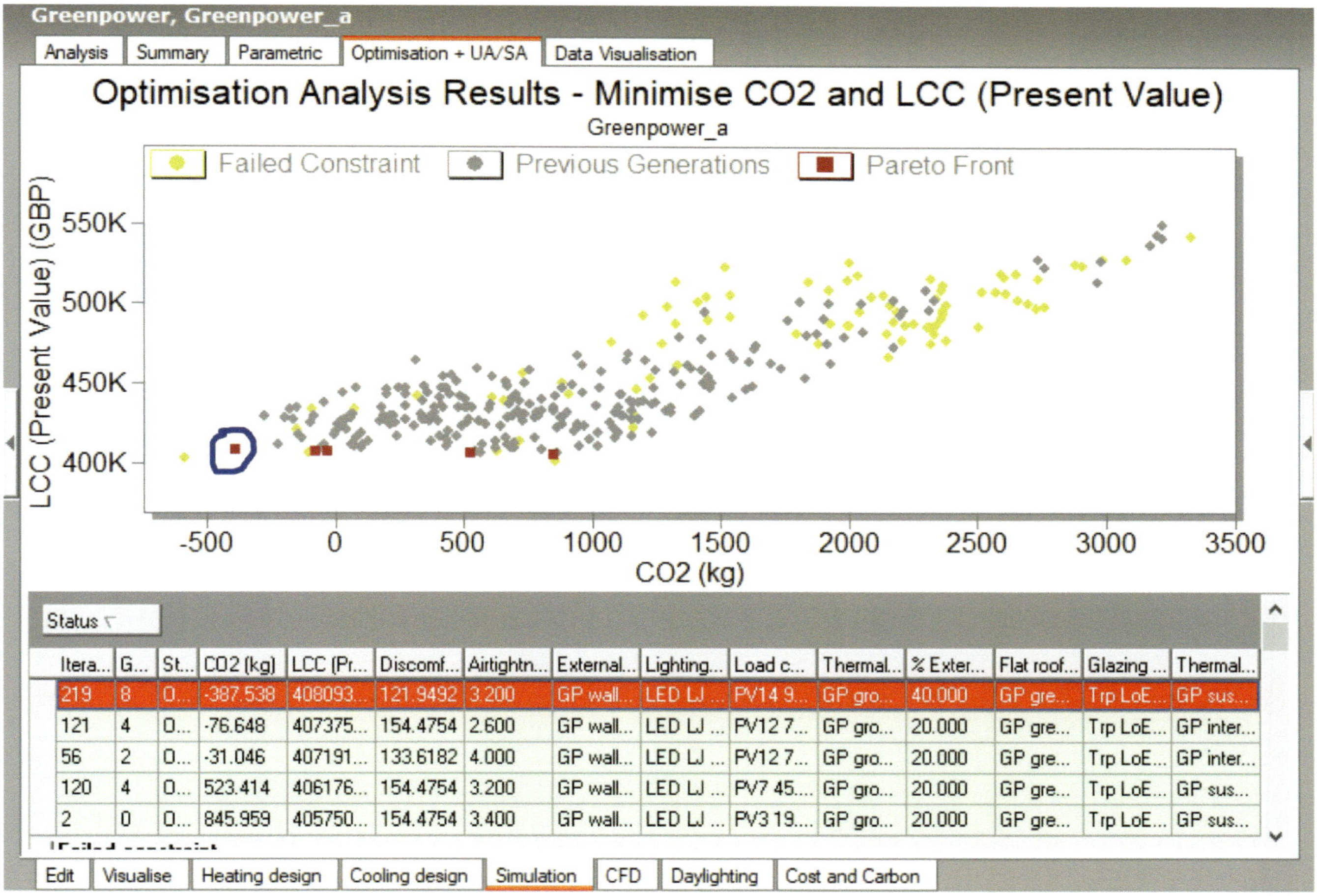

a) Better operational emissions and discomfort hours were achieved than in the single simulations

Figure 210   Results of initial Cost-CO$_2$ emissions optimisation with constraints in DesignBuilder   (Continued)

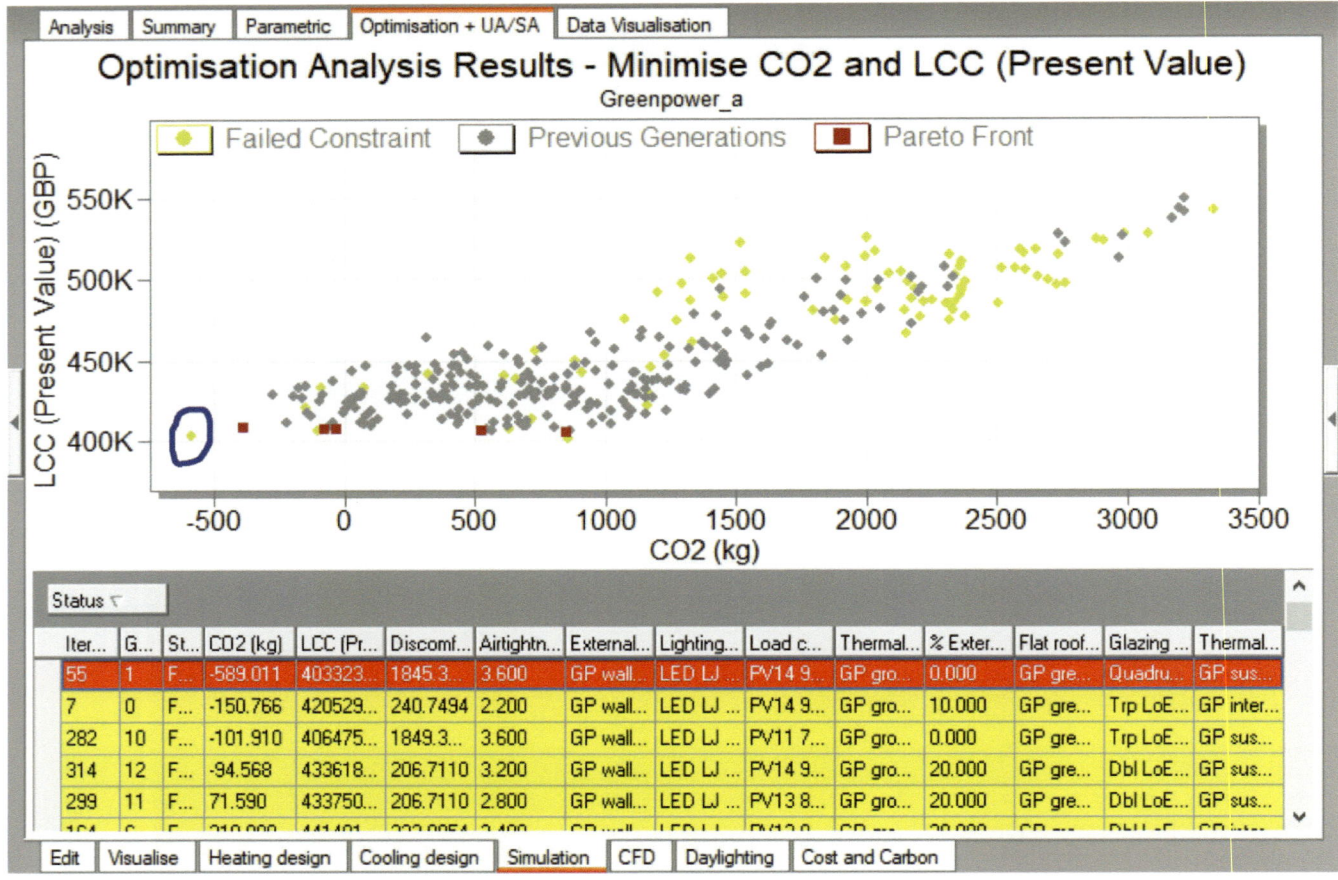

b) Lower operational emissions may be possible with adjusted window opening

Figure 210   (Continued)

TABLE 46   DESIGN VARIABLES AND VALUES CHOSEN FOR AMENDED MULTI-OBJECTIVE OPTIMISATION (AMENDMENTS AS HIGHLIGHTED)

Variable	Minimum value	Maximum value	Step	Number of cases
	or List			
Air tightness (m³/h/m²)	0.6	4	0.2	18
Wall construction U-value (W/m²·K)	0.10, 0.18, 0.26			3
Roof construction U-value:	0.10, 0.16, 0.25			3
Thermal mass – Ground floor U-Value (W/m²·K)	Floors (ground): U-Value=0.10, 0.16, 0.22			3
Thermal mass – Internal floors: material, thickness	Floors (internal): 25 mm timber, 100 mm concrete, 150 mm concrete			3
Glazing	Low emissivity double, triple, quadruple			3
External window opening	20%	100%	10%	9
Lighting power density	1 W/m²/100 lux Uncontrolled 1 W/m²/100 lux Controlled 2 W/m²/100 lux Uncontrolled 2 W/m²/100 lux Controlled 3 W/m²/100 lux Uncontrolled 3 W/m²/100 lux Controlled			6
PV (m²) – using all 14 clusters	84.422	90.916	6.494	2
	The size of solution space: product of the above			472,392

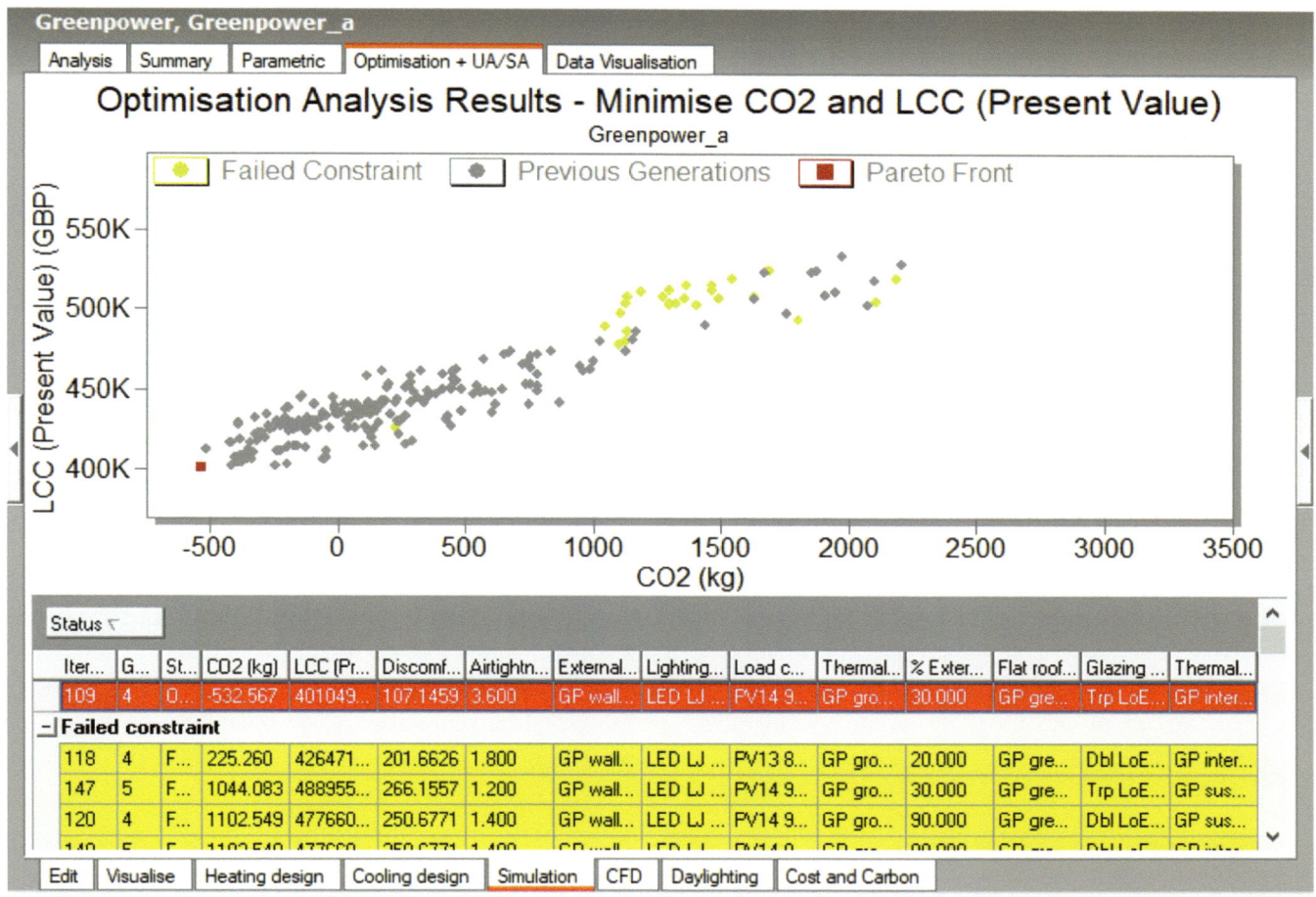

Figure 211   Results of the final Cost-CO₂ emissions optimisation with constraints in DesignBuilder

TABLE 47 ZERO CARBON SOLUTION EXTRACTED FROM THE NUMERICAL RESULTS FILE	
Pareto point	**I**
Iteration	109
Generation	4
kg CO₂/year	−532.6
LCC (£)	401,049
Discomfort (hrs)	107
Air tightness (m³/h/m²)	3.6
Wall *U*-value (W/m²·K)	0.10
Lighting power density (W/m²/100 lux)	1 W/m²/100 lux (Controlled)
PV Cluster/Area (m²)	PV14/90.916
Ground floor slab *U*-value (W/m²·K)	0.16
External window opening %	30
Green roof *U*-value (W/m²·K)	0.25
Glazing	Triple LoE Clr 3mm/13mm Argon
Thermal mass internal	100 mm concrete

it can be implemented in practice by selecting the model parameters from the optimum point as a design recipe for negative operational emissions of this building.

This solution will be used in combination with embodied emissions in Chapter 19, in order to determine when and how cumulative zero emissions, embodied and operational, can be achieved.

18.6 DISCUSSION

We can now reflect on the value of multi-objective optimisation in the design of building operational emissions. First, after following the single simulation design workflow in Figure 199, the model achieved negative operational emissions of -87.97 kgCO$_2$/year and 186.62 discomfort hours per year. After questioning whether this was the best solution and deploying multi-objective optimisation, we found a significantly better solution, with the performance of -387.54 kg CO$_2$/year and 121.95 discomfort hours per year. But there was a hint of an even better solution, with emissions of less than -500 kgCO$_2$/year, but that solution did not achieve the comfort constraint. After changing the ranges of some design variables, the model achieved an even better solution of -532.57 kgCO$_2$/year and 107.15 discomfort hours per year. Could this have been done with a series of single simulations? Potentially yes, but it would have taken a long time and it would have been close to guesswork. Multi-objective optimisation takes simultaneous influences of multiple design variables, where in some cases a reduction of the specification of one variable and an increase of the specification of another variable or a set of variables could lead to a better solution. Multi-objective optimisation runs millions of simulations, navigating through a large solution space by deploying genetic algorithms inspired by the biological paradigm of natural evolution. This is hard to do in a piecemeal fashion one step at a time using the single simulation workflow. Thus, multi-objective optimisation can squeeze every 'gram' of building performance and automate the way we design buildings. Given a choice between single simulations and multi-objective optimisation, we should in my view always go for the latter.

18.7 TASKS FOR SIMPLE SIMULATION EXPERIMENTS

Redesign a 20-foot shipping container into a building with negative operational emissions using the methods from this chapter. The container dimensions are: 20ft (6.096m) long × 8ft (2.438m) wide × 8ft 6in (2.591m) high and the material is corrosion-resistant steel. Use the following design variables, taking those through a wide enough range and in sufficiently small steps in order to ensure that operational zero carbon design can be found in that range:

• Airtightness
• Wall construction
• Roof construction
• Thermal mass
• Glazing
• Window to wall ratio
• Percentage of external window opening
• Lighting power density
• PV array size

Discuss the trade-off solutions with colleagues, using the resultant Pareto fronts and numerical result files.

18.8 SUMMARY OF THE METHOD AND DESIGN PRINCIPLES

The method for operational zero carbon design is summarised in Figure 199 for single simulations workflow and in Figure 204 for multi-objective optimisation workflow.

The summary of design principles is as follows:

• Develop an initial model and run a few single annual simulations, improving thermal efficiency, passive and active controls, and lighting efficiency, and adding renewable energy. Repeat the process if negative emissions and minimum discomfort hours have not been achieved.

- If negative emissions and minimum discomfort hours have been achieved, investigate if there are better solutions using multi-objective optimisation.
- Set design variables and their value ranges in sufficient increments, including renewable energy systems, to cover the solution space widely and in a sufficiently fine grid.
- Set an optimisation constraint using an appropriate number of discomfort hours.
- Run a Cost versus CO_2 optimisation using the chosen value of discomfort hours as a constraint.
- Use the best Pareto front points from this optimisation that have fulfilled the constraint and are in the negative part of the CO_2 emissions axis as the solution for operational zero carbon emissions.
- Use this solution in relation to embodied emissions in Chapter 19, in order to determine how and when cumulative zero emissions, embodied and operational, can be achieved.

18.9 SUMMARY OF OPTIMISATION SETTING UP PRINCIPLES

- Increasing the number of design variables will increase the solution space.
- Increasing the initial population size will increase the coverage of the solution space by the chromosomes (collections of genes), and therefore it will increase the number of points within the solution space from which the parallel search for the optimum solution will start. A population size equivalent to the number of parallel processes that can run on a simulation server can be specified.
- Increasing the number of generations will make the search more exhaustive, but it will also make it longer. Specifying a large number of generations may result in a lot of unnecessary computation after the optimum results have been reached.
- Specifying a small number of generations may result in an incomplete search for the optimum solution, and therefore give sub-optimum results.

CHAPTER 19

DESIGNING EMBODIED AND OPERATIONAL EMISSIONS TO TARGET A ZERO EMISSIONS YEAR FOR A BUILDING

In this chapter, we will apply the Zero Equation introduced in Chapter 1. Whether we represent this equation as an analogy of a jam jar and two spoons such as in Figures 4 and 5, or as a mathematical equation – Equation (1) and its derivatives in the form of Equations (2), (3) and (4), the objective will be the same: to find out when a building design will reach zero emissions.

As explained in Chapter 1, embodied emissions are the starting point in our analysis of zero carbon design. They set the starting point higher if conventional materials are used, or lower if bio-sourced materials, such as wood, straw, hemp, and others are used. Regardless of whether or not bio-sourced materials are used, a certain amount of conventional materials will be inevitable in each building, such as metal fasteners, nails, screws, tiles, glass, and others. This means that the initial embodied emissions will most likely be greater than zero.

Having accounted for the initial embodied emissions, there are two sets of operational emissions to be taken into account on an annual basis: positive emissions from using fossil fuels in running the building and positive emissions from periodic maintenance; and negative emissions from the operation of renewable energy systems installed in, or associated with the building.

Some simulation tools provide operational emissions as a combined number of positive and negative emissions, as shown in the previous chapter. Thus, deducting these combined operational emissions from the starting embodied emissions on an annual basis will result in reaching zero in a certain number of years, assuming that the net operational emissions are negative.

This chapter builds on the results of the previous chapter and retraces the key points from Chapter 5 to introduce the workflow for designing embodied and operational emissions that target a zero emissions year for a building. It uses the example building introduced in Section 18.2 in the previous chapter with additional assumptions that modify the design beyond the actual building in order to demonstrate the calculation of embodied emissions.

19.1 ESTABLISHING THE EMBODIED EMISSIONS

19.1.1 Embodied emissions from building materials and glazing

Following the completion of multi-objective optimisation in the previous chapter, we can now populate the model with values of design variables corresponding to a Pareto point of interest. This is achieved by selecting the Pareto point shown circled in blue in Figure 212, and using 'Apply selected design option' button, pointed by a blue arrow in the same figure.

After that is done, the 'Cost and Carbon' tab in Figure 212 leads to 'LCA and Embodied Carbon' tab in Figure 213. Full details are shown in Table 48. This simplifies the process introduced in Chapter 5, Tables 8 and 9, where each layer of each construction was analysed separately in order to calculate the mass of the materials and corresponding carbon emissions.

    DOI: 10.4324/9781003342342-23

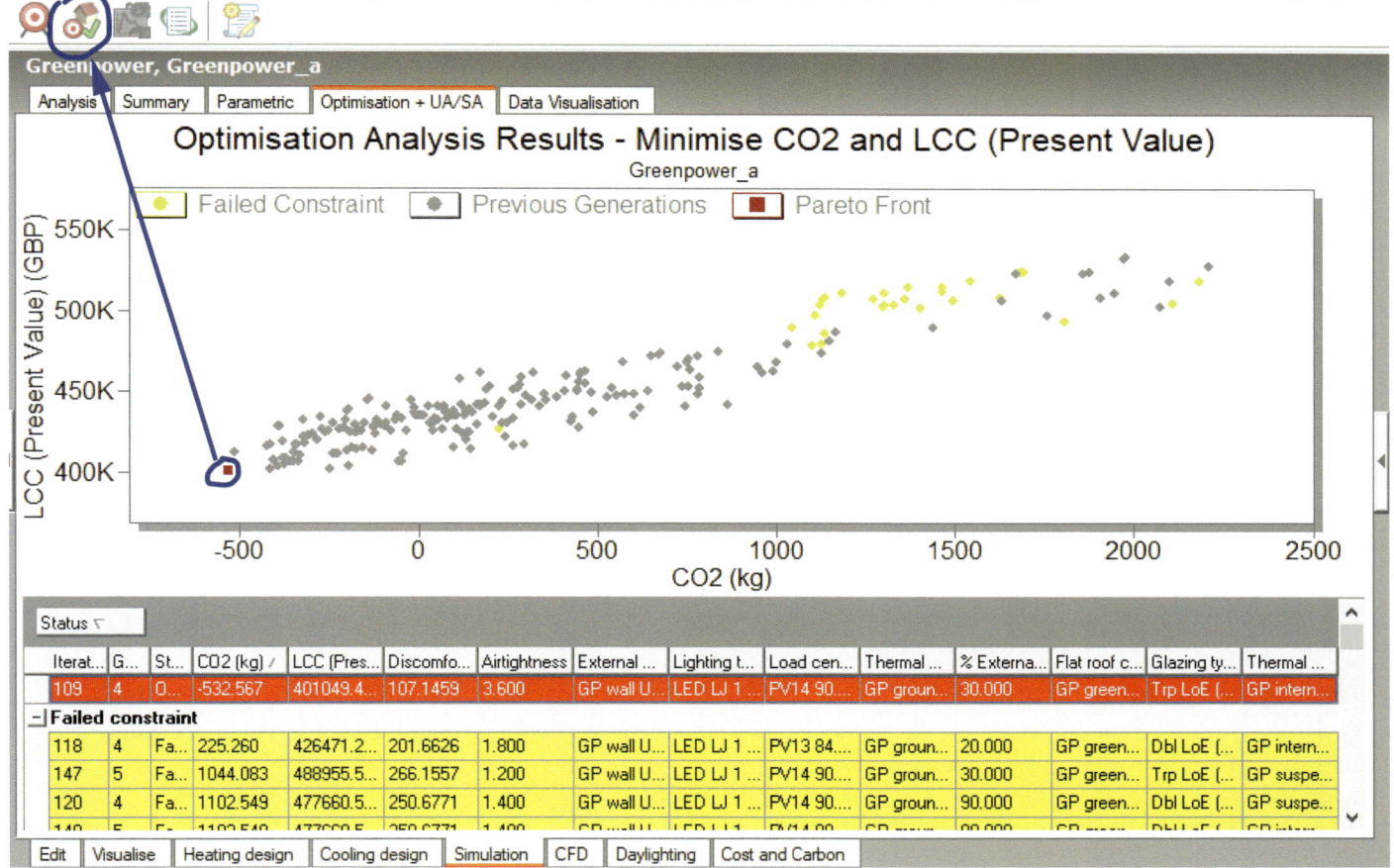

Figure 212   Applying selected design option to the model in DesignBuilder

Subsequently, embodied emissions in glulam arcs and columns are added as calculated in Table 49. The total emissions in construction materials are shown in Table 50.

19.1.2 Embodied emissions from HVAC systems, lighting, and electrical system

CIBSE TM65 (Harnot et al., 2021) introduces a calculation methodology for embodied carbon in building services. It refers to environmental product declarations (EPDs) as a standardised way of expressing embodied carbon in products. However, it also states that very few manufacturers provide this information. As a prerequisite for embodied carbon calculations is information on materials and their quantities, such information can be obtained from a BIM (building information model) (Kiamili et al., 2020). This enables the geometric data for pipes and ducts, fittings and accessories to be converted into material types and quantities, so that detailed life cycle assessment can be carried out. This level of detail is outside of the scope of this text, and instead we will use relatively simple methods to estimate embodied emissions in HVAC systems and lighting. Nevertheless, these simple methods will be based either on published material, or informed assumptions based on measurements.

Embodied emissions in the air-to-water heat pump. The first step in determining embodied emissions in the HVAC systems in this example is to calculate embodied emissions in the air-to-water heat pump installed in the building. The heat pump rating, which is essential for this calculation, is obtained from the simulation analysis of cumulative heating loads shown in Figure 214. As the selected heat pump has the capacity of 14 kW, it can be seen from this figure that it fulfils over 96% of hourly heating loads on an annual basis. Selecting a heat pump that fulfils all hourly heating loads would require the capacity of 86 kW, and would lead to a significant oversizing, considering that heating loads beyond the selected capacity occur only during a small number of annual hours.

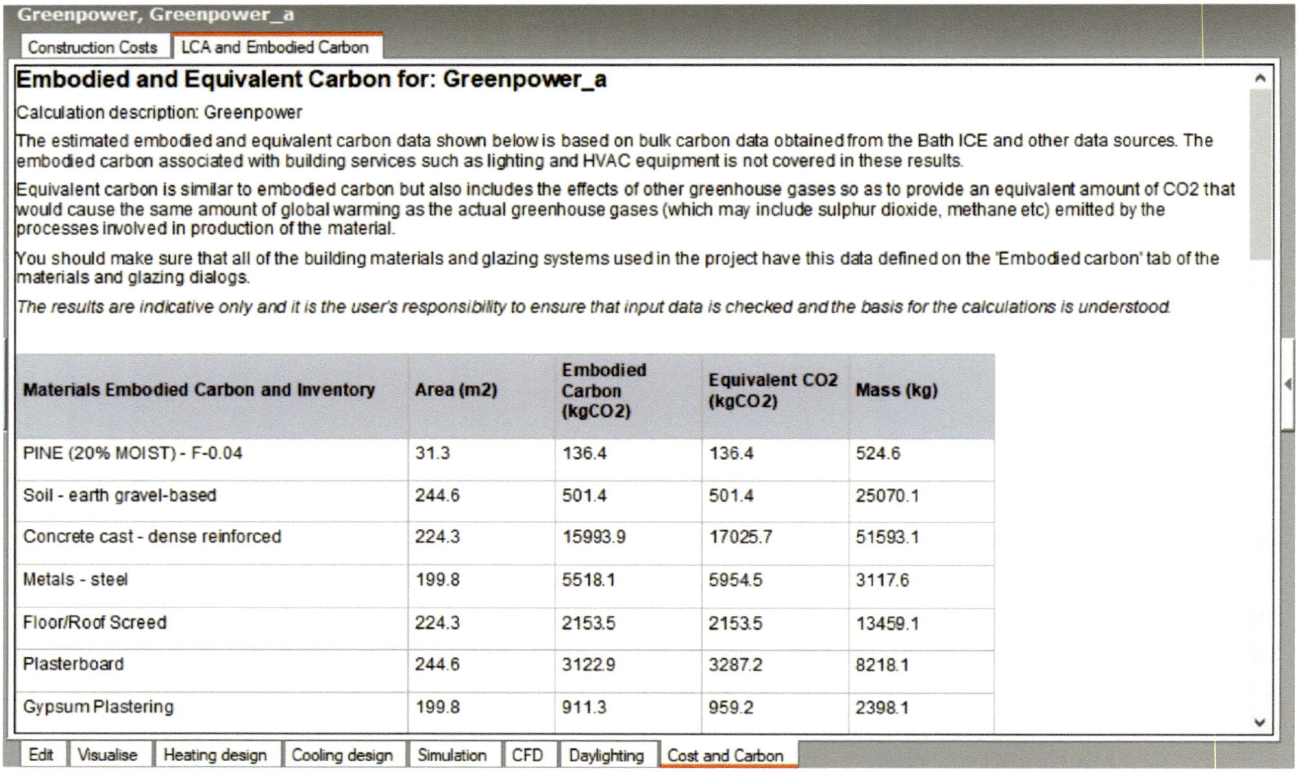

Greenpower, Greenpower_a

Construction Costs | LCA and Embodied Carbon

Embodied and Equivalent Carbon for: Greenpower_a

Calculation description: Greenpower

The estimated embodied and equivalent carbon data shown below is based on bulk carbon data obtained from the Bath ICE and other data sources. The embodied carbon associated with building services such as lighting and HVAC equipment is not covered in these results.

Equivalent carbon is similar to embodied carbon but also includes the effects of other greenhouse gases so as to provide an equivalent amount of CO2 that would cause the same amount of global warming as the actual greenhouse gases (which may include sulphur dioxide, methane etc) emitted by the processes involved in production of the material.

You should make sure that all of the building materials and glazing systems used in the project have this data defined on the 'Embodied carbon' tab of the materials and glazing dialogs.

The results are indicative only and it is the user's responsibility to ensure that input data is checked and the basis for the calculations is understood.

Materials Embodied Carbon and Inventory	Area (m2)	Embodied Carbon (kgCO2)	Equivalent CO2 (kgCO2)	Mass (kg)
PINE (20% MOIST) - F-0.04	31.3	136.4	136.4	524.6
Soil - earth gravel-based	244.6	501.4	501.4	25070.1
Concrete cast - dense reinforced	224.3	15993.9	17025.7	51593.1
Metals - steel	199.8	5518.1	5954.5	3117.6
Floor/Roof Screed	224.3	2153.5	2153.5	13459.1
Plasterboard	244.6	3122.9	3287.2	8218.1
Gypsum Plastering	199.8	911.3	959.2	2398.1

Edit | Visualise | Heating design | Cooling design | Simulation | CFD | Daylighting | Cost and Carbon

Figure 213 Obtaining embodied emissions from building materials and glazing in DesignBuilder

TABLE 48 LCA AND EMBODIED CARBON OUTPUT WITH CARBON STORAGE INCLUDED FOR TIMBER

Row	Materials embodied carbon and inventory	Area (m^2)	Embodied carbon ($kgCO_2$)	Equivalent CO_2 ($kgCO_2$)	Mass (kg)
1	PINE (20% MOIST) - F-0.04	31.3	−676.7	−676.7	524.6
2	Soil – earth gravel-based	244.6	501.4	501.4	25,070.1
3	Concrete cast – dense reinforced	224.3	15,993.9	17,025.7	51,593.1
4	Metals – steel	199.8	5,518.1	5,954.5	3,117.6
5	Floor/Roof Screed	224.3	2,153.5	2,153.5	13,459.1
6	Plasterboard	244.6	3,122.9	3,287.2	8,218.1
7	Gypsum Plastering	199.8	911.3	959.2	2,398.1
8	Gypsum Plasterboard	266.9	720.7	780.8	6,006.1
9	MW Glass Wool (rolls)	244.6	1,658.1	1,820.6	1,083.7
10	PUR Polyurethane Board (Diffusion open)	199.8	5,653.0	5,653.0	1,884.3
11	EPS Expanded Polystyrene (Standard)	224.3	3,235.2	4,218.7	1,294.1
12	Cast Concrete (Dense)	76.0	1,277.2	1,277.2	15,965.5
13	Bitumen/Felt Layers	244.6	3,534.3	3,534.3	2,079.0
14	0.75 in. Plywood/wood panels	244.6	−1,955.0	−1,955.0	2,102.2
	Sub-total		41,647.8	44,534.3	134,795.6

Row	Materials embodied carbon and inventory	Area (m²)	Embodied carbon (kgCO$_2$)	Equivalent CO$_2$ (kgCO$_2$)	Mass (kg)
TABLE 48 CONTINUED					
	Glazing Embodied Carbon and Inventory	Area (m²)	Embodied Carbon (kgCO$_2$)	Equivalent CO$_2$ (kgCO$_2$)	
	Trp LoE (e5=.1) Clr 3mm/13mm Arg	105.2	2,915	2,915	
	Window shading		5,261.7	5,261.7	
	Sub-total	105.2	8,176.7	8,176.7	

TABLE 49 EMBODIED EMISSIONS IN GLULAM ARCS AND COLUMNS

Description	Area	Thickness	Volume	Nos	Total volume	Density	Emissions per kg	Emissions per category
	m²	m	m³		m³	kg/m³	(kgCO$_2$/kg)	(kgCO$_2$)
	(1)	(2)	(3) = (1) · (2)	(4)	(5) = (3) · (4)	(6)	(7)	(8)= (5) · (6) · (7)
Column	1.49	0.12	0.18	5	0.89	750	−0.9	−604
Arc	12.06	0.24	2.90	5	14.48	750	−0.9	−9,772
							Sub-total	−10,376

TABLE 50 TOTAL EMBODIED EMISSIONS IN CONSTRUCTION MATERIALS

Description	Source	Embodied emissions (kgCO$_2$)
Embodied emissions in construction materials	Table 48	44,534
Embodied emissions in glazing	Table 48	8,177
Embodied emissions in glulam timber	Table 49	−10,376
Total materials		42,335

Finnegan et al. (2018) provide embodied emissions per functional unit (FU) of 1 kWh for air-water heat pump, however, in addition to embodied emissions this also includes 20 years of operational emissions and therefore it cannot be used in our calculation. Another study provides embodied emissions per unit of mass of air source heat pumps (Machnouk, 2021). Therefore, if we can find the mass of the installed heat pump that corresponds to 14 kW heating capacity, we will be able to obtain its embodied emissions by multiplying the mass by the emissions factor. As found in the manufacturer's online specifications, the installed heat pump (Ecodan_PUHZ-HW140YHA-BS from Mitsubishi Electric) has the mass of 148 kg. Taking the emissions factor of 10.8 kg CO$_2$/kg (Machnouk, 2021), embodied emissions in the air-to-water heat pump are calculated in Table 51.

Embodied emissions in underfloor heating system. The calculation of embodied emissions is based on estimating the length of the underfloor heating pipes on a 20 m × 11 m plan, and using a five-layer underfloor heating pipe of 20 mm outer diameter and 2 mm wall thickness. The five layers of the pipe wall are assumed to be: polyethylene, glue, aluminium, glue, polyethylene. The thickness of these layers was estimated from images of pipe cross-sections from various manufacturers web sites. As Figure 195b in Chapter 18 shows, there are 10 zone valves in the underfloor heating manifold. A sample valve was weighed, and the manifold was assumed to be made of modular components of equal weight as the valves. Corresponding materials and

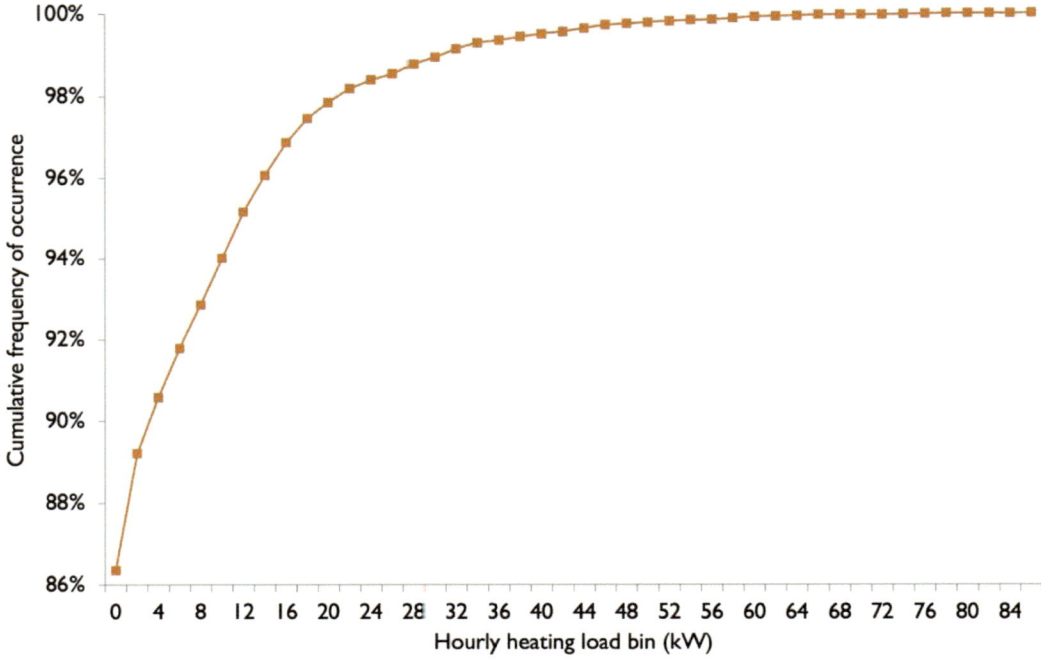

Figure 214   Heating capacity of air-to-water heat pump

their emissions are shown in each calculation table, where simple arithmetic operations between table columns are specified. These calculations are shown separately as follows:

• Embodied emissions in underfloor heating system pipes, Table 52.
• Embodied emissions in underfloor heating system manifold, Table 53.
• Embodied emissions in underfloor heating system valves, Table 54.

TABLE 51 EMBODIED EMISSIONS IN THE AIR-TO-WATER HEAT PUMP			
Description	Mass (kg)	Emissions factor (kg CO_2/kg)	Embodied emissions (kg CO_2)
Column	(1)	(2)	(3) = (1) · (2)
Air-to-water heat pump	148	10.8	1,598

Embodied emissions in LED lamps. The building contains 71 light fittings and we assume that they are all fitted with LED lamps. The LED lamp embodied emissions are obtained from a very thorough study by the US Department of Energy (Scholand & Dillon, 2012). Embodied emissions for an LED lamp of 12.5 Watts and 812 lumen (therefore 65 lm/W luminous efficacy) are obtained from Table 7-3 of that document (Scholand & Dillon, 2012), providing all stages of life cycle impacts of the 2012 LED lamp. Embodied emissions from the LED lamps in the building are calculated in Table 56 on this basis.

Embodied emissions in wiring, light fittings, and electric sockets. The reason for calculating embodied emissions in the light fittings and wiring separately from embodied emissions in LED lamps is to make it easier to account for maintenance emissions: LED lamps are replaced as part of periodic maintenance, and the wiring and the fittings are not. Electric wiring for electrical sockets, and the sockets follow similar calculation procedure and are therefore introduced here in the same category. These calculation procedures are shown in each table in the second row, where simple arithmetic operations between table columns are specified. An average length of wiring is assumed per each light fitting and power socket, bearing in mind that

TABLE 52 EMBODIED EMISSIONS IN UNDERFLOOR HEATING SYSTEM PIPES

Pipe length (m)	Pipe outer diameter (m)	Pipe inner diameter (m)	Pipe section area (m²)	Pipe volume (m³)	Materials	Proportion of the cross section	Embodied carbon (kg CO_2/kg)	Volume per material (m³)	Density (kg/m³)	Mass (kg)	Emissions per material proportion (kg CO_2)
(1)	(2)	(3)	$(4) = \pi \cdot [[2]^2 - (3)^2]/4$	$(5) = (1) \cdot (4)$	(6)	(7)	(8)	$(9) = (5) \cdot (7)$	(10)	$(11) = (9) \cdot (10)$	$(12) = (8) \cdot (11)$
					polyethylene	0.35	2.54	0.029	940	27.60	70.09
					glue	0.05	2.98	0.004	1190	4.99	14.87
					aluminium	0.20	6.67	0.017	2800	46.97	313.29
					glue	0.05	2.98	0.004	1190	4.99	14.87
742	0.02	0.016	1.13E–04	8.39E–02	polyethylene	0.35	2.54	0.029	940	27.60	70.09
										Total	**483.22**

TABLE 53 EMBODIED EMISSIONS IN UNDERFLOOR HEATING SYSTEM MANIFOLD

Manifold size (number of valves)	Valve mass (kg)	Total mass – (steel 100%) (kg)	Emissions factor – steel (kgCO_2/kg)	Emissions – steel (kgCO_2)
(1)	(2)	$(3) = (1) \cdot (2)$	(4)	$(5) = (3) \cdot (4)$
10	0.3	3	2.89	**8.67**

TABLE 54 EMBODIED EMISSIONS IN UNDERFLOOR HEATING SYSTEM VALVES

Number of valves	Valve mass (kg)	Total mass (kg)	Steel (95%)	Plastic (5%)	Emissions factor – steel (kgCO_2/kg)	Emissions factor – plastic (kgCO_2/kg)	Emissions – steel (kgCO_2)	Emissions – plastic (kgCO_2)	Emissions – total valves (kgCO_2)
(1)	(2)	$(3) = (1) \cdot (2)$	$(4) = (3) \cdot 0.95$	$(5) = 3 \cdot 0.05$	(6)	(7)	$(8) = (4) \cdot (6)$	$(9) = (5) \cdot (7)$	$(10) = (8) + (9)$
10	0.3	3	2.85	0.15	2.89	3.1	8.24	0.46	**8.70**

shorter and longer wiring will average out. The total number of power sockets is assumed to be the same as the number of light fittings. The wiring cross-section for lighting is assumed to be 1.5 mm² of copper per wiring core, and the cross section of wiring for power sockets is assumed to be 2.5 mm² per wiring core. Live and neutral cores are double insulated with their own insulation and an overall outer insulation, whilst earth core is insulated only by the overall outer insulation. Insulation thicknesses

TABLE 55   EMBODIED EMISSIONS IN UNDERFLOOR HEATING SYSTEM

Category	Source	Embodied emissions (kgCO$_2$)
Embodied emissions – pipes	Table 52	483.22
Embodied emissions – manifold	Table 53	8.67
Embodied emissions – valves	Table 54	8.70
Total embodied carbon – underfloor heating		**500.59**

TABLE 56   EMBODIED EMISSIONS IN LED LAMPS

Description	Embodied emissions per lamp (kg CO$_2$/lamp)	Number of lamps	Embodied emissions in LED lamps (kg CO$_2$)
Column	(1)	(2)	(3) = (1) · (2)
Raw materials	12.752		
Manufacturing	3.450		
Transport	0.052		
Total	16.254	71	1,154

TABLE 57   EMBODIED EMISSIONS IN WIRING FOR LIGHTING – COPPER CORES

Number of fittings	Wiring length per fitting (m)	Total wiring length (m)	Copper wiring diameter (mm)	Copper wiring cross section (mm²)	Copper wiring cores	Unit volume (m³)	Wiring density – copper (kg/m³)	Total copper wiring mass (kg)	Copper emissions (CO$_2$/kg)	Copper wiring emissions (kgCO$_2$)
(1)	(2)	(3) = (1) · (2)	(4)	(5)	(6)	(7) = (5) · (6)/10−6	(8)	(9) = (3) · (7) · (8)	(10)	(11) = (9) · (10)
71	10	710	1.38	1.50	3	4.50E−06	8900	28.44	2.71	77.06

are based on measurement, and weights of light fittings and sockets are based on weighing sample products. These calculations are shown separately for each category as follows:

- Embodied emissions in wiring for lighting – copper cores, Table 57.
- Embodied emissions in wiring for lighting – insulation for copper cores, Table 58.
- Embodied emissions in wiring for lighting – outer insulation, Table 59.
- Embodied emissions in LED fittings, Table 60.
- Embodied emissions in wiring for electric sockets – copper cores, Table 61.
- Embodied emissions in wiring for electric sockets – insulation for copper cores, Table 62.
- Embodied emissions in wiring for electric sockets – outer insulation, Table 63.
- Embodied emission in electric sockets, Table 64.

TABLE 58 EMBODIED EMISSIONS IN WIRING FOR LIGHTING – INSULATION FOR COPPER CORES

Number of fittings	Wiring length per fitting (m)	Total wiring length (m)	Insulation outside diameter (mm)	Core insulation inside diameter (mm)	Core insulation cross section area (mm²)	Insulated cores	Unit volume (m³)	PVC density (kg/m³)	Total insulation mass (kg)	PVC embodied emissions (kgCO₂/kg)	Emissions inner insulation cores kgCO₂
(1)	(2)	$(3) = (1) \cdot (2)$	(4)	(5) = [Table 57 Column (4)]	$(6) = \pi \times [(4)^2 - (5)^2]/4$	(7)	$(8) = (6) \cdot (7)/10^{-6}$	(9)	$(10) = (3) \cdot (8) \cdot (9)$	(11)	$(12) = (10) \cdot (11)$
71	10	710	3.00	1.38	5.57	2	1.11E−05	1379	10.90	3.1	33.78

TABLE 59 EMBODIED EMISSIONS IN WIRING FOR LIGHTING – OUTER INSULATION

Number of fittings	Wiring length per fitting (m)	Total wiring length (m)	Cable cross section (mm²)	Inner core areas (mm²)	Cable insulation cross section (mm²)	Unit volume (m³)	PVC density (kg/m³)	Total insulation mass (kg)	PVC embodied emissions (kgCO₂/kg)	Emissions outer insulation kgCO₂
(1)	(2)	$(3) = (1) \cdot (2)$	(4) = 8.6 mm × 5 mm	(5) = [Table 58 Column (4)]2 π/4 + [Table 57 Column(5)]	$(6) = (4) - (5)$	$(7) = (6)/10^{-6}$	(8)	$(9) = (3) \cdot (7) \cdot (8)$	(10)	$(11) = (9) \cdot (10)$
71	10	710	43.00	15.63	27.37	2.74E−05	1379	26.80	3.1	83.07

TABLE 60 EMBODIED EMISSIONS IN LED FITTINGS

Number of fittings	Fitting mass (kg)	Total mass of fittings (kg)	Steel (95%)	Copper (5%)	Emissions factor - steel (kgCO₂/kg)	Emissions factor – copper (kgCO₂/kg)	Emissions – steel (kgCO₂)	Emissions – copper (kgCO₂)	Emissions – total fittings (kgCO₂)
(1)	(2)	$(3) = (1) \cdot (2)$	$(4) = (3) \times 0.95$	$(5) = (3) \times 0.05$	(6)	(7)	$(8) = (4) \cdot (6)$	$(9) = (5) \cdot (7)$	$(10) = (8 + 9)$
71	0.4	28.4	26.98	1.42	2.85	2.71	76.89	3.8482	80.74

TABLE 61 EMBODIED EMISSIONS IN WIRING FOR ELECTRIC SOCKETS – COPPER CORES

Number of sockets	Wiring length per socket (m)	Total wiring length (m)	Copper wiring diameter (mm)	Copper wiring cross section (mm²)	Copper wiring cores	Unit volume (m³)	Wiring density – copper (kg/m³)	Total copper wiring mass (kg)	Copper emissions (CO₂/kg)	Copper wiring emissions (kgCO₂)
(1)	(2)	$(3)=(1)\cdot(2)$	(4)	(5)	(6)	$(7)=(5)\cdot(6)/10^{-6}$	(8)	$(9)=(3)\cdot(7)\cdot(8)$	(10)	$(11)=(9)\cdot(10)$
71	10	710	1.78	2.50	3	7.50E−06	8900	47.39	2.71	128.43

TABLE 62 EMBODIED EMISSIONS IN WIRING FOR ELECTRIC SOCKETS – INSULATION FOR COPPER CORES

Number of sockets	Wiring length per socket (m)	Total wiring length (m)	Insulation outside diameter (mm)	Core insulation inside diameter (mm)	Core insulation cross section area (mm²)	Insulated cores	Unit volume (m³)	PVC density (kg/m³)	Total insulation mass (kg)	PVC embodied emissions (kgCO₂/kg)	Emissions inner insulation cores kgCO₂
(1)	(2)	$(3)=(1)\cdot(2)$	(4)	$(5)=$ [Table 61 Column (4)]	$(6)=\pi\times[(4)^2-(5)^2]/4$	(7)	$(8)=(6)\cdot(7)/10^{-6}$	(9)	$(10)=(3)\cdot(8)\cdot(9)$	(11)	$(12)=(10)\cdot(11)$
71	10	710	3.50	1.78	7.12	2	1.42E−05	1379	13.93	3.10	43.20

TABLE 63 EMBODIED EMISSIONS IN WIRING FOR ELECTRIC SOCKETS – OUTER INSULATION

Number of fittings	Wiring length per fitting (m)	Total wiring length (m)	Cable cross section (mm²)	Inner core areas (mm²)	Cable insulation cross section (mm²)	Unit volume (m³)	PVC density (kg/m³)	Total insulation mass (kg)	PVC embodied emissions (kgCO₂/kg)	Emissions outer insulation kgCO₂
(1)	(2)	$(3)=(1)\cdot(2)$	$(4)=8.6$ mm · 5 mm	$(5)=$ [Table 62 Column (4)]²· π/4 + [Table 61 Column(5)]	$(6)=(4)-(5)$	$(7)=(6)/10^{-6}$	(8)	$(9)=(3)\cdot(7)\cdot(8)$	(10)	$(11)=(9)\cdot(10)$
71	10	710	69.30	21.73	47.57	4.76E−05	1379	46.57	3.10	144.38

TABLE 64 EMBODIED EMISSION IN ELECTRIC SOCKETS

Number of sockets	Socket mass (kg)	Total mass of sockets (kg)	Plastic (80%)	Copper (20%)	Emissions factor – plastic ($kgCO_2$/kg)	Emissions factor – copper ($kgCO_2$/kg)	Emissions – plastic ($kgCO_2$)	Emissions – copper ($kgCO_2$)	Emissions – total sockets ($kgCO_2$)
(1)	(2)	(3) = (1) · (2)	(4) = (3) × 0.8	(5) = (3) × 0.2	(6)	(7)	(8) = (4) · (6)	(9) = (5) · (7)	(10) = (8) + (9)
71	0.4	28.4	22.72	5.68	3.31	2.71	75.20	15.39	90.60

Assumptions about materials are shown in the corresponding tables and a summary of results of tables 57 to 65 is shown in Table 65.

TABLE 65 SUMMARY OF EMBODIED EMISSIONS IN WIRING, LIGHT FITTINGS, AND ELECTRIC SOCKETS

Description	Source	Embodied emissions ($kgCO_2$)
Wiring – lighting	Total of Table 56+Table 58+Table 59	193.91
Light fittings	Table 60	80.74
Wiring – sockets	Total of Table 61+Table 62+Table 63	316.01
Sockets	Table 64	90.60
Total – wiring, fittings, and sockets		**681.26**

Embodied emissions in the ventilation system. The calculations in this section are based on manufacturer's data regarding the mass of roof-mounted ventilation terminals and assumption about aluminium being the predominant material for the roof terminals and associated wall-mounted louvres. Information about the geometry of the wall-mounted louvres is derived from corresponding drawings and assumptions about material thickness. Table 66 shows corresponding calculations.

TABLE 66 EMBODIED EMISSIONS IN THE VENTILATION SYSTEM

Description	Mass per unit (kg)	Number of units	Total mass (kg)	Material embodied emissions ($kgCO_2$/kg)	Embodied emissions ($kgCO_2$)
	(1)	(2)	(3) = (1) · (2)	(4)	(5) = (3) · (4)
Passivent Airstruct Terminals 1700mm diameter Material: aluminium	246.00	2	492.00	6.67	3281.64
Louvres	9.32	2	18.65	6.67	124.38
Total					**3406.02**

19.1.3 Embodied emissions in the solar photovoltaic system

Embodied emissions in the solar PV system will be calculated in three scenarios, each corresponding to three different embodied emissions factors, which were available at the time of writing.

Using the model corresponding to the first Pareto point from Table 47 from the previous chapter and running an hourly simulation provides information about the maximum solar photovoltaic output. The fuel totals hourly grid output shows a maximum output of 7.762218 kW (with a minus sign indicating energy generation), occurring at 13:00 on 27th June (Figure 215).

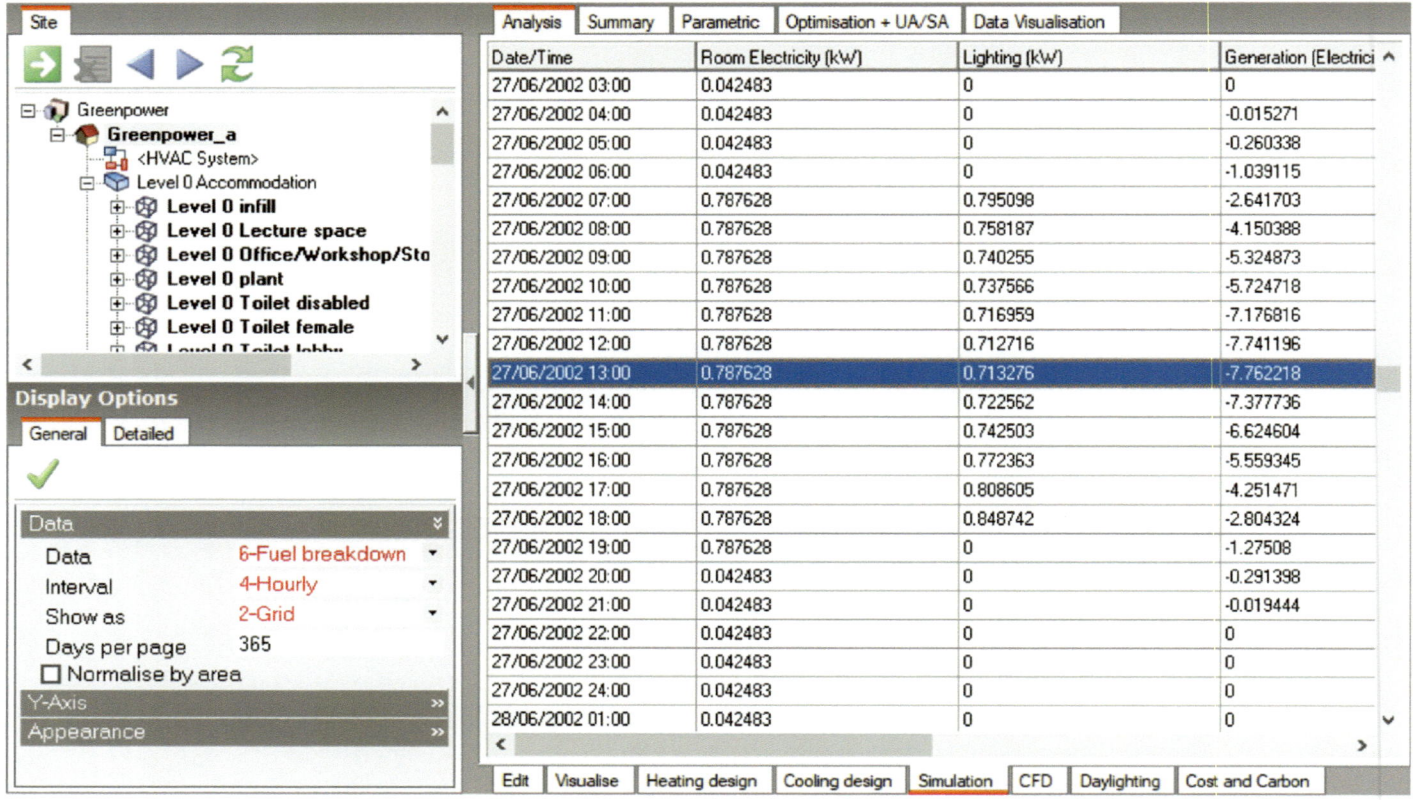

Figure 215   Maximum photovoltaic output

As shown in Table 10 in Chapter 5, 2,560 $kgCO_2$ of typically embodied emissions in 2021 is attributed to each kW_p (kilo-Watt-peak) of photovoltaic output (Circular Ecology, 2021) and in 2040 this embodied emissions are projected to be 325 $kgCO_2$/ kW_p. We will refer to these two sets of figures as Scenario 1 and 2.

Solar PV systems are rated on the basis of energy generation at peak performance under solar radiation of 1000 W/m², ambient temperature of 25°C, and clear sky (Sol Voltaics, 2022). Inspection of the weather data file used in this simulation showed that horizontal total radiation corresponding to the date and time of the maximum photovoltaic output was 774.3 W/m². As there were no occurrences of solar radiation of 1000 W/m², we need to convert the maximum PV output of 7.76 kW from the simulation into kW_p rating.

As a simplified way of conversion, we will normalise the simulation output by dividing 1000 W/m² by the solar radiation of 774.3 W/m² that caused the simulation output and multiplying it with the simulation output itself to obtain the kW_p rating. Subsequently, we will multiply the kW_p rating with the 2,560 $kgCO_2$ embodied emissions factor to obtain embodied emissions in the solar photovoltaic system. This is summarised in Table 67.

As PV embodied emissions in Scenario 1 (Circular Ecology, 2021) correspond to 2021 and in Scenario 2 (Worboys, 2021) these correspond to 2040, we will use the former for establishing the starting embodied emissions, and the latter for maintenance emissions due to PV system replacement.

TABLE 67 EMBODIED EMISSIONS IN THE SOLAR PV SYSTEM

Row	Description	Scenario 1 (1)	Scenario 2 (2)	Scenario 3 (3)
(1)	Max generation by solar PV system (kW)	7.76	7.76	7.76
(2)	Corresponding global horizontal radiation (Wh/m²)	774.30	774.30	774.30
(3)	Solar radiation for kW_p rating (W/m²)	1000	1000	1000
(4)	kW_p rating of solar PV system = (3) / (2) · (1)	10.0248	10.0248	
(5)	Embodied emissions factor ($kgCO_2/kW_p$)	2560	325	
(6)	PV area from Table 47 (m²)			90.916
(7)	Embodied emissions factor ($kgCO_2/$ m²) from EPD by Vindian Solar (2023)			0.7185
(8)	Embodied emissions in solar PV system for Scenarios 1 and 2 ($kgCO_2$) = (4) · (5)	**25,664**	**3,258**	
(9)	Embodied emissions in solar PV system for Scenario 3 ($kgCO_2$) = (6) · (7)			**65.32**

As embodied emissions in solar PV panels were generated at the time of panel production, can it be argued that the above approach is project specific and therefore invalid? The PV panels manufactured in a specific factory will have the same embodied emissions arising from their production wherever they are installed. However, as the kW_p rating of the panels is based on a normalised solar radiation that is not project specific, the above estimate of embodied emissions is also not project specific, but it follows published guidance on embodied emissions from solar PV based on the rating of the PV system.

Additionally, figures for embodied emissions in PV systems have become available from the first environmental product declaration for PV systems (Vindian Solar, 2023). We will derive embodied emissions per unit surface area of a PV system and refer to it as Scenario 3 (Table 67).

Two sets of cumulative embodied emissions will be ultimately calculated later in this chapter: one for combined Scenarios 1 and 2, and one for Scenario 3.

19.1.4 Emissions from the construction process: construction workers' travel, material deliveries, operation of site machines

In this section, we will replicate the workflow developed in Chapter 5. We will use the five phases of construction work as in Table 14 to replicate the calculations from Table 15 for emissions from travel of construction workers, from Table 17 for emissions from material deliveries and from Table 18 for emissions from operation of construction site machines. The total amount of materials for deliveries of 170,448 kg is combined from the mass of construction materials, the PV system, air source heat pump, glulam, and other entities discussed in this chapter, with a 15% contingency added.

It is assumed that the building was constructed by eight construction workers, and that they travelled in three person vans, which means that between two and three vans were used. The project was delivered by a local construction company, and therefore it is assumed that the construction workers travelled only locally within a radius of ten miles, on 20-mile day return trips. The source for vehicular carbon emissions used is the same as in Chapter 5 (VCA, 2016), however the reader is advised to check relevant sources for a corresponding location and the time of construction. This now enables the calculation of emissions from construction workers' travel, as shown in Table 68.

TABLE 68 CARBON EMISSIONS FROM THE TRAVEL OF CONSTRUCTION WORKERS

Phase (from Table 14)	No. of people	No. of vans	No. of days	Miles travelled per day	Miles travelled	km travelled	Emissions factor (gCO$_2$/km)	Total emissions – people travel (kgCO$_2$)
1–5	8	2.7	200	20	10,667	17,166	156.5	2,687
						Total emissions from people travel (kgCO$_2$) =		**2,687**

Deliveries are assumed to be done by 7.5 metric-ton trucks. The number of trucks is obtained by dividing the mass of materials delivered by 7,500kg (Table 69, column 4). Another assumption is that all deliveries are within 60-mile range, and thus each delivery consists of a 120-mile round trip (Table 69, column 5). The total number of miles travelled is obtained by multiplying the number of trucks by the number of miles travelled per day (Table 69, column 6). As data for truck emissions is available in gCO$_2$/km, the trips in miles are converted into kilometres (Table 69, column 7). Total emissions from material deliveries are then obtained by multiplying the kilometres travelled by the corresponding emissions per kilometre (Table 69, column 9).

TABLE 69 CARBON EMISSIONS FROM MATERIAL DELIVERIES

Phase (Table 14)	Percentage of materials per phase	Materials delivered (kg)	No of 7.5 t trucks	Miles travelled per day	Total tiles travelled	Total kilometres travelled	Emissions factor (gCO$_2$/km)	Total emissions - material deliveries (kgCO$_2$)
(1)	(2)	(3) = (2) · 208,254	(4) = (3)/7500	(5)	(6) = (4) · (5)	(7) = (5) · 1.609	(8) (Source: TheyWorkForYou, 2013)	(9) = (7) · (8)/1000
1	0%	0	0	0	0	0	327	0.00
2	40%	68,179	10	120	1200	1931.21	327	631.51
3	48%	81,815	11	120	1320	2124.33	327	694.66
4	12%	20,454	3	120	360	579.36	327	189.45
5	0%	0	0	30	0	0	327	0.00
	Total	**170,448**					**Total**	**1,515.62**

Emissions from the operation of site machines are calculated using the power rating of these machines and days spent on site, and applying published emissions factors (Heidari & Marr, 2015). The emissions from the operation of site machines are summarised in Table 70.

TABLE 70 CARBON EMISSIONS FROM THE OPERATION OF SITE MACHINES

Phase	Hours of operation per day (hr/day)	Days of operation	Power (kW)	Emissions factor (g CO$_2$/(kW.hr))	Total emissions – site machines (kgCO$_2$)
(1)	(2)	(3)	(4)	(5)	(6) = (2) · (3) · (4) · (5) /1000
2	8	4	42	138	185
3	8	2	42	138	95
			Total emissions from the operation of site machines (kgCO$_2$) =		**278**

Total emissions from construction of 4,480 kgCO$_2$ are obtained by adding the totals from Tables 68–70.

19.1.5 Embodied emissions from maintenance

Embodied emissions from maintenance occur only after a number of years. The reason for calculating these at the start in this example is to find out whether these would influence the time of reaching total zero. As can be seen from Chapter 5, Table 19, the most significant emissions from maintenance are those from the replacement of renewable energy systems. In addition, we will also calculate emissions from replacement of some of the HVAC systems, such as the air-to-water heat pump and emissions from replacement of LED lamps. Thus, embodied emissions from maintenance are calculated in Table 71.

TABLE 71	EMBODIED EMISSIONS FROM MAINTENANCE				
Description	Embodied emissions (kgCO$_2$)	Source	Service life (years)	Replacement frequency in 60-year cycle	Total embodied emissions from maintenance (kgCO$_2$)
Column	(1)	(2)	(3)	(4)	(5)
Replace air-to-water heat pump	1,598	Table 51	20	2	3,196
Replace solar PV system	3,258	Table 67, row (6) column (2)	30	1	3,258
Replace LED bulbs	1,154	Table 56	10	5	5,770

19.2 COMBINING EMBODIED AND OPERATIONAL EMISSIONS TO ACHIEVE ZERO TOTAL CUMULATIVE EMISSIONS

In this section, we will apply the Zero Equation in order to combine embodied and operational emissions and find out when the total emissions will reach zero.

The starting point are embodied emissions from the day the building is completed. These consist of emissions in materials (Table 50), construction process (Table 68, Table 69, Table 70), solar PV (Table 67), air-to-water heat pump (Table 51), under-floor heating (Table 55), lighting (Tables 56 and 65) and ventilation (Table 66). The starting emissions are then reduced annually by the negative operational emissions from Table 47 and increased when maintenance events occur as per Table 71. The result of combined embodied and operational emissions is shown in Figure 217. The maintenance emissions introduce vertical spikes in the cumulative emissions lines.

As it can be seen from this figure, the negative operational emissions from the first Pareto point in Table 47 of −532.6 kgCO$_2$, corresponding to 90.92 m² of PV, are not sufficient to achieve zero cumulative emissions within the 60-year lifecycle, shown as red line in Figure 217. This is where we will use the Zero Equation to find out the magnitude of negative operational emissions from the PV system required to achieve zero cumulative emissions by a specified year.

As the emissions of −532.6 kgCO$_2$/year obtained from the Pareto point are a balance between positive emissions from using fossil fuels and negative emissions arising from the operation of the PV system, we first need to re-run the simulation model without the PV in order to find the magnitude of the positive emissions. This is done by disabling 'Include electric load centres' in the Generation tab (Figure 216a), which subsequently shows the PV disabled in Figure 216b. The resulting operational emissions from the use of fossil fuels were 964 kgCO$_2$/year. Therefore, the operational emissions from the PV on its own are −533 −964 = −1497 kgCO$_2$/year.

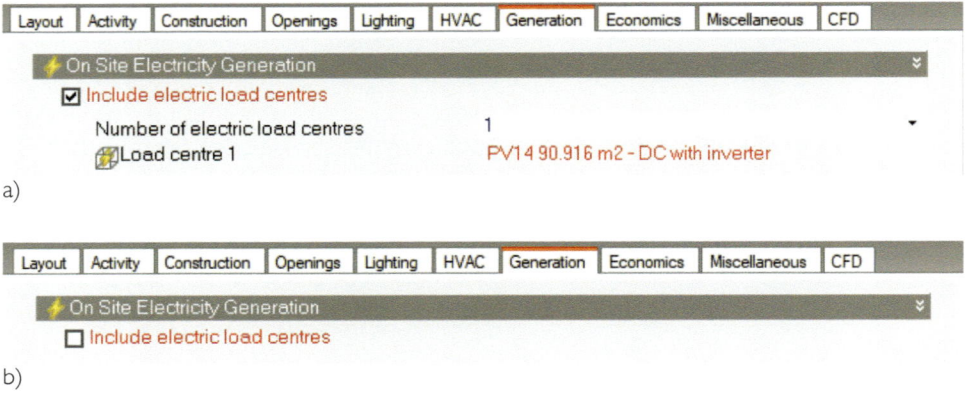

a)

b)

Figure 216   Disabling the PV in the simulation model in DesignBuilder

19.2.1 Summary of embodied emissions

TABLE 72 SUMMARY OF EMBODIED EMISSIONS			
	Insulation panel wall (kgCO$_2$)	Hempcrete wall (kgCO$_2$)	Source
Materials	42,335	14,042	Table 50 and Section 19.2.3 for hempcrete
Construction	4,480	4,480	Table 68, Table 69, Table 70
Solar PV	25,664	25,664	Table 67
Heat pump	1,598	1,598	Table 51
Underfloor heating	501	501	Table 55
Lighting	1,154	1,154	Table 56
Ventilation	3,406	3,406	Table 66
Wiring, fittings, and sockets	681	681	Table 65
Totals	**79,819**	**51,526**	
Totals less Solar PV	**54,156**	**25,863**	

19.2.2 Applying the Zero Equation in combined Scenarios 1 and 2

To find out what is required for this building to achieve total zero emissions and by when, we apply the Zero Equation (Equation 4), which is replicated here for convenience:

$$f = \frac{CE_{CO2} + P_{CO2} \times t}{BN_{CO2} \times t - BVE_{CO2}} \tag{4}$$

where:

CE_{CO2} – Constant embodied emissions, arising from materials, construction process, HVAC system, and maintenance (kgCO$_2$)
P_{CO2} – Positive operational emissions that occur from year $t = 1$ (kgCO$_2$/year)
t – time in years
BN_{CO2} – Base level of negative operational emissions that occur from year $t = 1$ (kgCO$_2$/year)
BVE_{CO2} – Base level of variable embodied emissions, arising for from changing the size of a renewable energy system, such as a
            PV system, to target a zero emissions year (kgCO$_2$)

f – Base level multiplying factor for renewable energy system, such as a PV system, so that $f \times BN_{CO2} = N_{CO2}$ and $f \times BVE_{CO2} = VE_{CO2}$. This factor is used to target a zero emissions year by changing the size of a renewable energy system.

N_{CO2} – Negative operational emissions arising from renewable energy that occur from year $t = 1$ (kgCO$_2$/year)

VE_{CO2} – Variable embodied emissions, arising from changing the size of a renewable energy system, such as a PV system, whilst targeting a zero emissions year (kgCO$_2$)

We first set time to $t = 40$ years from the date of construction, which sets the year to 2050, and calculate all other terms of this equation as follows:

- CE_{CO2} = starting embodied emissions + embodied emissions from maintenance = $54,156 + (1,154 \times 3 + 1,598) = 59,216$ kgCO$_2$

- $P_{CO2} \times t = 964$ kgCO$_2$/year $\times$ 40 years = 38,560 kgCO$_2$

- $BN_{CO2} \times t = 1,497$ kgCO$_2$/year $\times$ 40 years = 59,880 kgCO$_2$

- $BVE_{CO2} = (25,664 + 3,258)$ kgCO$_2$ = 28,922 kgCO$_2$ from initial embodied emissions from the PV plus PV replacement emissions after 30 years.

Substituting CE_{CO2}, $P_{CO2} \times t$, $BN_{CO2} \times t$, and BVE_{CO2} in Equation (4), we calculate f as follows:

$$f = \frac{59,216 + 38,560}{59,880 - 28,922} = 3.158 \tag{4a}$$

Therefore, in order to achieve total zero emissions in 40 years from the date of construction, we need to increase the negative operational emissions by a factor of 3.158, which means increasing the surface of the PV array by the same amount. Thus, the base level PV surface obtained from the first Pareto point in Table 47 of 90.92 m² will now be increased to: 90.92 m² $\times$ 3.158 = 287.15 m². The result is shown with the blue line in Figure 217. As can be seen in this figure, zero cumulative emissions are reached exactly in 2050, as calculated.

19.2.3 Using bio-sourced materials to lower embodied emissions in combined Scenarios 1 and 2

What would be the effect of using bio-sourced material for external walls, instead of thermal insulation panels? To answer this, we replace the materials in rows 4, 7, and 10 in Table 48 with hemp-lime (hempcrete) wall. The equivalent thickness of this wall that achieves the same U-value as the thermal insulation panel wall is 0.729 metres. Combining this with 199.8 m² of the external wall area and negative embodied emissions of −108 kgCO$_2$/m³ (Bevan & Woolley, 2008, p. 81) results in embodied emissions reduction from hempcrete of $0.729 \times 199.8 \times (-108) = -15,726$ kgCO$_2$. We also need to take out the embodied emissions from the conventional materials that hempcrete is replacing on lines 4, 7, and 10 in Table 48, namely: $5954.5 + 959.2 + 5653 = 12,566.7$ kgCO$_2$. Therefore, the overall reduction resulting from introducing hempcrete is −28,293 kgCO$_2$. Applying this adjustment in Equation (4) leads to:

$$f = \frac{59,216 - 28,293 + 38,560}{59,880 - 28,922} = 2.244 \tag{4b}$$

Therefore, the base level PV surface obtained from the first Pareto point in Table 47 of 90.92 m² will need to be increased by factor 'f' to 204.04 m² when using hempcrete as a replacement for insulation panel walls. The result is shown as the green line in Figure 217. Even though the embodied emissions in the materials are much lower in the case of hempcrete wall (green line) than in the base case (red line), the increased embodied emissions in the PV increase the overall starting embodied emission slightly above the base case (blue line). Despite that increase in the initial embodied emissions, the base level multiplying factor for the PV system area ensures that zero cumulative emissions are reached in 2050, as shown in Figure 217.

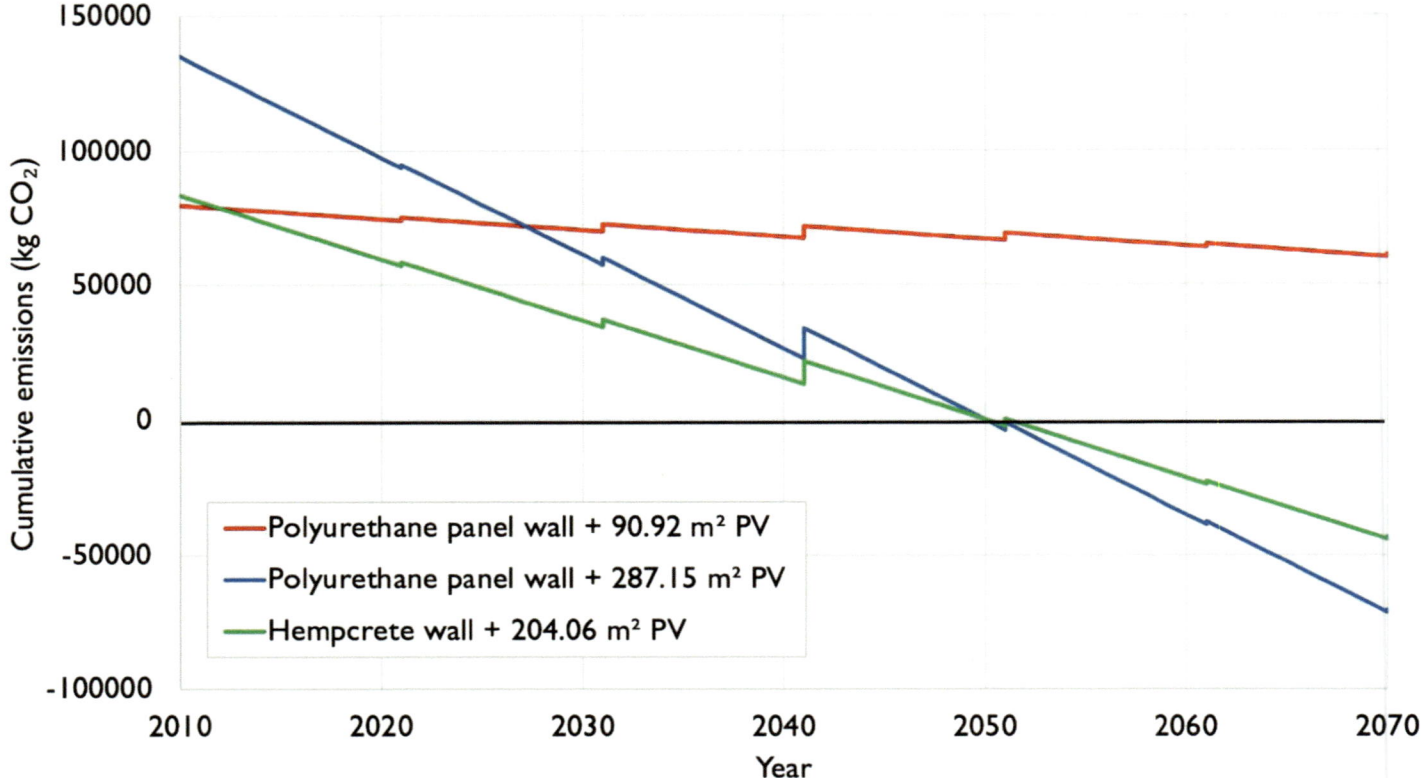

Figure 217   Cumulative emissions: embodied plus operational in combined Scenarios 1 and 2 – the vertical spikes are due to maintenance emissions

The two examples above illustrate the power of the Zero Equation: after constant and variable embodied emissions are calculated, and after positive and negative operational emissions are identified, the Zero Equation ensures that zero cumulative emissions are reached with certainty and by a specified year.

19.2.4 Applying the Zero Equation in Scenario 3

We will now re-run the calculations using the Zero Equations for Scenario 3 from Table 67, where embodied emissions in the PV system are much lower and based on an environmental product declaration. We will apply the same embodied emissions factor for the starting and for the maintenance embodied emissions. With reference to the Zero Equation in its formulation as Equation (4), we use the following values:

- CE_{CO2} = starting embodied emissions + embodied emissions from maintenance = 54,156 + (1,154 × 3 + 1,598) = 59,216 $kgCO_2$

- $P_{CO2} \times t$ = 964 $kgCO_2$/year × 40 years = 38,560 $kgCO_2$

- $BN_{CO2} \times t$ = 1,497 $kgCO_2$/year × 40 years = 59,880 $kgCO_2$

- BVE_{CO2} = (65.32 + 65.32) $kgCO_2$ = 130.64 $kgCO_2$ from initial embodied emissions from the PV plus PV replacement emissions after 30 years.

Substituting CE_{CO2}, $P_{CO2} \times t$, $BN_{CO2} \times t$ and BVE_{CO2} in Equation (4), we calculate f as follows for conventional building materials:

$$f = \frac{59,216 + 38,560}{59,880 - 131} = 1.636 \qquad \text{(4a)}$$

Applying the same reduction from introducing hempcrete as a replacement for conventional materials in the previous section, the Zero Equation becomes as follows:

$$f = \frac{59,216 - 28,293 + 38,560}{59,880 - 131} = 1.163$$

(4b)

The results are illustrated in Figure 218.

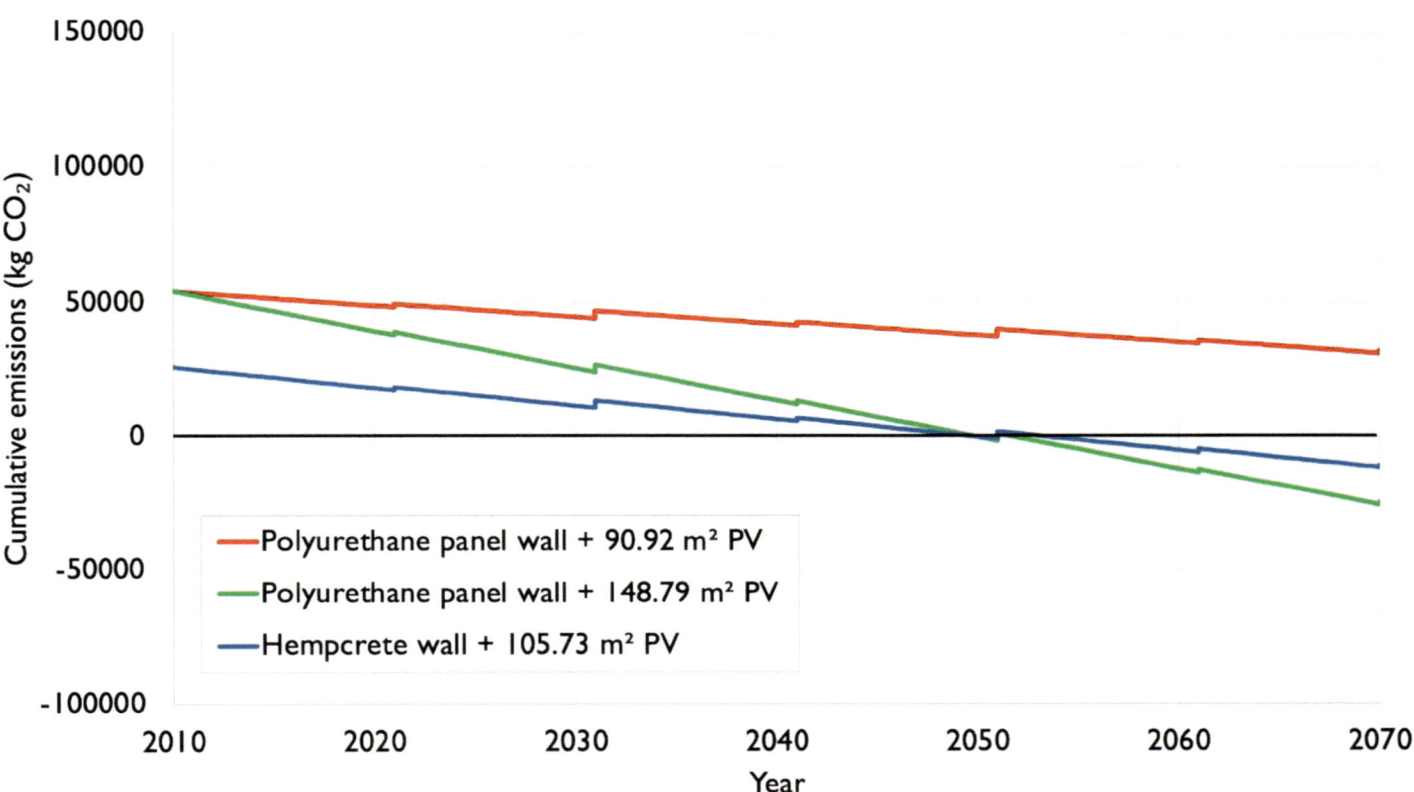

Figure 218   Cumulative emissions: embodied plus operational in Scenario 3 using the same vertical axis scale as in Figure 217

19.2.5 Discussion

As we have seen, having negative operational emissions is a necessary but not a sufficient condition for achieving zero cumulative emissions, as shown with the red line for combined Scenarios 1 and 2 in Figure 217, and in Figure 218. In order to achieve zero cumulative emissions with certainty and by a specified year, we use the Zero Equation to find out the magnitude of negative operational emissions required and to adjust the size of renewable systems accordingly. In addition to increasing negative operational emissions, this adjustment will increase initial embodied emissions from the renewable system. The results of two cases, one with conventional materials (blue line), and one with hemp-lime (hempcrete, green line) are shown in Figure 217 for combined Scenarios 1 and 2 and in Figure 218 for Scenario 3.

Although the application of the Zero Equation provides a method for achieving zero cumulative emissions with certainty and by a specified year, the differences between the results in Figures 217 and 218 arise because of the uncertainties between different embodied emissions factors for PV systems. Whilst the combined Scenarios 1 and 2 use the current/recent embodied emissions and projected future emissions per kW_p, the embodied emissions factor in Scenario 3 is derived from a recent, and seemingly the first environmental product declaration for a particular PV product at the time of writing. Although there is a range of possible embodied emissions figures resulting from these different emissions factors, the workflow established by the application of the Zero Equation remains the same. As embodied emissions factors for the PV, as well as other materials and products, become more established, applying the Zero Equation will lead to more accurate results. Regardless of these differences, the point made by this analysis is that the available roof space on the building in this particular example is not sufficient for achieving zero cumulative emissions by year 2050, and that space for an additional PV area is required.

Using non-intensive ways to reducing embodied emissions

Locally sourced materials lead to lower transportation costs. Locally based workforce has shorter travel to work and that causes less carbon emissions for transport. Using manual labour instead of heavy construction site machines where appropriate will also result in lower emissions. The analysis in Section 19.1.4 already assumes local transport of construction workers, and a minimum number of construction site machines and their hours of operation. But what would be the impact of scaling up the total emissions from construction of 4,480 $kgCO_2$ by a factor of 10 to represent longer transportation/travelling distances and more intensive construction methods? Such increase would result in the increase of embodied emissions from construction from 5.6% to 37.3% and a new total of initial embodied emissions of 120,143 $kgCO_2$. This illustrates the importance of locally sourced materials, locally based construction workers, and a minimum use of heavy construction site machines.

The analysis in this chapter shows how to achieve total cumulative emissions with certainty and by a specified year. It is based on a number of assumptions, and the accuracy and certainty of these calculations are dependent on these assumptions. As the industry evolves and as information about embodied emissions becomes more accurate and more readily available, this will lead to an increased accuracy of zero cumulative emissions calculations.

Regardless of the assumptions and details of emissions factors, the workflow presented in this chapter provides the reader with a method for replicating the calculations in their own projects. The application of Zero Equation provides a structured approach for achieving true zero carbon building designs whilst targeting a specific year and simultaneously accounting for operational and embodied emissions.

19.3 TASKS FOR SIMPLE EXERCISES

Task 1:

- Use a simulation tool of your choice to create the geometry of a shipping container. The container dimensions are: 20ft (6.096m) long × 8ft (2.438m) wide × 8ft 6in (2.591m) high and the material is corrosion resistant steel. Set the longer side of the container on the East-West axis.
- Find a manufacturer's specification of a solar PV panel.
- Populate the container roof with as many panels as can be fitted, using south orientation and 35° inclination.
- Run annual simulation using a weather data location of your choice.
- Calculate kW_p rating and embodied emissions of the PV system on the shipping container roof by following the example from Table 67.

Task 2:

- Add a south-facing window of the size of your choice and a door on the north-facing. side to the shipping container from Task 1.
- Add 300 mm of thermal insulation of your choice to the shipping container.
- Calculate embodied emissions by consulting the workflows from this chapter and from Chapter 5.
- Combine embodied and operational emissions and determine the year when the total emissions will reach zero.
- Change thermal insulation material to a bio-sourced material of your choice, such as straw, hemp-lime or other, and repeat the process from step 2 of this task.
- Discuss the results with your colleagues.

19.4 SUMMARY OF DESIGN PRINCIPLES

As we have seen in this chapter, the calculation of combined embodied and operational emissions involves numerous assumptions that may influence the accuracy of results. Whilst different assumptions will lead to different outcomes, the methodology in this chapter provides a workflow for achieving total zero emissions with certainty in the context of these assumptions.

The workflow developed in this and in the previous chapter is as follows:

- Run multi-objective optimisation of a building design until Pareto points with negative CO_2 emissions emerge.
- Populate the model with values of design variables corresponding to a negative CO_2 Pareto point of interest.
- Obtain embodied emissions from construction materials and glazing from this model.
- Calculate embodied emissions from HVAC systems and lighting.
- Calculate embodied emissions from the solar photovoltaic system and other renewable energy systems if applicable.
- Calculate embodied emissions from the construction process: construction workers travel, material deliveries, and operation of site machines.
- Calculate embodied emissions from maintenance.
- Draw a chart of combined embodied and operational emissions to establish whether or not zero cumulative emissions are achieved and by when.
- Apply the Zero Equation in order to adjust the requirement for renewable energy systems and corresponding negative operational emissions for reaching zero cumulative emissions by a specified year.
- Investigate the application of non-intensive ways to reduce emissions.
- Investigate the effects of bio-sourced materials in reducing emissions.
- Repeat the process until a satisfactory outcome is achieved.

CHAPTER 20

POST-OCCUPANCY MONITORING AND PERFORMANCE EVALUATION

Post-occupancy evaluation (POE) helps us complete the feedback loop in our design, as it gives us comprehensive information about the design and on lessons learnt for future designs. POE can take several forms as follows:

- One-off tests
  - co-heating
  - air-tightness testing
  - thermal imaging
  - observation of building envelope
- Instrumental monitoring
- Experimental measurement of physical parameters of buildings
- Energy performance analysis
- Troubleshooting of building operation and performance
- Calibration of dynamic simulation models
- Occupant comfort and behaviour studies

20.1 ONE-OFF TESTS

20.1.1 Co-heating tests

Co-heating tests are designed to measure the overall conductive and ventilation heat loss. They are carried out in unoccupied buildings in which a sufficient temperature difference between the inside and the outside is maintained with thermostat-controlled portable electric heaters over a period of between one and three weeks (Figure 219). As the building enters into a steady state after a few days of heating, the overall fabric and ventilation heat loss becomes proportional to the heat loss coefficient and inversely proportional to the temperature difference. The heat loss coefficient is calculated by rearranging the heat loss equation

$$Q = UA \times \Delta T \tag{58}$$

and solving it for UA as follows:

$$UA = \frac{Q}{\Delta T} \tag{59}$$

where

UA – overall heat loss coefficient (W/K)
Q – heat input (W)
ΔT – temperature difference between inside and outside (K).

DOI: 10.4324/9781003342342-24

Figure 219   Co-heating test with thermostat-controlled portable electric heaters with oscillating base

The co-heating test needs to be carried out in winter when a sufficient temperature difference between inside and outside of typically 10°C can be maintained. Fluctuation of external temperature and solar radiation can make the results of the co-heating test unusable. The co-heating test is therefore best conducted during overcast weather conditions with relatively stable outside temperatures. In order to minimise fluctuations of temperature differences, the results of co-heating tests are analysed using daily average values of heat input Q and temperature difference ΔT. When the former is plotted against the latter and a regression line is fitted to the plotted points, the slope of the line will represent the heat loss coefficient UA (Figure 220). In this particular case, the slope of the line and therefore the overall heat loss coefficient is UA = 69.3 W/K.

However, co-heating tests in general are conducted under unrealistic circumstances, attempting to maintain steady-state conditions in the building whilst trying to pick a period of steady-state conditions outside. In reality, buildings never operate under steady-state conditions. Additionally, conducting a co-heating test without capturing the dynamics of the building behaviour is something of a lost opportunity. It will be demonstrated later in this chapter how the pre-conditioning period of a co-heating test can be considered to be a dynamic heating test. The results of that test will be the building time constant, the overall heat loss coefficient, and the effective thermal capacitance, in other words, a lot more than from a co-heating test.

20.1.2 Air-tightness tests

Air-tightness tests are carried out in order to establish building infiltration characteristics called air permeability. A blower door and a fan are inserted into the frame of an existing door (Figure 221), and the fan pressurises or depressurises the building, depending on the test method. The air flow rate through the building is monitored for different fan speeds and air permeability is subsequently calculated by dividing the flow rate by the building envelope area.

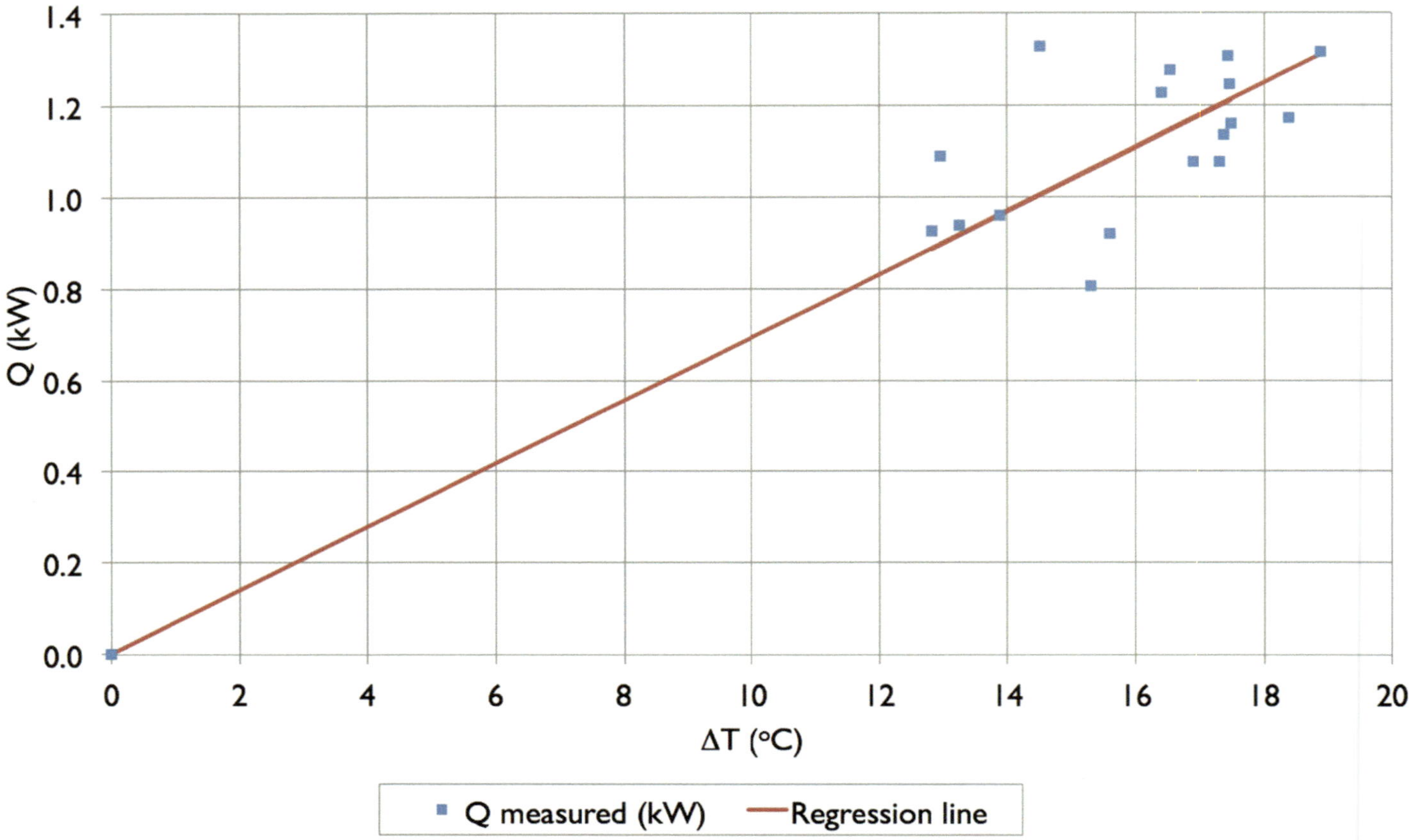

Figure 220   Co-heating test results and fitted regression line

Figure 221   A blower door with a fan (left) and instrumentation for air-tightness testing (right)

Sometimes it is not possible to complete air-tightness tests due to considerable air leaks in the building. These normally occur due to gaps between pipes or ducts and the building envelope, or due to poorly sealed interfaces between building components such as windows and walls. Figure 222 shows some of the air leakage sources found during air-tightness tests.

a)   A hole in the ceiling for a cable run

b)   A hole in the window frame

c)   Air leakage through a loft hatch

d)   Air leakage under patio doors

e)   Air leakage through wall electricity sockets

f)   Air leakage around unsealed pipe runs

Figure 222   Sources of air leakage found during air-tightness tests

Air-tightness tests are mandatory under UK building regulations; however, voluntary or repeat tests can also be carried out to assess the need for improvement of air-tightness or to assess the quality of improvements that have been carried out. The ventilation heat loss arising from building air permeability and its effect on heat losses and internal temperatures is discussed in Chapters 8 and 12.

20.1.3 Thermal imaging

Thermal imaging is a method of qualitative investigation of the building envelope. It can show relative differences between building components. It can also reveal defects in the envelope, such as missing insulation and thermal bridges, and can help with directing repair work. Figure 223 shows a thermal image of the back of the Birmingham Zero Carbon House (ZCH), of which a detailed case study is in Chapter 21. The figure shows that heat losses through the windows of the ZCH (bottom right) are higher around the edges than in the centre of the windows. The heat loss from the edges of windows of the ZCH is comparable to the heat loss from the walls of the house next door (left). There is a sharp vertical edge between the ZCH wall and the next door house wall, indicating much lower heat loss from the ZCH as a consequence of 350 mm thermal insulation that achieves U-value of 0.11 W/(m²K).

The passage tunnel between the two houses (bottom middle) glows with heat in the thermal image, and a closer inspection revealed that the source of heat loss is a thermal bridge in the tunnel, coming from the next door house (Figure 224).

Figure 223   *Thermal image of the back of the Birmingham Zero Carbon House (right) and the next door house (left) with a passage tunnel between the two*

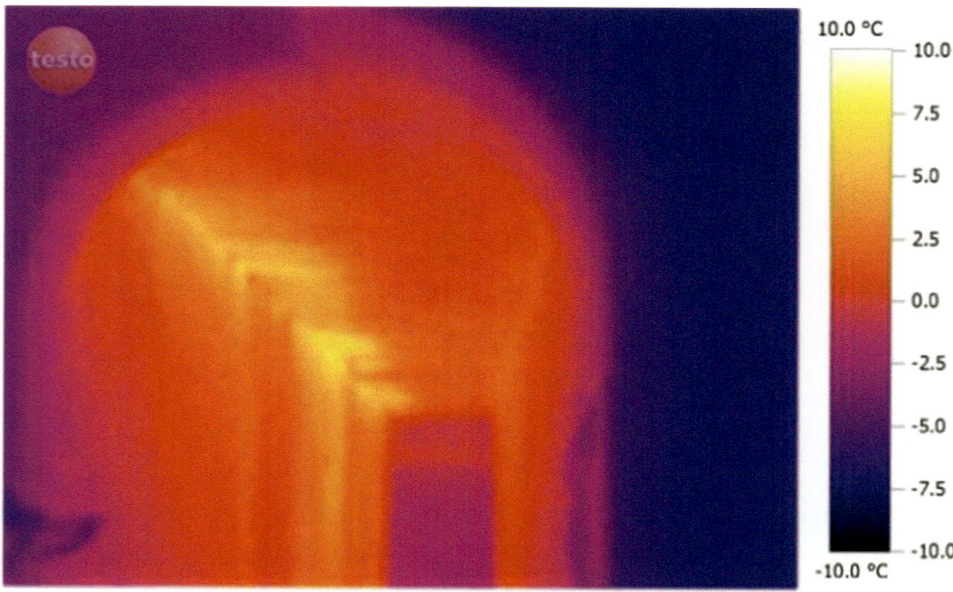

Figure 224   *Heat losses through a thermal bridge on the interface between vertical wall and first floor slab*

An example of missing insulation is shown in Figure 225, taken from the south-facing side of a passive solar house in Rowheath Solar Village in Bournville. The 100 mm cavity was initially filled with polystyrene beads, but after satellite TV installers drilled a hole the insulation had leaked out, showing angle of repose typical of granular materials descending from a higher to a lower point, a 'snow slope' looking shape inside the cavity between two windows. This figure also shows missing insulation above and to a smaller extent below the window frames.

Thermal imaging can also help to understand the effect of thermal reflective blinds as shown in Figure 226. This figure shows two semi-detached houses with the boundary marked by the vertical rainwater pipe running in the middle of this image. The house on the left has reflective thermal blinds drawn and the house on the right does not. There is a surface temperature difference of approximately 3°C between the respective windows of the two houses and the consequent heat loss from the house on the right is higher. The image also shows a patch of missing insulation between the windows of the two houses, manifested with brighter colours signifying higher heat loss.

Figure 225   Missing insulation from the cavity wall

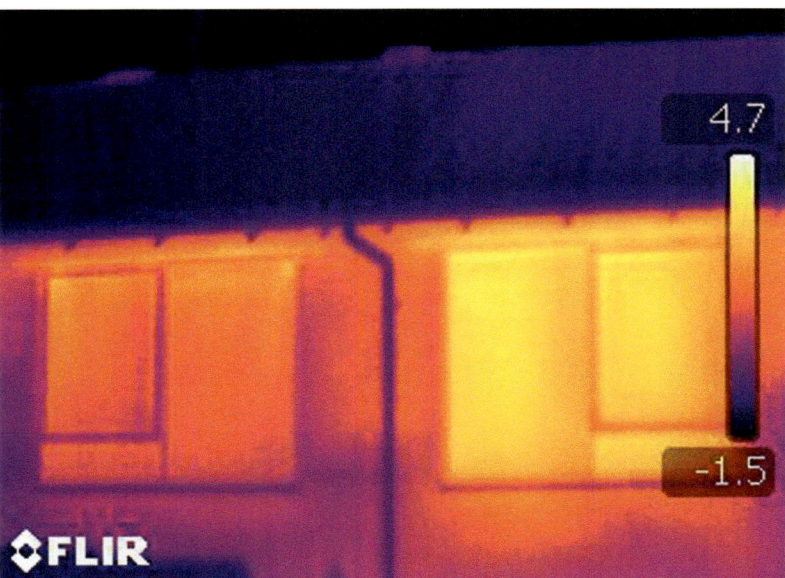

Figure 226   The effect of reflective thermal blinds on heat losses in two semi-detached houses

20.1.4 Observation of building envelope

Observation of the building envelope can reveal problems, such as gaps and cracks that need to be rectified and can help with learning how to overcome similar problems in future designs.

Figure 227 shows cracks observed in the building envelope in the Solar Demonstration House from Figure 230, several months after completion. Whilst conventional sealants cracked at interfaces of different materials (red arrows), silicon sealant remained intact (green arrow).

a) Cracks between sunspace walls and the floor (red arrows)

b) Cracks between window frame and the opening in the wall sealed with conventional putty (red arrow) and absence of cracks between the sunspace frame and the wall sealed with silicon (green arrow)

Figure 227   Cracks observed in the building envelope

This prompted an analysis of the temperature differences between day and night on various building surfaces. It was found that temperature differences between day and night on the sunspace wall surface were greater than 15°C during 33% of the total annual hours; and on the sunspace floor surface, temperature differences between day and night were greater than 15°C during 45% of the total annual hours. This difference of temperatures caused differential expansion between different materials and consequent cracks in the building envelope. Unlike conventional sealants, the silicon sealant showed high durability and resistance to cracks, but applying this sealant alone will not help unless expansion tolerances are designed for building components, especially those facing south.

20.2 INSTRUMENTAL MONITORING

The objective of this type of monitoring is to determine the energy consumption, carbon emissions, and comfort conditions in buildings, typically over a period of two years of occupation. The main challenges of this type of monitoring are to maintain uninterrupted data recording whilst making regular downloads of data and dealing with any equipment maintenance and malfunction. As heat transfer in buildings does not change much over a time interval of five to six minutes, this is the recommended time step for recording the monitored parameters. All heat inputs and outputs need to be monitored in order to be able to analyse building energy performance.

Figure 228a shows a large scale monitoring system as deployed at Bournville Solar Village; Figure 228b shows a much smaller portable monitoring system as deployed at a monitoring project in the Black Country in the West Midlands region of England; Figure 228c shows wireless monitoring system as deployed at the Birmingham ZCH.

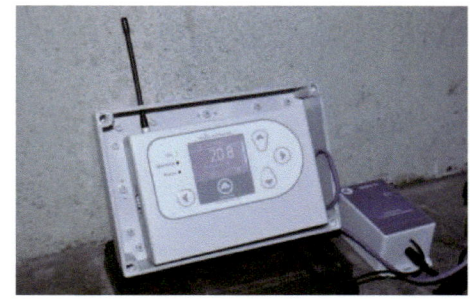

a) Fixed large scale monitoring system handling 250 measuring points

b) Portable battery operated monitoring system

c) Wireless monitoring system with web based remote access

Figure 228   Different types of monitoring systems

20.2.1 Signal filtering, data processing, and analysis

Temperature sensors installed in building walls and in the ground are susceptible to low voltage electric currents that can be found there. This especially applies to thermocouple thermometry, where the noise introduced by the electric current from the surroundings can interfere with the signal and can make the results of monitoring unusable. This problem can be overcome by hardware based signal filtering in the monitoring equipment, but the cost of these measures can be very high and impractical if implemented retrospectively. In one of such cases, I developed a software equivalent of a hardware filtering device, by programming the filter in machine code and applying it to 1,000 samples in each reading. Figure 229 shows a sample of data set before and after filtering. The resultant filter was integrated into the monitoring system, and due to its high speed as result of its machine code application it did not have a detrimental effect on the regular time interval of monitoring of circa 250 sensors, each sampled 1,000 times and filtered, with a sampling and recording intervals of six minutes.

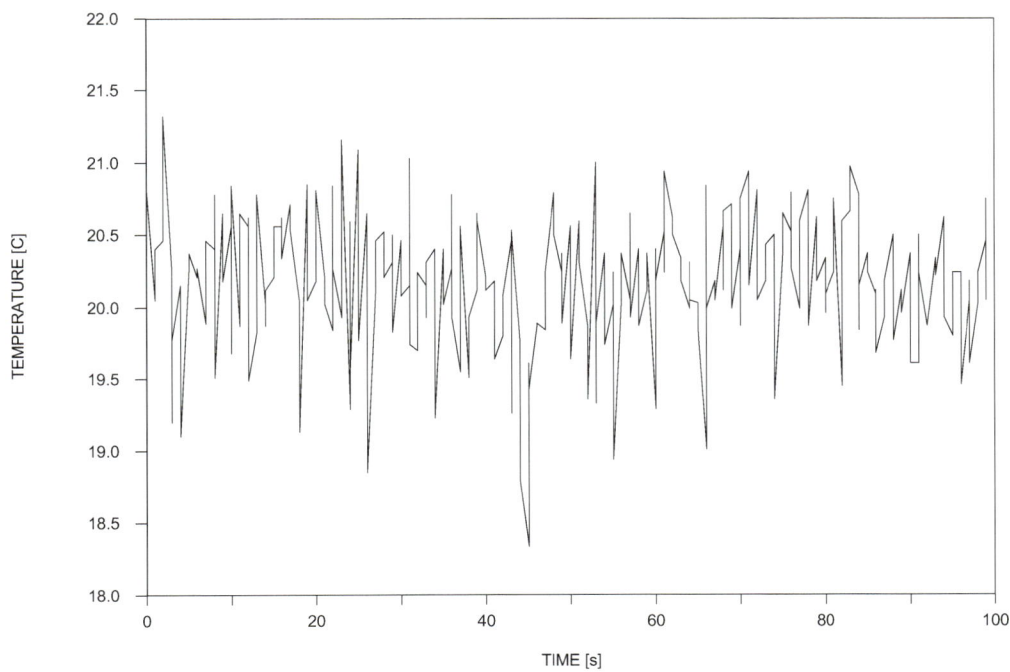

a) Effects of signal noise in a temperature channel at 1 sample per reading without filtering

Figure 229   A sample data set before and after filtering in machine code

(Continued)

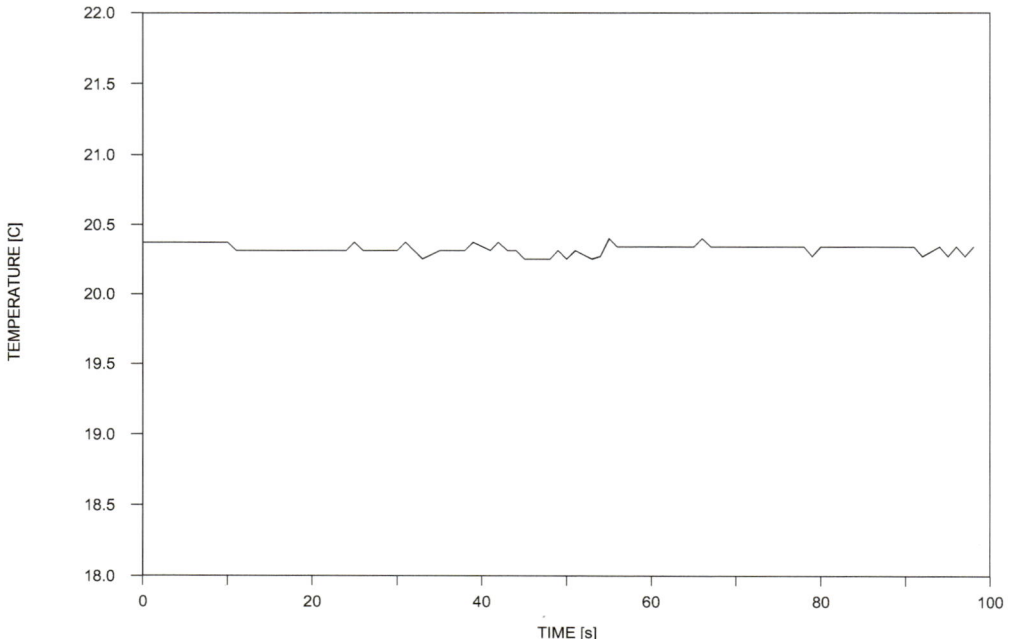

b) Effects of signal noise in a temperature channel at 1000 samples per reading after filtering

Figure 229   (Continued)

20.3 EXPERIMENTAL MEASUREMENT OF BUILDING PHYSICS PARAMETERS

20.3.1 Measuring thermal mass – building time constant, heat loss coefficient, and effective thermal capacitance

Physical parameters of buildings in use are never the same as their theoretical equivalents. Published properties of building materials and components are determined in laboratories under controlled conditions. The operational use of buildings subjects the materials and components to different working regimes. Properties of materials will also depend on the exact composition of the material and the moisture content, and both are likely to differ from theoretical conditions.

During my PhD study at Birmingham University, I conducted instrumental monitoring and analysis of energy performance of the Rowheath Solar Village in Bournville, Birmingham. A Solar Demonstration House within the solar village (Figure 230) was equipped as a monitoring lab, with circa 250 sensors, measuring solar radiation on different planes, ambient air temperature, relative humidity and numerous temperatures in rooms, walls, and mechanical systems in this house and in several neighbouring houses (Figures 231 and 232).

The Demonstration House was unoccupied during a three year monitoring period and was available for experimentation. One of the experiments I conducted was a dynamic heating test, in order to investigate the parameters of the building's transient thermal behaviour (Figure 233). The results of these experiments are complementary to the explanation of thermal mass in Chapter 11, and a good introduction into dynamic behaviour of buildings.

The specific objective of the dynamic heating test was to obtain a building time constant t_c, overall conductance-area product, and effective thermal capacitance. By definition, the time constant is the time required for a system response to reach 63.3% of its final value after an initial step input.

Figure 230   Solar Demonstration House in Bournville, Birmingham

Figure 231   Weather station in Rowheath Solar Village, measuring horizontal total and diffuse radiation (pyranometers without and with a shading ring), air temperature, and air humidity

Figure 232   Data acquisition and signal conditioning equipment for various sensors in Rowheath Solar Village

Here is how the dynamic behaviour parameters were obtained from these tests. To start, the heat balance for a building can be expressed as

$$C \times \frac{dT_{room}}{dt} = -Q_{loss} + Q_{solar} + Q_{internal} \qquad (60)$$

where C is the effective thermal capacitance of the building. In the absence of solar gain Q_{solar} (thermal blinds were down during the tests, the sky was overcast and ambient temperatures were relatively constant), Equation (60) can be rearranged as

$$\frac{dT_{room}}{dt} + \frac{UA}{C} \times T_{room} = \frac{UA}{C} \times T_{amb} + \frac{Q_{internal}}{C} \qquad (61)$$

where

UA – overall conductance-area product
T_{amb} – difference between ambient air temperature and the initial room temperature: $T_a - T_{r,o}$
T_{room} – difference between the room temperature and the initial room temperature: $T_r - T_{r,o}$.

The solution of the differential Equation (61) can be expressed as

$$T_{room} = \frac{T_{amb} + Q_{internal}}{UA} \times \left(1 - e^{-\frac{t \times UA}{C}} \right) \qquad (62)$$

Equation (62) can then be rewritten as

$$T_{room} = a \times \left(1 - e^{-b \times t}\right)$$ (63)

where

$$a = \frac{T_{amb} + Q_{internal}}{UA}$$

$$b = \frac{UA}{C}$$

and the time constant is then determined as

$$t_c = \frac{1}{b} = \frac{C}{UA}$$ (64)

This is the method for simultaneous calculation of building thermal properties from dynamic heating tests I developed in my doctoral thesis (Jankovic, 1988). It is based on fitting the curve from Equation (63) to raw test data in Figure 233a. When the fitted curves are obtained from this process (Figure 233b), the three parameters from Equation (64) are also obtained. In the particular case the following values were obtained for the Solar Demonstration House:

t_c = 44.5 hours

UA = 256.2 W/K

C = 41.1 MJ/K.

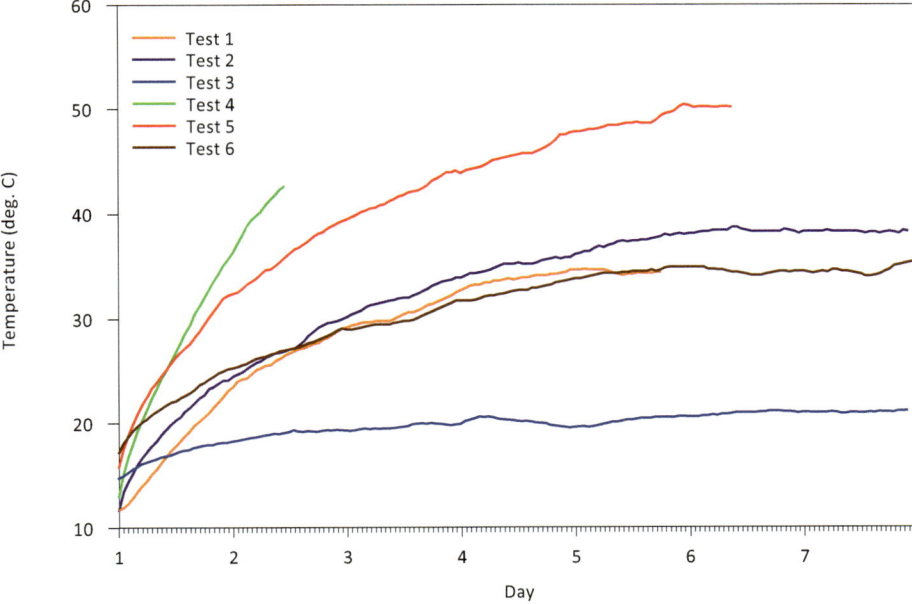

a) Room temperatures recorded during the tests (this may be a good time to confess that I did indeed heat the building to about 50°C for research purposes)

Figure 233   Dynamic heating tests conducted in the Bournville Demonstration House

(Continued)

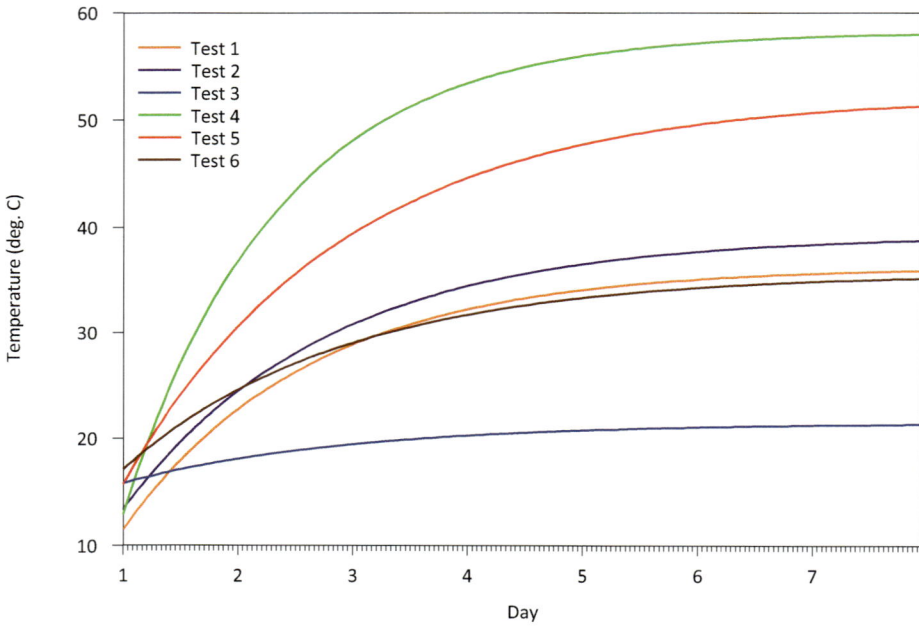

b) Curves fitted to room air temperatures data

Figure 233   (Continued)

The above results show that it takes almost two days for this particular building to significantly change its temperature, by 63% of its final value. The results of this work were used to explain various aspects of thermal mass design in Chapter 11.

The result for UA value above is consistent with the UA value of 249.6 W/K obtained from a co-heating test in this building.

20.3.2 Effective solar aperture and conductance-area product

In addition to the measurement of the steady-state properties of the building fabric using the co-heating tests, other methods that utilise day-to-day data from monitoring have been developed for the simultaneous calculation of U-values and effective solar apertures (effective surface areas of glazing that admit solar gain into the building).

The method used in this particular example is based on work by Jankovic (1988). The heat balance equation

$$Q_{aux} + Q_{int} + Q_{sol} = Q_{loss}$$

(65)

where

Q_{aux} – auxiliary heating energy
Q_{int} – internal heat gain from people and appliances
Q_{sol} – solar energy input
Q_{loss} – conductive and infiltration heat loss of the building

is first rearranged as suggested by Siviour (1981)

$$Q_{aux} + Q_{int} + A_{eff} \times I = \left(UA_c + UA_v\right) \times \left(T_r - T_a\right)$$

(66)

where

UA_c – overall conductance-area product
UA_v – an equivalent of UA_c but prescribed for infiltration
$Q_{sol} = A_{eff} \times I$
A_{eff} – effective solar aperture

I – solar irradiance

T_r – internal room temperature

T_a – external air temperature

Subsequently, both sides of the heat balance equation written in this form are divided by the temperature difference between room air and external air ($T_r - T_a$) and the term with the solar input is moved to the right hand side as follows:

$$\frac{Q_{aux} + Q_{int}}{T_r - T_a} = UA_V + UA_C - A_{eff} \times \frac{I}{T_r - T_a} \qquad (67)$$

If terms ($Q_{aux} + Q_{int}$)/($T_r - T_a$) and (I/($T_r - T_a$)) are known and if the former is plotted versus the latter then the plot will be a straight line with the slope of $-A_{eff}$ and the intercept of ($UA_c + UA_v$).

If points related to the two known terms in the above equation are then subjected to linear regression analysis, the result will be the values of the effective solar aperture Aeff and the sum of the overall conductance-area product and its infiltration equivalent ($UA_c + UA_v$).

In practice, however, the application of this method is not straightforward. In this particular case additional filtering of data entries for the above equation was required before the method could be used for the calculation of the building thermal properties. Analysis of six passive solar houses was carried out on an hourly basis with daily total values calculated as entries.

An example of the plot of data pairs (($Q_{aux} + Q_{int}$)/($T_r - T_a$), I/($T_r - T_a$)) for House A is shown in Figure 234a. The existence of scattered data pairs makes it difficult to obtain meaningful results from the regression analysis. Scattering of results occurs as result of dynamic heat transfer, occupant behaviour that was not directly monitored and possible noise in the instrumental monitoring channels.

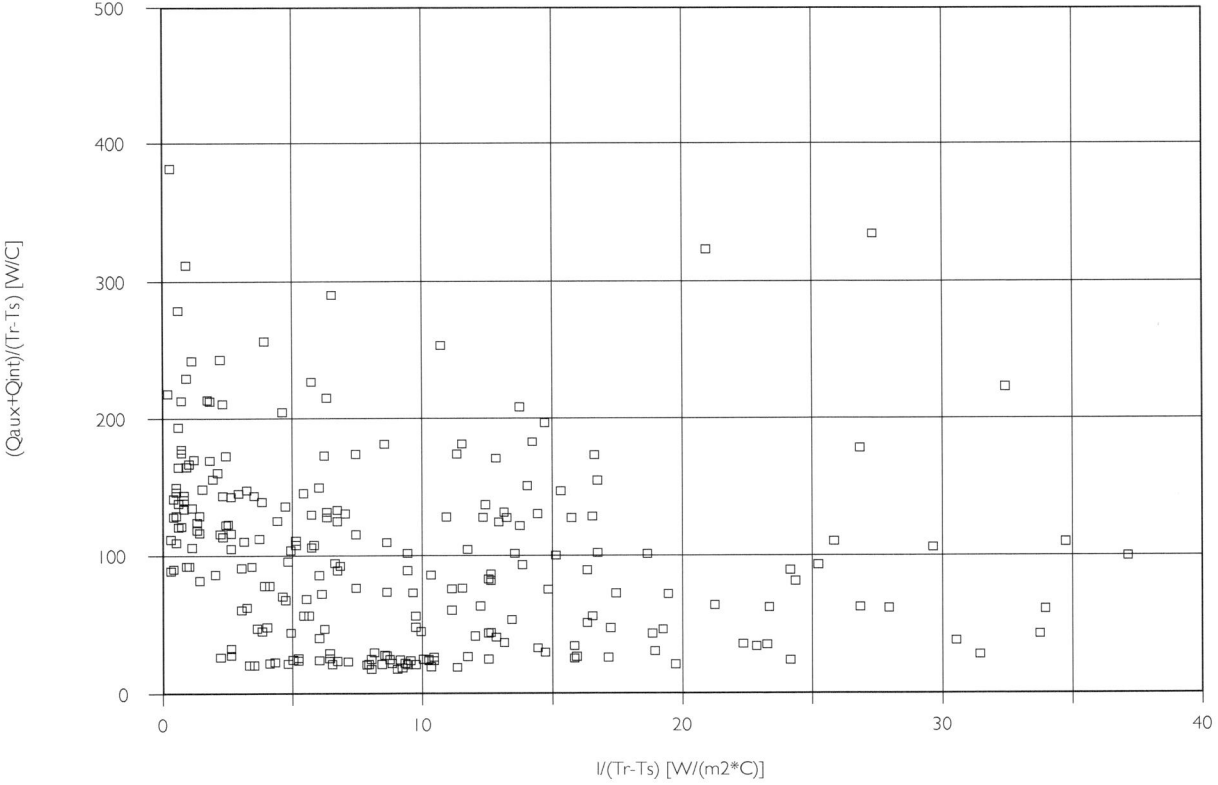

a) Scatter-plot of rearranged heat balance equation before filtering

Figure 234   An example of measurement of the overall UA value and effective solar aperture

(Continued)

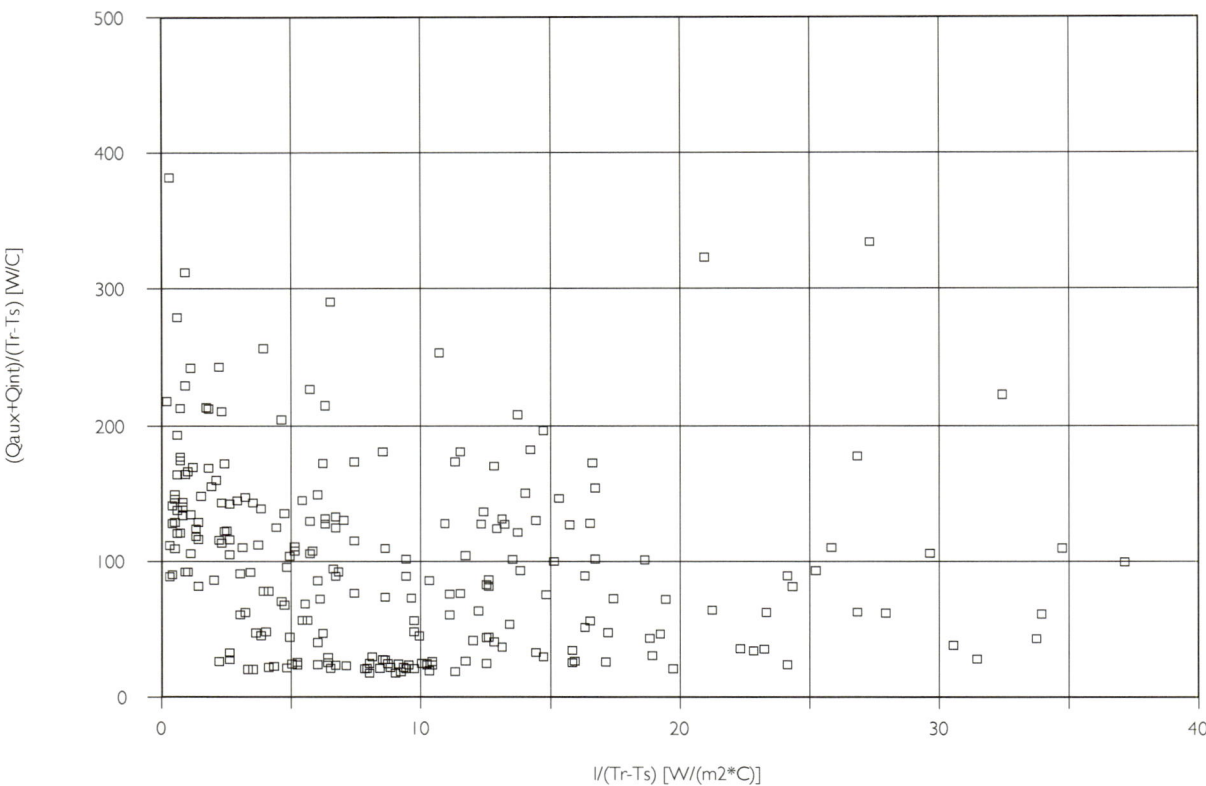

b) Scatter-plot of rearranged heat balance equation after filtering

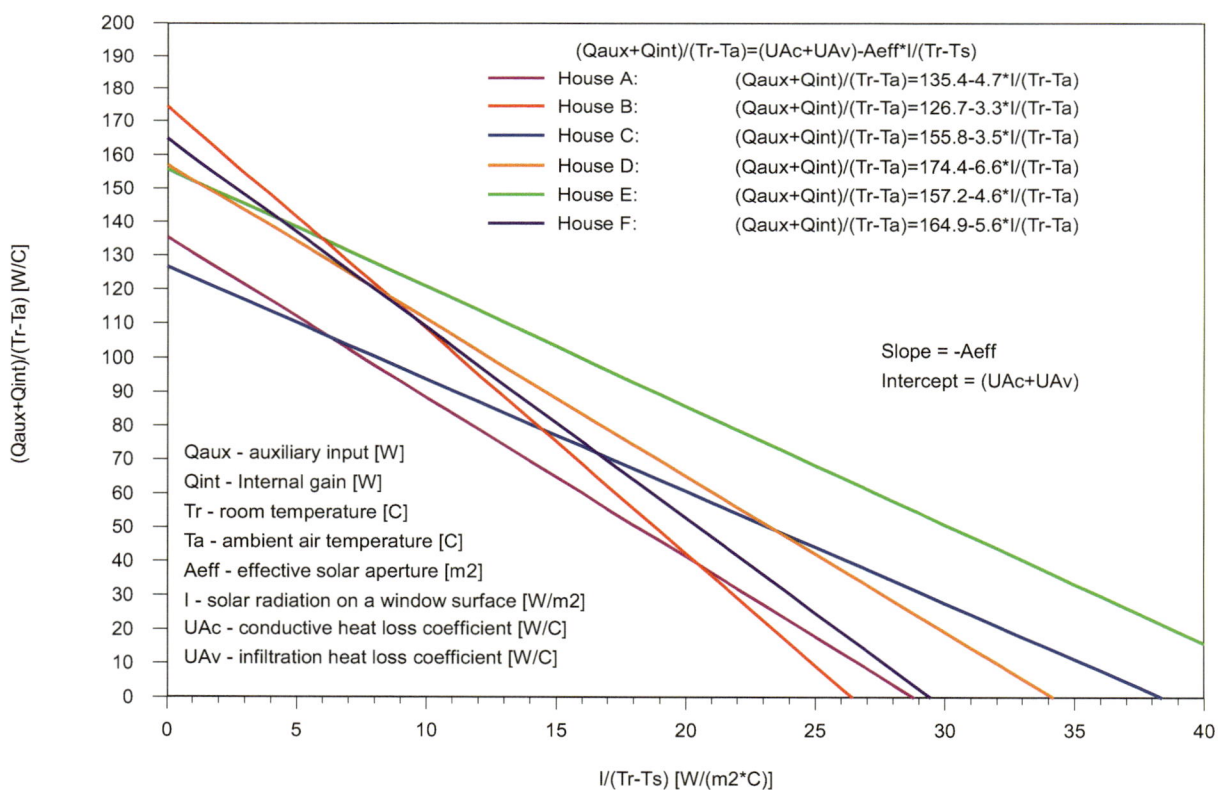

$$(Q_{aux}+Q_{int})/(T_r-T_a)=(UA_c+UA_v)-A_{eff}*I/(T_r-T_s)$$

House A: $(Q_{aux}+Q_{int})/(T_r-T_a)=135.4-4.7*I/(T_r-T_a)$
House B: $(Q_{aux}+Q_{int})/(T_r-T_a)=126.7-3.3*I/(T_r-T_a)$
House C: $(Q_{aux}+Q_{int})/(T_r-T_a)=155.8-3.5*I/(T_r-T_a)$
House D: $(Q_{aux}+Q_{int})/(T_r-T_a)=174.4-6.6*I/(T_r-T_a)$
House E: $(Q_{aux}+Q_{int})/(T_r-T_a)=157.2-4.6*I/(T_r-T_a)$
House F: $(Q_{aux}+Q_{int})/(T_r-T_a)=164.9-5.6*I/(T_r-T_a)$

Slope = -A_{eff}
Intercept = (UA_c+UA_v)

Q_{aux} - auxiliary input [W]
Q_{int} - Internal gain [W]
T_r - room temperature [C]
T_a - ambient air temperature [C]
A_{eff} - effective solar aperture [m2]
I - solar radiation on a window surface [W/m2]
UA_c - conductive heat loss coefficient [W/C]
UA_v - infiltration heat loss coefficient [W/C]

c) Calculation of effective solar aperture and overall conductance-area product for six buildings based on scatter-plot data

Figure 234   An example of measurement of the overall UA value and effective solar aperture

To overcome this problem, additional filtering of data set was carried out. The area of the graph was divided into 40 rectangles (Figure 234a), and all data points were eliminated from the data set if their density per rectangle was less than a certain number. This number was determined iteratively, by trial and error.

Additionally, data points from the lower left half of the lower left rectangle were eliminated regardless of their density as they represented low energy input during low solar radiation, which was believed to be typical for the night time energy balance. The result of filtering is shown in Figure 234b, which helped to carry out a regression analysis.

Results of similar analysis for all six houses are shown in Figure 234c and in Table 73 where the effective solar aperture A_{eff} and the building heat loss coefficient $(UA_c + UA_v)$ were obtained from the regression line equations.

TABLE 73 EFFECTIVE SOLAR APERTURES AND HEAT LOSS COEFFICIENTS FOR SIX ANALYSED HOUSES			
House	Design type	UA (W/C)	A_{eff} (m²)
A	B4	135.4	4.7
B	B3	126.7	3.3
C	B4	155.8	3.5
D	B4	174.4	6.6
E	A4	157.2	4.6
F	B4	164.9	5.6

The differences between the six houses occurred as a result of slight design variations, as built properties, and occupant lifestyle and behaviour.

The gross window area for each analysed house was 11.6 m² (13.7 m² for House D). The effective solar apertures were, however, found to be much smaller and different for each house. The difference was as much as 71% in the case of House B. This could be due to the different absorption coefficients of walls, carpets, and furniture in the houses. House B had the smallest effective solar aperture (Table 73) and it was found that the interior of this house had exceptionally light colours and therefore lower solar absorption than the interiors of the other analysed houses.

20.4 ENERGY PERFORMANCE ANALYSIS

Energy performance analysis will be explained on an example of analysis carried out for six passive solar houses in Rowheath Solar Village. The starting point was the heat balance equation

$$Q_{aux} + Q_{int} + Q_{sol} = Q_{loss} + Q_{res} \tag{68}$$

where the terms of the equation were calculated as follows:

$$Q_{aux} = Q_{gas,measured} \times (1.06 - e^{-(Tr - Ta)/(Tr - Ta)max})$$

$$Q_{int} = 0.7 \times Q_{el,measured} + Q_{occ} \times 12hr \times (daypeople + totalpeople) \times days$$

$$Q_{sol} = A_{eff} \times I$$

$$Q_{loss} = (UA_c + UA_v) \cdot (T_r - T_a)$$

$$Q_{res} = Q_{aux} + Q_{int} + Q_{sol} - Q_{loss}$$

where

$Q_{gas,measured}$ – monthly measured gas consumption
T_r – monthly average room air temperature
T_a – monthly average ambient air temperature
$(1.06 - e^{-(Tr - Ta)/(Tr - Ta)max})$ – efficiency of gas boilers (Jankovic, 1988)
0.7 – overall coefficient of heat gain from electrical
appliances; this coefficient could not be known exactly
and the error introduced in this way was 4% if, for
example, the coefficient had been 0.5 instead of 0.9
$Q_{el,measured}$ – monthly measured electricity consumption
Q_{occ} – heat gain per occupant seated at rest assumed as 100 W per occupant
daypeople – number of daytime occupants, obtained from interviews (Table 75)
totalpeople – total number of occupants (Table 75)
days – number of days in a month

A_{eff} – effective solar aperture, as calculated in the previous section

I – measured monthly average solar radiation on vertical south-facing surface

$(UA_c + UA_v)$ – building heat loss coefficient due to conduction and constant infiltration,
as calculated in the previous section

Q_{res} – residual of the heat balance.

Residuals of the heat balance equation were handled in two ways: positive residuals were attributed to variable infiltration (window opening and operation of extract fans) and negative residuals were attributed to measured internal temperatures being higher than actual temperatures. The reason was that room air temperatures were measured in living rooms only, and bedroom temperatures, which were not measured, were lower. Negative residuals Q_{res} occurred only in winter months.

Variable infiltration rates were calculated from positive residuals Q_{res} of the heat balance equation

$$N_{var} = \frac{(Q_{res} \times 3.6)}{(\rho \times V \times c \times (T_r - T_a))} \text{ if } Q_{res} > 0 \tag{69}$$

where

ρ – density of air
c – specific heat of air

If the residuals Q_{res} were negative, then the corrections for differences between the room air and the ambient air temperatures were calculated as

$$\Delta T = \frac{Q_{res}}{UA_c + UA_v} \text{ if } Q_{res} < 0 \tag{70}$$

and the measured temperature differences were then corrected as

$$(T_r - T_a)_{corrected} = (T_r - T_a)_{measured} - \Delta T \tag{71}$$

Degree days were then calculated on the basis of the actual monthly average temperature differences between the room air temperatures and the ambient air temperatures as follows:

$$DD = (T_r - T_a) \times days \qquad (72)$$

20.4.1 Energy performance results

The above analysis subsequently enabled energy performance results to be produced. Total heat losses in six houses are shown in Figure 235. Most of the heat losses are clustered together and the inconsistency of heat losses in House E was explained from residents' interview (see Table 76) as a consequence of a six weeks vacation period at the beginning of the year.

Total delivered heating energy Q_{aux} is shown in Figure 236. House E again appears to be different in comparison with other houses. This is explained again with the six weeks' vacation together with high thermostat setting (Table 76).

Solar energy delivered to the houses is shown in Figure 237. It follows the pattern of solar radiation on a vertical south-facing surface and the differences between the delivered solar energy in different houses are due to different effective solar apertures (see Table 73).

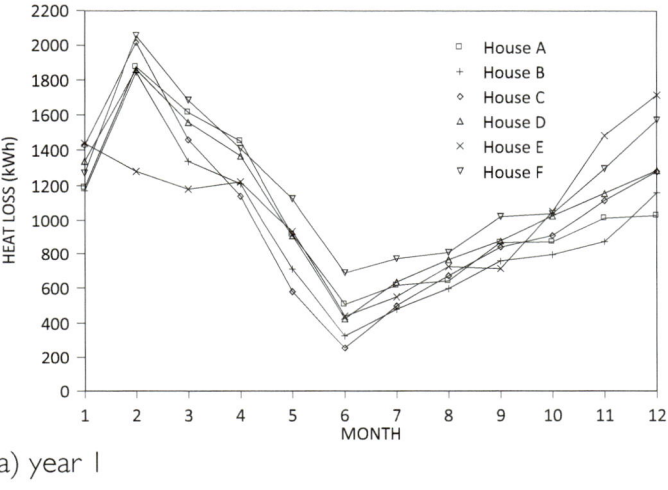

a) year 1   b) year 2

Figure 235   Total heat losses

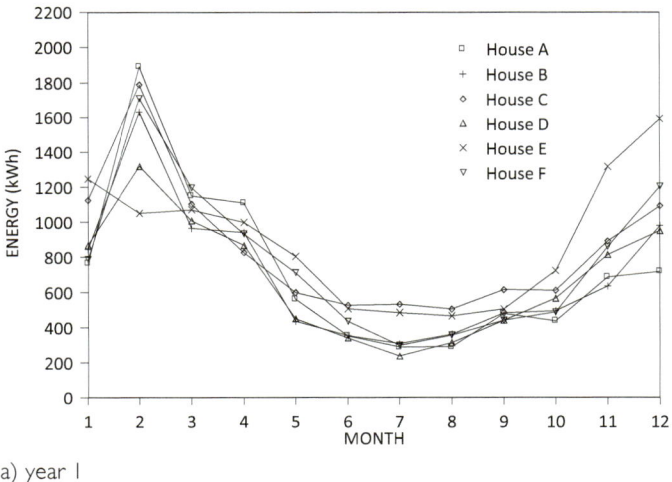

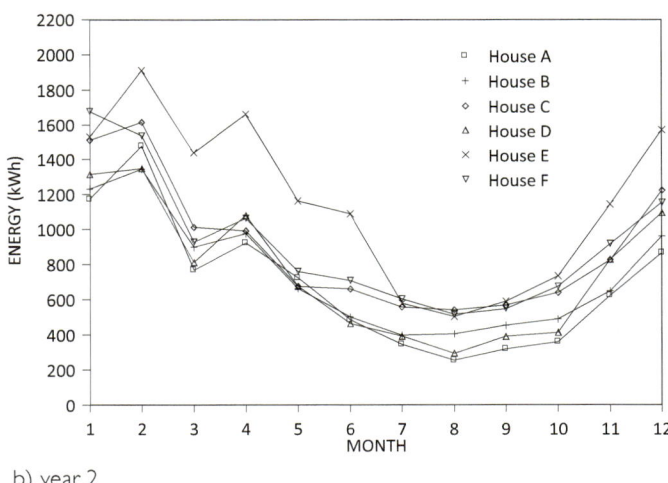

a) year 1   b) year 2

Figure 236   Total delivered heating energy

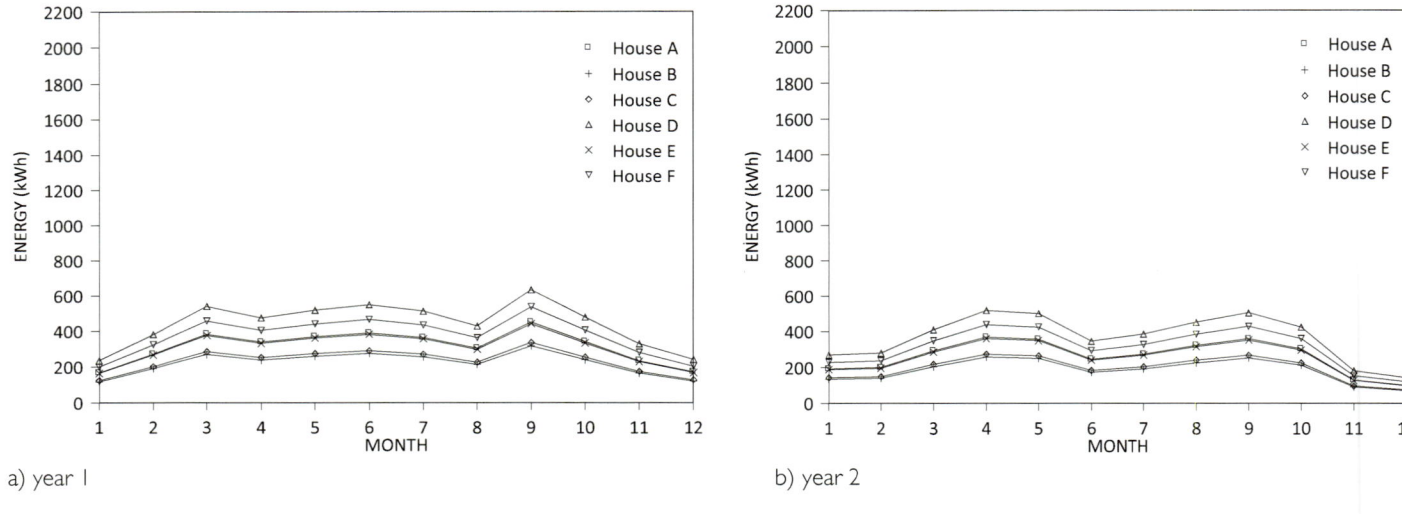

a) year I

b) year 2

Figure 237   Total delivered solar energy

Fractions of total heat loss that are offset by solar energy (solar fractions) are shown in Figure 238. House D has the largest solar fraction as result of the largest effective solar aperture (Table 73). The summary of two-year average solar fractions is shown in Table 74.

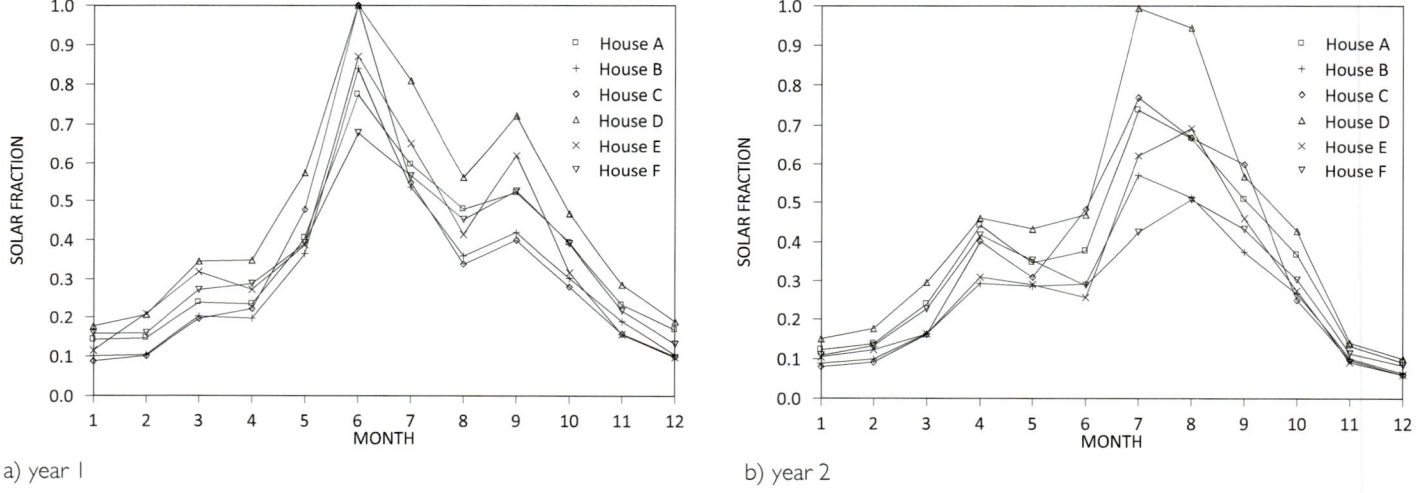

a) year I

b) year 2

Figure 238   Solar fractions achieved

TABLE 74 BIENNIAL AVERAGE SOLAR FRACTIONS	
House	**Solar fraction**
A	0.36
B	0.29
C	0.33
D	0.45
E	0.33
F	0.32

Calculated variable infiltration rates are shown in Figure 239. The highest variable infiltration rates occurred in House C. From Table 73, we can see that House C had the largest number of occupants and consequently the highest requirement for fresh air, which is consistent with the findings in Figure 239.

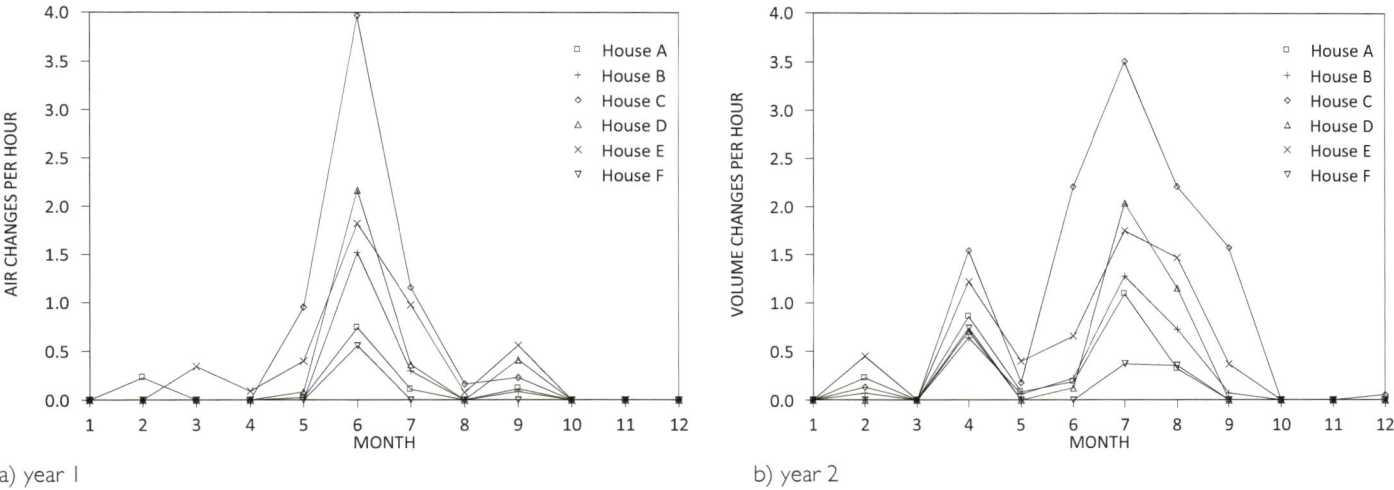

a) year 1   b) year 2

Figure 239   Variable infiltration rates

Two-year averages of the total energy consumption are shown in Figure 240. Total gas consumption is between 9.1 and 14.1 MWh per year and total gas and electricity consumption between 13.3 and 17.2 MWh per year. House E is the highest energy user and House C the lowest. This is a consequence of a combination of factors, including heat loss coefficients, effective solar apertures, slightly different design types, and different occupants' energy use patterns.

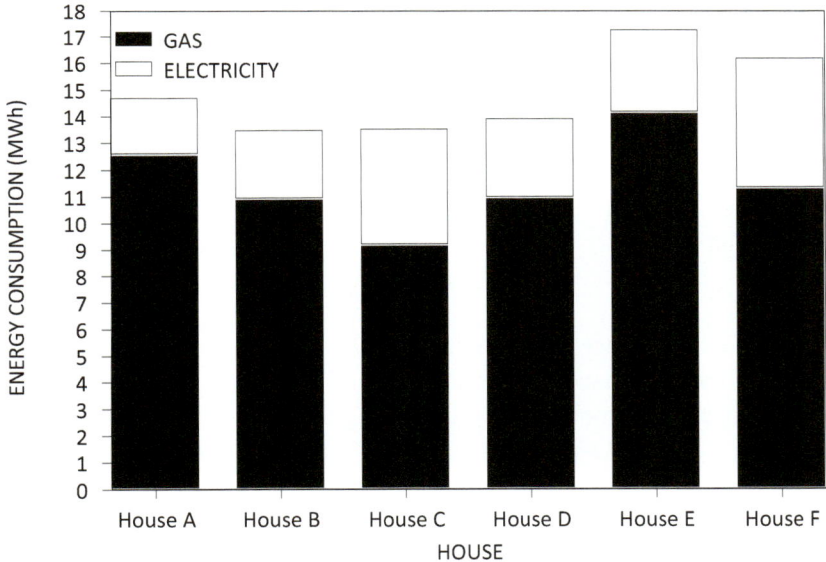

Figure 240   Two-year average of the total energy consumption

20.5 TROUBLESHOOTING OF BUILDING OPERATION AND PERFORMANCE

Monitoring is a useful tool for troubleshooting incorrect building operation. Figure 241 shows results of the monitoring of an office building in which there were complaints about thermal comfort. The data clearly shows internal air temperatures above the comfort level and prolonged times of operation of the heating system beyond normal office hours. By conducting a dynamic simulation of the same building using temperatures obtained through monitoring, it was found that that the times of operation of the heating system could be reduced considerably, by starting the heating at 8:30 in the morning and switching it off at 12:00 noon.

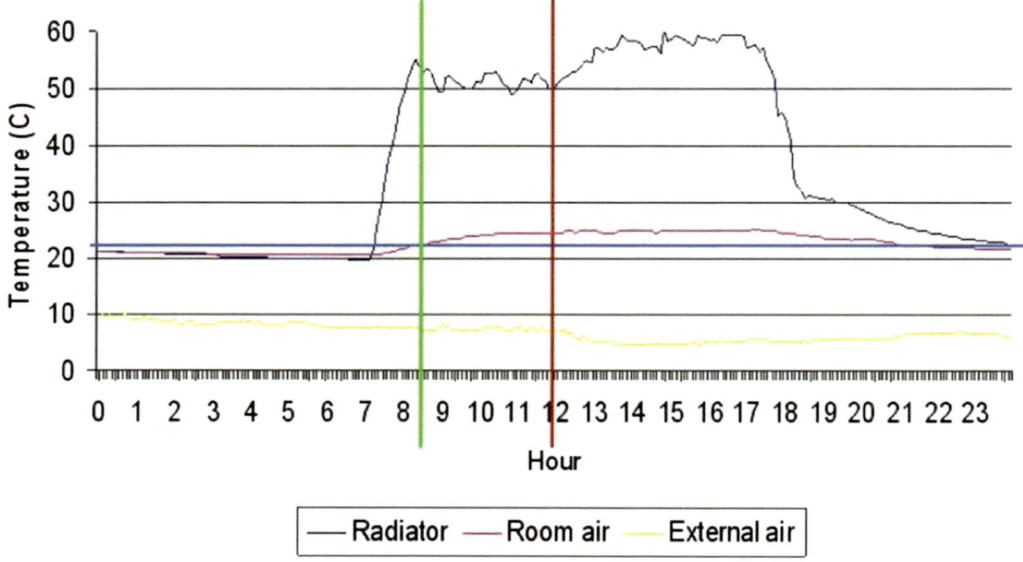

Figure 241   *Results of monitoring of an office building in which there were complaints about thermal comfort*

These recommendations were implemented in this office building, resulting in increased comfort and in a 30% reduction of energy consumption.

20.6 CALIBRATION OF DYNAMIC SIMULATION MODELS

Although dynamic simulation models replicate all aspects of building thermal performance using fundamental principles of physics, experience shows that inexperienced users can generate results that are considerably inaccurate. Simulation models contain a number of built-in default values of parameters (values used if not specified by the user). This ensures that the model runs rather than crashes, but it causes inaccuracy if the default parameter values do not correspond to the actual building and its environment.

However, it is not just user inexperience that can cause model inaccuracy. In 'Co-heating tests' and 'Measuring thermal mass – Building time constant, heat loss coefficient and effective thermal capacitance' in this chapter, the results of empirical calculation of the overall heat loss coefficient were reported using two different methods: the co-heating test method and the dynamic heating test method. Both methods produced similar results, giving the UA value of 249.6 and 256.2 W/K, within 2.5% discrepancy between the two. The theoretical UA value for the same building, calculated on the basis of properties of building materials was 157 W/K, representing between 59% and 63% difference from the two empirical values. In 'Effective solar aperture and conductance-area product' we saw that effective window size (effective solar aperture) can be nearly 70% different from the actual window size. This indicates the magnitude of differences between the buildings in use and buildings 'on paper', and justifies the need for improving the accuracy of simulation models using data from monitoring. The importance of monitoring projects as an invaluable resource for improvements of simulation models and of their accuracy is emphasised by these findings.

The process of improving the accuracy of a simulation model will be referred to as 'calibration'. Temperature differences between monitored and simulated performance will be an indicator of the accuracy of the simulation model. During the calibration process libraries of weather data files cannot be used, but the weather file needs to be synthesised using parameters obtained through monitoring.

The process of calibration of simulation models is demonstrated in detail in Section 21.1.5, explaining a manual method of calibration, and in Section 22.1.2, explaining an automated method of calibration using multi-objective optimisation and evolutionary computing. Advanced methods for reducing simulation performance gap are introduced in Section 22.1.1.

20.7 OCCUPANT COMFORT AND BEHAVIOUR STUDIES

Occupant studies are conducted for two main reasons

1   to establish occupants' perception of and the degree of their satisfaction with the internal environment
2   to provide complementary information for instrumental monitoring in order to explain discrepancies between expected and actual building performance

As we have seen from the first part of this chapter, identical building types have noticeable differences in energy performance. Occupant studies represent an invaluable source of information that helps to explain these differences.

Questionnaires are well established instruments for collecting data for occupant behaviour studies. A seven point scale established by Fanger (1972, 1988) is a good way of collecting occupants' perception of internal conditions (Figure 242a). Qualitative descriptions of occupants' responses are quantified by converting them into corresponding numbers and are subsequently used for various analyses.

However, the seven point scale is restrictive as it limits occupants' choices to a discrete number of possibilities. This can be overcome by a visual-analogue scale (Kildesø et al., 1999), as shown in Figure 242b. Wyon, one of the co-authors of this work, who worked with Fanger in the early days of thermal comfort research, found that the visual-analogue scale gives the occupant a much wider range of choices and makes it almost impossible to give the same response twice even under the same internal conditions. Positions of occupants' responses, represented as lines crossing the long horizontal line, are digitised into numbers, and these numbers are then used in analysis of occupants' responses.

Low extreme	Much lower than neutral	Slightly lower than neutral	Neutral	Slightly higher than neutral	Much higher than neutral	High extreme
-3	-2	-1	0	+1	+2	+3

a) The seven point scale – the occupant is asked to mark one of the seven points on the scale that corresponds to their perception of the tested parameter

Low extreme   High extreme

b) The visual-analogue scale – the occupant is asked to mark the horizontal line in the position that corresponds to their perception of the tested parameter

Figure 242   Occupant response scales

Questionnaires can also be simpler, as shown in Figure 243, where occupants are given three choices between low, neutral, and high. The pie charts on the right represent the number of responses to each choice as a fraction of the total number of responses.

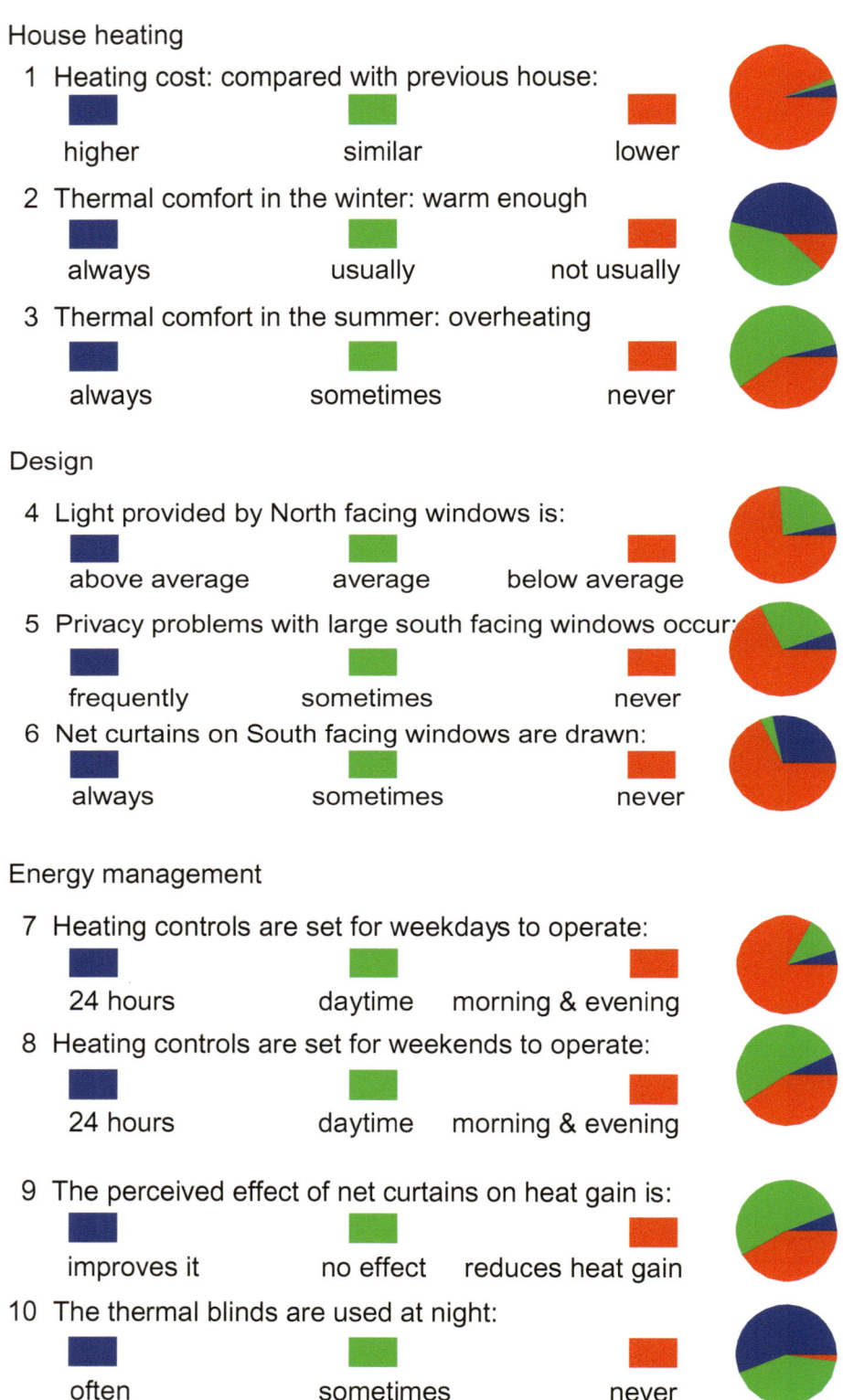

House heating

1 Heating cost: compared with previous house:

higher	similar	lower

2 Thermal comfort in the winter: warm enough

always	usually	not usually

3 Thermal comfort in the summer: overheating

always	sometimes	never

Design

4 Light provided by North facing windows is:

above average	average	below average

5 Privacy problems with large south facing windows occur:

frequently	sometimes	never

6 Net curtains on South facing windows are drawn:

always	sometimes	never

Energy management

7 Heating controls are set for weekdays to operate:

24 hours	daytime	morning & evening

8 Heating controls are set for weekends to operate:

24 hours	daytime	morning & evening

9 The perceived effect of net curtains on heat gain is:

improves it	no effect	reduces heat gain

10 The thermal blinds are used at night:

often	sometimes	never

Figure 243   Extract from results of an occupant survey of 90 households obtained through questionnaires

Questionnaires often need to be followed up by interviews with a smaller sample of occupants, in order to obtain further details, for instance of the number of appliances in the house (Table 75), energy use patterns (Table 76) or clarification of certain responses.

TABLE 75 EXAMPLE OF INTERNAL HEAT GAINS INFORMATION OBTAINED THROUGH INTERVIEWS						
Heat gains information	**House A**	**House B**	**House C**	**House D**	**House E**	**House F**
Occupants	2	3	5	2	3	2
Daytime occupants	0	1	2	0	2	0
Washing machine	1	1	1	1	1	1
Electric cooker	1	0	1	1	0	0
Refrigerator	1	1	1	1	1	1
Tumble drier	1	1	1	0	1	0
Gas cooker	0	1	0	0	1	1
Freezer	0	1	1	1	1	1
Electric shower	0	0	1	1	0	0

TABLE 76 EXAMPLE OF OCCUPANTS' ENERGY USE INFORMATION OBTAINED THROUGH INTERVIEWS						
	House A	**House B**	**House C**	**House D**	**House E**	**House F**
Thermostat location	Hall	Hall	Hall	Hall	Hall	Hall
Thermostat setting	20	20	18	22	20	18
Thermostat altered	Yes 25	No	Yes	Yes	Yes 25	Yes
Timer on/off	05:30–07:00 18:00–22:00	Manual	07:00–23:00	06:00–07:30 16:15–22:00	06:00–08:00 16:00–22:30	6:30–08:00 17:00–23:00
Summer off/winter on	Jun/Oct	Not known	May/Oct	Apr/Oct	May/Oct	May/Oct
House vacated	No	Jun 3 weeks	No	Jul	1 Jan–13 Feb	No
Heating system	Ok	Ok	Ok	Ok	Ok	Ok

Depending on their responses to the questionnaires prior to the interviews, there may be opportunities to help occupants improve some aspects of their energy management or to increase their understanding of related issues. For instance, around 50% of occupants' responses to question 9 in Figure 243 indicated their misunderstanding that net curtains had no effect on solar gain from south-facing windows, and a small minority even thought that net curtains increased solar gain. A simple explanation during the interview therefore helped the corresponding occupants to learn about the effect of net curtains and thus improve the energy performance of their houses. Interviews can also help occupants through advice on the heating control strategy.

Some occupants believed that running the heating system intermittently, rather than permanently, leads to both good thermal comfort and energy savings. This may well be true in internally insulated lightweight buildings, but is not true in externally insulated heavyweight buildings. The thermal response of the occupant's building can be explained to them in a way that can help them to optimise control strategies.

Collecting comprehensive information about level of occupancy, energy use patterns, and energy consciousness, as well as information about thermal, visual, and ventilation comfort is invaluable for establishing the level of success of each individual design and for explaining discrepancies between expected and actual results of instrumental monitoring.

20.8 SUMMARY

- One-off tests
  - co-heating tests
    - need to be conducted in winter months during overcast weather and relatively stable external temperature
    - result in overall heat transfer coefficient
    - are conducted under unrealistic circumstances as buildings almost never operate in steady-state conditions
    - are a missed opportunity as analysis of the heating up period during the test can result in more thermal behaviour parameters than the co-heating test alone
  - air-tightness testing
    - will help identify sources of air leakage
    - larger sources of leakage may need to be sealed before the test can give meaningful results
    - air leakage needs to be eliminated at the design stage, with designed openings for pipes and ducts, and by specifying flexible and durable sealants
  - thermal imaging
  - needs to be conducted at night in cold weather ideally after the building is brought to a steady state
  - can reveal missing insulation from cavity walls, thermal bridges, and other sources of heat losses
  - observation of building envelope
  - can reveal cracks that occur as a result of differential thermal expansion
  - such cracks can be eliminated by designing interfaces of different materials with expansion tolerances and by specifying flexible sealants between them
- Instrumental monitoring
  - This is an indispensable way of evaluating building energy performance.
  - Monitoring needs to be conducted over a period of two years of occupation.
  - Monitoring needs to measure all energy inputs and outputs in order to calculate energy balances.
  - Parameters need to be recorded at regular intervals of five to six minutes (10–12 readings per hour).
  - It is important to determine noise to signal ratio in monitored sensors and to eliminate the noise using hardware or software filters.
- Experimental measurement of physical parameters of buildings
  - Physical parameters of buildings in use can be considerably different from theoretical parameters.
  - Experimental measurement can be carried out using the dynamic heating test method, rearranged heat balance equation method, or co-heating method.
- Energy performance analysis
  - All energy inputs and outputs need to be monitored.
  - In order to ensure parity between results in different monitoring projects, energy performance needs to be presented in terms of monthly average summaries.
- Troubleshooting of building operation and performance
  - Short-term monitoring, lasting between one day and one week can help determine causes of operational problems in buildings and help to recommend improvements.
- Calibration of dynamic simulation models
  - Simulation models use built-in default values of certain parameters to enable them to work if these parameters are not specified by the user. This can lead to considerable discrepancies between the model performance and the building performance.
  - Theoretical physical parameters of building materials as used by simulation models can be radically different from parameters in buildings in use.
  - Calibration helps to increase the accuracy of simulation models and also to increase confidence in the models.

• Occupant comfort and behaviour studies
  • Help establish the success of a design.
  • Complement monitoring results and help explain discrepancies of the performance of similar buildings used by different occupants.
  • Questionnaires and interviews are the instruments used in occupant behaviour studies.
  • Discrete choice scales or analogue visual scales are used to obtain occupants' perception of parameters that influence their satisfaction with the internal environment.

PART 5

PRACTICE, RESEARCH, AND POLICY DEVELOPMENT SUPPORT

CHAPTER 21

PRACTICE: CASE STUDIES

This chapter introduces a number of case studies of zero carbon buildings around the world. A very detailed case study of a retrofit house in England, the Birmingham Zero Carbon House (ZCH), is followed by shorter case studies of several buildings. The case studies demonstrate a range of unique ways to zero carbon design, showing a variety of approaches in different climates. All of the approaches demonstrate the underlying desire to make better buildings for people, and have pushed the boundaries of integration of design and technology. The variety of solutions to zero carbon design problems will provide the reader with an enhanced design toolbox to stimulate their own design thinking.

21.1 CASE STUDY 1: BIRMINGHAM ZERO CARBON HOUSE, ENGLAND

In this section, I will examine a somewhat unusual zero carbon project – a Birmingham ZCH. I will first give an overview of the design, followed by a description of my own IES simulation model. Subsequently the carbon performance of the model, the analysis of comfort conditions, and the life cycle cost analysis will be used as criteria for determining the degree of success of the project. This will be followed by the results of monitoring over a five-year period and by calculations of combined embodied and operational carbon emissions to determine cumulative emissions over the life cycle.

The Birmingham ZCH is an unusual implementation of zero carbon design. We usually associate zero carbon design with new buildings, but this building was originally built 170 years ago. It was recently renovated and extended by its owner, architect John Christophers, into a residence that uses only renewable energy (Figure 244).

The house has achieved level 6 (the highest level) of the UK Code for Sustainable Homes and has won the Architecture Award from the Royal Institute of British Architects in 2010. I will here examine the underlying design principles and evaluate its performance.

Plans and sections of the ZCH are shown in Figure 245 through to Figure 248.

21.1.1 Building energy efficiency

First, here is the outline of building energy efficiency measures implemented in ZCH.

Response to site context. The original building is set in a dense housing area and it is overshadowed by a house next to it (Figure 245). In response to this limitation, the garden facade of the extended part of the house has been angled towards the south-west, from where it captures solar radiation (Figures 246 and 247). A deciduous ash tree in the garden (Figure 245) provides shading in summer months and allows exposure to solar radiation at lower incidence angles in winter, spring and autumn.

Building geometry. The design is based on a shallow plan depth (Figures 246 and 247) with roof lights (Figure 248) and openable windows that enable the utilisation of natural daylight and natural ventilation.

Figure 244   Birmingham Zero Carbon House from the street side (left) and from the garden side (right) (photos with permission from Martine Hamilton Knight/STO)

Figure 245   ZCH site plan (courtesy of John Christophers)

zero carbon house

1. Existing neighbouring house
2. Sunspace /living (phase 2)
3. Kitchen
4. Dining
5. Store
6. WC/shower
7. Wood-burning stove
8. Floor hatch to cellar
9. Double-height living
10. Garage
11. Hall
12. Bicycle store
13. Lobby
14. Coats
15. Shelves
16. Living
17. Side passageway
18. Existing neighbouring house

Figure 246   ZCH ground-floor plan (courtesy of John Christophers)

Thermal insulation. The *U*-values of thermal elements, which are considerably better than UK Part L Building Regulations 2010, are given in Table 77.

Thermal insulation in the external walls is based on recycled newspaper. This material, named Warmcel, is fixed to the walls using adhesives and rendered using 'glaster' — a mixture of lime and ground recycled glass. In Chapter 8, I explained the difference between inside and outside insulation, and in Chapter 11 implications of positioning insulation inside or outside. From these explanations we learnt that external insulation contributes much more to building energy efficiency as it maximises the effect of thermal mass. However, installing 350 mm of insulation on the outside facade in a well-established 19th century street was not possible due to planning and aesthetic reasons. As a compromise, insulation was installed on the existing internal wall surface on the street side, and on the new external wall surface on the garden side.

Airtightness. Airtight membranes that are vapour-permeable and moisture-sensitive are used throughout and they achieve airtightness of N50 = 0.57 h^{-1}, equivalent to Q50 = 0.99 m³/(h.m²), both at 50 Pa airtightness test pressurisation.

Thermal mass. A 200 mm dense clay blockwork on the garden side, exposed internal brickwork retained from the original building (Figure 249), and 75 mm rammed earth floors (Figure 250) on all three levels constitute this building's thermal mass.

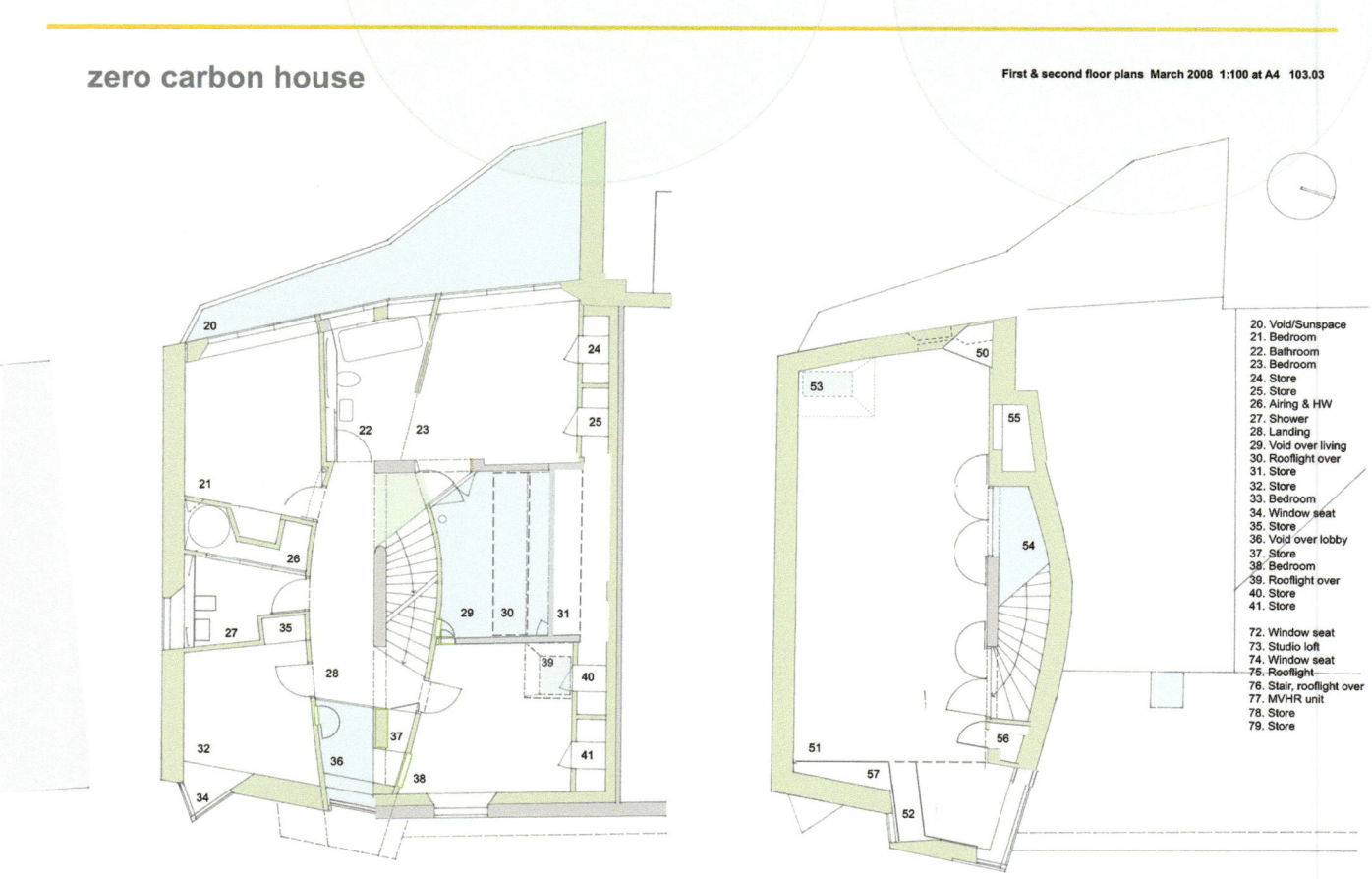

20. Void/Sunspace
21. Bedroom
22. Bathroom
23. Bedroom
24. Store
25. Store
26. Airing & HW
27. Shower
28. Landing
29. Void over living
30. Rooflight over
31. Store
32. Store
33. Bedroom
34. Window seat
35. Store
36. Void over lobby
37. Store
38. Bedroom
39. Rooflight over
40. Store
41. Store

72. Window seat
73. Studio loft
74. Window seat
75. Rooflight
76. Stair, rooflight over
77. MVHR unit
78. Store
79. Store

Figure 247   ZCH first- and second-floor plan (courtesy of John Christophers)

Passive solar gain. The building utilises a direct gain system, with a large amount of south-west glazing (Figure 244-right), high density rammed earth floors (Figure 250) and internal brickwork partitions (Figure 249) that absorb solar energy. This is also evident from Figure 251 which shows solar illumination of the rammed earth floor deep into internal space.

Natural ventilation. Openable windows at high-level drive stack effect that enables night-time cooling of thermal mass (Figure 252).

Mechanical ventilation with heat recovery. As result of high airtightness, mechanical ventilation with a cross flow heat recovery heat exchanger is used (Figure 253), with 93% heat recovery efficiency.

Natural daylight. The building is a shallow plan with a number of openable windows and roof lights. High-level light is funnelled into the building using uneven mirror lining internally around the window frame that creates changing illumination patterns throughout the day (Figure 254).

Electrical lighting. The house utilises manually dimmable T5 fluorescent triphosphor lamps with high luminous efficacy (Figure 255-left).

Figure 248   ZCH sections (courtesy of John Christophers)

TABLE 77 *U*-VALUES OF BUILDING ELEMENTS IN ZERO CARBON HOUSE	
Element	**W/(m²K)**
Exposed floor	0.14
Exterior walls	0.11
Roof	0.08
Doors	0.38
Triple glazing	0.65

Additional heating. A wood-burning stove provides water heating to the domestic hot water system and bathroom heating, as well as to space heating in the central core of the house via the vertical flue pipe, from which adjacent rooms receive heat through internal openings (Figure 255).

Other features. There are a few other features that contribute to saving of natural resources and reuse of materials that are worth referring to. The house itself has been recycled by utilising most of its original structure, including bricks and timber, roof, floorboards, and drains. Stairs and window seats are made of reclaimed 200-year-old Canadian honeydew maple. Handrails are made of hemp, and door handles are also reclaimed. Thermal insulation is made of recycled newspaper, with internal rendering made of lime and ground recycled glass. Earth for rammed earth floors comes from the site ground works and load bearing

Figure 249   Exposed internal brickwork (left and right) and rammed earth floor (right) illuminated by solar radiation (photos with permission from Martine Hamilton Knight/Builtvision)

Figure 250   Two views of the top floor studio with rammed earth floor providing thermal mass and high-level windows providing natural daylight (photos with permission from Martine Hamilton Knight/Builtvision)

walls are made of unfired clay blocks. Rainwater is collected into a 2.5 m³ tank in the cellar and used for flushing toilets, running the washing machine, and supplying a dedicated tap in the kitchen. Grey water from the bath on the first floor is used for watering plants in the garden. Water consumption is generally reduced through special low-use water taps and low capacity dual-flush toilet tanks.

21.1.2 Renewable energy
There are three renewable energy systems in the house:

• a high-efficiency wood-burning stove rated at 8 kW, with 70% output towards the water heating system and 30% output towards space heating (Figure 256a)
• a solar photovoltaic system, consisting of 35.6 m² multi-crystalline panels (Figure 256b)
• a solar thermal system, consisting of 8.8 m² evacuated tube collectors (Figure 256c)

21.1.3 Development of the simulation model
Details from the previous two sections were used to create a dynamic simulation model in IES Virtual Environment.

Figure 251   Double height space living room showing solar radiation illuminated rammed earth floor; exposed brickwork; flue from the wood-burning stove for space heating; high-level internal openings for natural ventilation and heating; high-level glazing for natural daylight; and T5 low energy lamps (photo with permission from Martine Hamilton Knight/Builtvision)

Figure 252   High-level openable glazing above stairs and daylight patterns (courtesy of John Christophers)

The model geometry was created from architectural drawings which were available in PDF format. These were first converted into DXF format, and although the conversion resulted in the right proportions it created a completely wrong scale. To correct the scale, key dimensions were taken on every floor of the house, and DXF drawings were rescaled accordingly. These rescaled drawings were then imported into IES Virtual Environment, and internal spaces were created by tracing the DXF outline of each floor and subdividing the floors with partitions.

Construction types were then created using the specifications obtained for the ZCH. Insulation thickness in the walls, roof, and floors was varied iteratively until U-values from the specifications were obtained. Windows U-values were obtained in a similar way, by varying the thermal resistance of cavities.

Occupancy profiles for living rooms, kitchen, bedroom, and bathrooms were created specifically for the house, as general library profiles were found not to be appropriate. Although manual switching of lights is installed throughout the house, occupants operate the lights in the living rooms and kitchen in response to the available daylight. Hence the corresponding automatic variation profiles were created as daylight sensitive.

To take into account daylight sensitivity, daylight sensors were created in Radiance IES part of the simulation model, and the sensor readings were generated for use in the thermal simulation.

Figure 253   Mechanical ventilation heat recovery unit in the Zero Carbon House

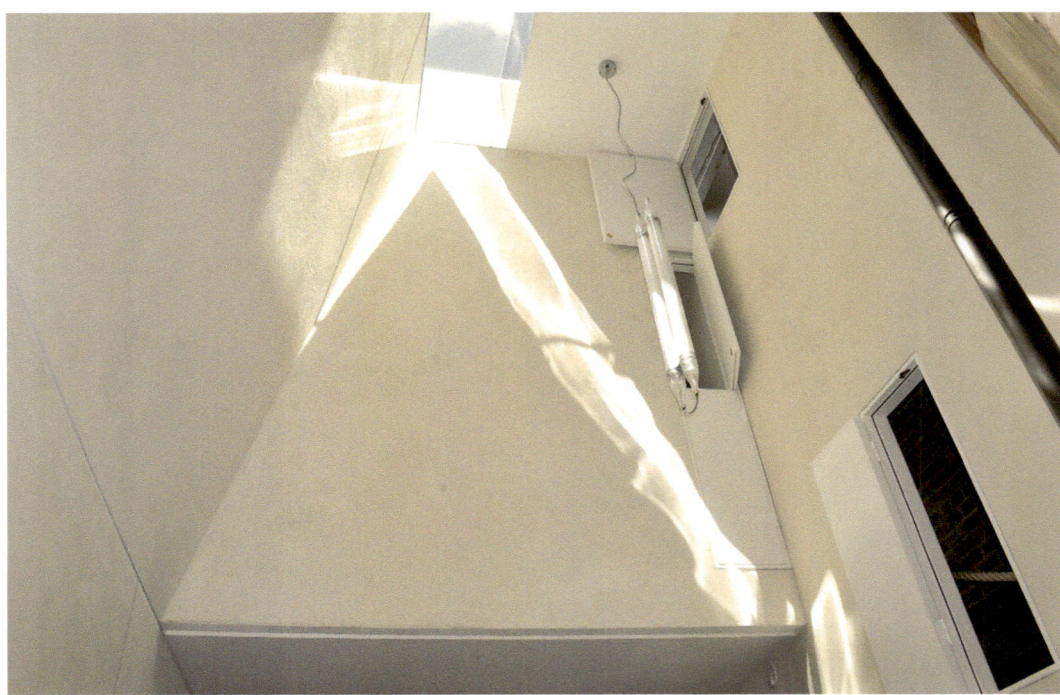

Figure 254   Daylight patterns created by mirror-lined roof lights

Figure 255   Internal openings in the Zero Carbon House with T5 lamps shown on the left

a) High efficiency wood burning stove   b) Solar photovoltaic array   c) Evacuated tube solar thermal collector

Figure 256   Renewable energy systems in the Zero Carbon House

Thermal condition templates were then created for different room types, specifying room conditions, heating and ventilation systems, internal gains and air exchanges.

A heating system type was set to biomass with mechanical ventilation and heat recovery. Domestic hot water heating was set to the same system as space heating, with additional heating from the solar thermal system.

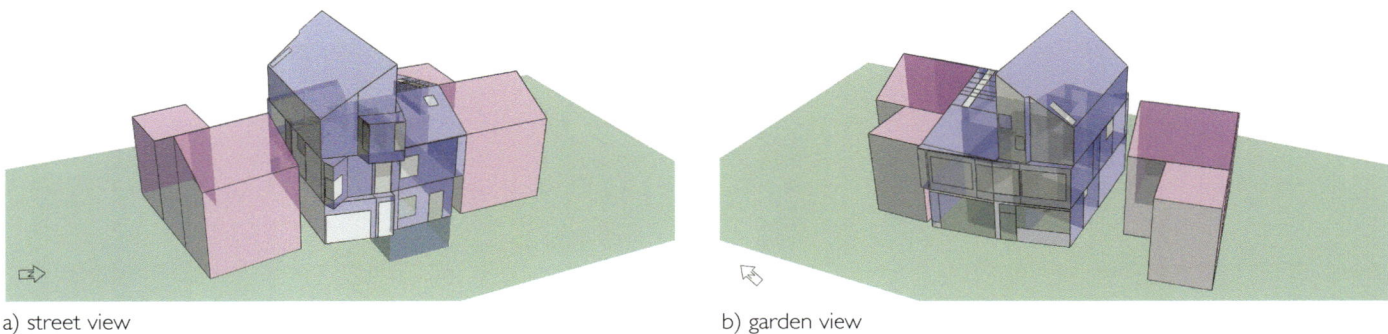

a) street view   b) garden view

Figure 257   Geometry of the IES simulation model of the Zero Carbon House

Both solar PV and solar thermal systems were set to 110° azimuth and 35° slope and with surface areas and efficiencies corresponding to the actual system specifications.

After all of the above specifications were implemented in the model, the model was completed (Figure 257) and ready to run. However, the first run attempt failed due to geometry related errors. It appeared that the simulation tool could not handle the simultaneous horizontal curvature of the second-floor staircase wall and the sloping glazed roof above it. A number of 'non-planar' surface errors were thrown which interrupted the run at the start. These problems were rectified by deleting the original staircase and re-entering a new one with the glazed roof set to horizontal position. This discrepancy in the glazed roof geometry was unlikely to cause any significant inaccuracy of the model performance because of its relatively small size, especially after a calibration of the model explained below.

21.1.4 Energy production and consumption records

Before instrumental monitoring commenced, there were only manual energy production and consumption records available for the ZCH. Nevertheless this information was invaluable for the purpose of the calibration of the simulation model, which will be described in the next section.

Biomass heating energy had to be estimated from the records of how many times the wood-burning stove was used (Table 78).

TABLE 78 USE OF BIOMASS IN THE ZERO CARBON HOUSE IN 2010–2011 (INFORMATION COURTESY OF JOHN CHRISTOPHERS)	
Month	Number of times the stove was used
Nov	4
Dec	12
Jan	10
Feb	9

To convert this information into energy figures the dimensions of the wood-burning stove firebox were obtained and an utilisation factor (a fraction of the full volume filled by wood logs) was estimated. Based on the internal dimensions of the firebox of 400 × 300 × 280 mm, and applying calorific value and density of ash tree logs, we estimated that a single use of the wood stove filled with ash tree logs, it was estimated that a single use of the wood stove filled with ash tree logs generates 50.72 kWh (Table 79).

TABLE 79 ENERGY GENERATED BY THE WOOD STOVE FILLED WITH ASH TREE LOGS				
Wood stove fire box capacity (m³)	Fire box utilisation factor	Calorific value of ash tree logs (kWh/kg)	Density of ash tree logs (kg/m³)	Energy per full stove (kWh)
0.0336	0.85	4.44	400	50.72

Subsequently this enabled us to calculate the total biomass energy use in the ZCH as 1.36 MWh, as shown in Table 80.

TABLE 80 BIOMASS ENERGY USE IN THE ZERO CARBON HOUSE		
Month	**Number of times the stove was used**	**Actual energy use (MWh)**
Nov	4	0.203
Dec	12	0.609
Jan	10	0.507
Feb	9	0.457
	Total	1.775

Electricity generation and import records are shown in Table 81, from which we can learn that a surplus of 47% of electricity was generated, despite using electricity to top up space heating during the coldest December on record.

TABLE 81 ELECTRICITY GENERATION AND IMPORT RECORDS IN THE ZERO CARBON HOUSE (INFORMATION COURTESY OF JOHN CHRISTOPHERS)					
Date	**Meter**	**Meter reading**	**Generated electricity (kWh)**	**Imported electricity (kWh)**	**Notes**
29/09/2009	Generation meter	0.0			Still building site
	Import meter	Unknown			
10/10/2009	Generation meter	100.9			Extra electric heat used to help building dry out
	Import meter	2,658.0			
31/03/2010	Generation meter	892.4			Start of feed-in tariff
	Import meter	6,367.0			
10/07/2010	Generation meter	2,827.9	1,935.5		
	Import meter	6,611.0		244.0	
06/09/2010	Generation meter	3,655.7	827.8		
	Import meter	6,744.0		133.0	
07/12/2010	Generation meter	4,343.0	687.3		
	Import meter	7,471.0		727.0	
06/01/2011	Generation meter	4,376.1	33.1		Back up heat for coldest December on record
	Import meter	8,443.0		972.0	
01/02/2011	Generation meter	4,464.9	88.8		
	Import meter	8,684.0		241.0	
01/03/2011	Generation meter	4,579.0	114.1		
	Import meter	8,931.0		247.0	
31/03/2011	Generation meter	4,922.2	343.2		12 months figures: 147% of used electricity generated
	Import meter	9,113.0		182.0	
	Total 31 March 2010 to 31 March 2011		4,029.8	2,746.0	

We will now use the recorded energy production and consumption to calibrate the simulation model.

21.1.5 Calibration of the simulation model

The calibration of the simulation model was subsequently performed using the annual energy records from the previous section.

First, the calibration in respect of heating energy consumption was carried out, with target error of 1% between actual and simulated energy consumption. During the course of calibration, internal temperatures were adjusted iteratively in the model until simulated annual energy consumption totals converged towards the actual annual energy totals within less than the target error. This process is shown in Table 82. Results from an initial model are shown in column 4. This model, which was based on 20°C internal temperatures, shows considerably higher heating energy consumption than the actual consumption in column 3. The relative error between the simulated and actual values, was calculated as the following fractional difference:

$$(\text{simulated} - \text{actual})/\text{actual}$$

TABLE 82	CALIBRATION OF THE SIMULATION MODEL FOR HEATING ENERGY CONSUMPTION							
Month	Number of times the stove was used	Actual energy use (MWh)[a]	Simulated energy use Ti = 20°C (MWh)	Simulated energy use Ti = 19°C (MWh)	Simulated energy use Ti = 18°C (MWh)	Simulated energy use Ti = 18.5°C (MWh)	Simulated energy use Ti = 18.2°C (MWh)	Simulated energy use Ti = 18.4°C (MWh)
1	2	3	4	5	6	7	8	9
Jan	10	0.507	0.831	0.717	0.604	0.660	0.627	0.649
Feb	9	0.457	0.347	0.251	0.161	0.205	0.178	0.196
Mar	0	0.000	0.000	0.000	0.000	0.000	0.000	0.000
Apr	0	0.000	0.000	0.000	0.000	0.000	0.000	0.000
May	0	0.000	0.000	0.000	0.000	0.000	0.000	0.000
Jun	0	0.000	0.000	0.000	0.000	0.000	0.000	0.000
Jul	0	0.000	0.000	0.000	0.000	0.000	0.000	0.000
Aug	0	0.000	0.000	0.000	0.000	0.000	0.000	0.000
Sep	0	0.000	0.000	0.000	0.000	0.000	0.000	0.000
Oct	0	0.000	0.000	0.000	0.000	0.000	0.000	0.000
Nov	4	0.203	0.479	0.366	0.263	0.313	0.283	0.303
Dec	12	0.609	0.811	0.698	0.586	0.642	0.608	0.631
Total		1.775	2.469	2.032	1.614	1.821	1.696	1.779
Relative error		–	39.053%	14.471%	−9.074%	2.547%	−4.461%	0.198%

[a] From Table 80.

The model was subsequently run with different internal temperatures, changing the set temperature in smaller steps as the relative error decreased (columns 7–9, Table 82) until the error was reduced to less than 1%. The simulation, with internal temperatures set to 18.4°C and with the heating profile set to November to February from 16:00 to 20:00 hours, shows a considerably higher accuracy than the initial model, with a relative error of less than 0.2%, satisfying the target error criterion of less than 1% (Table 82, column 9). In this way the calibration of the model in respect of heating energy consumption was completed.

In the next step, the calibration of the simulation model in terms of PV electricity generation and total electricity use was carried out using the same target error of 1%. This process is documented in Table 83.

TABLE 83 CALIBRATION OF THE SIMULATION MODEL FOR PV-GENERATED ELECTRICITY

	Actual*	17.00 % PV efficiency	15.00 % PV efficiency	16.50 % PV efficiency	15.50 % PV efficiency	15.25 % PV efficiency
1	2	3	4	5	6	7
PV Generated (MWh)	4.03	4.4951	3.9663	4.3629	4.0985	4.0324
Relative error PV Generated (%)		11.54	1.58	8.26	1.70	0.06

The starting PV system efficiency was as quoted by the manufacturer, and internal heat gains and auxiliary energy were set to low (Table 83, column 3). Subsequently, the PV system efficiency was reduced and auxiliary energy increased until the relative error between the simulated and actual energy was reduced to well below the target error of 1% (Table 83, column 7).

The model achieved in this way is calibrated for heating energy consumption, PV electricity generation, and total electricity use. This model is therefore considered to be the closest representation of the performance of ZCH, and will be used in the remainder of this chapter as the basis for further analysis.

21.1.6 Development of a reference model

An essential ingredient of the life cycle cost analysis that will be carried out as part of this case study is the choice of a reference case against which improvements are measured. The most relevant reference case for a retrofit implementation is the original building before the retrofit took place. A simulation model of the original building was therefore developed as specified below, to be used as a reference for improvement comparison.

The original building had no thermal insulation and *U*-values of thermal elements that were set in the model are given in Table 84.

TABLE 84 *U*-VALUES OF THE THERMAL ELEMENTS IN THE ORIGINAL BUILDING IN ZERO CARBON HOUSE

Element	W/(m²K)
Exposed floor	0.89
Exterior walls	2.04
Roof	2.14
Doors	2.19
Single glazing	5.66

Airtightness was set to the equivalent of 0.5 air changes per hour, corresponding to the air permeability of 11 m³/(h.m²) at 50 Pa. Thermal mass comprised of dense brick uninsulated external walls and internal partitions. Openable windows enabled natural ventilation and the shallow plan with tall windows provided a degree of natural daylight. Electrical lighting was based on low efficiency incandescent lamps and additional space heating on gas fires.

The geometry of the simulation model of the reference building is shown in Figure 258.

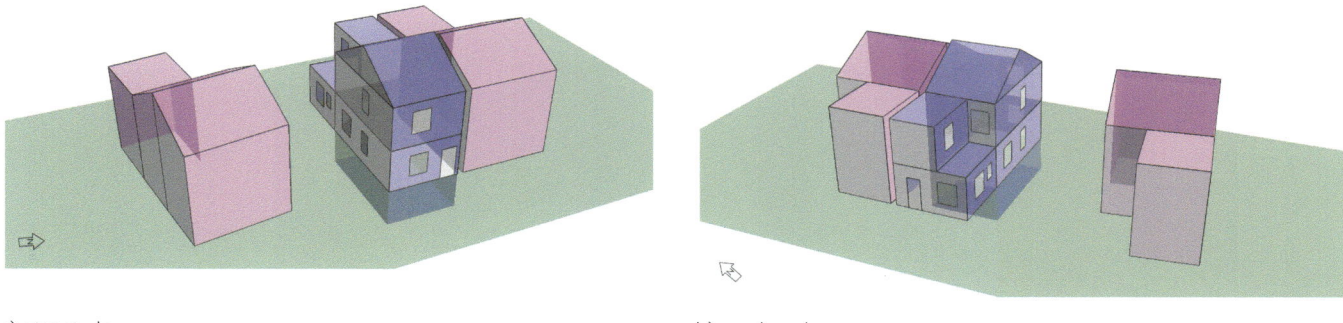

a) street view   b) garden view

Figure 258   Geometry of the IES simulation model of the original house

21.1.7 Energy and carbon performance

A summary of simulation results for the two models, the original house and ZCH are shown in Table 85.

Description	Units	Original house	Zero carbon house
Total heating energy	MWh	61.6	1.31
Total electricity used	MWh	14.22	2.82
Total electricity generated	MWh		−4.03
Total electricity exported	MWh		−1.21
DHW energy with ST	MWh		3.91
DHW energy without ST	MWh	10.45	6.46
Solar thermal	MWh		−2.54
CO_2 emissions	$kgCO_2$	17,102	−527

TABLE 85   SIMULATION RESULTS COMPARISON BETWEEN ZERO CARBON HOUSE AND THE ORIGINAL HOUSE

As Table 85 shows, the building achieves carbon-negative performance, with negative emissions of −527 kg of CO_2 per annum, which is a huge improvement from the reference building which generated emissions of over 17 tonnes of CO_2 per annum.

21.1.8 Life cycle benefit

Life cycle cost calculations will now be developed to establish the financial viability of Birmingham ZCH. These calculations are based on a comparison between the reference model and the ZCH model summarised in Table 85.

A BRIEF INTRO INTO LIFE CYCLE BENEFIT ANALYSIS

The total cost of construction and operation of the building over a chosen period is referred to as life cycle cost. Zero carbon projects will have additional costs in comparison with the cost of a conventional building that will make these projects zero carbon. At the end of a chosen period, say, 25 years, that additional cost would most likely be paid off by energy savings and any renewable energy incentives, thus generating surplus. That surplus will be referred to as life cycle benefit. In general, we need to maximise life cycle benefit or minimise life cycle cost.

To start with, a concept of time-dependent value of money needs to be introduced. If one borrows £1 at 15% annual interest rate, its value at the end of the first year will be £1/(1 + 0.15) = £0.87; at the end of second year £1/(1 + 0.15)^2 = £0.76; at the end of fifth year £1/(1 + 0.15)^5 = £0.50 and so on. To repay £1 borrowed today, the borrower needs to pay

£0.13 more at the end of the first year, £0.24 more at the end of the second year, or £0.50 more at the end of the fifth year. The parameter that helps to calculate the present value of an investment after several years is called net present value factor (NPVF) or present worth factor (PWF) and it is calculated as follows:

$$NPVF = \frac{1}{(1+d)^n}$$

where:

d – market discount rate or borrowing interest rate
n – number of periods in years.

The above calculation, however, does not take inflation into account. For instance, with an inflation rate of 3%, a cost of £1 will be £1 × (1 + 0.03) at the end of the second year, £1 × (1 + 0.03)^2 at the end of the third year, and £1 × (1 + 0.03)$^{n-1}$ at the end of the nth year.

As both inflation rate and discount rate influence the value of money simultaneously, a parameter that combines both is needed in order to calculate future value of an investment. Taking inflation into account, NPVF from the above equation becomes:

$$NPVF = \frac{(1+i)^{n-1}}{(1+d)^n}$$

where:

i – inflation rate
d – market discount rate or interest rate on borrowed money
n – number of periods in years.

To calculate life cycle benefit for a period of N years, we first need to list all future payments and returns, discount them to the present value using NPVF, and add all these discounted values over the entire period N. The life cycle benefit is the sum of all future expenditure and income over N years, discounted to the present value.

One of the challenges in the life cycle calculations was to establish the initial investment, as the house was extended as well as upgraded. One of the ways of establishing the initial investment would be to calculate the difference between the overall construction costs of £150/sq.ft and deduct the typical house construction costs of £120/sq.ft as established by the UK Homes and Communities Agency. In this way the difference of £30/sq.ft would be equal to extra investment of £323/m². This figure multiplied by the useful area of 204 m² gives the total extra investment cost of £65,875. This cost however contains the cost of building the extended envelope, and it therefore does not give us realistic information for the life cycle cost.

Another way of establishing the initial investment would be to obtain costs of upgrading thermal insulation in the original building only, plus the costs of the three renewable energy systems. We will refer to this cost as 'Initial Investment A'. However, as the simulation results of the calibrated model refer to the performance of the upgraded building with a much larger envelope than the original building, we also need to take into account the cost of insulating the upgraded building. We will refer to this cost as 'Initial Investment B'.

We can then be confident that the true life cycle cost will be in the range between the costs calculated using Initial Investment A and Initial Investment B.

The life cycle cost analysis for Initial Investment A is shown in Table 87 with assumptions for cases A and B shown in Table 86. The summary of life cycle benefit for cases A and B is shown in Table 88.

As energy prices have varied considerably since the house was retrofitted and are expected to vary into the future, a chart of energy price variations has been assembled from several sources (BEIS, 2022; Statista, 2022a,b) and shown in Figure 259. These prices will be used in the life cycle cost calculations, payback period, and return on investment.

TABLE 86	LIFE CYCLE COST CALCULATION ASSUMPTIONS			
	Assumptions	Case A	Case B	Units
1	Inflation rate	3.00%	3.00%	%
2	Mortgage rate[a]	3.05%	3.05%	%
3	Electricity price	See Figure 259		£/kWh
4	Gas price	See Figure 259		£/kWh
5	Biomass price	0.000	0.000	£/kWh
6	Feed-in tariff – generation	0.361	0.361	£/kWh
7	Feed-in tariff – export	0.030	0.030	£/kWh
8	Initial investment[b]			
9	Insulation cost	6,011	10,286	£
10	Windows cost	9,678	18,654	£
11	Wood-burning stove cost	5,548	5,548	£
12	Solar thermal cost	11,123	11,123	£
13	PV cost	14,985	14,985	£
14	Initial investment cost	47,345	60,596	£

[a]Actual rate in the particular case.
[b]Including installation cost.

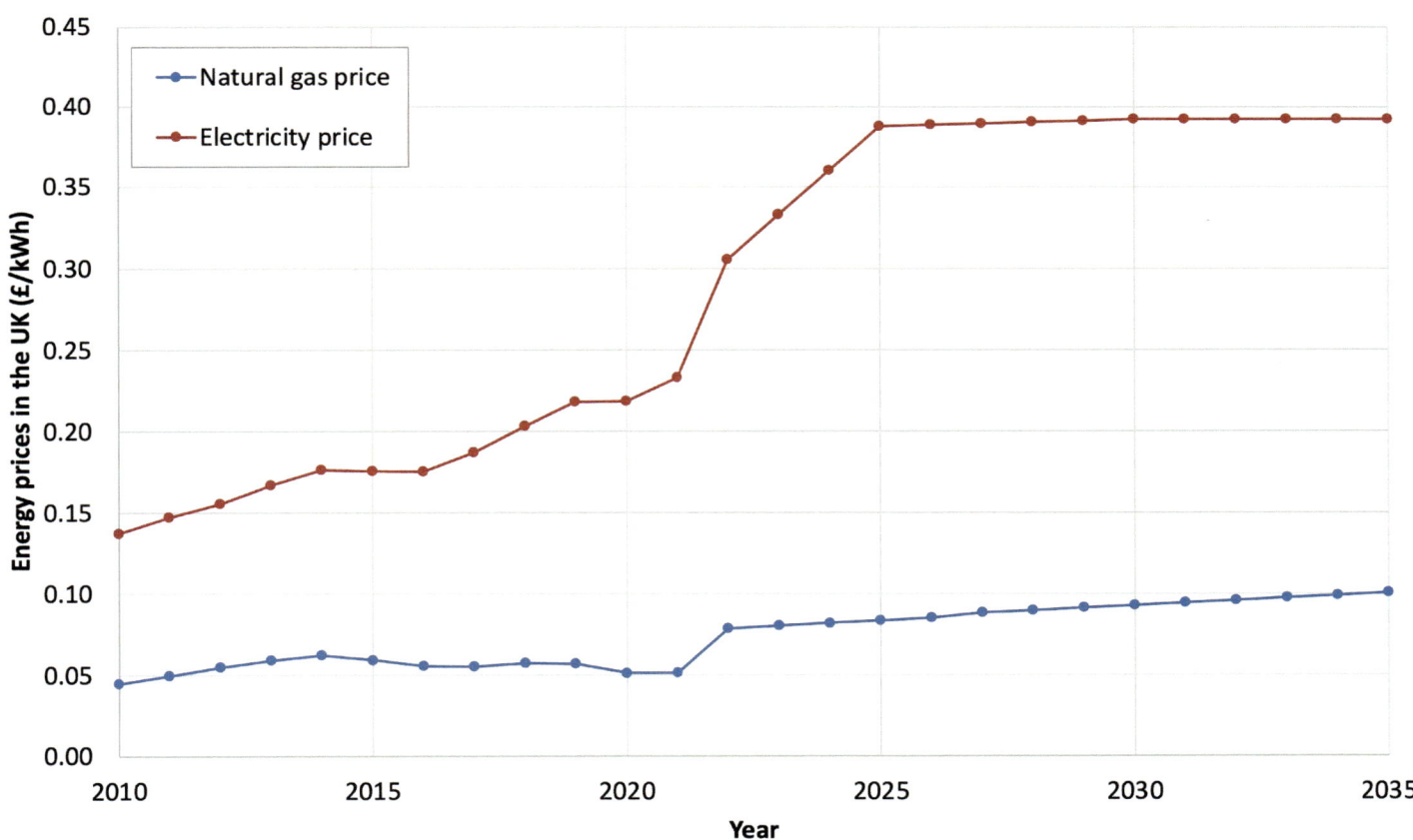

Figure 259   Variation of energy prices in the past and the future

TABLE 87 LIFE CYCLE COST CALCULATIONS FOR THE ZERO CARBON HOUSE

Year	Initial extra investment (£)	Heating energy savings = ref model – zch model[a] (MWh)	Electricity energy savings = ref model – zch model (MWh)	Heating energy savings = ref model – zch model[a] (£)	Electricity energy savings = ref model – zch model (£)	Energy savings (£)	Feed-in tariff income (£)	Total income from savings and feed-in tariff (£)	N = Year – 2009	NPVF[b]	Net present value (£)	Cumulative net present value (£)
	1	2	3	4	5	6	7	8	9	10	11	12
2009	-47,345								0		-47,345	-47,345
2010		61.6	11.4	2,772.91	1,565.22	4,338.13	1,491.13	5,829.26	1	0.97	5,657	-41,688
2011		61.6	11.4	3,074.62	1,678.95	4,753.57	1,491.13	6,244.70	2	0.97	6,057	-35,631
2012		61.6	11.4	3,396.43	1,774.15	5,170.58	1,491.13	6,661.71	3	0.97	6,458	-29,173
2013		61.6	11.4	3,656.08	1,905.42	5,561.50	1,491.13	7,052.63	4	0.97	6,834	-22,339
2014		61.6	11.4	3,830.70	2,011.64	5,842.34	1,491.13	7,333.47	5	0.97	7,103	-15,236
2015		61.6	11.4	3,656.99	2,003.62	5,660.61	1,491.13	7,151.74	6	0.97	6,923	-8,313
2016		61.6	11.4	3,441.23	1,999.11	5,440.34	1,491.13	6,931.47	7	0.97	6,707	-1,606
2017		61.6	11.4	3,402.83	2,133.89	5,536.72	1,491.13	7,027.85	8	0.97	6,797	5,190
2018		61.6	11.4	3,530.83	2,318.27	5,849.10	1,491.13	7,340.23	9	0.97	7,095	12,286
2019		61.6	11.4	3,516.20	2,486.62	6,002.81	1,491.13	7,493.94	10	0.97	7,240	19,526
2020		61.6	11.4	3,165.13	2,490.62	5,655.75	1,491.13	7,146.88	11	0.97	6,902	26,428
2021		61.6	11.4	3,170.61	2,655.46	5,826.08	1,491.13	7,317.21	12	0.97	7,063	33,491
2022		61.6	11.4	4,852.83	3,484.17	8,337.00	1,491.13	9,828.13	13	0.96	9,482	42,973
2023		61.6	11.4	4,949.89	3,797.30	8,747.18	1,491.13	10,238.31	14	0.96	9,873	52,846
2024		61.6	11.4	5,046.94	4,110.42	9,157.36	1,491.13	10,648.49	15	0.96	10,263	63,109
2025		61.6	11.4	5,144.00	4,423.55	9,567.55	1,491.13	11,058.68	16	0.96	10,654	73,762
2026		61.6	11.4	5,241.05	4,433.49	9,674.55	1,491.13	11,165.68	17	0.96	10,751	84,514
2027		61.6	11.4	5,435.17	4,443.43	9,878.60	1,491.13	11,369.73	18	0.96	10,943	95,456

TABLE 87 CONTINUED

Year	Initial extra investment (£)	Heating energy savings = ref model – zch model[a] (MWh)	Electricity energy savings = ref model – zch model (MWh)	Heating energy savings = ref model – zch model[a] (£)	Electricity energy savings = ref model – zch model (£)	Energy savings (£)	Feed-in tariff income (£)	Total income from savings and feed-in tariff (£)	N = Year – 2009	NPVF[b]	Net present value (£)	Cumulative net present value (£)
2028		61.6	11.4	5,532.22	4,453.37	9,985.60	1,491.13	11,476.73	19	0.96	11,040	106,497
2029		61.6	11.4	5,629.28	4,463.31	10,092.59	1,491.13	11,583.72	20	0.96	11,138	117,634
2030		61.6	11.4	5,726.34	4,473.25	10,199.59	1,491.13	11,690.72	21	0.96	11,235	128,869
2031		61.6	11.4	5,823.39	4,473.25	10,296.65	1,491.13	11,787.78	22	0.96	11,323	140,192
2032		61.6	11.4	5,920.45	4,473.25	10,393.70	1,491.13	11,884.83	23	0.96	11,411	151,603
2033		61.6	11.4	6,017.51	4,473.25	10,490.76	1,491.13	11,981.89	24	0.96	11,498	163,101
2034		61.6	11.4	6,114.56	4,473.25	10,587.82	1,491.13	12,078.95	25	0.96	11,586	174,687
										Life cycle cost	174,687	

[a]ZCH uses own grown biomass, hence the entire heating energy in the reference model is a saving.
[b]NPVF – Net Present Value Factor.

It is worth mentioning that these results are slightly different from the results published in the first and second editions of this book. The reason is that variable energy prices are used in this third edition, as sufficient information of this variability has become available after the house was retrofitted.

As it can be seen from Table 87, column 12, cumulative net present value goes from negative to positive between years 2016 and 2017, or between the seventh and eighth year since the start. In order to determine the payback period more accurately than it can be derived from Table 87 a graph of both cases A and B are drawn in Figure 260.

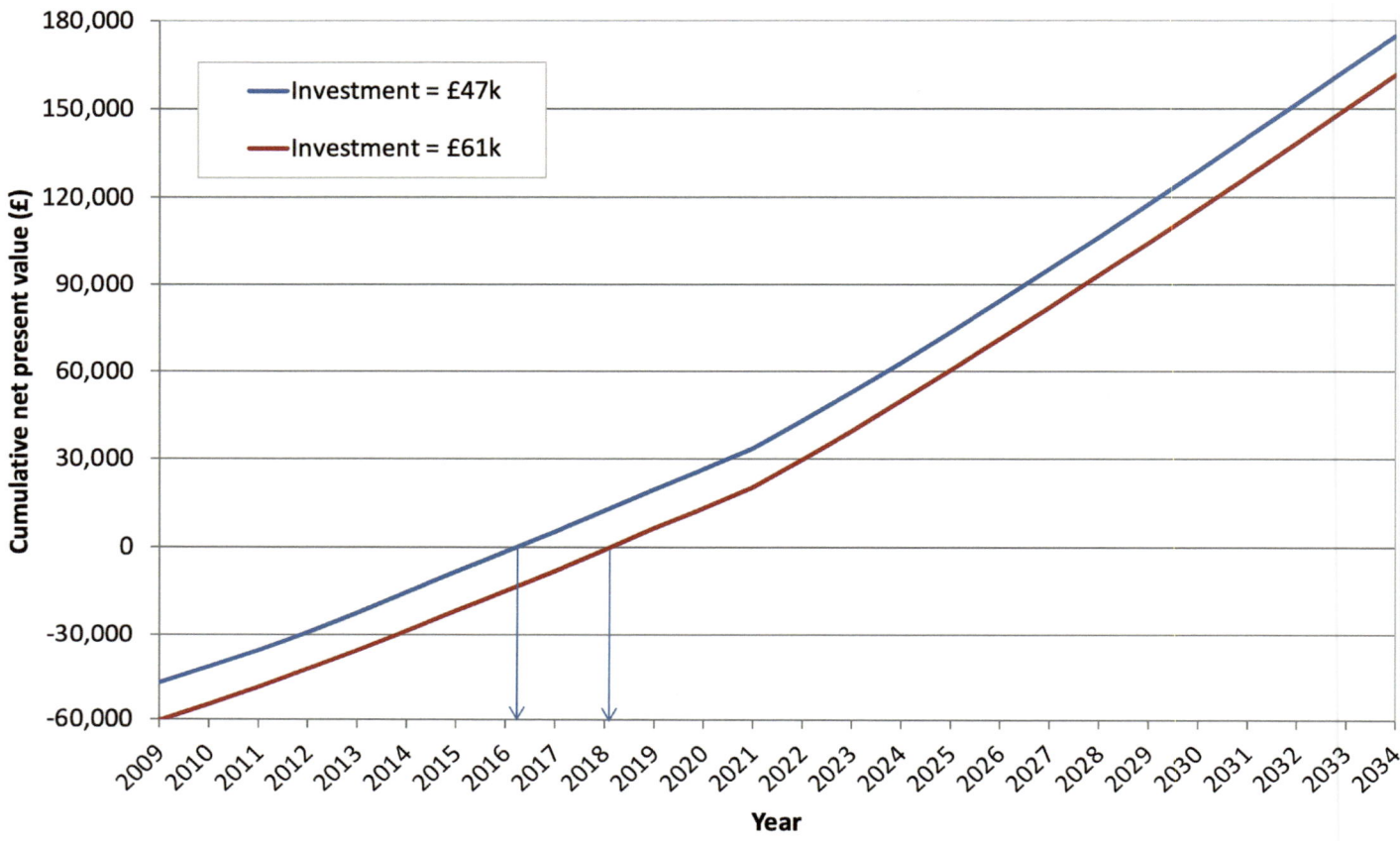

Figure 260   Life cycle benefits

The payback period for both cases is denoted with vertical arrows pointing to the time axis. These payback periods as well as lifecycle benefit and return on investment are summarised in Table 88.

TABLE 88	SUMMARY OF LIFE CYCLE COST ANALYSIS FOR TWO DIFFERENT INITIAL INVESTMENTS			
Case	Initial investment (£)	Life cycle benefit (£)	Payback period (years)	Return on investment (%)
A	47,345	174,687	7.2	269
B	60,596	161,436	9.1	166

As we can see from the life cycle cost analysis, in the case of Initial Investment A which takes into account the upgrade of insulation and windows in the original building only, payback is achieved 7.2 years after the retrofit and by the end of the 25-year cycle the building will generate £174,687 income for the owner. In the case of Initial Investment B which takes into account cost of insulation and windows in the upgraded building, the payback is achieved in 9.1 years after the completion and the building will generate £161,436 for the owner after 25 years. The actual payback period and the life cycle cost will be somewhere between these two sets of figures.

The return on investment (ROI) is calculated as follows:

$$ROI = (Gain - Cost) / Cost$$

and the results are shown in Table 88. If our assumption for the long-term inflation rate of 3% over the life cycle period and the cost estimates were correct, which one can never be absolutely sure of, the return on investment for ZCH would be 269% in case A and 166% in case B. Even if we take the lower of the two figures, the investment into zero carbon seems to pay back very well.

We also need to be aware that the life cycle cost and payback period are sensitive to changes of the inflation rate and the mortgage interest rate (Figure 261). For instance, if the inflation rate changes to 2% and mortgage interest rate remains at 3.05%, the payback period will be 7.4 years and the income at the end of the 25-year period will be £ 147,477. If the inflation rate is 3% and mortgage interest rate is 5%, the payback period will be 7.9 years and the income at the end of the 25-year period will be £ 122,876. And if the inflation rate is 10% and mortgage interest rate is 3.05%, the payback period will be 6 years and the income at the end of the 25-year period will be £ 554,254.

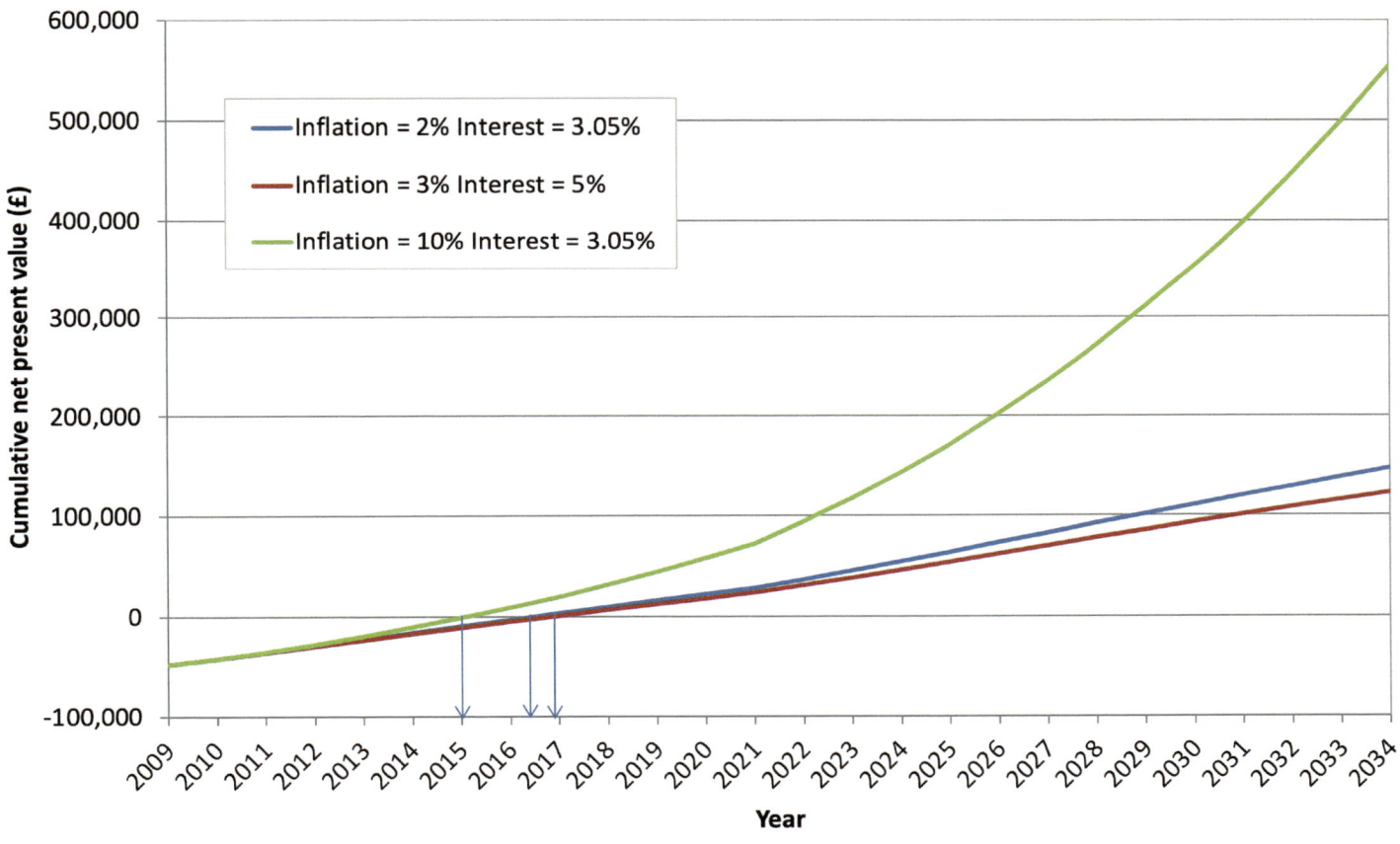

Figure 261   Alternative life cycle benefit depending on inflation and interest rates

However, regardless of exact figures, we can conclude that the project is financially viable. The sensitivity analysis from Figure 261 gives us an idea of the range of alternatives.

21.1.9 Occupant survey and analysis of thermal comfort
Thermal comfort in ZCH was assessed through occupant study and through simulation results. The occupant study was based on a questionnaire followed by an interview and email correspondence to elaborate the answers and to obtain an in-depth understanding of the building use and perceived performance. Results of the occupant study are shown in Tables 89 and 90.

We can see that most of the scores in Table 89 are high and the overall performance of the house is rated by its occupants as 92% excellent. We can also see that the lowest score of 36% relates to manual dimming of electrical lights. This clearly shows that automatic dimming would have been a better option, as frequent changes of available daylight throughout the day make manual dimming impractical. Mechanical ventilation seems to be more in use than natural ventilation, so that there is a 64%/36% split between the former and the latter.

TABLE 89 RESULTS OF OCCUPANT STUDY – SCALES FROM 0% TO 100%		
Question	Score (%)	High end scale description
Windows/doors/internal openings are opened or closed to capture or conserve heat	89	Regularly
Electrical lights are switched on/off in response to available daylight	85	Regularly
Electrical lights are dimmed in response to available daylight	36	Regularly
Natural ventilation is used to replace the operation of MVHR during mild external temperatures	64	Regularly
When electrical lights and electrical devices are not used, they are switched off, including devices on standby	91	Regularly
How would you rate the space heating system performance?	70	Excellent
How would you rate the water heating system performance?	71	Excellent
How would you rate the mechanical ventilation system performance?	82	Excellent
How would you rate the solar thermal system performance?	87	Excellent
How would you rate the PV system performance?	90	Excellent
How would you rate the overall performance of the house, including comfort and energy?	92	Excellent

The thermal comfort results from the occupant survey are shown in Table 90. The scale has been derived from the visual analogue scale used in the questionnaire. The mid-point of the scale represents thermal neutrality or 0%. The low extreme of this sale of −100% represents a perception of being too cold, and the high extreme of this scale represents a perception of being too hot. Using the results from Table 90, we can conclude that the occupants of the ZCH were 3% above the neutral point in summer and 10% below the neutral point in winter.

TABLE 90 RESULTS OF OCCUPANT STUDY – THERMAL COMFORT SCALES FROM −100% TO +100%		
Question	Score (%)	Scale description
What is your perception of your personal thermal comfort in the house in summer?	3	−100% = too cold 0% = neutral +100% = too hot
What is your perception of your personal thermal comfort in the house in winter?	−10	−100% = too cold 0% = neutral +100% = too hot

If the visual analogue scale is converted into Fanger's 7-point scale (see 'Measurement of comfort' in Chapter 17), we can see that every 33.3% on the former corresponds to each point below or above zero on the latter. Fanger's '1 = slightly warm' therefore corresponds to the range from 0% to +33.3% on the visual analogue scale. Fanger's '−1 = slightly cold' corresponds to the range from −33.3% to 0% on the visual analogue scale. When we convert the score of −10% on the visual analogue scale the Fanger's scale by dividing −10 by 33.3 we get the Fanger's predicted mean vote (PMV) of −0.3 and the corresponding predicted percentage of dissatisfied of 6.87 using Equation (54) for PPD from 'Measurement of comfort' in Chapter 17.

Conversely, a comfort score of +3% on the visual analogue scale converted to Fanger's scale gives PMV of +0.09 and PPD of 5.17 (Table 91).

TABLE 91 SUMMARY OF OCCUPANT THERMAL COMFORT IN ZERO CARBON HOUSE			
	PMV (−3 to +3)	PPD (%)	Discrepancy from neutrality (%)
Summer	0.09	5.17	0.17
Winter	−0.30	6.87	1.87

We now need to remember that the best PPD score for a thermally neutral vote of PMV = 0 is 5% (see Figure 188 in Chapter 17). In other words, the best we can do as designers is to have 5% thermally dissatisfied people in a building, whilst 95% will be thermally satisfied. The PPD scores of 6.87% in winter and 5.17% in summer differ from thermal neutrality score (the best score) of 5% by 1.87% in winter and by 0.17% in summer (Figure 262). We can therefore safely say that the ZCH occupants perceive their thermal environment as very comfortable.

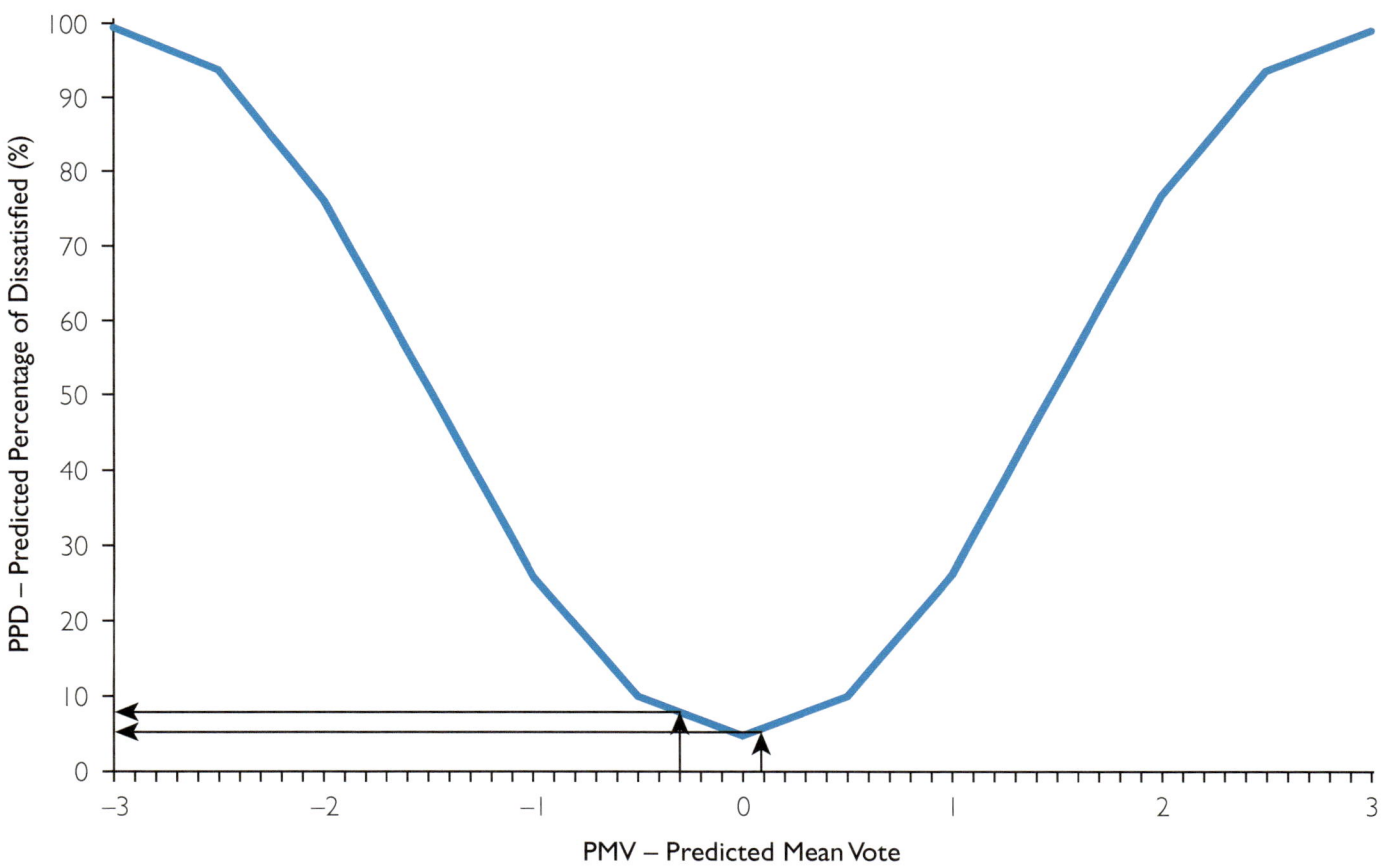

Figure 262   Predicted mean vote and predicted percentage of dissatisfied in Zero Carbon House

Thermal comfort will now be analysed using results of a dynamic simulation. A frequency of occurrence analysis of hourly temperatures is summarised in Table 92.

TABLE 92 ANALYSIS OF FREQUENCY OF OCCURRENCE OF HOURLY TEMPERATURES IN THE KITCHEN/DINING ROOM

Temperature interval end	Frequency of occurrence (hours)	Frequency of occurrence (%)	Cumulative frequency of occurrence (hours)	Cumulative frequency of occurrence (%)
15	21	0.2	21	0.2
16	327	3.7	348	4.0
17	992	11.3	1,340	15.3
18	1,015	11.6	2,355	26.9
19	707	8.1	3,062	35.0
20	312	3.6	3,374	38.5
21	531	6.1	3,905	44.6
22	697	8.0	4,602	52.5
23	2,730	31.2	7,332	83.7
24	934	10.7	8,266	94.4
25	253	2.9	8,519	97.2
26	108	1.2	8,627	98.5
27	49	0.6	8,676	99.0
28	34	0.4	8,710	99.4
29	21	0.2	8,731	99.7
30	15	0.2	8,746	99.8
31	9	0.1	8,755	99.9
32	5	0.1	8,760	100.0
	Total above 28°C	1.0		

We can see from this analysis that there are two frequency peaks: at 18°C and at 23°C. Temperatures below 18°C occur mostly at night when the heating is turned off and they are not a reason for concern. Temperatures above 23°C occur in summer months, and higher temperatures above 28°C occur 1% of the total time. Internal winter temperatures vary between 18°C and 21°C, although they are more frequent towards the lower end of this range, as evident from Table 92.

From this analysis, which is consistent with the occupant survey, we can conclude that the ZCH provides a comfortable environment for its occupants. We can also see that the occupants frequently keep winter temperatures near 18°C (18.4°C is the calibration temperature for the simulation model as per Table 82), which shows their commitment for energy conscious living.

21.1.10 Post-occupancy monitoring – five years on

At the time of writing this section ZCH has been monitored for five years. This section reports on the results of monitoring and compares the actual and predicted zero carbon status. It contains extracts from my collaborative conference work on comparing ZCH with the Passivhaus standard (Christophers and Jankovic, 2017).

21.1.10.1 Hourly data and thermal comfort

Details of temperatures and relative humidity recorded during a 12-month period are shown in Figure 263. It was at first slightly surprising to see that internal temperatures fluctuate from 15°C in winter to 25°C in summer. A question whether the house occupants experienced discomfort levels as result of winter temperatures was raised. The temperature information was in contradiction with the findings from the occupant survey reported earlier in this text, in which it appeared that the occupants' perception was that they were close to thermal neutrality, both in summer and winter.

For that reason, an adaptive clothing algorithm from the section 'Improving thermal comfort assessment through simulated adaptive clothing behaviour' in Chapter 17 was applied and the predicted percentage of dissatisfied (PPD) was calculated for each half-hourly interval. The adaptive clothing algorithm looks at PMV in the previous half-hourly period, and increases the clothing level for PMV < 0 and reduces it for PMV > 0 in the current half-hourly period. As a result, the clothing level varies on a continuous basis throughout the year (Figure 264), as it would do in real life where people adjust their clothing to match internal conditions, and it effectively represents adaptive clothing behaviour. Using adaptive clothing, the calculated PPD level does not go over 10% (Figure 264), which is consistent with the findings from the occupant survey (Figure 262), and the close proximity to thermal neutrality throughout the year. This effectively explains how the occupants regulate their thermal comfort throughout the year.

The occupants have a choice to heat the house to higher temperatures in winter if they wish to. This is especially so because it does not cost them anything to do that: all of the electricity comes from the PV system and the wood for the stove comes from fallen branches in the garden. The finding that they prefer to adapt to the internal conditions with their clothing, rather than with heating, indicates that they have developed a level of adaptation to thermal conditions that resonates with nature, and gives an example how it can be done.

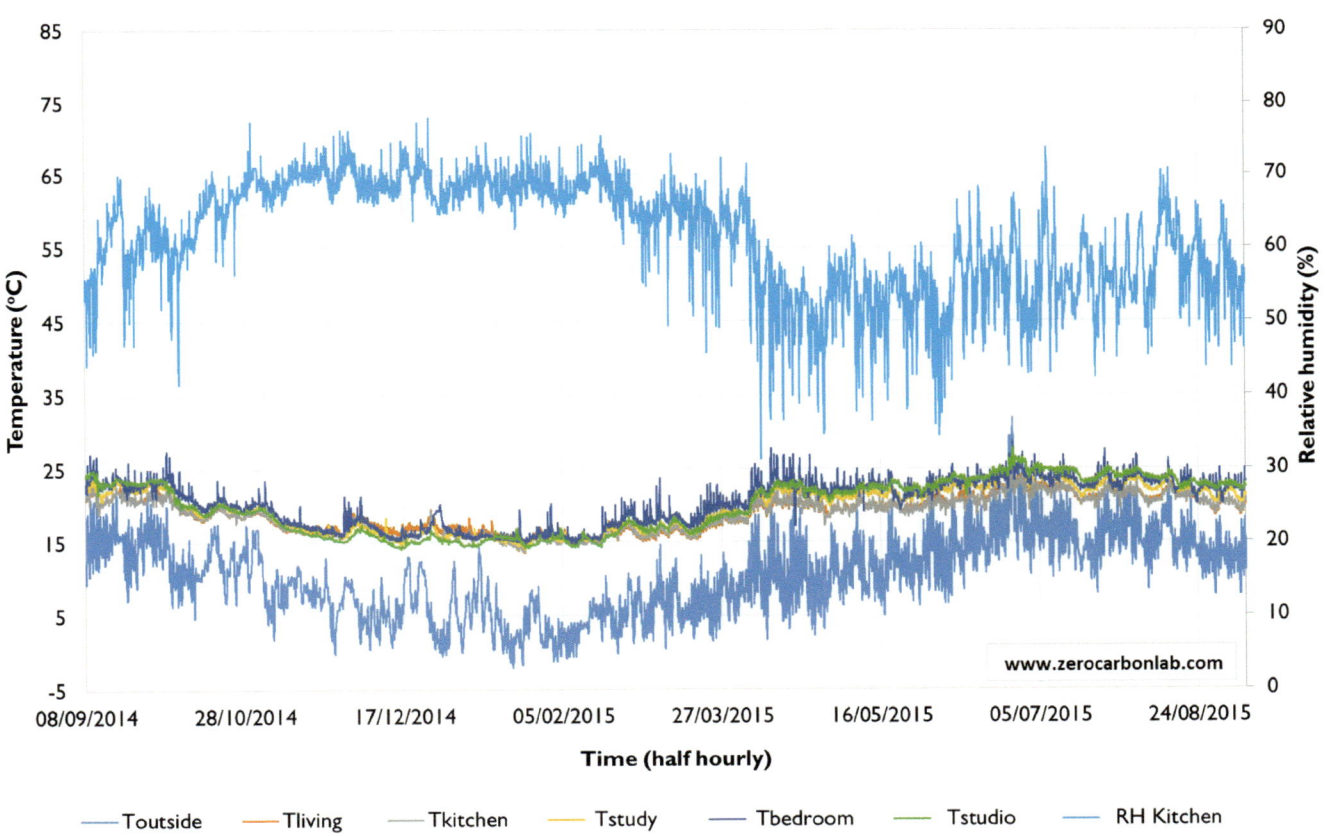

Figure 263   Temperature and relative humidity

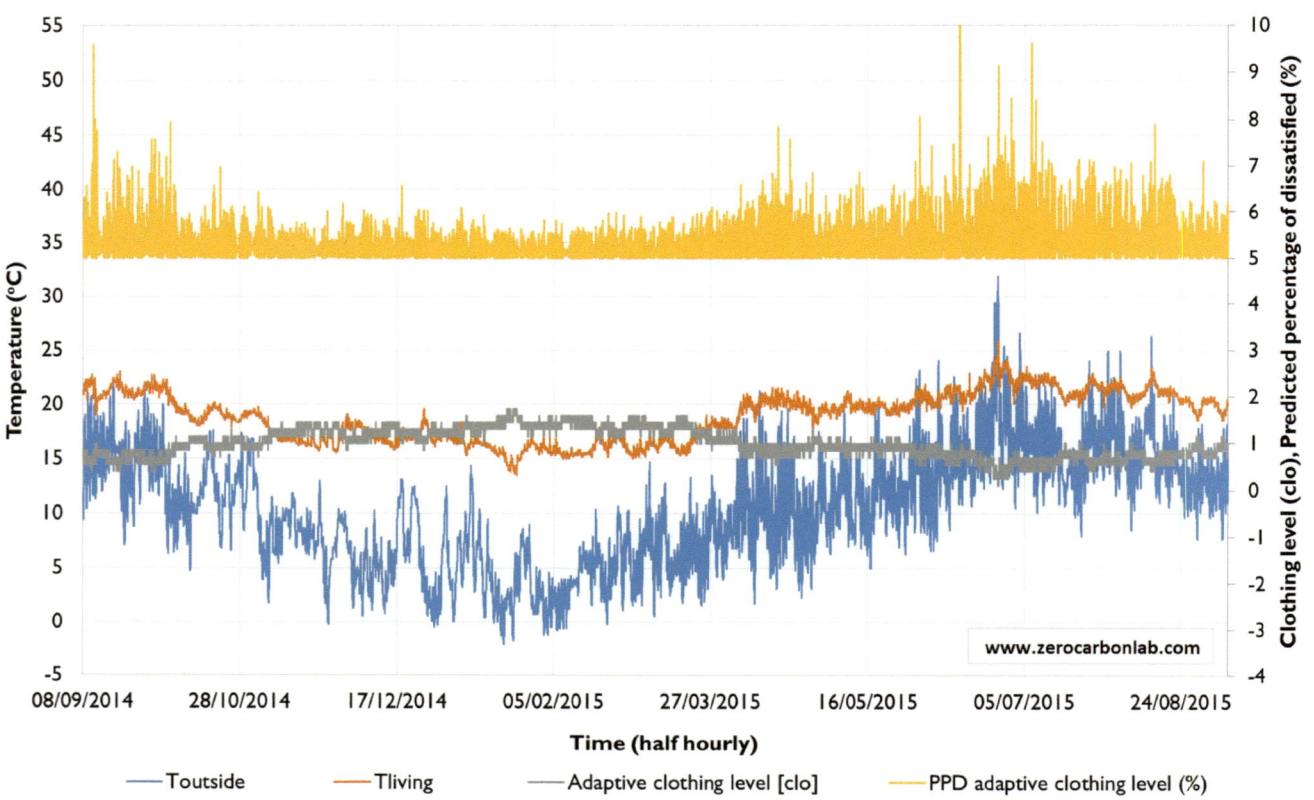

Figure 264   *Thermal comfort levels resulting from adaptive clothing behaviour*

21.1.10.2 Energy and carbon performance

Thermal energy input into the house is shown in Figure 265 in the context of external air temperature. As it can be seen from this figure, the wood-burning stove is used infrequently during the coldest weather when the output from the solar thermal system is the lowest.

During the coldest period when external air temperature was below 0°C, the output temperatures from the solar hot water system were between 30°C and 50°C (Figure 266). The solar thermal output followed the pattern of solar radiation falling on the roof surface with a degree of seasonal time delay. This can be attributed to the seasonal changes in the outside air temperatures and corresponding higher heat losses from the solar thermal system in winter and mid-season, as well as to a lower use of solar thermal energy and consequent higher output temperatures in summer months.

The magnitude of electricity energy imported from the grid (Figure 267) is inversely proportional to the solar radiation on the roof surface (Figure 266), whilst the surplus energy exported to the grid exceeds the energy imported from the grid (Figure 267).

Energy balances have been calculated from monitored data over a period of five years and summarised per year in Table 93. Emissions for electricity import and export are assumed to be time-dependent (Department for Business, Energy & Industrial Strategy, 2021) as result of gradual decarbonisation of the electrical grid and are shown in Figure 268.

The above results have been compared with the Passivhaus standard applicable to new buildings and EnerPHit standard applicable to refurbished buildings (Passivhaus Trust, 2023) and summarised in Table 94.

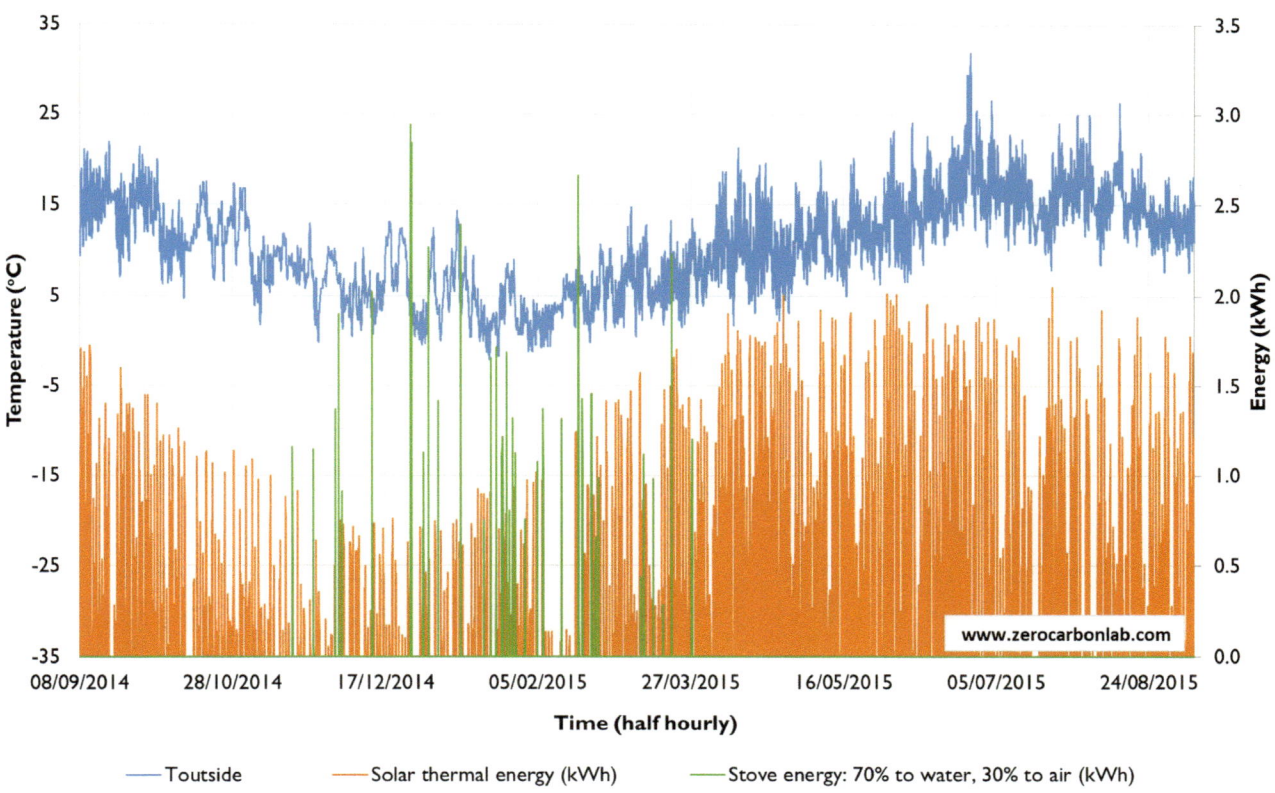

Figure 265   Wood-burning stove and solar thermal inputs

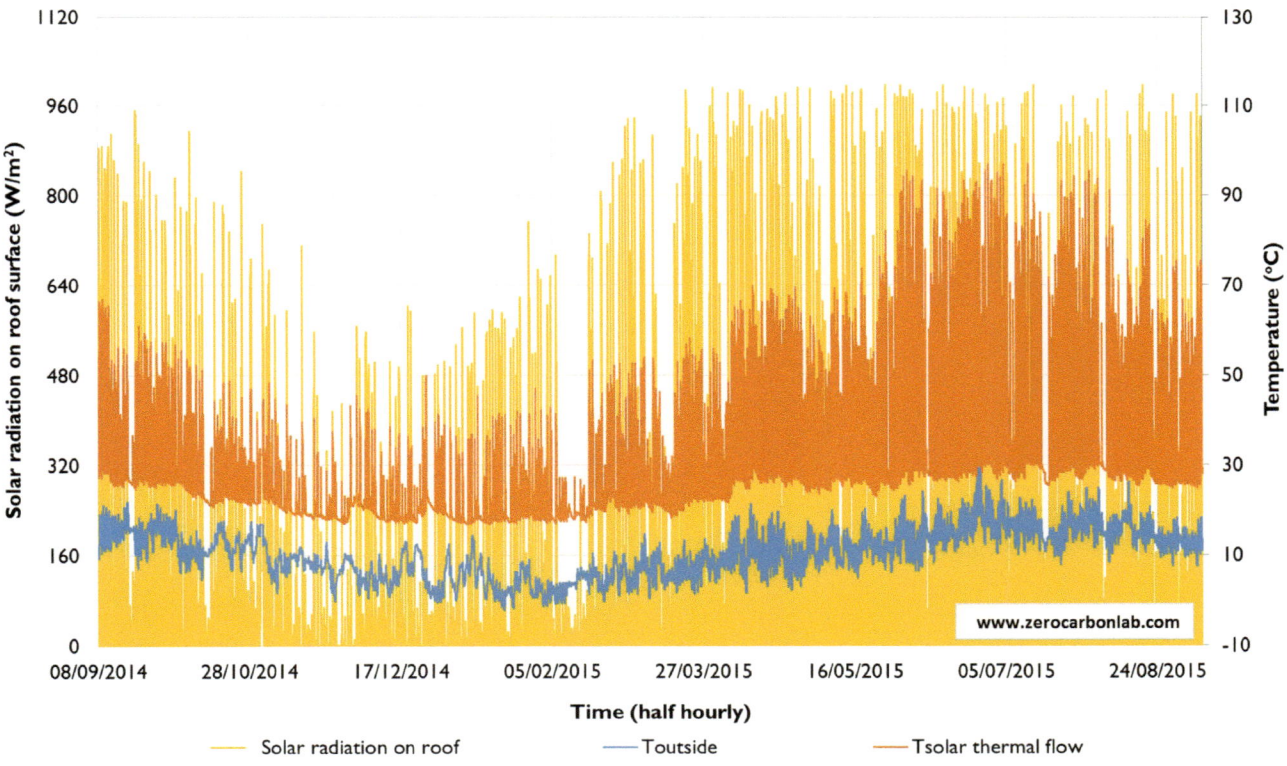

Figure 266   Solar radiation and solar thermal output

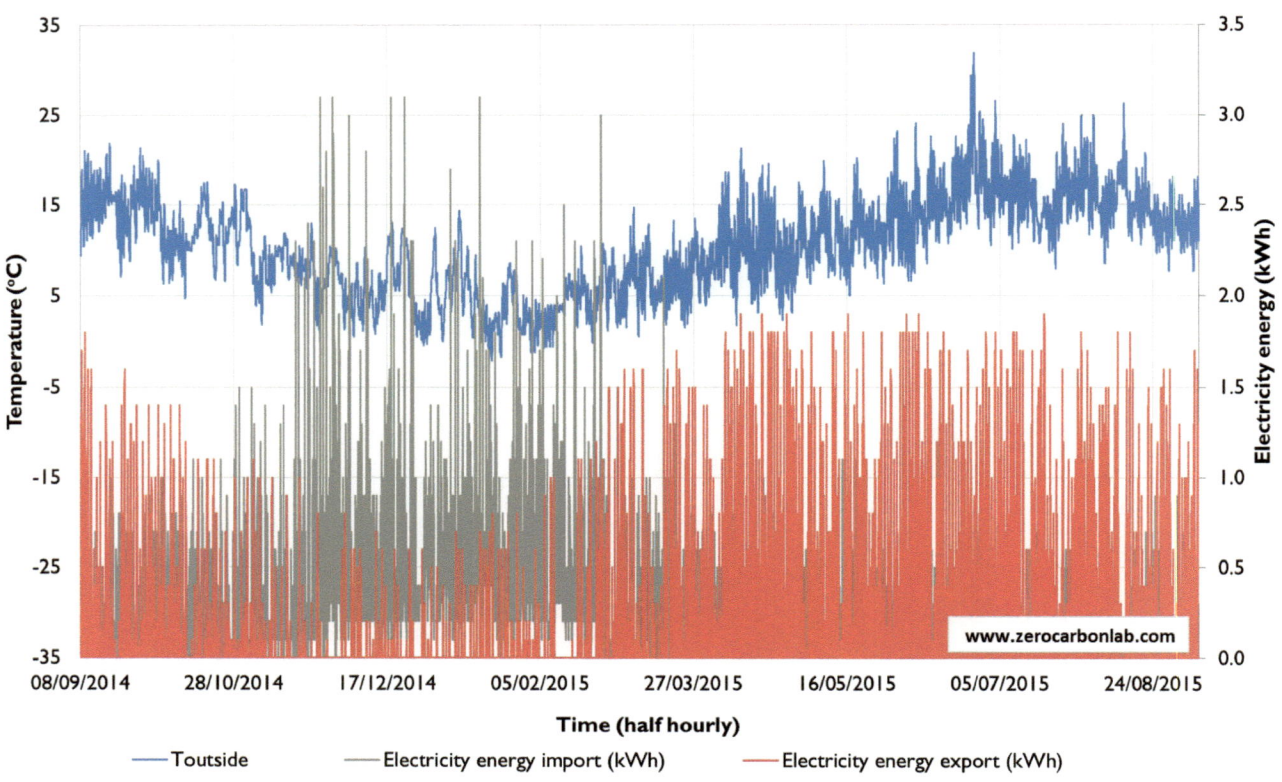

Figure 267   Electricity energy import and export

TABLE 93	SUMMARY OF ANNUAL ENERGY BALANCES AND OPERATIONAL EMISSIONS				
Row	**Description**	**kWh /year**	**Emissions factor (kgCO$_2$ /kWh)**	**Emissions (kgCO$_2$ /year)**	**Notes**
1	Savings from retrofit			−17,102	From Table 85
2	Biomass energy used	217	0.029	6.30	217 kWh/year was obtained from instrumental monitoring
3	Solar PV: Export	3,303	Time dependent – see note	Time dependent between −1,509 and −21	Emission factors and therefore the emissions are time dependent as per Department for Business, Energy and Industrial Strategy (2021) – see Figure 268
4	Grid electricity: Import	2,989	Time dependent – see note	Time dependent between 1,489 and 21	
5	Solar PV: Net electricity export	314 '	Time dependent – see note	Time dependent between −20 and −0.18	
6	Solar thermal	3,707	Solar thermal replaces electricity heating for DHW and is already taken into account in solar PV import figure above (2,989 kWh/year). Without solar thermal the electricity import would have been 3,707 + 2,989 = 6,696 kWh/year.		

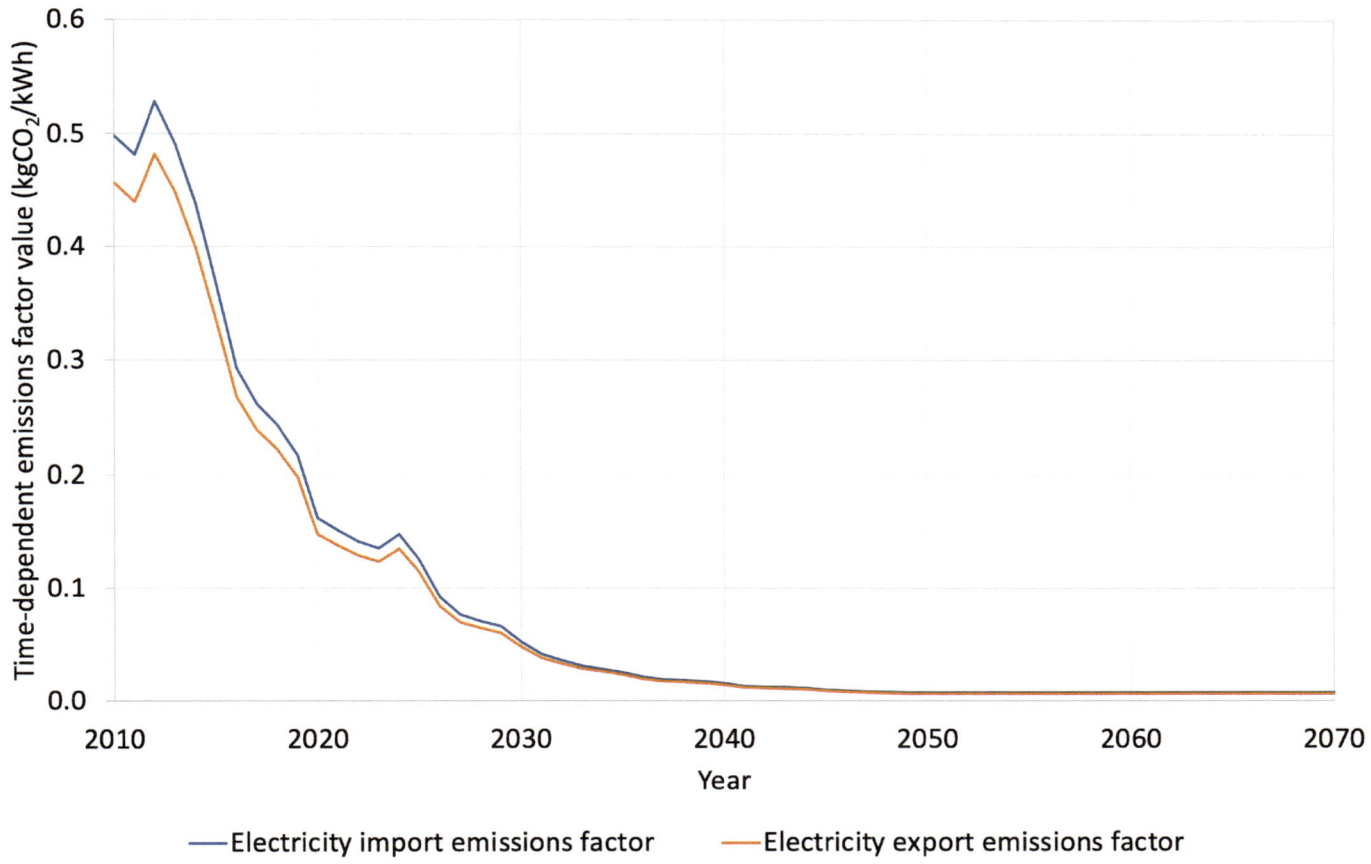

Figure 268   *Time-dependent emissions factor for electricity import and export*

TABLE 94 COMPARISON WITH PASSIVHAUS AND ENERPHIT STANDARDS			
	ZCH	**Passivhaus**	**EnerPHit**
Specific heating energy demand (kWh/m²/year)	1.18	≤15	≤25

As it can be seen from this table, the ZCH significantly outperforms the Passivhaus standard and even more so the EnerPHit standard. This can be better understood taking into account that Passivhaus/EnerPHit performance metrics exclude renewable energy from the design calculations. However, ZCH is beyond the design stage and renewable energy is an everyday reality in it. Thus renewable energy cannot be separated from the overall energy balance, resulting in a significantly better performance than the other two benchmarks.

However, extended Passivhaus standards take into account renewable energy generation (Passive House Institute, 2023). ZCH compares well with these new Passivhaus Classic, Plus, and Premium standards. From the point of view of renewable energy generation calculated to be 63.4 kWh/m²/year, ZCH is between Passivhaus Plus (>60 kWh/m²/year) and Premium (>120 kWh/m²/year) standards, and therefore considerably better than Classic standard (0 kWh/m²/year renewable energy generation) as already reported in Table 94. In the context of renewable energy demand calculated to be 29.7 kWh/m²/year), ZCH is better than Passivhaus Premium standard (≤30 kWh/m²/year), and therefore even better than Plus and Classic standards (≤45 kWh/m²/year and ≤60 kWh/m²/year, respectively). Details of these comparisons are discussed by Christophers and Jankovic (2017).

21.1.11 Cumulative embodied and operational emissions

In this section, we utilise the results of embodied emissions calculations from Chapter 5, Table 23, and integrate them with measured operational emissions (Table 93) in order to find out how combined emissions vary with time and the time they reach zero. The total of embodied and operational emissions is shown in Figure 269. This figure also shows a comparison of emissions with and without the pre-retrofit reduction – the emissions attributed to the house before the retrofit, shown in Table 93, Row 1, as −17,102 $kgCO_2$/year.

The starting point in Figure 269 is 41,781 $kgCO_2$, being a combined total of embodied emissions in materials of 39,825 $kgCO_2$ and emissions from the construction process of 1,956 $kgCO_2$, as shown in Chapter 5, Table 23. This starting value is then being deducted annually by the amount of operational emissions from Table 93. These operational emissions vary with time due to the time-dependent emissions factor for electricity import and export in Figure 268. At the start of the 60 years period, emissions from biomass heating of 6.30 $kgCO_2$/year (Table 93, Row 2) are suppressed and reversed by the negative emissions from the net electricity export of −20 $kgCO_2$/year (Table 93, Row 5). Towards the end of the 60 years period, the negative emissions from net electricity export of −0.18 $kgCO_2$/year (Table 93, Row 5) can no longer suppress/reverse emissions from biomass heating of 6.30 $kgCO_2$/year (Table 93, Row 2).

The savings arising from the emissions attributed to the house before the retrofit of −17,102 $kgCO_2$/year (Table 93, Row 1) reduce the combined emissions rapidly, so that the house reaches zero cumulative emissions between 2012 and 2013, just over two years after the retrofit (Figure 269, green line). However, if the emissions attributed to the original house are not taken into account, then the cumulative emissions increase continuously, as result of the relative differences between the biomass emissions and the negative emissions from net electricity export (blue line).

Emissions from maintenance in relation to repainting the house externally and repainting the ceilings (Table 19, Rows 1 & 3) are included in the combined emissions at intervals of 20 years, therefore in year 2030 and 2050. However, due to their relatively low magnitude in comparison with the starting emissions, these maintenance emissions are not easily visible in Figure 269. The maintenance emissions due to replacement of the solar PV system (1,638 $kgCO_2$) and replacement of the solar thermal system (5,821 $kgCO_2$), carried out 30 years after the retrofit (Table 19, Rows 4 & 5), cause a slightly noticeable 'kink' in both red and blue lines in Figure 269, due to their higher magnitude than the other maintenance emissions.

End of life emissions of 4,476 $kgCO_2$ (Table 23, second last row) also cause noticeable 'kink' in both red and blue lines in Figure 269 at the end of the life cycle of 60 years.

The total cumulative emissions of −970,377 $kgCO_2$ are calculated at the end of the lifecycle of 60 years.

21.1.12 Discussion

The above analysis confirms that combined embodied and operational emissions in Birmingham ZCH achieved zero just over two years after the retrofit. The financial and payback analysis, although depending on variable inflation rates, mortgage rates, costs of materials and systems, installation costs, and other factors, clearly shows significant financial viability and payback period between six and nine years. This comes in the context of excellent thermal comfort and award-winning architecture.

The occupants of the ZCH have gone considerably further in reducing energy consumption by heating the house to around 18°C in winter and thus demonstrating their resource-conscious living, giving an important example of how one can live in more harmony with the environment. Or simulation analysis confirmed that this lower winter temperature was not the reason why the house is carbon-negative, as it remains carbon-negative with much higher internal temperatures.

This design can and should be used as an example of what can be done with the existing building stock. The methods for achieving building energy efficiency and implementing renewable energy do not differ from the corresponding methods for new buildings discussed throughout this book and demonstrated in Chapters 17 and 19. It is clear from this analysis that dynamic simulation is the key ingredient necessary in order to design with confidence.

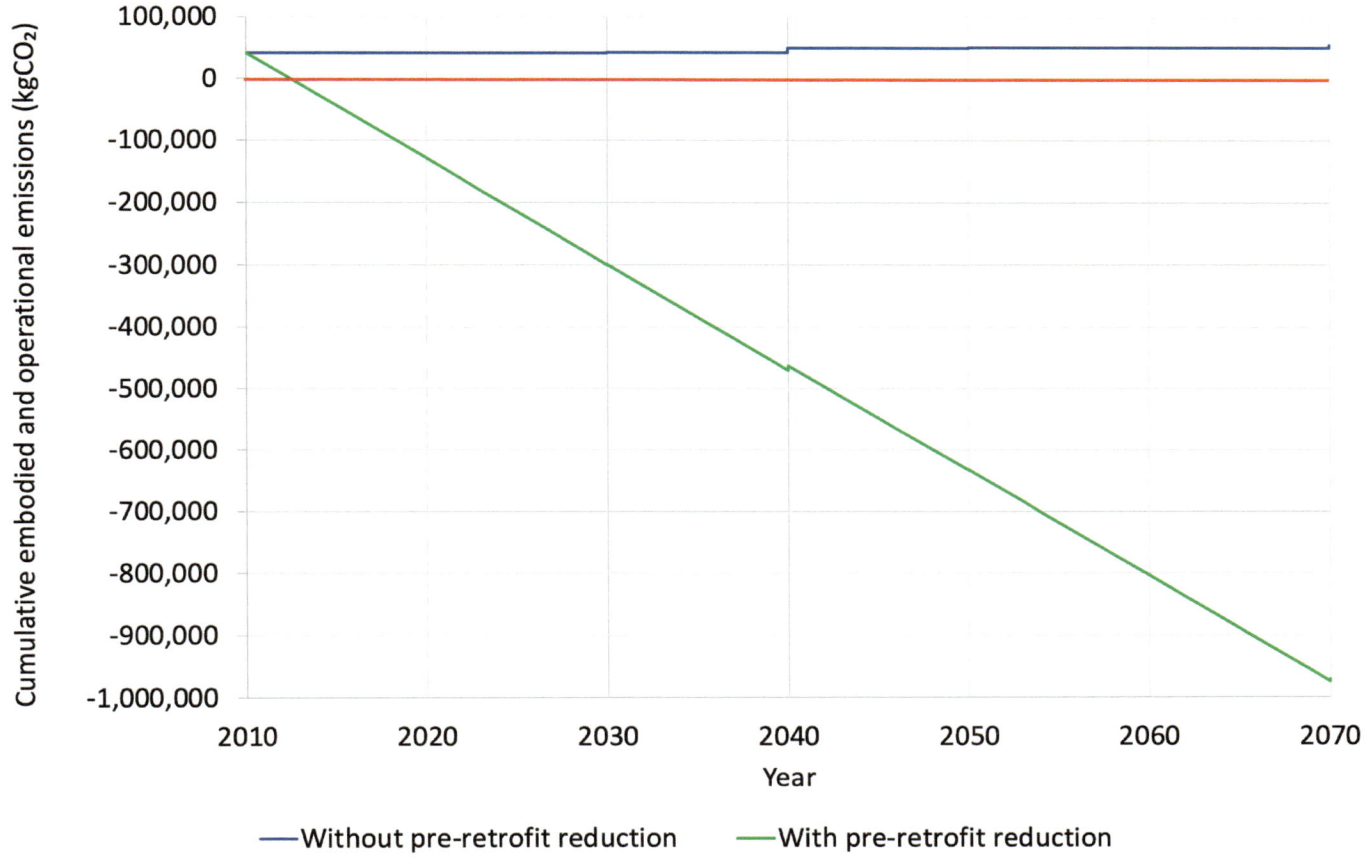

—Without pre-retrofit reduction   —With pre-retrofit reduction

Figure 269   The total of embodied and operational emissions

We therefore have the methods, tools and technology for designing zero carbon retrofit conversions of existing buildings, and we now need new stimulating measures, new business models and access to finance to facilitate the change from old inefficient to new efficient retrofit applications. In addition to being a design tool, dynamic simulation can be used for convincing stakeholders of the environmental and financial potential of retrofit projects.

To retrofit a 170-year-old house into a ZCH truly represents a remarkable design achievement, and this project has been used widely for the education and inspiration of a wide range of organisations and individuals. As John Christophers points out, 'It can be done, it should be done and we ought to be doing more of it'.

ACKNOWLEDGEMENTS

I am grateful to John Christophers and his family for the long-term collaboration since the first edition of this book, and for generous access to the house for the purpose of instrumental monitoring, performance analysis and preparation of this case study.

21.2 CASE STUDY 2: FLOATING OFFICE ROTTERDAM, THE NETHERLANDS

21.2.1 Design concept

The Floating Office Rotterdam (Figure 270) was designed by Powerhouse Company in Rotterdam and it opened on 6 September 2021. It is the largest floating office in the world. The structure is based on timber that can be reused at the end of the building life cycle. The building floats on water and adapts to changing water levels and water level increases due to climate change. This is particularly important for the Netherlands in the context of −2 metres ground altitude as referred to in the next section. The building uses water from the Maas River for cooling of the upper floors and for heating of the lower floor via a water source heat pump. A large photovoltaic array on the roof provides significant contribution to the energy requirements.

According to information provided by Powerhouse Company, the building is carbon-neutral and self-sufficient. The building is home to the Global Center on Adaptation, a Rotterdam based NGO that provides knowledge and advice to countries and organisations in the field of climate change. The building is also home to its designers, the Powerhouse Company.

Figure 270   Floating Office Rotterdam (photo by Marcel I Jzerman with permission from Powerhouse Company)

21.2.2 The climate

The building is located in Rotterdam in the Netherlands. The nearest location for the EPW weather data file is for latitude 52.30N, longitude 4.77E, and altitude of −2 m, from WMO station number 062400. Although this weather data file is for the Amsterdam area, it is 56 kilometres (35 miles) in a straight-line distance northeast from Rotterdam and therefore it is reasonably representative of the site climate. The following data (Table 95) is derived from the corresponding weather data file 'NLD_Amsterdam.062400_IWEC.epw', available from DOE (2023):

TABLE 95 FLOATING OFFICE ROTTERDAM: SUMMARY OF WEATHER CONDITIONS													
Month	1	2	3	4	5	6	7	8	9	10	11	12	Year
Maximum air temperature (°C)	12	10	17	19	25	33	28	31	23	25	15	10	33
Average air temperature (°C)	4	4	5	8	13	15	17	17	14	11	6	4	10
Minimum air temperature (°C)	−5	−8	−3	−1	3	5	10	8	7	−1	−5	−2	−8
Maximum relative humidity (%)	100	100	100	100	100	99	100	100	100	99	100	100	100
Average relative humidity (%)	87	87	83	87	74	74	85	80	85	83	89	89	84
Minimum relative humidity (%)	55	52	35	42	30	32	46	33	53	55	49	66	30
Total sunshine hours (direct solar irradiance >120 W/m²)	64	91	143	134	204	176	202	179	124	88	70	43	1,518

21.2.3 The site

The building is located on the water in Rijnhaven harbour on the Maas River, at the address of Antoine Platekade 1006, 3072 ME Rotterdam, Netherlands. It is south-east facing, giving good solar exposure to a large solar photovoltaic array on the building roof (Figure 271).

Figure 271   Floating Office Rotterdam site

21.2.4 The building

Timber is the main construction material used in this building (Figure 272). In addition to a possibility of recycling and reuse after the end of the lifecycle, this material offers significant carbon storage that lowers the starting embodied emissions of this building. The aim of demountable and recyclable design has led to a grid construction on three levels, with a restaurant, a large outdoor space and a floating swimming pool on the ground floor, and with offices on the first and second floors (Figure 273).

21.2.5 Thermal insulation

In addition to thermal insulation properties of thick timber walls, the building is triple-glazed with floor to ceiling glazing on the main south-east facade whilst it is partially glazed on the smaller side facades. The green roof on the northwest side provides thermal inertia in summer and winter, making the internal conditions more comfortable. It also buffers the rainfall drainage into the river, contributing to the prevention of a rapid water level rise. The ecological value of the green roof is in that it introduces new green surface into the area that otherwise would not have had such surface, thus improving biodiversity on the site. Details of thermal insulation of building components are shown in Table 96.

21.2.6 Solar gain and solar shading

The floor to ceiling triple glazing on the main south-east facade and partial glazing on the two side facades provide significant exposure to solar gain. The balconies provide overhang shading to lower floors, and internal movable shading with reflective surface provides additional control of the solar gain (Figure 275).

21.2.7 Natural daylight

There is a significant amount of natural daylight entering the building through floor to ceiling triple glazing, where glare is controlled by overhanging balconies and internal movable shading with reflective surfaces (Figure 276).

Figure 272   Floating Office Rotterdam: The construction process showing an extensive use of timber (top: photo by Sebastian van Damme; bottom: proto by Marcel IJzerman) with permission from Powerhouse Company

(Continued)

Figure 272   (Continued)

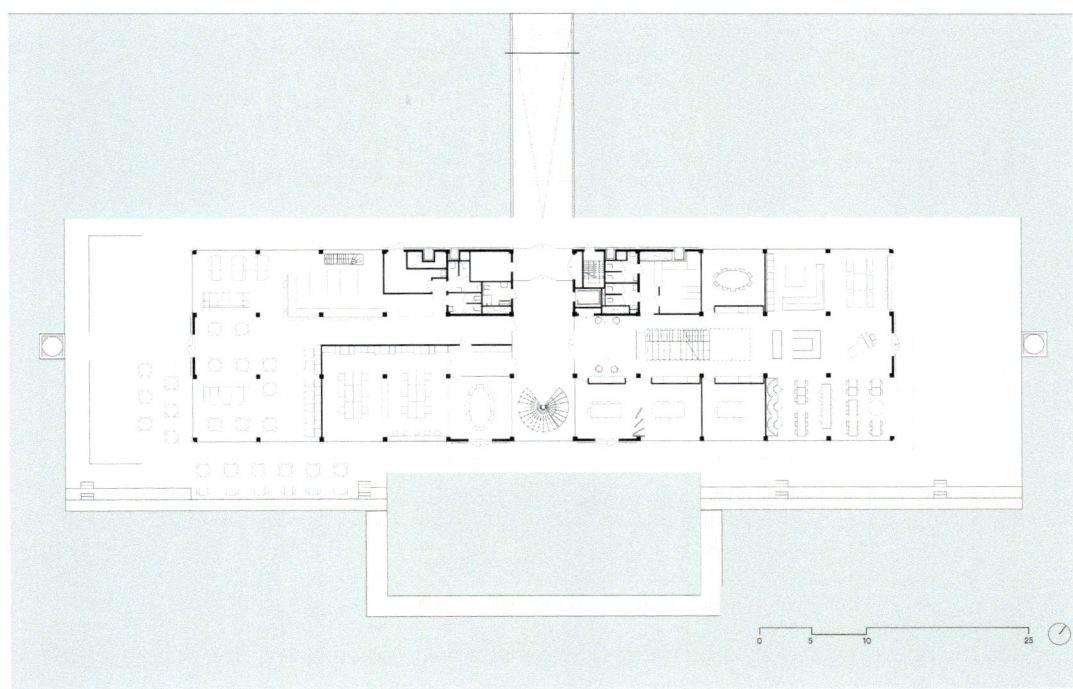

a) Ground floor

Figure 273   Floating Office Rotterdam: floor plans (with permission from Powerhouse Company)

(Continued)

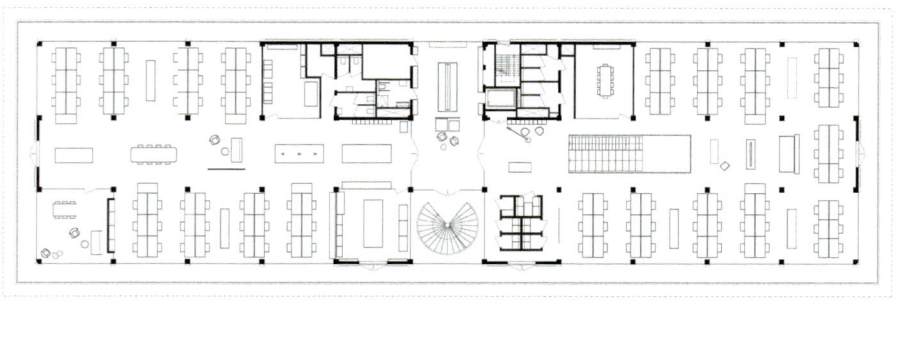

b) First floor

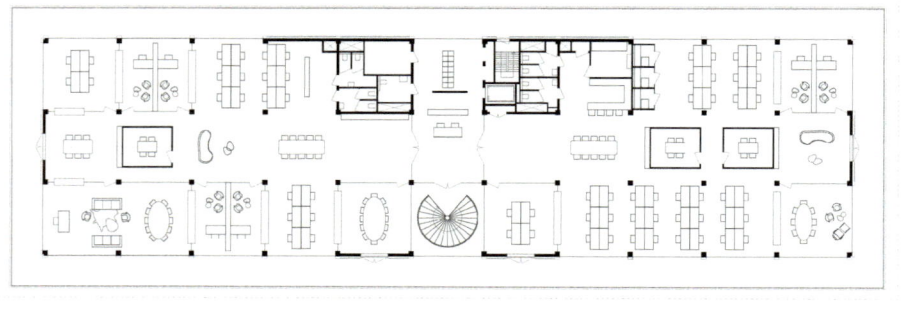

c) Second floor

Figure 273   (Continued)

TABLE 96 *U*-VALUES IN FLOATING OFFICE ROTTERDAM (COURTESY OF POWERHOUSE COMPANY)	
Description	*U*-value (W/m²K)
Windows	0.97
Wooden facade	0.22
Roof	0.17
Ground floor	0.27

21.2.8 Heating and cooling

All heating is provided by a water heat pump, using the Rijnhaven harbour water as a heat source, and delivered via underfloor heating on the lowest floor. Comfort cooling is achieved by circulating canal water through the ceiling heat exchanger panels on each floor.

21.2.9 Renewable energy

The building is fitted with a photovoltaic array of 800 m² surface area on the south-east facing roof, 135° azimuth angle (clockwise from north) and 10° inclination from horizontal (Figure 274). An energy simulation carried out for this photovoltaic array with an hourly timestep using the weather data file referred to above under 'The climate' heading shows energy production from the sun of 127 MWh per annum.

Figure 274 Floating Office Rotterdam: the south-east roof is fitted with photovoltaic panels providing solar electricity for the building and the north-west roof is a green roof providing thermal inertia and giving an ecological dimension to the building (photo by Mark Seelen, with permission from the Powerhouse Company)

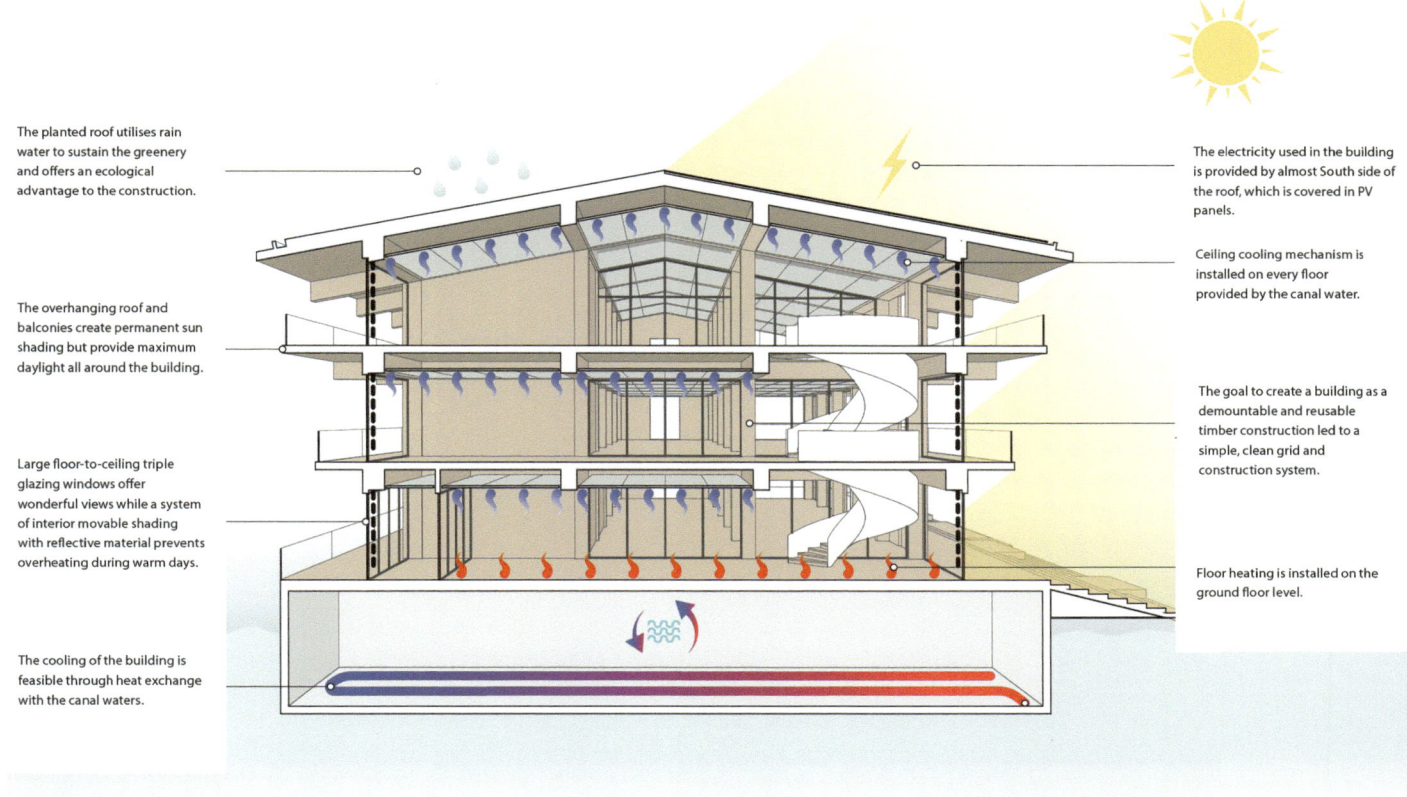

The planted roof utilises rain water to sustain the greenery and offers an ecological advantage to the construction.

The overhanging roof and balconies create permanent sun shading but provide maximum daylight all around the building.

Large floor-to-ceiling triple glazing windows offer wonderful views while a system of interior movable shading with reflective material prevents overheating during warm days.

The cooling of the building is feasible through heat exchange with the canal waters.

The electricity used in the building is provided by almost South side of the roof, which is covered in PV panels.

Ceiling cooling mechanism is installed on every floor provided by the canal water.

The goal to create a building as a demountable and reusable timber construction led to a simple, clean grid and construction system.

Floor heating is installed on the ground floor level.

Figure 275 Floating Office Rotterdam: environmental design diagram (with permission from Powerhouse Company)

Figure 276   Floating Office Rotterdam: interior views showing significant amount of daylight (photos by Sebastian van Damme with permission from Powerhouse Company)

21.2.10 Embodied and operational emissions

Embodied emissions from building materials are calculated on the basis of the volumes of materials provided by Powerhouse Company, applying emissions factors from ICE database (Jones & Hammond, 2019) and taking into account carbon storage in timber – the negative emissions resulting from carbon sequestration (Table 97).

TABLE 97 EMBODIED EMISSIONS IN MATERIALS AND PV IN FLOATING OFFICE ROTTERDAM				
Description	Volume (m³)	Density (kg/m³)	Material emissions (kgCO$_2$ /kg)	Building embodied emissions (kgCO$_2$)
Wood				
CLT	868.5	750	−1.2	−781,650
Glulam	493.7	750	−0.9	−333,248
Chipboard	90.9	800	−1.12	−81,446
Finishes	145.3	800	−1.12	−130,189
Other				
Glass	61.4	2,500	1.75	268,625
Steel	7.9	7,800	1.99	122,624
Concrete	1,737.5	2,380	0.148	612,017
Gypsum board	78	640	0.39	19,469
Tiles	5.1	1,700	0.7	6,069
Insulation				
Expanded polystyrene	725	15	3.29	35,779
Stone wool	142	40	1.12	6,362
Embodied emissions from materials				−255,589

Embodied emissions from the PV system are calculated in Table 98. Considering the variation of figures for PV embodied emissions from: 2,560 $kgCO_2/kW_p$ corresponding to 2021 (Circular Ecology, 2021); embodied emissions projections of 325 $kgCO_2/kW_p$ corresponding to 2040 (Worboys, 2021), and 0.7185 $kgCO_2/m^2$ from (Vindian Solar, 2023), and the uncertainty of where exactly we are between these embodied emissions figures, Table 98 presents calculations in three scenarios, one for each of the three corresponding figures.

TABLE 98 EMBODIED EMISSIONS IN PV IN FLOATING OFFICE ROTTERDAM			
Description	Row	Value	Source
Max generation (kW)	(1)	115	
Corresponding global horizontal radiation (W/m²)	(2)	854	
Solar radiation for kW_p rating (W/ m²)	(3)	1,000	
kW_p rating	(4)	134	
PV area (m²)	(5)	800	
Emissions ($kgCO_2/kW_p$) – Scenario 1	(6)	2,560	Circular Ecology (2021)
Emissions ($kgCO_2/kW_p$) – Scenario 2	(7)	325	Worboys (2021)
Emissions ($kgCO_2/m^2$) – Scenario 3	(8)	0.7185	Derived from Vindian Solar (2023)
PV embodied emissions – Scenario 1 ($kgCO_2$)	(9) = (4) × (6)	343,966	
PV embodied emissions – Scenario 2 ($kgCO_2$)	(10) = (4) × (7)	43,668	
PV embodied emissions - Scenario 3 ($kgCO_2$)	(11) = (5) × (8)	575	

Operational emissions are calculated from energy consumption and energy production figures obtained from Powerhouse Company (96,700 kWh/year and 109,500 kWh/year) and applying emissions factors for consumption and production for the Netherlands (0.37434 $kgCO_2/kWh$ and 0.45172 $kgCO_2/kWh$) from Carbon Footprint (2022).

Results of cumulative embodied and operational emissions calculations are shown in Figure 277.

As it can be seen from this figure, the embodied emissions start from negative values in Scenarios 2 and 3 and from positive values in Scenario 1.

In order to find the time when zero cumulative emissions will be reached in Scenario 1, we will apply the Zero Equation in its formulation as Equation (2), rewritten below from Chapter 1:

$$t = \frac{E_{CO2}}{N_{CO2} - P_{CO2}}$$

(2)

where

E_{CO2} – Embodied emissions in materials, in this case −255,589 $kgCO_2$ (Table 101) plus PV embodied emissions of 343,966 $kgCO_2$ from Scenario 1 (Table 98)

P_{CO2} – Positive operational emissions that occur from year $t = 1$, in this case 96,700 kWh/year × 0.37434 $kgCO_2/kWh$ = 36,199 $kgCO_2$/year

N_{CO2} – Negative operational emissions arising from renewable energy that occur from year $t = 1$, in this case 109,500 kWh/year × 0.45172 $kgCO_2/kWh$ = 49,463 $kgCO_2$/year

t – time in years.

Substituting the above values in Equation (2) gives the following time in years when zero cumulative emissions will be achieved in Scenario 1:

$$t_1 = \frac{-255,589 + 343,966}{49,463 - 36,199} = 6.7 \qquad (2a)$$

In the other two scenarios, zero cumulative emissions are achieved on the completion date of the building.

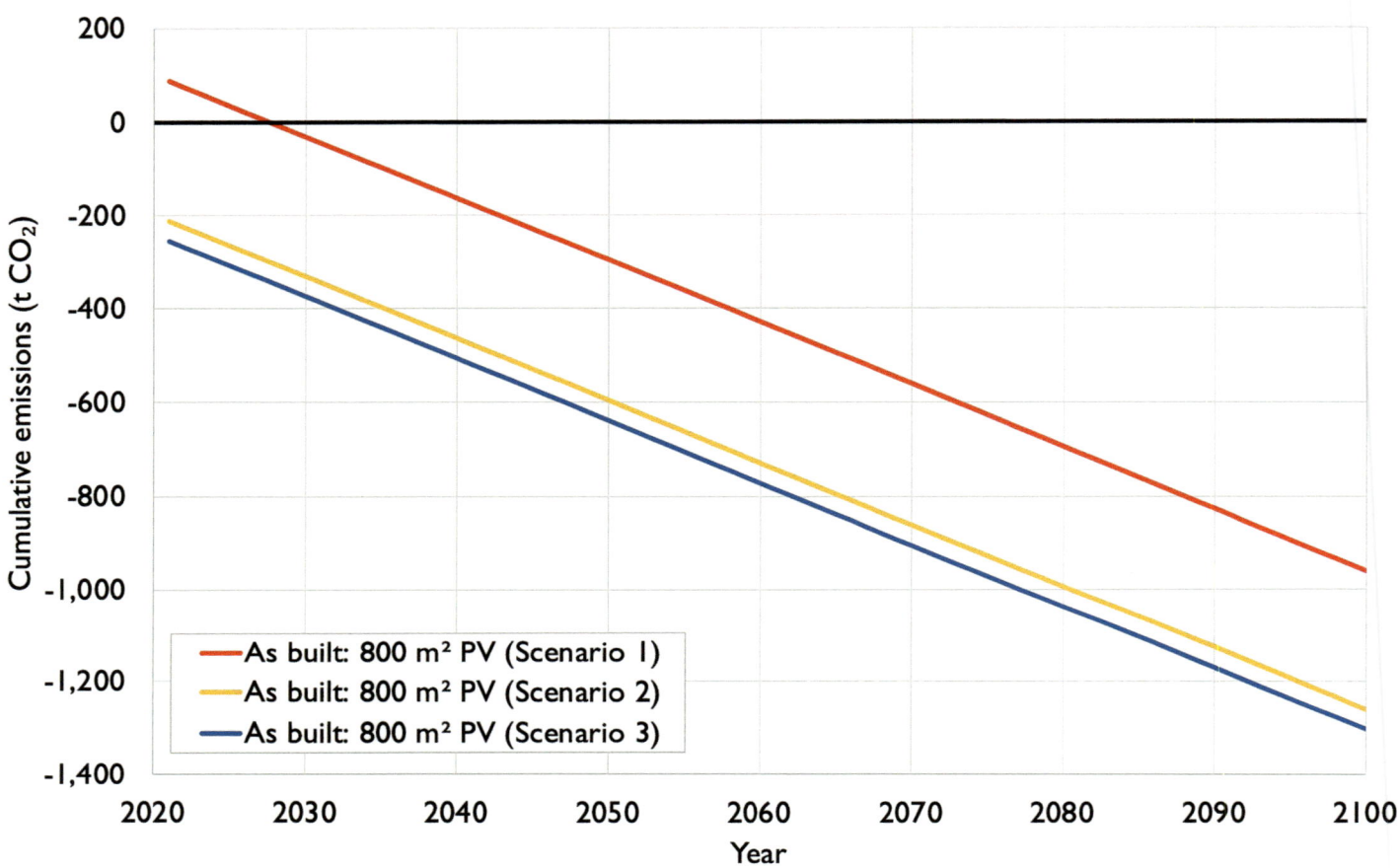

Figure 277   Floating Office Rotterdam: combined embodied and operational emissions

21.2.11 Lessons learnt

The Floating Office Rotterdam is an inspiring building with exceptional environmental credentials. It is one of the best I have come across. The reusable timber construction holds a significant amount of carbon storage (negative embodied emissions) and the floating construction with a large PV array and the water source heat pump provide exceptional environmental credentials. In Scenario 1, zero cumulative emissions are achieved in 6.7 years. In Scenarios 2 and 3, zero cumulative emissions, embodied and operational, were achieved at the completion date of the construction process.

ACKNOWLEDGEMENTS

The case study has been prepared with both technical information and a written consent provided by Powerhouse Company. Photographs are credited to the respective authors and both the photographs and architectural drawings are used with permission from Powerhouse Company. The support from Powerhouse Company in the preparation of this case study is gratefully acknowledged.

21.3 CASE STUDY 3: ASHRAE NEW GLOBAL HEADQUARTERS, USA

ASHRAE is a global professional society in the field of sciences of heating, ventilation, air conditioning, refrigeration, and related fields. It is an industry leader in research, standards writing, publishing, certification, and continuing education and it is dedicated to serve humanity and promote a sustainable world. Since its foundation in 1894, it has grown to more than 50,000 members in over 130 countries. ASHRAE has supported the built environment profession through numerous handbooks, such as *ASHRAE Handbook of Fundamentals 2021* (ASHRAE, 2021a), and has established numerous standards that influence the built environment globally, such as for instance ANSI/ASHRAE Standard 55-2000 – Thermal Environmental Conditions for Human Occupancy (ASHRAE, 2021b) and ANSI/ASHRAE/IES Standard 90.1-2022 – Energy Standard for Sites and Buildings Except Low-Rise Residential Buildings (ASHRAE, 2022). At the start of COVID-19 pandemic in 2020, the Society was the first to draw attention to the airborne infection transmission with ASHRAE Position Document on Infectious Aerosols (ASHRAE, 2020), believed to have saved many lives.

This case study is an abbreviated digest of an extensive work by ASHRAE on retrofitting its new headquarters, which was opened on 18th November 2021. Whilst most of the information in this case study has been generously provided by ASHRAE, and considering that several case studies of this building have been previously published elsewhere, the main contribution of this case study is in the analysis of timing for achieving zero cumulative emissions, combined from embodied and operational emissions.

21.3.1 Design concept

In early 2020, ASHRAE started a renovation of an existing 1978 building for its new headquarters, after a local hospital expressed interest in acquiring their previous headquarters in Atlanta, Georgia, for its expansion. Commensurate with ASHRAE's international standing in the field of sustainability, the retrofit of the new headquarters was to become a showcase of net zero performance, to be used as a learning tool for the society's worldwide membership and the built environment profession in general (Figure 278). It was designed to comply with ASHRAE standards and to be an exemplar for the industry.

Figure 278   ASHRAE New Global Headquarters (with permission from ASHRAE)

21.3.2 The climate

The building is located northeast of Atlanta, GA. The nearest location for the EPW weather data file is for latitude 33.65 N, longitude 84.43 W, and altitude of 315 m, from WMO station number 722190. This weather data is for Atlanta, approximately 29 km (18 miles) southwest from the site location in a straight-line distance, and therefore it is representative of the site climate. The following data (Table 99) is derived from the corresponding weather data file 'USA_GA_Atlanta.722190_TMY2.epw', available from DOE (2023):

TABLE 99 ASHRAE NEW GLOBAL HEADQUARTERS: SUMMARY OF WEATHER CONDITIONS													
Month	**1**	**2**	**3**	**4**	**5**	**6**	**7**	**8**	**9**	**10**	**11**	**12**	**Year**
Maximum air temperature (°C)	19	22	25	30	31	34	36	33	30	32	23	21	36
Average air temperature (°C)	5	7	12	16	21	24	25	24	23	17	11	7	16
Minimum air temperature (°C)	−9	−11	−1	−1	12	15	14	16	14	3	−3	−8	−11
Maximum relative humidity (%)	100	100	97	93	100	100	100	100	100	100	100	100	100
Average relative humidity (%)	69	61	62	62	64	64	74	75	79	70	69	68	68
Minimum relative humidity (%)	13	21	28	29	33	36	37	22	18	22	13	13	21
Total sunshine hours (direct solar irradiance >120 W/m²)	153	168	220	286	287	277	288	288	217	255	192	161	2,792

21.3.3 The site

The building is located at address of 180 Technology Parkway, Peachtree Corners, Georgia, USA. It is situated on the edge of Technology Park Lake that is just to the south side from the building (Figure 279), creating opportunities for additional energy input from low incidence sun angles reflected from the water surface in winter, and for using the lake as a heat pump source.

Figure 279   ASHRAE New Global Headquarters site (image from before the retrofit)

21.3.4 The building

Before the retrofit, the building was poorly insulated and had excessive glazing, resulting in significant heat gains in summer and heat losses in winter (Figure 280a). In comparison, the retrofitted building has improved thermal insulation and reduced amount of glazing (Figure 280b), making the occupant comfort better and energy consumption lower.

a) Before the retrofit

b) After the retrofit

Figure 280   ASHRAE New Global Headquarters: before and after the retrofit

UPPER / ENTRY LEVEL　　　　　**MIDDLE LEVEL**

Figure 281　ASHRAE New Global Headquarters: final floor plans (with permission from ASHRAE)

The building footprint consists of two blocks: a larger block to the right and smaller block to the left. These are linked by an atrium, making an overall T-shape (Figure 281). The entry level consists mainly of open plan offices, working pods and small meeting rooms, and the middle level in addition has a range of different sizes of training rooms, conference rooms, catering facilities, and the library. A large lobby on this floor provides an informal breakout space and offers opportunities for displays by ASHRAE sponsors.

21.3.5　Thermal insulation

The condition of external envelope before the retrofit was poor: failed double-glazed units and an extensive air leakage equivalent to a 10 m² hole in the building facade (Pearson, 2021). Due to relatively mild winters and hot summers, as corroborated by the summary of weather conditions in Table 99, it was found to be more critical to reduce summer heat gains whilst achieving a moderate reduction of winter heat losses. Hence, 90 mm of expanded polystyrene was added to external walls, whilst 100 mm of roof insulation was placed internally, and it was found that adding more thermal insulation would lead to diminishing returns (Pearson, 2021). The improvements of the thermal envelope also included a reduction of the window-to-wall (WWR) ratio explained in the next section, and improvements in airtightness. Details of thermal insulation before the retrofit (where known) and after the retrofit are shown in Table 100.

TABLE 100　ASHRAE NEW GLOBAL HEADQUARTERS: SUMMARY OF *U*-VALUES		
	Before retrofit	**After retrofit**
Description	*U*-value (W/m²K)	*U*-value (W/m²K)
Windows		2.63
Roof lights	N/A	3.94
External walls	0.58	0.33
Ground contact floor		0.19
Ground overhang exposed floor		0.42
Roof		0.18

21.3.6　Solar gain, solar shading and natural daylight

The existing building before the retrofit had window-to-wall ratio of nearly 80%, where most of the double-glazed units had failed. The high WWR was the source of a significant heat gain and heat loss. Analysis conducted by ASHRAE retrofit team investigated WWR options from 33.5% to 75.4% and found that the optimum WWR for East and West facades was 38% and for North and

South facades was 29%. The application of these WWRs reduced the heat gain and heat loss significantly, whilst still providing sufficient daylight levels. Thus, 57% of the upper level and 23% on the middle level receive more than 300 lux of daylight.

The vaulted glazing in the atrium roof in the original building (Figure 280a) was causing overheating in summer, and it was replaced by a flat opaque roof (Figure 280b), creating more space for the PV system on the roof.

As can be seen from the diagram in Figure 282, the external shades control high incidence solar radiation. Light-coloured ceilings ensure reflection of daylight into the depth of the plan, and the new skylights on the top floor provide daylight directly into the depth of the plan.

21.3.7 Ventilation, heating, and cooling

Ventilation, heating, and cooling are illustrated in Figure 282. Natural and mechanical ventilation are achieved by operable windows, atrium exhaust, and bidirectional ceiling fans. A dedicated outside air system (DOAS) provides heating and cooling using radiant ceiling panels. Enthalpy heat recovery and demand-controlled ventilation are deployed to save energy. Fan-assisted night flush is used for to reduce daytime cooling load. A high-efficiency plant consists of air-to-water and water-to-water heat pumps. The latter uses the return water from the radiant chilled panels as a heat source. There is a potential to use the nearby lake as a heat pump heat source.

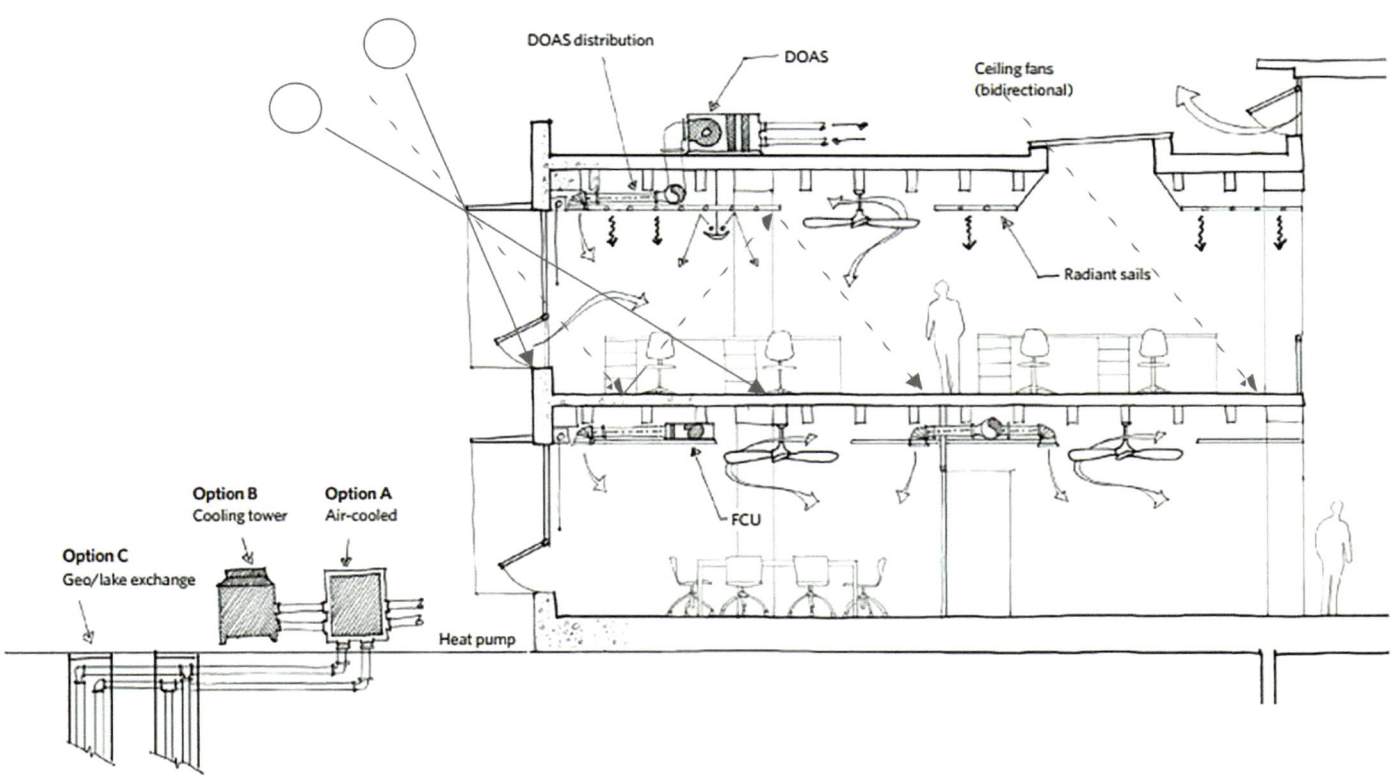

Figure 282   ASHRAE New Global Headquarters environmental design diagram (with permission from ASHRAE)

21.3.8 Renewable energy

The building is powered by three separate PV systems (Figure 283):

• Roof installed 187 kW$_p$ system
• Ground installed 65 kW$_p$ adjacent to the building
• Ground installed 81 kW$_p$ in an unused part of the carpark

This makes the total of 333 kW$_p$ installed. The system was designed under conflicting constraints: on the one hand, a maximum output from the system was sought, by achieving maximum utilisation of the roof and adjacent areas; on the other hand, preserving established trees and staying within the maximum size cap set by the local utility provider placed limits on the system size. The PV control system mitigates the effects where partial shading occurs as the sun angle changes and it enables monitoring of performance of each individual panel.

Figure 283   ASHRAE New Global Headquarters photovoltaic systems (with permissions from ASHRAE)

Performance of the PV system over the past twelve months from the time of writing this text (April 2023) is shown in Figure 284. Thus, 251,578kWh produced by the PV system (Figure 284a) was used for own electricity consumption (Figure 284b), whilst 216,292 kWh was surplus production (Figure 284). This makes the total energy produced over the past twelve months to be 467,870 kWh (467.87 MWh). A different monitoring dashboard for ASHRAE PV production recorded the output of 447.9 MWh for year 2022. Although the two respective time windows (the past 12 months and the 2022 calendar year) had different weather conditions, the two corresponding annual production figures are only 4.4% different and therefore show mutual correspondence.

21.3.9 Embodied and operational emissions and emissions savings

From information received from ASHRAE, it appears that the aspiration for this project was for the power generated on site to be equal to the power consumed on site plus to provide a surplus for export within the limitations set by the local utility provider. The above figures show that this aspiration was achieved and that the power generated was well in excess of the power consumed. We will now investigate combined embodied and operational emissions to calculate the requirements for achieving total zero emissions.

Embodied emissions from building materials were calculated using a combination of a simulation model of the retrofitted building by Buckley (2023) and OneClickLCA (2022a). The simulation model containing detailed specification of constructions for all building surfaces was exported to OneClickLCA twice: the first time as a complete building and the second time as a building without concrete in walls, slabs and roof as this represented the effect of reuse of the existing building. The results for the complete building are shown in Figure 285a, and for the building without concrete are shown in Figure 285b. The two sets of upfront carbon values from these two figures were subtracted to obtain an embodied emissions saving due to reusing the

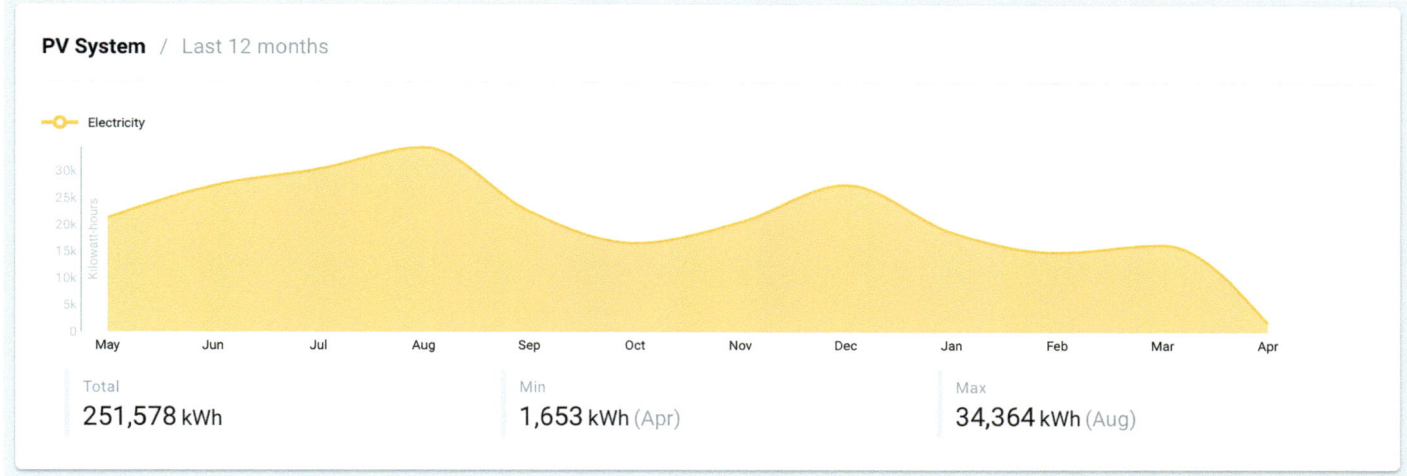

a) PV system generation for own electricity use

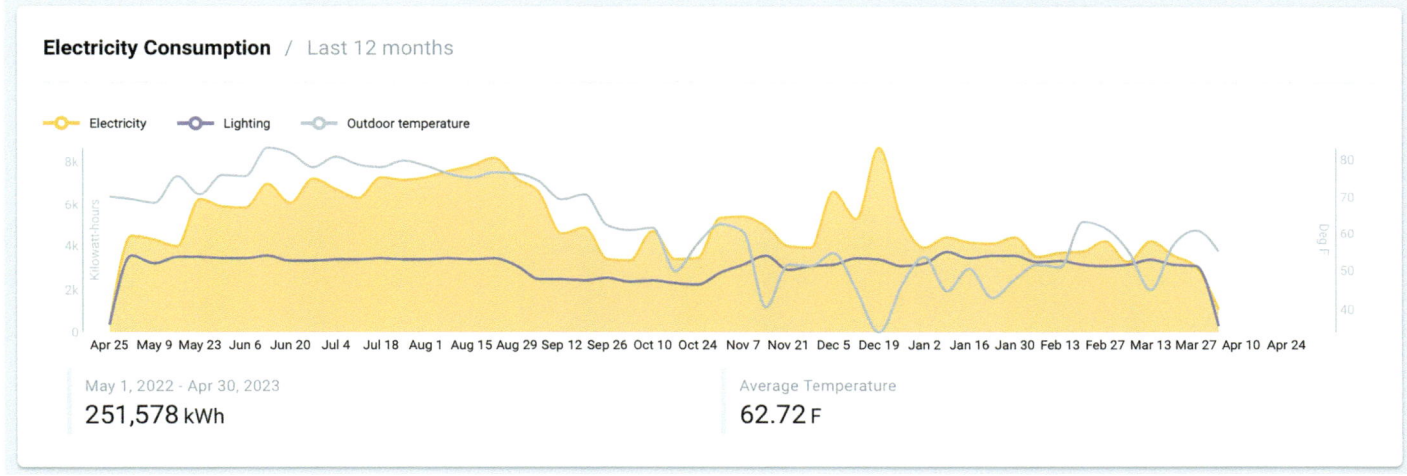

b) Electricity consumption

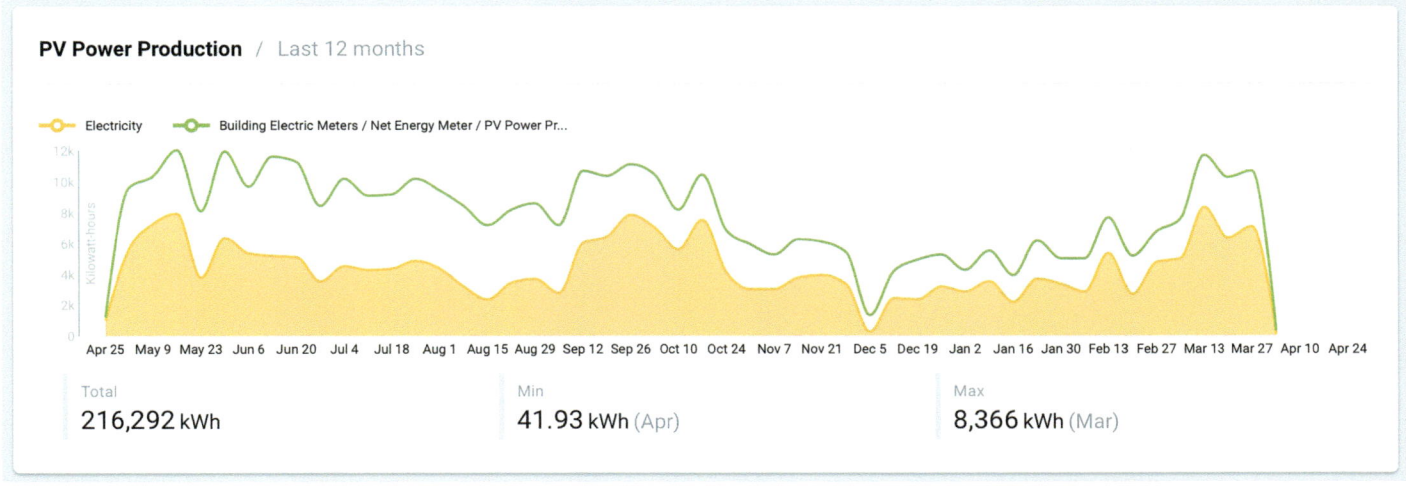

c) PV production for export

Figure 284   ASHRAE New Global Headquarters energy generation and consumption from the building performance dashboard (with permission from ASHRAE)

existing building and the results are summarised in Table 101. Although OneClickLCA results also show operational carbon, this figure was not used as more accurate operational emissions figures were available from the ongoing monitoring data, as already introduced in the previous section.

	Result category	Carbon emissions kg CO$_2$e	Biogenic carbon kg CO$_2$e bio ⑦	Carbon savings from materials reuse kg CO$_2$e	Carbon savings from exported energy kg CO$_2$e	Carbon offsets kg CO$_2$e	Net Carbon kg CO$_2$e	
➕ A1-A5	Upfront carbon	899 448	0	−10 187			889 261	Details
➕ B1-B7	Operating carbon	2 923 456		−25 870			2 897 587	Details
➕ C-D	End of life	69 026	0	0			69 026	Details
	Total	**3 891 930**	**0**	**−36 057**			**3 855 874**	
	Results per denominator							
	Gross Internal Floor Area (ASHRAE) 6196.0 m^2	628	0	−6			622	

a) complete building

	Result category	Carbon emissions kg CO$_2$e	Biogenic carbon kg CO$_2$e bio ⑦	Carbon savings from materials reuse kg CO$_2$e	Carbon savings from exported energy kg CO$_2$e	Carbon offsets kg CO$_2$e	Net Carbon kg CO$_2$e	
➕ A1-A5	Upfront carbon	210 064	0	0			210 064	Details
➕ B1-B7	Operating carbon	3 039 685		0			3 039 685	Details
➕ C-D	End of life	66 289	0	0			66 289	Details
	Total	**3 316 038**	**0**	**0**			**3 316 038**	
	Results per denominator							
	Gross Internal Floor Area (ASHRAE) 6196.0 m^2	535	0	0			535	

b) building without materials reused in retrofit

Figure 285   ASHRAE New Global Headquarters embodied emissions results from OneClickLCA

TABLE 101 ASHRAE NEW GLOBAL HEADQUARTERS TOTAL EMBODIED EMISSIONS FROM BUILDING MATERIALS	
Description	**Embodied emissions (kgCO$_2$)**
Upfront carbon	889,261
Saving from reusing existing building	−679,197
End of life	69,026
Total embodied emissions from materials	279,090

Embodied emissions in the PV system were calculated in three different scenarios, taking into account different sources of emission factors for PV. This is summarised in Table 102.

TABLE 102 ASHRAE NEW GLOBAL HEADQUARTERS EMBODIED EMISSIONS IN THE PV SYSTEM

Description	Row	Value	Source
PV rating (kWp)	(1)	333	Derived from ASHRAE.org (2021)
PV area (m²)	(2)	2,461.56	Derived from Buckley (2023)
Emissions kgCO2/kWp (Scenario 1)	(3)	2,560	Circular Ecology (2021)
Emissions kgCO2/kWp (Scenario 2)	(4)	325	Worboys (2021)
Emissions kgCO2/m² (Scenario 3)	(5)	0.7185	Derived from Vindian Solar (2023)
PV embodied emissions – Scenario 1 (kgCO$_2$)	(6) = (1) × (3)	852,480	
PV embodied emissions – Scenario 2 (kgCO$_2$)	(7) = (1) × (4)	108,225	
PV embodied emissions – Scenario 3 (kgCO$_2$)	(8) = (2) × (5)	1,769	

Operational emissions are calculated using information from the building performance dashboard extract, shown in Figure 284. As the PV system generation for own electricity use (Figure 284a) fulfils annual electricity consumption (Figure 284b), it is only the PV production for export of 216,292 kWh/year (Figure 284c), multiplied by the emissions factor of 0.3902 kgCO$_2$/kWh that contributes to negative operational emissions of −84,397 kgCO$_2$/year.

Operational emissions savings from the retrofit are calculated in Table 103. This table shows annual reduction of emissions attributed to replacement of an energy-inefficient building before the retrofit by an energy-efficient building after the retrofit.

TABLE 103 ASHRAE NEW GLOBAL HEADQUARTERS OPERATIONAL EMISSIONS SAVINGS FROM RETROFIT

Description	Row	Value	Source
Existing building energy use intensity (EUI) (kBTU/ft²/year)	(1)	36	(Scoggins et al., 2021)
Existing building EUI (kBTU/year)	(2) = (1) × 66,700 ft²	2,401,200	Floor area from (ASHRAE.org, 2021)
Retrofitted building EUI (kBTU/ft²/year)	(3)	21	EUI from (ASHRAE.org, 2021)
Retrofitted building EUI (kBTU/year)	(4) = (3) × 66,700 ft²	1,400,700	
Existing building EUI (kWh/m²/year)	(5) = (1) × 3.15459	114	Resultant unit conversion using conversion factor from unitconverters.net (2023)
Existing building EUI (kWh/year)	(6) = (5) × 6,196 m²	703,650	
Retrofitted building EUI (kWh/m²/year)	(7) = (3) × 3.15459	66	
Retrofitted building EUI (kWh/year)	(8) = (7) × 6,196 m²	410,463	
Energy saving (kWh/year)	(9) = (8) − (6)	−293,188	
Emissions prior to retrofit (kgCO$_2$/year)	(10) = (9) × 0.3902 kgCO$_2$/kWh	−114,402	Emissions factor from US EPA (2023)

We will now calculate the time when the building will reach zero cumulative emissions by applying the Zero Equation in its formulation as Equation (2), rewritten below from Chapter 1:

$$t = \frac{E_{CO2}}{N_{CO2} - P_{CO2}} \tag{2}$$

where

E_{CO2} – Embodied emissions in materials, in this case 279,090 kgCO$_2$ (Table 101) plus PV embodied emissions scenarios from Table 102 (852,480; 108,225; or 1,769 kgCO$_2$)

P_{CO2} – Positive operational emissions that occur from year $t = 1$ (kgCO$_2$/year), in this case nil

N_{CO2} – Negative operational emissions arising from renewable energy that occur from year $t = 1$ (kgCO$_2$/year), in this case $-84,397$ kgCO$_2$/year calculated above, and $-114,402$ kgCO$_2$/year of operational emissions prior to retrofit (Table 103)

t – time in years

Substituting the above values in Equation (2) gives the following time in years when zero cumulative emissions will be achieved in Scenario 1:

$$t_1 = \frac{279,090 + 852,480}{84,397 + 114,402} = 5.7 \tag{2a}$$

Similar calculations for Scenarios 2 and 3 are as follows:

$$t_2 = \frac{279,090 + 108,225}{84,397 + 114,402} = 1.9 \tag{2b}$$

$$t_3 = \frac{279,090 + 1,769}{84,397 + 114,402} = 1.4 \tag{2c}$$

Therefore, in the worst-case scenario, ASHRAE New Global Headquarters will achieve zero cumulative emissions in 5.7 years, and in the best-case scenario in 1.4 years after the completion of retrofit in 2021. A chart of cumulative embodied and operational emissions in Figure 286 reflects these results and shows emission reductions through to 2080. A limitation of this analysis is that it only uses embodied emissions in building materials and the PV systems, but not in the mechanical systems in the building. However, even with this limitation, the new ASHRAE Global Headquarters is set to achieve zero total emissions well before 2030.

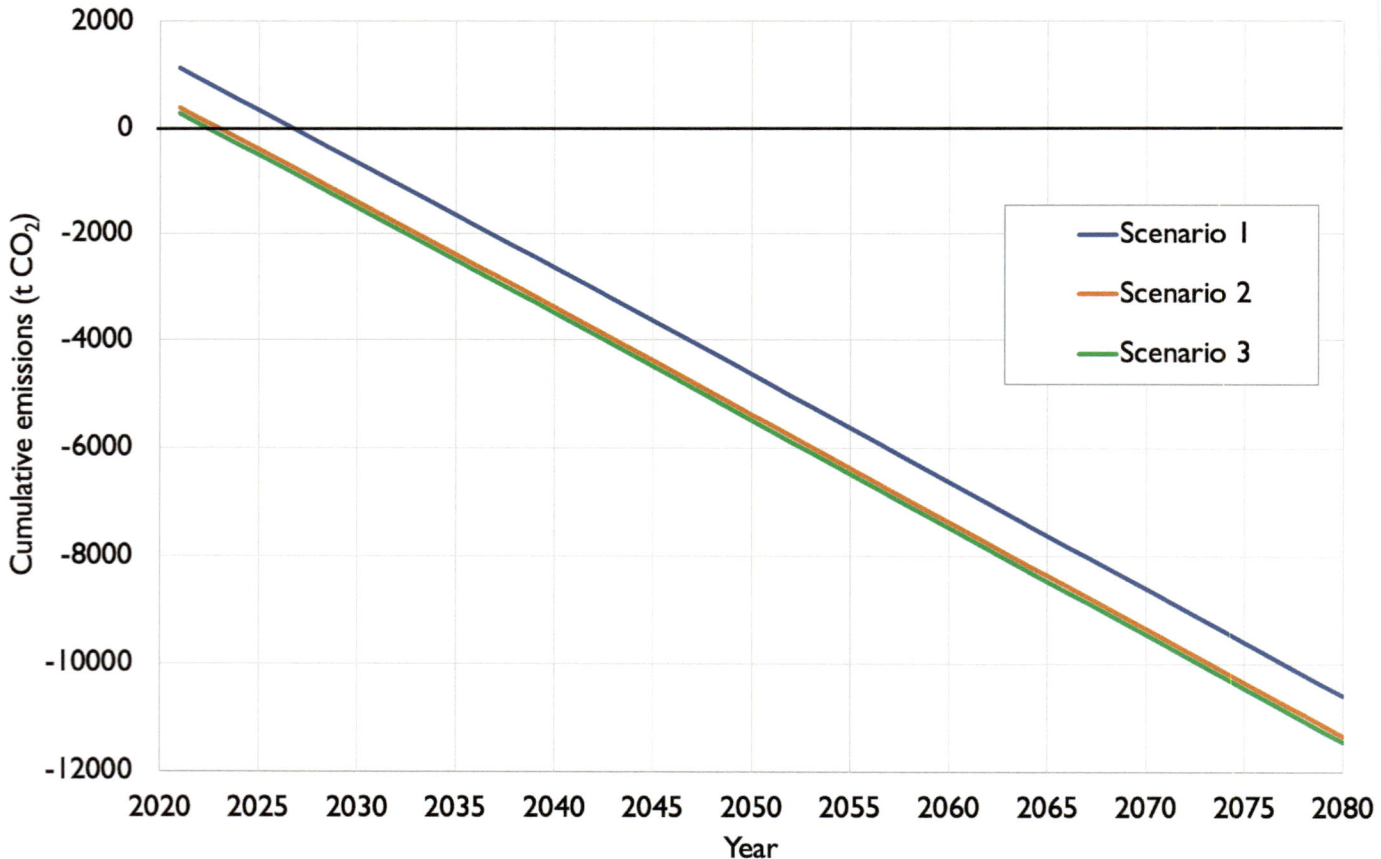

Figure 286   ASHRAE New Global Headquarters: combined embodied and operational emissions

21.3.10 Lessons learnt

ASHRAE New Global Headquarters is an exemplar building, which ASHRAE wishes to use as a learning tool for its worldwide membership and for the built environment profession in general. The savings from retrofitting an existing 1978 building into an energy-efficient building are significant. The new headquarters has a PV system that fulfils its entire annual energy consumption and provides a significant electricity production for export. When these are combined with embodied emissions, our Zero Equation shows that the building will achieve zero cumulative emissions in 1.4–5.7 years from the date of completion of retrofit, depending upon the applicable embodied emissions factors in the PV system.

ACKNOWLEDGEMENTS

The case study has been prepared with both technical information and a written consent provided by ASHRAE. Embodied emissions in building materials were derived from a simulation model provided with written consent from Integrated Environmental Solutions Ltd. All support in the preparation of this case study is gratefully acknowledged.

21.4 CASE STUDY 4: POWERHOUSE TELEMARK, NORWAY

21.4.1 Design concept

The Powerhouse Telemark building was developed as part of exploration of new approaches to sustainable construction of office buildings. The objective was to reduce annual energy consumption by 70%, whilst producing more renewable energy than consumed by the building. The building interiors are designed to enable building users to create flexible workspaces, compatible with increasing trends of hybrid onsite and remote offsite working. The building was competed in 2020 and it obtained BREEAM Excellent certification, creating an example for future sustainable workspaces. Powerhouse Telemark is a collaboration between Snøhetta, Skanska, Asplan Viak, and the client Recreate. The building is one of the series of Powerhouse projects, developed in compliance with a Powerhouse definition. The essence of this definition and the concept is that Powerhouse buildings are energy-producing/energy-positive buildings that generate more energy over a 60-year lifecycle than the total amount of energy required to produce the materials, build, operate, and dispose the building.

21.4.2 The climate

The building is in Porsgrunn, Telemark in Norway. The nearest location for the EPW weather data file is for latitude 59.90N, longitude 10.62 E, and altitude of 17 m, from WMO station number 014880. This weather data is for Fornebu near Oslo in Norway, 101 kilometres (63 miles) north-east from the site location in a straight-line distance, and therefore it is reasonably representative of the site climate. The following data (Table 104) is derived from the corresponding weather data file 'NOR_Oslo.Fornebu.014880_IWEC.epw', available from DOE (2023):

TABLE 104 POWERHOUSE TELEMARK: SUMMARY OF WEATHER CONDITIONS													
Month	1	2	3	4	5	6	7	8	9	10	11	12	Year
Maximum air temperature (°C)	5	11	14	21	23	27	28	25	20	15	9	9	28
Average air temperature (°C)	−4	−1	1	5	12	15	17	17	11	7	2	−2	7
Minimum air temperature (°C)	−14	−10	−13	−3	3	5	8	8	−2	−2	−6	−17	−17
Maximum relative humidity (%)	100	100	100	99	96	98	99	100	98	100	100	100	100
Average relative humidity (%)	94	81	69	65	61	64	69	73	71	79	88	76	74
Minimum relative humidity (%)	50	25	23	21	16	16	27	26	31	15	38	18	15
Total sunshine hours (direct solar irradiance >120 W/m²)	52	74	127	156	232	257	242	185	98	84	46	29	1,582

21.4.3 The site

The building is located at the address of Dokkvegen 11, 3920 Porsgrunn, Norway. The main facade is south-east facing, with other facades facing north-east, north-west and west (Figure 287). It is near the front of Porsgrunnselva fjord (Figure 288), within Herøya industry park, with good access to the road network.

Figure 287   Powerhouse Telemark: aerial view showing significant roof PV installation (with permission from Snøhetta)

21.4.4 The building

The building has a distinct form, with a 45° angle on the east facade, giving it an unusual appearance and attractive feature in the industrial park setting (Figure 288). It consists of 11 floors, occupying 7,917 m² of heated floor area, and it contains office and co-working spaces, meeting spaces, a staff restaurant, conference rooms and a roof terrace with a view of the fjord. Instead of the usual reception desk in office buildings, this building has a 'barception' desk, where the visitors can help themselves to refreshments when entering the building. Standardisation of interior design and materiality was at the forefront of design thinking, in order to minimise unnecessary waste when tenants change. Thus, there are standardised partitions, lighting, kitchenettes, washrooms, and flooring throughout.

21.4.5 Thermal insulation

The building is well insulated to Passivhaus standard (The Passivhaus Trust, 2022) with triple glazing throughout. Thermal properties of building components are shown in Table 105.

Measured airtightness is 0.37 1/h at 50 Pa pressure difference.

Figure 288   Powerhouse Telemark: a view of the building in its context near the front of Porsgrunnselva fjord (with permission from Snøhetta)

TABLE 105 *U*-VALUES IN POWERHOUSE TELEMARK (WITH PERMISSION FROM SNØHETTA)	
Description	**_U_-value (W/m²K)**
Windows (including frame)	0.80
External walls	0.17
Roof	0.12
Ground floor	0.11

21.4.6 Solar gain, solar shading, and natural daylight

The north-east and the north-west facades are the most exposed to low angle solar radiation and are protected with external shading made from wooden balusters (Figure 289). Solar gain is captured into thermal mass of the building, consisting of concrete slabs and it provides short term heat storage between the day and the evening.

The building has been designed to maximise the use of natural daylight and reduce the use of electrical lighting. High-level windows, sloping surfaces, light-coloured spaces, and loose furniture diffuse the natural daylight, creating a pleasant working environment (Figure 290).

21.4.7 Ventilation, heating, and cooling

Heating and cooling are provided to border zones of each floor via water loops, linked to ground source heat pumps with bore holes 350 metres deep. In the staff restaurant, ventilation ducts are integrated into seating benches, thus combining the interior

a) External view   b) Internal view

Figure 289   Powerhouse Telemark: solar shading (with permission from Snøhetta)

and the HVAC design. Elsewhere, displacement ventilation is implemented with vertical air outlets that release air at low speed and velocity at floor level.

21.4.8 Renewable energy

The building has extensive PV array on its 24° angle sloping roof (Figure 287). An additional PV system is integrated in the south-east facade (Figure 291), and another PV system is on the carport and on the bicycle parking. These PV systems produce estimated 248,326 kWh of solar electricity per year (Table 106).

21.4.9 Embodied and operational emissions and energy

The choice of material in the design and construction of the building was based on low embodied emissions. Thus, the timber used in the building was locally sourced, and other low embodied emissions materials, such as environmental concrete and gypsum were used.

At the time of design development, Snøhetta's Powerhouse Definition was based on primary energy calculations. Thus, primary energy balances over the building lifetime of 60 years are shown in Table 107.

As it can be seen from this table, the building achieves energy neutrality over its lifetime. However, what would be required for this building to achieve cumulative zero emissions by 2050?

To answer this question, we need to find embodied energy per kWp of solar PV, instead of embodied emissions per kWp used in previous analyses and case studies in this book, multiply that by the primary energy factor obtained from Snøhetta, and apply Zero Equation on the figures in Table 107. That will provide a multiplication (increase) factor for the PV system that achieves cumulative zero energy and therefore cumulative zero emissions by 2050.

Embodied energy in PV systems of 1,306 MJ/kWp (362.8kW/kWp) is obtained from the analysis by Todde et al. (2018). The Powerhouse Telemark PV system of 308 kWp, as per Table 106, will therefore have embodied energy of 308 kWp × 362.8 kW/kWp = 111,735.5 kW. Applying the primary energy factor of PEF = 2.4 to this figure consistently with the figures in Table 107 gives the equivalent primary embodied energy in the PV system of 268,165 kWh.

Figure 290   *Powerhouse Telemark: natural daylight and flexible work spaces (with permission from Snøhetta)*

Figure 291 Powerhouse Telemark: south-east facade-integrated PV system (with permission from Snøhetta)

TABLE 106 POWERHOUSE TELEMARK: PHOTOVOLTAIC SYSTEMS (WITH PERMISSION FROM SNØHETTA AND SKANSKA)			
Description	**Surface area (m²)**	**Installed capacity (kW$_p$)**	**Annual electricity production in 2020[a] (kWh/year)**
PV on the entire roof	808	179	166,544
PV in south-east facade	373	63	35,428
PV on carport	252	55	39,302
PV on bicycle parking	49	11	7,052
Total	1,482	308	248,326

[a]Annual production is estimated to increase over the building lifetime 2020–2080 to circa 280,000 kWh/year.

TABLE 107 POWERHOUSE TELEMARK: PRIMARY ENERGY BALANCES (WITH PERMISSION FROM SNØHETTA AND SKANSKA)

Row	Description	Value[a] (kWh/m²years[b])
(1)	Power generation solar cells	83.8
(2)	Bound energy[c] (embodied energy)	−44.5
(3)	Energy use in operation	−38.1
(4)	Total energy balance over lifetime	1.2

[a]Value is calculated using an average primary energy factor of 2.4 for the period 2020–2080.

[b]'years' refers to the expected building's lifetime of 60 years.

[c]Design document obtained from Snøhetta and Skanska and translated from Norwegian refers to energy embodied in materials as 'bound' energy.

Subsequently, a multiplying factor for the as built PV array that makes this project achieve zero emissions by 2050 is calculated by applying our Zero Equation, expressed below in terms of energy as Equation (73), instead of the Zero Equation expressed as Equation (4), initially developed in terms of CO_2 emissions, as follows:

$$f = \frac{CE_e + P_e \times t}{BN_e \times t - BVE_e} \tag{73}$$

where:

CE_e – Constant embodied energy, arising from materials, in this case 44.5 kWh/m²/year (Table 107, row 2) × 7,917 m² (floor area) × 30 years = 21,138,390 kWh. As it seems that this figure only included the construction materials but not the energy embodied in the PV system, the above calculated figure of 268,165 kWh for PV embodied energy is added to get a total of 21,406,555 kWh

P_e – Positive operational energy that occurs from year $t = 1$, in this case 38.1 kWh/m²/year (Table 107, row 3) × 7,917 m² (floor area) = 301,637.7 kWh/year

BN_e – Base level of negative operational energy that occurs from year $t = 1$, in this case 83.8 kWh/m²/year (Table 107, row 1) × 7,917 m² (floor area) = 663,444.6 kWh/year

BVE_e – Base level of variable embodied energy, arising for from changing the size of the PV system to target a zero emissions year, in this case 268,165 kWh as calculated above

t – Time in years, in this case 30 years from the date of construction in 2020 until 2050

f – Multiplication (increase) factor for the base level PV system.

By substituting the above values in Equation (73), we obtain:

$$f = \frac{21,406,555 + 301,637.7 \times 30}{(663,444.6 \times 30) - 268,165} = 1.55 \tag{73a}$$

Therefore, increasing the existing PV system output by a factor of 1.55 would enable the building to reach zero emissions by 2050. Upgrading 308 kW$_p$ of PV by this factor would require additional 169.4 kW$_p$ to be added to this project to achieve total zero energy and therefore total zero emissions by 2050 (Figure 292).

21.4.10 Lessons learnt

Powerhouse Telemark is a remarkable energy-producing building, which during its lifetime of 60 years achieves energy neutrality and therefore carbon emissions neutrality. The design shows how the high quality of architectural design, the quality and amenity of interior spaces, and the energy neutrality can be brought together as an exemplar for designing office buildings from

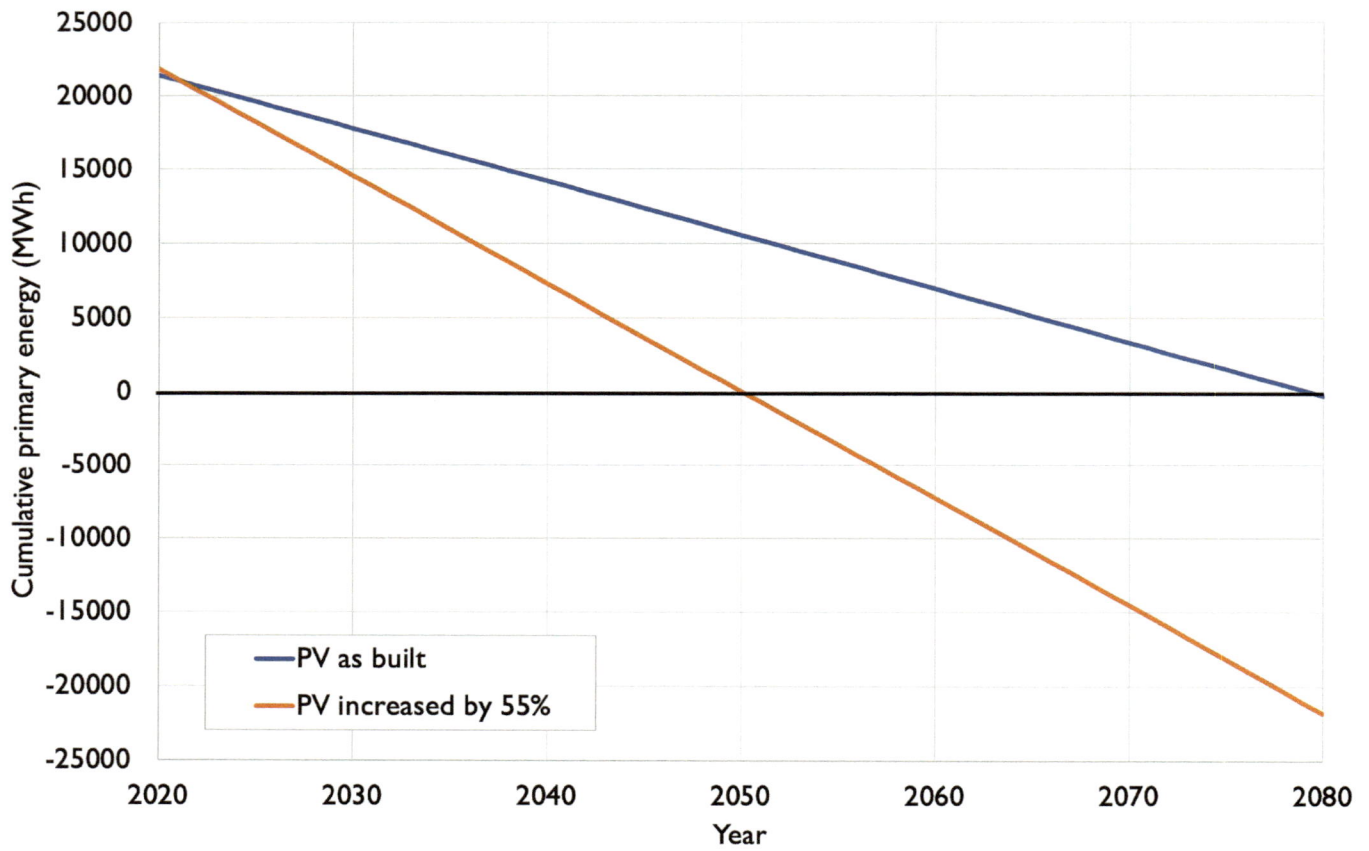

Figure 292   Powerhouse Telemark: combined embodied and operational energy

this point on and into the future. An interesting aspect of this project is the approach of expressing embodied emissions as embodied energy, or 'bound' energy, as translated from the Norwegian language. Whilst most of the information in this case study has been provided by Snøhetta and Skanska, the main contribution of this case study is in the identifying the requirements for this building to become carbon-neutral by 2050, 30 years ahead of its thermal neutrality at the end of the 60 years lifecycle. In doing so, the Zero Equation developed earlier in this book in terms of CO_2 emissions was re-expressed in terms of energy. The result of application of the Zero Equation is that a 55% of increase of the PV systems output, which could potentially be achieved by installing another PV system within a proximity of the building, would make this building energy and carbon-neutral by 2050. Such an increase would super-charge the energy-producing capacity of this excellent building.

ACKNOWLEDGEMENTS

This case study has been prepared with a written consent provided by Snøhetta. Whilst most of the information has been provided by Snøhetta, Skanska, has been very helpful in providing technical details for this case study. The support from Snøhetta and Skanska in the preparation of this case study is gratefully acknowledged.

21.5 CASE STUDY 5: THE ENTOPIA BUILDING, UNITED KINGDOM

This case study is an abbreviated digest of the extensive work on design and retrofit of the Entopia Building, transformed from offices, in a 1939-built former telephone exchange in the historic city centre of Cambridge, into a new home for the University of Cambridge Institute for Sustainability Leadership. Whilst most of the information in this case study has either been generously provided by the Institute or published elsewhere and referenced here, the main contribution of this case study is in the assessment of timing for achieving zero cumulative emissions.

21.5.1 Design concept

The aim of the Entopia Building is to demonstrate the way for the radical changes required to achieve a sustainable world. The design was developed with targets of BREEAM Outstanding (BRE Group, 2022), EnerPHit Classic (Passivhaus Institut, 2022a) and WELL Gold certification (IWBI, 2022), additionally focusing on circularity, bio-based materials and embodied emissions. This led to a transformation of a former 1939 telephone exchange building into a contemporary energy-efficient building. The retrofit strategy was based on the fabric-first approach to the EnerPHit standard (Passivhaus Institut, 2022a), prioritising energy demand reduction over using energy from sustainable sources. In order to keep the original appearance of the external facade, the building was insulated internally with a new insulation layer using bio-based materials. These materials included wood fibre insulation, lime and cork plaster and internal studwork for holding thermal insulation and represented 48% of all materials by volume. This approach was closely linked with careful planning for the whole life carbon reduction and also informed by moisture risk assessment. The interior of the building contains products designed for reuse, repair and disassembly, facilitating a circular economy approach (Architype, 2023).

21.5.2 The climate

The building is situated in Cambridge, UK. The nearest location for the EPW weather data file is for latitude 52.196 N, longitude 0.158 E, and altitude of 101 m, from WMO station number 03683. This weather data is for London Stansted Airport, approximately 36 kilometres (22.4 miles) south from the site location in a straight-line distance, and therefore it is representative of the site climate. The following data (Table 108) is derived from the corresponding weather data file 'cntr_Cambridge_TRY.epw', available from Robinson (2019):

TABLE 108 THE ENTOPIA BUILDING: SUMMARY OF WEATHER CONDITIONS													
Month	1	2	3	4	5	6	7	8	9	10	11	12	Year
Maximum air temperature (°C)	12	14	22	18	27	26	28	28	24	24	19	15	28
Average air temperature (°C)	4	4	5	9	12	14	16	17	14	11	6	5	10
Minimum air temperature (°C)	−5	−6	−4	−3	1	1	7	5	3	2	−4	−5	−6
Maximum relative humidity (%)	100	100	100	100	100	100	100	100	100	100	100	100	100
Average relative humidity (%)	86	86	78	78	76	73	81	77	80	87	82	90	81
Minimum relative humidity (%)	42	45	35	23	33	28	37	31	33	49	43	58	23
Total sunshine hours (direct solar irradiance >120 W/m²)	70	86	133	198	297	303	276	250	201	139	83	68	2,104

21.5.3 The site

The building is located in the historic centre of Cambridge at address 1 Regent Street, Cambridge, CB2 1GG, United Kingdom (Figure 293). Its footprint is approximately T-shaped with the main facade facing south-west. The rooftop terrace provides a suitable space for a PV system, which also serves as a canopy for a seating area.

21.5.4 The building

The Entopia Building is a six-storey (partial basement plus ground and four floors) building in a conservation area of the historic city of Cambridge. The building is home to a new international headquarters of the University of Cambridge Institute for Sustainability Leadership. Its purpose is to provide office space as well as incubator space for start-ups and facilitate collaboration activities. The structure is a concrete frame, except for extension to east wing, which is steel frame with hollow core concrete planks. The original building had uninsulated walls and single glazed sash windows (Figure 294a). The retrofitting included adding internal insulation in order to preserve the facade appearance, and a replacement of the original windows with triple glazing (Figure 294b). As result of the retrofit, the building is insulated to EnerPHit standard (Passivhaus Institut, 2022a), and it now requires 15% of the energy requirement of the original building (Passivhaus Trust, 2022).

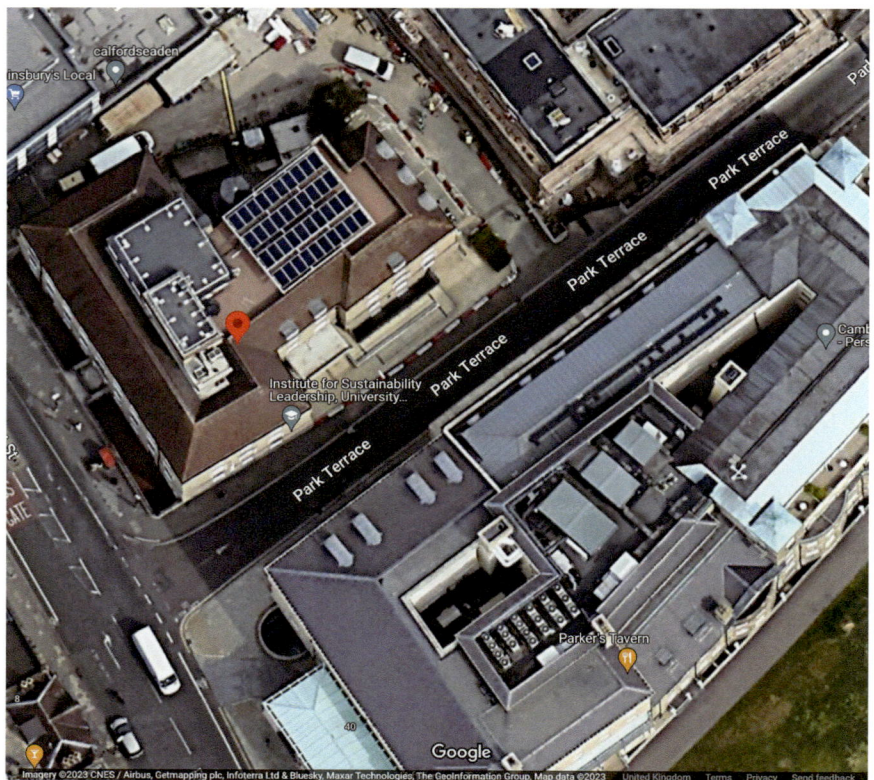

Figure 293   *The Entopia Building site*

a) Before the retrofit

b) After the retrofit

Figure 294   *The Entopia Building*

Source: Cronshaw (2022)

21.5.5 Thermal insulation

In order to preserve the facade, internal insulation was added to the existing masonry walls as part of the retrofit. The internal insulation consists of lime and cork plaster, acting also as the airtightness line, with wood fibre insulation and independent studwork with a plasterboard finish. In some instances, a liquid applied membrane was also used for airtightness line. The original sash windows were replaced by new triple-glazed windows (Figure 294b), placed as insets into the new interior wall. This resulted in a 35% reduction of heat demand (Blaylock et al., 2022). A summary of U-values is shown in Table 109. Airtightness was 1.33 m³/h/m² @ 50 Pa and 0.605 ach @ 50 Pa.

TABLE 109 THE ENTOPIA BUILDING U-VALUES (COURTESY OF ARCHITYPE)	
Description	U-value (W/m²K)
Windows	0.82
External walls to ambient	0.32
External walls in contact with ground (excluding ground effect)[a]	0.27
External walls in contact with ground (including ground effect)[a]	0.12
Roof	0.21
Ground floor (excluding ground effect)[a]	0.39
Ground floor (including ground effect)[a]	0.17

[a]Derived from PHPP (Passivhaus Institut, 2022b).

21.5.6 Natural daylight and solar gain

Natural daylight (Figure 295) has been improved by changing the original sash windows and removing the original transoms and mullions. This resulted in a 60% increase in the glazed area (Blaylock et al., 2022), and a 77% increase of natural daylight (Cronshaw, 2022). The solar gain received through the increased glazing surface (Figure 295) is absorbed in the building thermal mass, and either used to offset heating demand during the heating season or is purged with night-time cooling using external air during summer months.

Figure 295   The Entopia Building interior view (photo by Jack Hobhouse/Architype, with permission from Architype)

21.5.7 Electrical lighting

The LED lighting system contains 350 lights reused from a refurbishment project in London. These lights were retested and the original warranty honoured by the supplier. As the fittings for these lights did not fully match the intended exposed ceiling installation, additional end plates were added to overcome this issue and to provide up-lights too.

21.5.8 Heating, cooling and mechanical ventilation

Space heating is provided by an air source heat pump and distributed through a mechanical ventilation system throughout the building. Additional electric radiators are provided locally for topping up the heating in areas of higher heat demand. The concrete frame building structure contains a significant amount of thermal mass, from which heat build-up is removed by night-time cooling with external air, operated by an air handling unit in summer months. Cooling for most of the building is a combination of natural ventilation, mechanical ventilation with summer bypass and night purge, and peak looping by the reversible heat pump in the central air handling unit. Areas with high heat gain, such as meeting rooms, are equipped with direct expansion fan coil units, operated above 28°C in summer. The fan coils also providing space heating in winter. As the building complies with EnerPHit airtightness criteria of $n_{50} \leq 1.0$ h^{-1}, a high-efficiency MVHR system is installed to maintain the required levels of external air supply and to minimise the use of heating whilst maintaining thermal comfort. Heating energy has been completely transitioned to electricity, removing the need for and the connection to gas supply.

21.5.9 Renewable energy

A solar PV system is installed as a canopy on the rooftop terrace. In addition to providing 5.4 kWh/m^2/year of the gross internal area (5.4 kWh/m^2/year × 2,939 m^2 = 15,870.6 kWh/year), the PV canopy provides a pleasant seating area on the rooftop terrace (Figure 296).

Figure 296   The Entopia Building rooftop terrace PV canopy (photo by Jack Hobhouse/Architype, with permission from Architype)

21.5.10 Embodied and operational emissions

In this section, we calculate the time when the building will achieve zero cumulative emissions, using combined embodied and operational emissions obtained from the project documentation. As a first step, building performance figures per square metre of the floor area (Blaylock et al., 2022; Holbrook, 2022) were multiplied by total the floor area to obtain absolute CO_2 emissions figures. The results of the calculations are shown in Table 110, including 382,070 kgCO$_2$e of upfront carbon and 26,451 kgCO$_2$e/year of operational emissions.

TABLE 110 THE ENTOPIA BUILDING EMISSION CALCULATIONS

Description	Gross internal area[a] (m²)	Embodied emissions[c] (kgCO₂e/m²)	Operational emissions[a] (kgCO₂e/m²/year)	Embodied emissions (kgCO₂e)	Operational emissions (kgCO₂e/year)	Emissions prior to retrofit[b] (kgCO₂e/year)
Column	(1)	(2)	(3)	(4) = (1) × (2)	(5) = (1) × (3)	(6) = −(5) × 100/15
Value	2,939	130	9	382,070	26,451	−176,340

[a]Source: Holbrook (2022).

[b]Source: Blaylock et al. (2022) and communication with Architype.

[c]Derived from Passivhaus Trust (2022).

As result of the retrofit, this building requires 15% of the energy requirement of the original building (Passivhaus Trust, 2022). Assuming that CO_2 emissions are proportional to energy requirements, what were the 100% of CO_2 emissions before the retrofit, if 15% of these emissions after the retrofit are 26,451 $kgCO_2$e/year, as shown in Table 110 column (5)? To answer this, we write the following:

$$26,451\,kgCO_2e / year \rightarrow 15\%$$

$$\uparrow X\,kgCO_2e / year \rightarrow 100\% \uparrow \qquad (74)$$

From the above, using the vertical arrows to indicate the direction of divisions, we write the following proportion:

$$\frac{X}{26,451} = \frac{100}{15} \qquad (75)$$

and we find the value that corresponds to 100% emissions as follows:

$$X = 26,451 \times \frac{100}{15} = 176,340\,kgCO_2e / year \qquad (76)$$

This is shown with a negative sign in Table 110 column (6), representing replacement of emissions prior to retrofit.

In the next step, the time when the building will reach zero cumulative emissions is calculated using the Zero Equation in its formulation as Equation (2), rewritten below from Chapter 1 as follows:

$$t = \frac{E_{CO2}}{N_{CO2} - P_{CO2}} \qquad (2)$$

where

E_{CO2} – Embodied emissions in materials, in this case 382,070 $kgCO_2$ (Table 110, column (4))
P_{CO2} – Positive operational emissions that occur from year $t = 1$ in this case 26,451 $kgCO_2$/year (Table 110, column (5))
N_{CO2} – Negative operational emissions arising from replacement of emissions prior to retrofit from year $t = 1$, in this case −176,340 $kgCO_2$/year (Table 110, column (6))
t – time in years

Substituting the above values in the Zero Equation (2), we obtain the time in years when the building will achieve zero combined emissions as follows:

$$t = \frac{382,070}{176,340 - 26,451} = 2.5\ \text{years} \qquad (2a)$$

As it can be seen from these calculations, the significant savings from retrofitting the 1939 building, combined with the use of bio-sourced materials and low operational emissions, result in cumulative embodied and operational emissions reaching zero in 2024, 2.5 years after the completion of retrofit. A chart of combined embodied and operational emissions is shown in Figure 297.

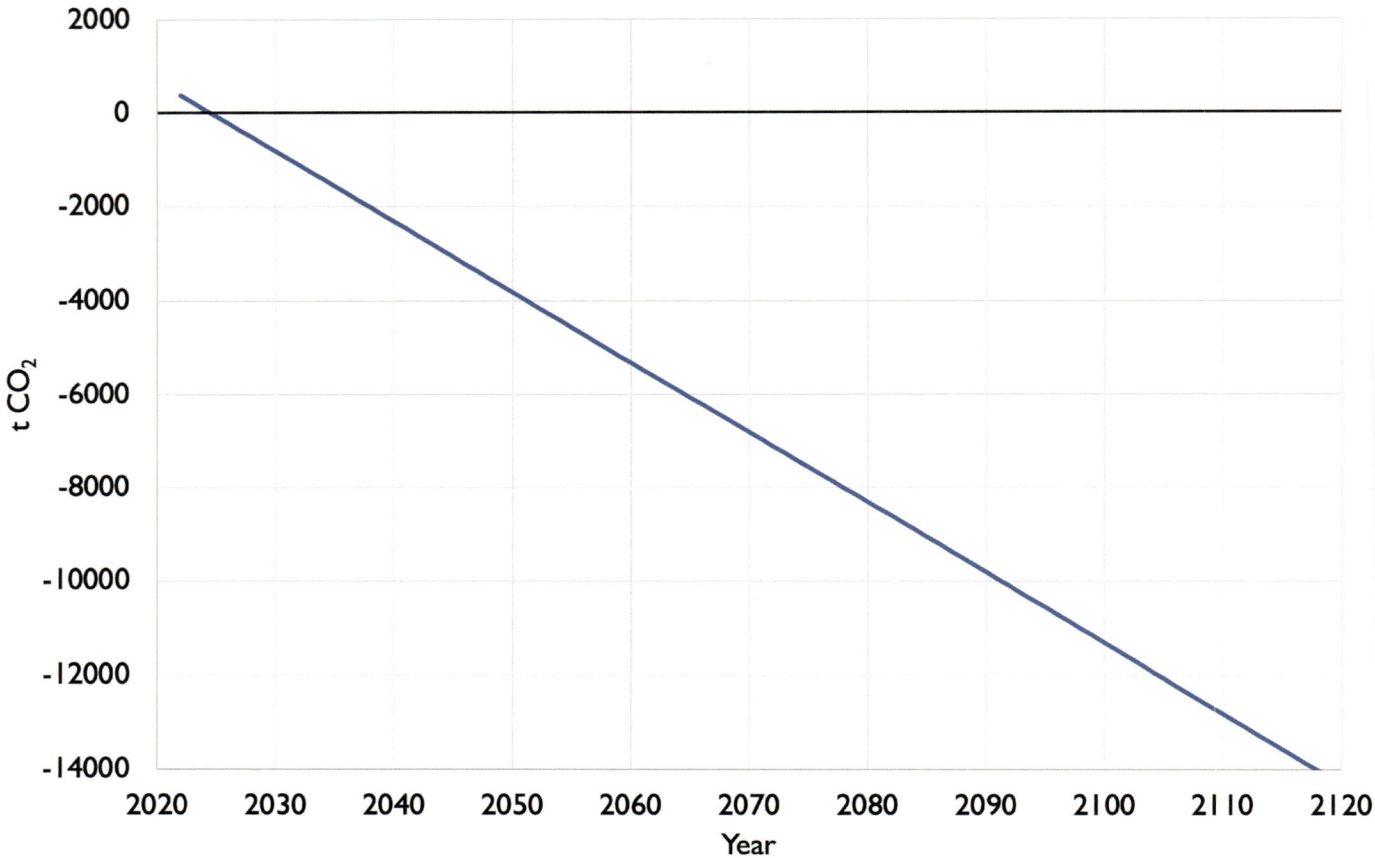

Figure 297   The Entopia Building combined embodied and operational emissions

21.5.11 Lessons learnt

Several causal relationships became apparent during the preparation of this case study. First, the fabric-first approach not only created better internal conditions for the occupants, but it also resulted in a lower specification and lower capital cost of the mechanical systems, and consequent lower embodied emissions. Second, the extensive use of bio-based materials for thermal insulation has lowered the overall embodied emissions to a level from where they can be effectively reduced to zero, in combination with operational emissions savings. Third, the size of the PV system, although moderate comparing with the building size, provides a good fit for the overall reduction of carbon emissions, so that zero cumulative emissions will be reached 2.5 years after the completion of retrofit. Fourth, the replacement of the original building and its original emissions leads to a rapid reduction of the cumulative emissions, embodied and operational, down to zero, achieved in 2.5 years after the completion of retrofit. Fifth, due to the completion of retrofit being close to the time of writing this text, information available for this analysis was based on design data rather than on in-use data. However, the main message coming from this building is loud and clear: Entopia is an exceptional building that provides an exemplar pathway to a sustainable world.

ACKNOWLEDGEMENTS

The case study has been prepared with a written consent provided by the University of Cambridge Institute for Sustainability Leadership. Whilst most of the information has been provided by the University of Cambridge Institute for Sustainability Leadership, Architype, and Max Fordham have been very helpful in providing technical details and comments for this case study. The support from the University of Cambridge Institute for Sustainability Leadership, Architype, and Max Fordham in the preparation of this case study is gratefully acknowledged.

21.6 MAIN FINDINGS FROM CASE STUDIES

The main findings from the case studies are

- retrofit projects benefit from reduced embodied emissions as materials retained from the building prior to retrofit do not constitute new emissions;
- retrofit projects benefit from reduced operational emissions through replacement of emissions from the building prior to retrofit;
- bio-sourced materials, such as timber, cork, and wood fibre, reduce embodied emissions through their carbon storage;
- the Zero Equation, in addition to its use in the context of carbon emissions, can also be applied in the context of energy.

21.7 SUMMARY OF DESIGN PRINCIPLES

Zero carbon retrofit:

- Utilise as much of the existing building as possible.
- Utilise recycled or reclaimed materials or components.
- Utilise site material from ground works.
- Remove any render or thermal insulation where appropriate to reveal internal thermal mass.
- Adjust retrofit solutions to the overall context (i.e., external insulation on the street side may not be possible without a radical change of the building character).
- Upgrade thermal insulation, installing it externally wherever it is possible without changing the overall character of the building.
- Use triple glazing with inert gas filling to minimise heat losses.
- Increase airtightness using vapour-permeable and moisture-sensitive membranes.
- Increase thermal mass by adding high density materials to the interior (i.e. rammed earth floors).
- Implement passive solar systems (i.e., direct gain).
- Design for natural ventilation using high- and low-level openings.
- Implement mechanical ventilation with high-efficiency cross flow heat recovery.
- Keep existing shallow plan if applicable.
- Insert light wells/passive stacks if the original building is deep plan.
- Implement mirrors to scoop daylight and direct it into internal spaces.
- Use high-efficiency dimmable lighting with photo-electric response.
- Design renewable energy systems as part of the overall building design.
- Integrate the above features using a dynamic simulation model.
- Evaluate thermal comfort.
- Calculate life cycle benefit.
- Calculate the time when zero cumulative emissions will be achieved, taking into account embodied and operational emissions.
- Calculate additional renewable energy requirements if zero cumulative emissions are not achieved by 2050.

Zero carbon design:

- Use shelter planting on the north side and deciduous trees on the south-facing side.
- Use green plants on site to reduce heat island effect.
- Design the building form in response to the prevailing wind to enhance air circulation around the building.
- Design the building form to follow the sun path.
- Use compact plan to reduce heat exchange with the environment.
- Use shallow plan to enhance natural daylight and natural ventilation.
- Design thermal zones perpendicular to the north-south axis.
- Design the building layout to give occupants opportunities for internal migration in search of more comfortable conditions.

- Use larger windows on the south side and smaller windows on the north side for predominantly heated buildings, and vice versa for predominantly cooled buildings (applies to the northern hemisphere and vice versa for the southern hemisphere).
- Use higher levels of thermal insulation in cool climates and lower levels in warm climates.
- Design a standby building operation in the 'absence mode', putting the measures in place for prevention of overheating or freezing.
- Use high thermal mass to smooth out fluctuations of internal temperatures.
- Use natural ventilation and night-time free cooling to precondition the building for the following warm day.
- Use correctly designed solar shading to prevent excessive solar gain in summer and admit solar radiation in winter.
- Use reflective coating on the outside of the building to reduce solar absorption.
- Design natural daylight and integrate electrical lighting with natural daylight using daylight sensitive controls.
- Design the whole building to be a passive stack chimney.
- Design an automated shading control that follows daily changes in solar radiation on different sides of the building.
- Use solar panels as shading devices.
- Insert light wells/passive stacks if the building is deep plan.
- Implement mirrors to scoop daylight and direct it into internal spaces.
- Use high-efficiency dimmable lighting with photo-electric response.
- Design renewable energy systems as part of the overall building design.
- Use battery storage for PV-generated electricity.
- Use reclaimed materials, such as cooking oil, for fuel.
- Combine renewable systems in innovative ways (heat pumps and MVHR, etc.).
- Use energy stored in the ground for heating and cooling via earth tubes.
- Use advanced materials, such as PCM.
- Integrate solar systems into building components (PV in transparent roofs, PV as shading devices, etc.).
- Integrate MVHR into design.
- Use dynamic simulation models for design development and integration.
- Ensure that the tradesmen working on site understand the relationship between the quality of work and building performance.
- Carry out continuous monitoring and post-occupancy evaluation.
- Design for gradual reduction of cumulative emissions, consisting of embodied and operational emissions, to reach zero by a target year using the Zero Equation.
- Calculate additional renewable energy requirements if zero cumulative emissions are not achieved by the target year.

CHAPTER 22

WHERE THE FIELD IS GOING – RESEARCH AND POLICY DEVELOPMENT SUPPORT

22.1 RESEARCH

In this section, I present material from my personal research. This is not to say that all of the research in this field hinges on my personal research, but it is to give the reader an insight into the direction of some research topics relevant to the subject of this book. The work discussed here contains extracts from my relevant publications, which will be referenced accordingly.

22.1.1 Using reverse modelling to reduce simulation performance gap

Whilst working on design simulations for a building to be constructed from hemp-lime bio-composite material (hempcrete) I came across data from monitoring of a hempcrete test cell at the University of Bath. This was useful to have at the start of design work with this new material, and I created a simulation model of the same test cell to see how simulated and measured values corresponded. To my disappointment, I found considerable discrepancies between monitored and simulated values, to the extent that the results of simulation were completely unusable, as shown in Figure 298. In this figure, 'Ext_Temp' represents monitored external air temperature; 'Ext_RH' represents monitored external relative humidity; 'Int_Tem' represents monitored internal air temperature; 'Int_RH' represents monitored internal relative humidity; 'IES Int_Tem' represents simulated internal temperature using IES Virtual Environment; and 'IES Int_RH' represents simulated internal relative humidity using IES Virtual Environment. As can be seen from this figure, both monitored internal temperature and relative humidity are stable, whilst their simulated equivalents show significant fluctuations. How can this information be used to help future designs?

This problem made me dig deep into my previous research work, going back to the time when I did my doctoral research at the University of Birmingham. A part of my research at that time involved the creation of building simulation models from monitored data using so-called 'Fourier series'. I called these models 'reverse models'.

All models discussed in the book so far can be categorised as forward models. They are assembled on the basis of fundamental principles of heat transfer, using information about building geometry, material properties, and assumptions about occupancy, heat gains, air tightness, etc. Governed by a numerical solver and driven by weather data, the model produces outputs, such as internal room temperatures, energy consumption, and others.

Reverse modelling is a process of creating a dynamic model of a building from a relationship between measured inputs and outputs. For instance, if driving functions such as solar radiation, external and internal air temperature are known from monitoring, then creating a functional relationship between these inputs and outputs constitutes reverse modelling.

Using appropriate tools, reverse models are simpler to build and are far more accurate than the forward models. One of these tools that I worked with at the time was Fourier series. A useful fact that I learnt at the time was that a Fourier transform of a

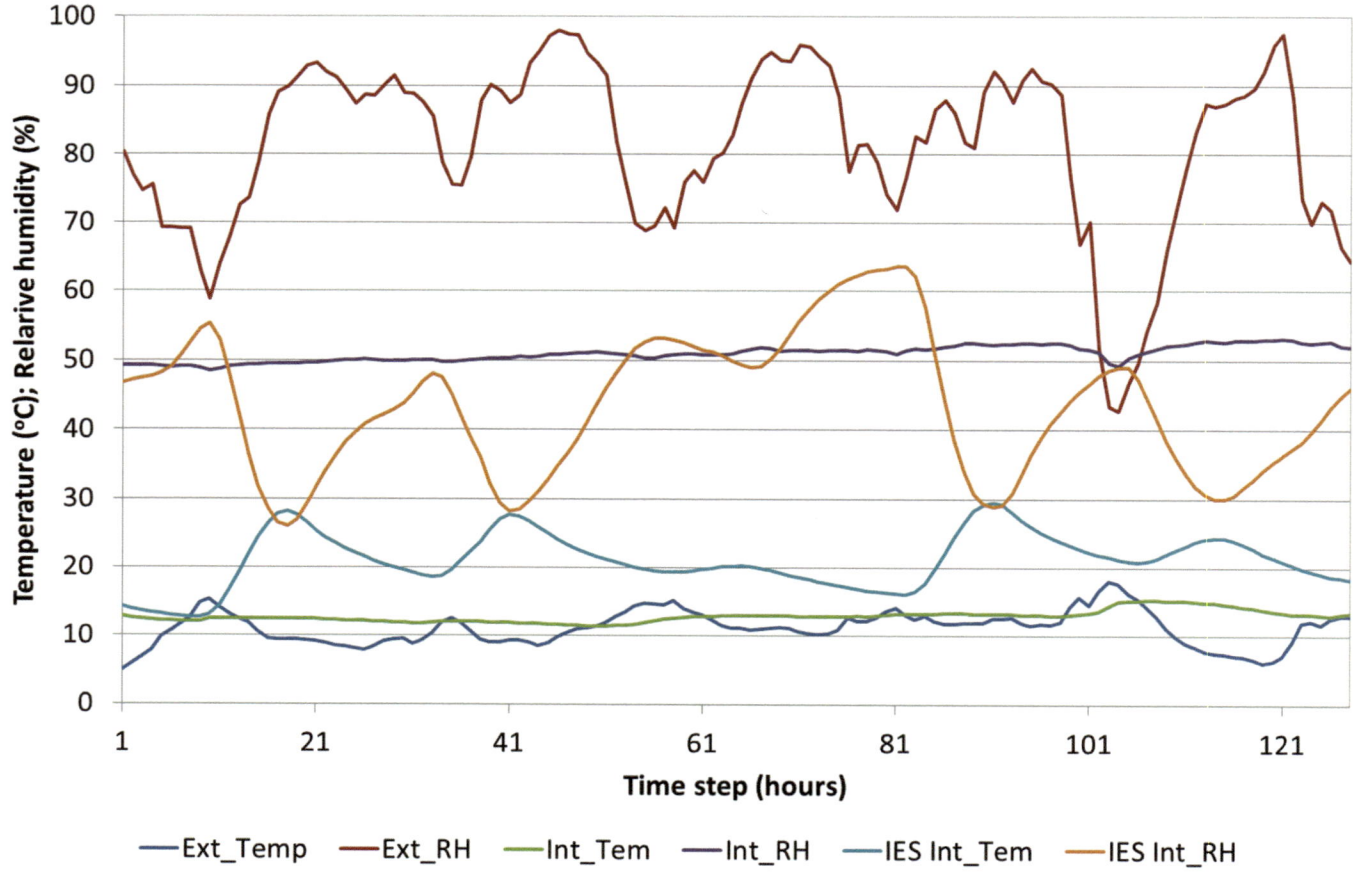

Figure 298 *Measured and simulated conditions in experimental building constructed from hemp-lime bio-composite (from Jankovic (2016))*

series of values, such as temperature readings in a building over a period of time, can be represented as a formula with an exceptionally high accuracy. The same applies to external conditions, for instance, solar radiation. Taking such two formulae together and manipulating them analogously to simple arithmetic operations of division and multiplication, could help to establish the relationship between these two time series.

This is how the solution to the problem of discrepancy between monitored and simulated values in Figure 298 started to emerge. If we can say that two buildings, one already built and one to be built from the same material and having similar geometry are generally similar, then we can write the following expression:

$$\frac{\text{Monitored data}_{\text{existing building}}}{\text{Simulated data}_{\text{exisiting building}}} = \frac{\text{Monitored data}_{\text{non-existing building}}}{\text{Simulated data}_{\text{non-existing building}}} \qquad (77)$$

After some 'obvious transformations', as my maths teacher used to say, we can re-write the above as:

$$\text{Simulated data}_{\text{non-existing building}} \times \frac{\text{Monitored data}_{\text{existing building}}}{\text{Simulated data}_{\text{exisiting building}}} = \text{Monitored data}_{\text{non-existing building}} \qquad (78)$$

It follows that creating Fourier series of the entities on the left from the equals sign in Equation (78) will enable the calculation of the values to the right of the equals sign, namely data equivalent to monitoring a building that does not yet exist! Here is how this is done.

22.1.1.1 Fourier transform

In his publication 'The Analytical Theory of Heat', originally published in 1878, Joseph Fourier (Fourier, 2009) demonstrated that every periodic function can be represented with a combination of a series of sine and cosine waves with varying amplitudes and frequencies using a discrete Fourier transform (DFT):

$$f(x) = \frac{a_0}{2} + \sum_{k=1}^{n}(a_k \sin(kx) + b_k \cos(kx)) \tag{79}$$

where

$f(x)$ – periodic function of x
$a_{0,1,\ldots,k}$ – weighting factors
n – number of harmonics.

Diurnal cycles of solar radiation and external air temperatures are periodic functions of time that drive building energy flows, and thus building response to these driving functions is also periodic. Therefore, both the driving functions and building response can be expressed with the Fourier series.

The above representation of the Fourier series can be rearranged in the form of:

$$f(x) = c_{av} + \sum_{k=1}^{n} c_{max,\,k} \cos\left(\frac{2\pi k}{n}x - \varphi_k\right) \tag{80}$$

where

c_{av} – the average level of periodic fluctuation
$c_{max,\,k}$ – maximum amplitude of the kth harmonic
n – number of harmonics
φ – phase angle of the kth harmonic

By calculating and inserting building time lag and decrement factor into the above equation, a simplified building simulation model can be obtained as follows:

$$f(x) = c_{av} + \sum_{k=1}^{n} d_{b,\,k}\ c_{max,\,k} \cos\left(\frac{2\pi k}{n}x - \varphi_k - \varphi_{b,k}\right) \tag{81}$$

where

$d_{b,k}$ = building decrement factor for the kth harmonic
$\varphi_{b,k}$ = building time lag for the kth harmonic

However, this manual approach is limited to only several harmonics and will lead to significant inaccuracies. Adding more harmonics increases the accuracy of resultant models. However, automating this approach to obtain more harmonics in the DFT is computationally intensive.

A breakthrough in the application of the Fourier transform was introduced through Danielson-Lanczos Lemma in the form of a Fast Fourier Transform or FFT (Weisstein, 2023). Whereas the computational intensity of the DFT is proportional to $O(N^2)$, the computational intensity of the FFT is proportional to $O(N\log_2 N)$. In practice, this respectively means a difference between weeks and seconds of CPU time (Press, 2007).

22.1.1.2 Creating and evaluating a Fourier transform-based reverse model

As explained earlier, a reverse model involves the creation of a relationship between driving functions acting on a building and corresponding consequences – building performance outputs.

In order to create a reverse model using Fourier transforms, we will first create a Fourier transform of an individual driving function, say of a combination of solar radiation and external air temperature added together, and a Fourier transform of a consequence of this driving function, say free-running internal air temperature in the building. Dividing the latter by the former creates a reverse model of the building in the form of a Fourier filter.

There are a few details that need to be borne in mind in the above process. The original time series of the combination of hourly or sub-hourly values of solar radiation and outside air temperature is a function of time. Creating a Fourier transform of this series of time-dependent numbers converts them into frequency-dependent numbers – numbers that have the form of $x + iy$, where 'i' is the imaginary unit defined as $i^2 = -1$. Effectively, a Fourier transform converts a function from a time domain to a frequency domain. Division of functions is also referred to as 'deconvolution' of two functions. Deconvolution in the case of two Fourier transforms involves division of complex numbers.

Having obtained a reverse model of a building in the form of a Fourier filter, we can use it to simulate that building. When the Fourier filter, which is effectively a response function of the building to external inputs, is multiplied by a Fourier transform of a combination of solar radiation and external air temperature added together, the result is a Fourier transform of the free-running internal air temperature in the building. Using the process of an inverse Fast Fourier Transform, or iFFT, the internal temperature is transformed into a time domain. This process of multiplying the Fourier filter by another Fourier transform (effectively multiplying two Fourier transforms) and obtaining an inverse Fourier transform is called 'convolution' of two functions.

The use of the FFT for reverse modelling is justified on the basis of the accuracy of this method. Whereas the DFT is limited to only several harmonics for practical computational reasons, the FFT has as many harmonics as the number of points in the source time series, having a direct impact on the accuracy of the reverse model.

I will demonstrate this by creating a reverse model of EnergyPlus model outputs and comparing the accuracy of the reverse model and the original model. Firstly, an EnergyPlus model of a simple box was created, and its inputs and outputs were obtained in the form of time-dependent series of numbers, representing solar radiation, external air temperature, and internal air temperature over the entire year in ten-minute intervals. Therefore, there are six values in each hour, and the total number of data points per variable is 8,760 hours × 6 points per hour = 52,560 points. One of the limitations of the FFT is that the number of points in the data series has to be a power of two. As 52,560 is not a power of two, the time series has to be padded to the next nearest power of two, namely to 65,536 data points. This is achieved by simply replicating data points from timestep one into timestep 52,561, timestep two into timestep 52,562, and so on, until 65,536 is reached.

Subsequently, a reverse model was built by creating a deconvolution of the FFT of the free-running internal temperature and the FFT of a combination of ambient temperature plus direct normal solar radiation. The free-running temperature was then re-built as a convolution of the Fourier filter and the FFT of the combination of ambient temperature plus direct normal radiation. A root mean square error (RMSE) was calculated between the original EnergyPlus free-running internal temperature and reverse-modelled free-running internal temperature, resulting in RMSE = 0.000000003°C. This means that the reverse model is almost identical to the original over 52,562 data points, representing the whole year of hourly data in ten-minute intervals. This demonstrates exceptional accuracy of the reverse model based on Fourier transforms and justifies the use of this method for reverse modelling.

Traditionally, a combination of solar radiation and external air temperature was used to determine a combined effect of solar radiation and external air temperature on a building. The calculation of this so-called sol-air temperature includes the outside surface conductance and the solar radiation absorption coefficient and contains a degree of uncertainty in the form of a 'remainder' that covers radiation temperature exchange from the wall surface (Jones, 2020). It is therefore obvious that there are limitations and uncertainties associated with sol-air temperature. The limitations are due to the outside surface conductance and solar radiation absorption being theoretical values that can potentially be inaccurate, with the additional limitation of dealing with heat transfer through solid walls only. The uncertainties are due to the 'remainder' that is ignored in the sol-air temperature calculation, as well as due to the potential inaccuracy in the values of the outside surface conductance and the solar radiation absorption coefficient. Additionally, sol-air temperature only applies to heat transfer through solid walls and not through glazing.

The approach of a simple addition of solar radiation and external air temperature used here overcomes the limitations and uncertainties of the sol-air temperature. As the FFT will resonate with different frequencies of the two driving functions, and will effectively separate their influences, this makes it possible to combine external air temperature and solar radiation through a simple addition in reverse modelling as described here.

22.1.1.3 Method for reducing simulation performance gap

The essence of the method is to investigate whether a reverse model of a relationship between monitored and simulated temperatures in a test building (building A) could be used as a filter to morph simulated temperatures of another building built from the same material into monitored temperatures of that building (building B). In this particular case, both A and B were existing buildings which were also monitored. If the method proves to be accurate within certain limitations as to the type and size of the building, then an interesting possibility would arise if building B does not exist but is on a 'drawing board'. In that case, the result of the method would be a representation of monitored temperatures of a building that does not exist.

Figure 298 shows significant discrepancies between monitored and simulated results from a test building named Hempod at the University of Bath (building A). In the first step, FFT transforms of monitored and simulated temperatures and relative humidity of that building were obtained, and a deconvolution between the two sets of time series constituted the reverse models of temperature and relative humidity in the form of two respective Fourier filters. This initial analysis was based on 128 data points selected when building A was in free-running mode and weather conditions were stable.

The Fourier filters developed in this way were subsequently applied to an existing terraced house (building B) built from hemp-lime composite material (Figure 299).

Figure 299   A two-bedroom house used in this analysis, built from hemp-lime composite material (from Jankovic (2016))

House B was under continuous monitoring of temperature and relative humidity over a period of two years. Airtightness tests and a co-heating test were carried out during the monitoring period, in order to obtain accurate performance characteristics for the simulation model developed in IES VE. It was fortunate for the development of this method that house B was unoccupied during the monitoring period. Thus the house received only minimum heat input.

In preparation for the simulation of house B, an EPW weather data file for the particular location was obtained and partially modified. Technical specification of EPW weather files (DOE, 2023) was consulted in this process in order to determine positions of columns corresponding to specific weather data parameters. Dry bulb temperature, relative humidity, and global horizontal radiation were used directly from monitored data and inserted into the EPW file, replacing the existing data. Dew point temperature was derived from monitored data using published calculations (Vaisala, 2022). Direct normal radiation and diffuse horizontal radiation were calculated using monitored data and formulae published by Duffie and Beckman (Duffie & Beckman, 2013). All calculated values were then inserted into relevant columns in the EPW file, replacing the existing values.

Dynamic simulation with the IES VE model was subsequently carried out and the Fourier transforms of simulated internal air temperature and relative humidity were obtained. The two Fourier filters obtained from reverse modelling of building A were subsequently applied to these Fourier transforms of IES simulation outputs. The results are shown in Figures 300 and 301, where vertical green and red lines denote the start and end of the period without heat input in house B. As building A was in free-running mode, the period between the green and red vertical lines was used for comparison of accuracy of the results.

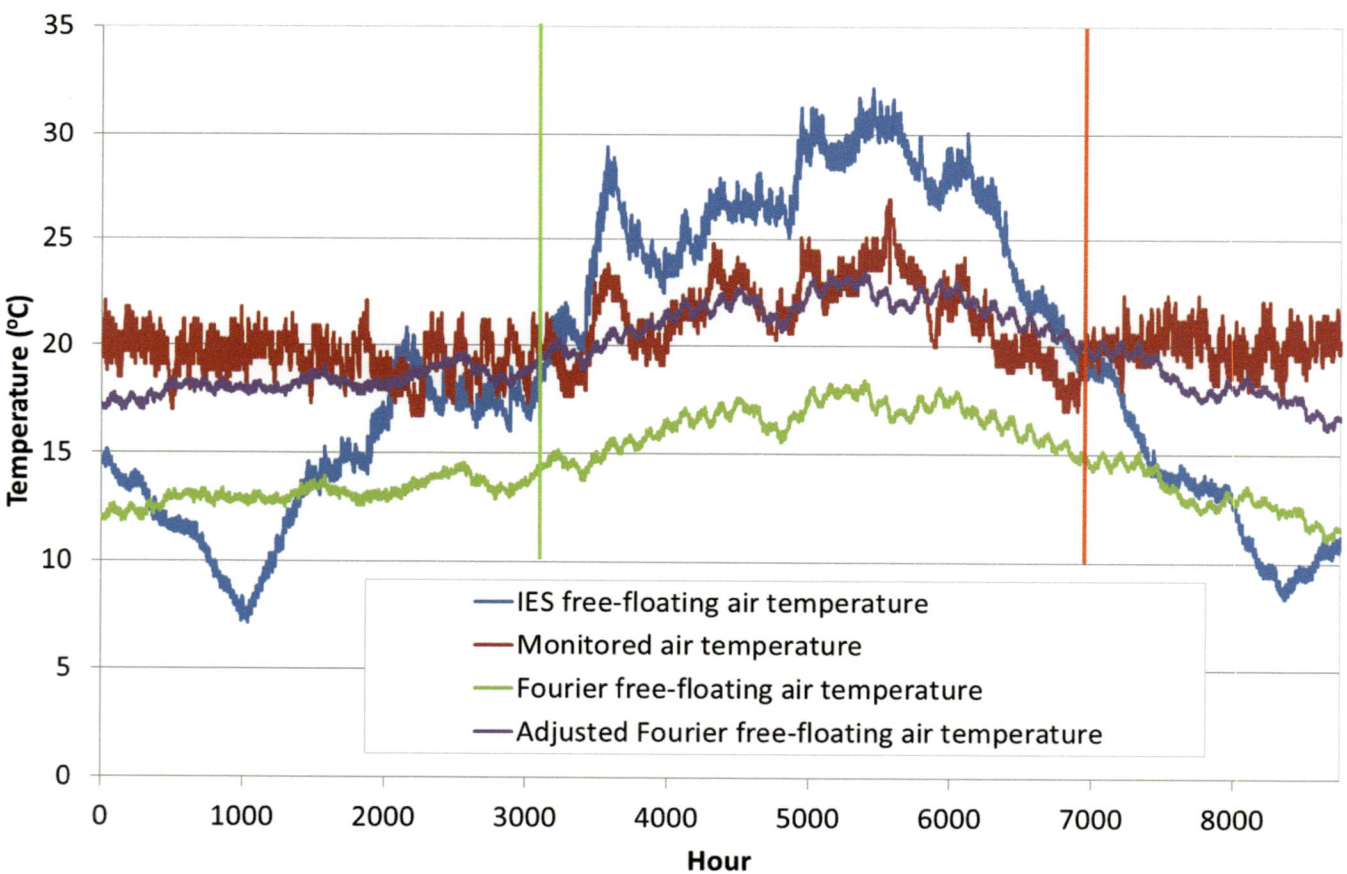

Figure 300   Comparison between IES temperature, Fourier-filtered temperature, and monitored air temperature of the living room (from Jankovic (2016))

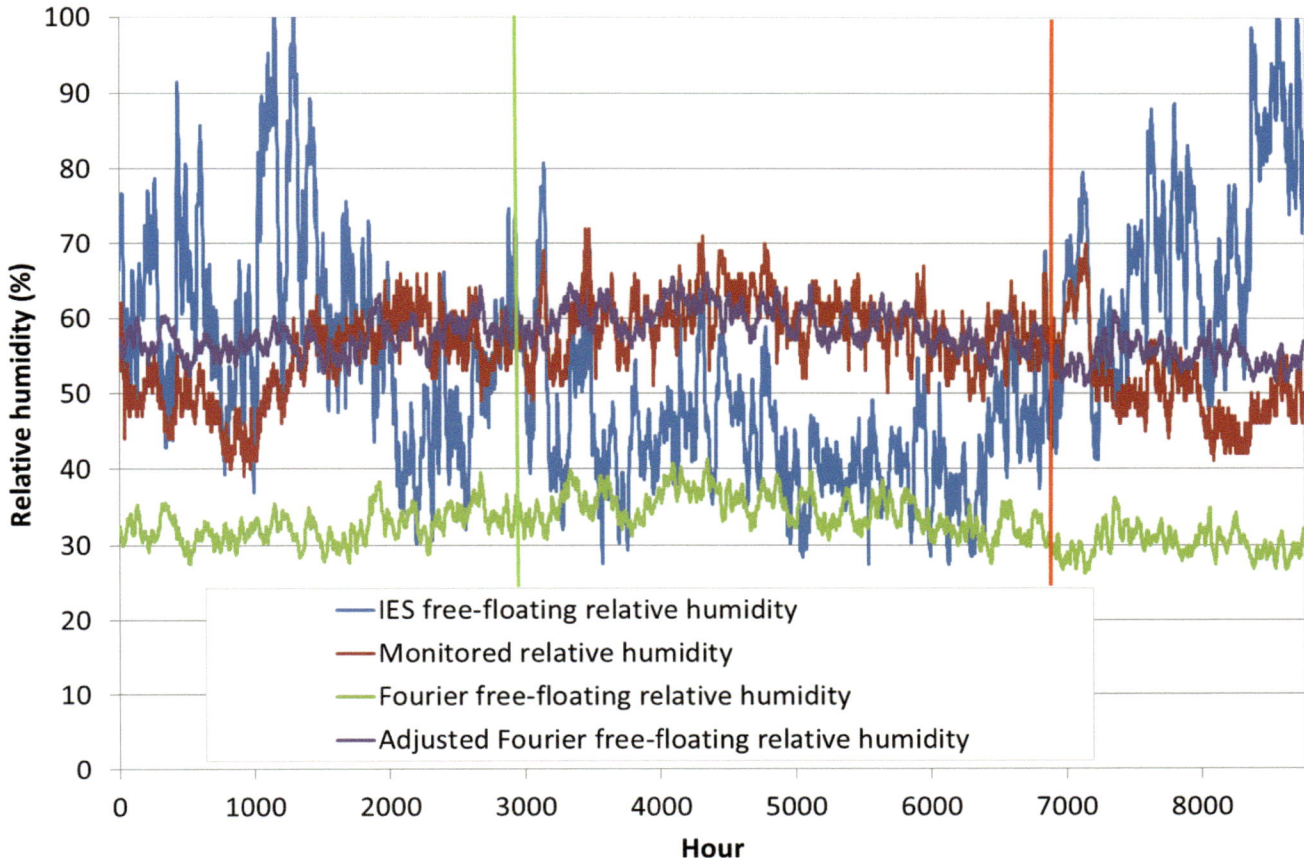

Figure 301   Comparison between IES relative humidity, Fourier-filtered relative humidity, and monitored relative humidity of the living room (from Jankovic (2016))

RMSE as defined in Equation (82) was used to assess the accuracy of results:

$$RMSE = \sqrt{\frac{\sum_{I}^{N}(simulated\text{-}actual)^2}{N}}$$ (82)

The analysis shows that the RMSE between the IES-simulated temperature and the monitored temperature during the unheated period was 4.89°C, and the RMSE between the Fourier-filtered temperature and monitored temperature was 1.48°C. Similarly, the RMSE between the IES-simulated relative humidity and the monitored relative humidity was 16.89%, and the RMSE Fourier-filtered relative humidity and the monitored relative humidity was 4.12%. The magnitude of these errors is proportional to the performance gap. The reduction of RMSE in the Fourier-filtered cases demonstrates significant reduction of the performance gap.

22.1.1.4 Summary

This analysis therefore demonstrates that a Fourier filter obtained from an experimental building (building A) and applied to the results of the simulation of another building (building B) has considerably reduced the simulation performance gap of that other building, both in terms of air temperature and relative humidity. If building B did not exist but was being designed, the results of this method could have been considered to represent data equivalent to monitored values from a non-existing building. These findings can help to design better buildings with more confidence, and thus reduce investment and operational costs. I discuss details of this method, results, and design implications of these findings in a separate technical paper (Jankovic, 2016).

22.1.2 Advanced methods for calibration of simulation models

In Sections 3.9.2 and 21.1.5, I explained a simple manual method for calibration of simulation models. The manual method requires the user to compare the discrepancy between simulated and measured output, apply their judgement to the model parameters that need to be changed to reduce discrepancy, run a single simulation, and repeat the process until the error is reduced to an acceptable level. The manual method can be very time-consuming, as well as limited to a small number of candidate solutions for a calibrated model, and the reduction of the discrepancy will depend on the user's expertise.

Calibration process can be automated using optimisation methods, in which thousands of values of the uncertain parameters of the model can be tested and evaluated. In this section, I explain how this was done in a live 'RetrofitPlus' project (Figure 302), in which the baseline model for the building to be retrofitted had to be established. The baseline model was subsequently used as a starting point for design simulations for retrofit.

Figure 302   Two semi-detached houses to be retrofitted (from Jankovic and Basurra (2017))

In this example, four simulation tools are used: (1) IES VE to create the initial model from the building survey and export it as gbXML; (2) DesignBuilder, to import gbXML model as 'BIM' model and create EnergyPlus IDF – Input Data File; (3) EnergyPlus, to modify the IDF file and make it suitable for multi-objective optimisation in JEPlus+EA; and (4) JEPlus+EA to create bespoke optimisation process, in which the target values are real heating and electricity consumption obtained from the energy bills from the house in question. These steps are now explained in more detail below.

1   The reason for using IES VE was the involvement of my students, who had the appropriate software licence and skills to create the simulation model from a building survey (Figure 303). IES has a gbXML export capability, which was used to pass the model on to DesignBuilder (Figure 304).

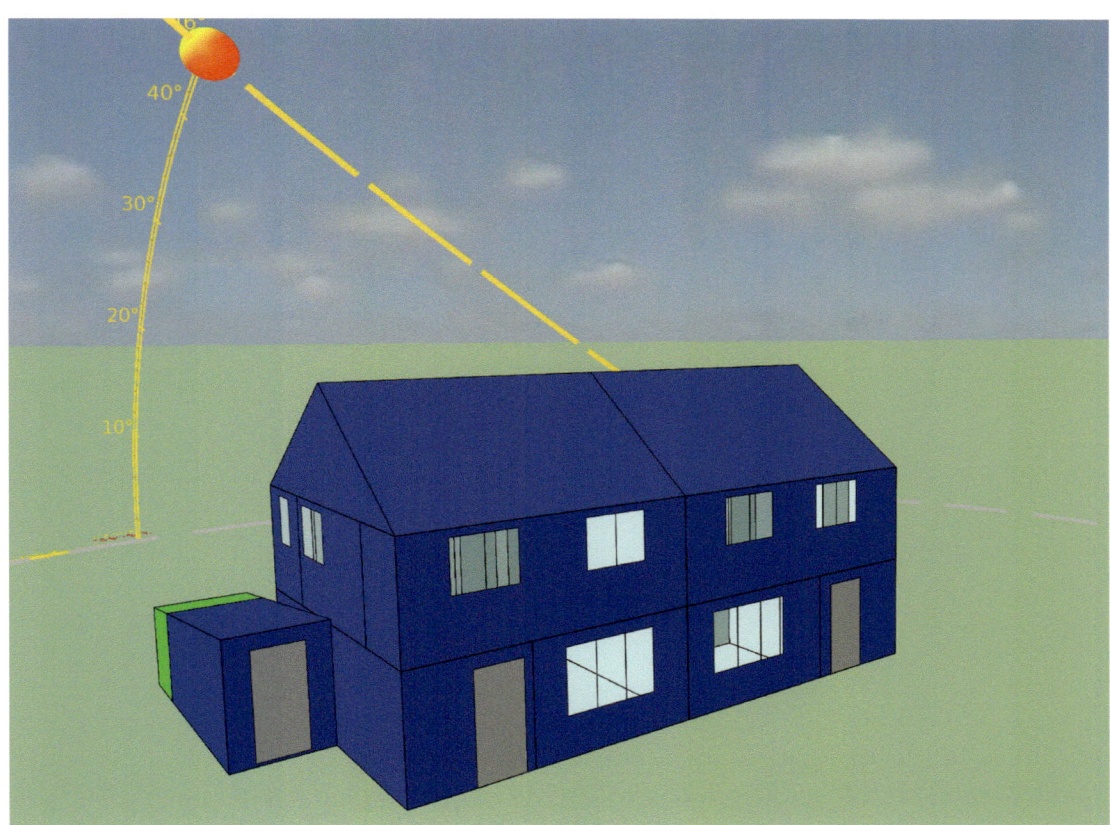

Figure 303   IES VE model created from the building survey (from Jankovic and Basurra (2017))

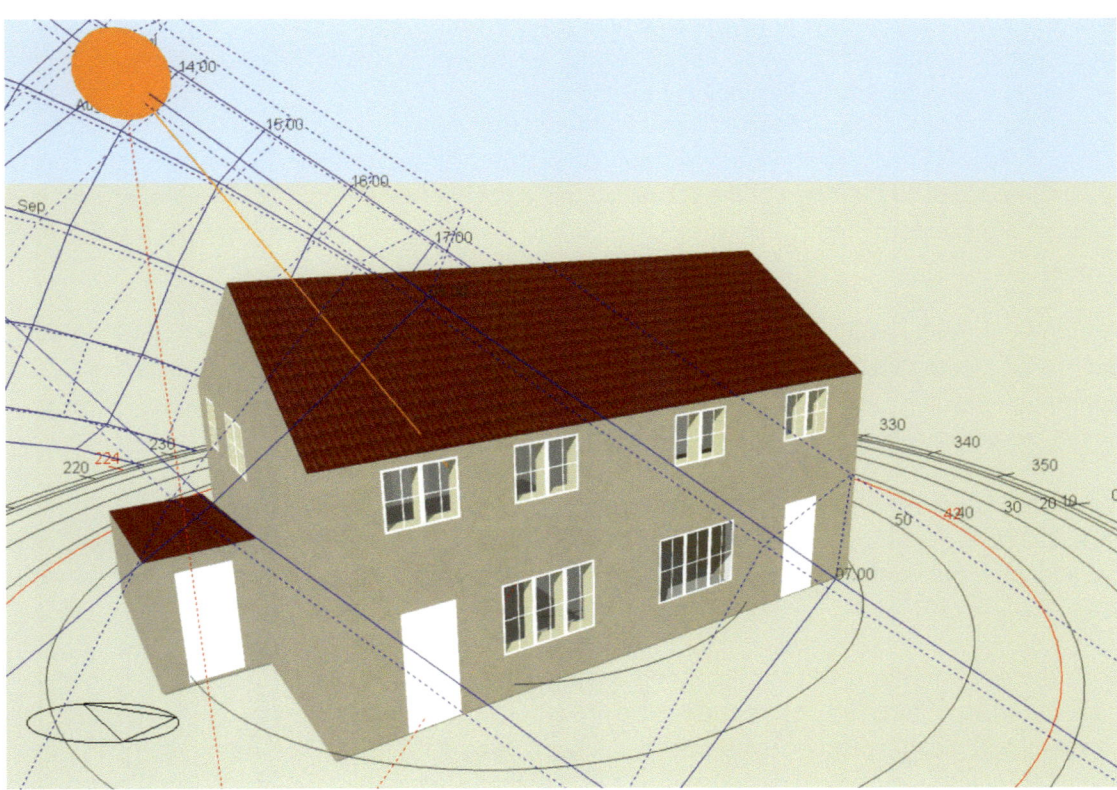

Figure 304   DesignBuilder model created from a gbXML import from IES VE (from Jankovic and Basurra (2017)

2 The reason for using DesignBuilder was its excellent workflow for editing building models and the underlying EnergyPlus simulation engine. Although DesignBuilder has its native file format, it is also capable of exporting EnergyPlus IDF file.

3 The IDF file created by DesignBuilder requires some modifications in EnergyPlus IDF Editor (Figure 305) before it can be used for multi-objective optimisation. Firstly, the results reporting frequency needs to be changed from hourly to annual. Secondly, DesignBuilder generates infiltration data for each zone as Flow/Zone in m³/s, which makes it difficult to modify infiltration simultaneously in all zones in the parametric simulations; infiltration is therefore changed into AirChanges/Hour, and different air flow figures in m³/s for individual zones were changed into a single figure of h⁻¹. Similar changes had to be made for similar reasons for LightingLevel, from W to W/m², and for EquipmentLevel from W to W/m².

The above changes in the IDF file enable the insertion of search tags in JEPlus and deal with all zones in a generalised way, rather than using specific hard-coded air flow rates, etc. for each individual zone. In each of the above steps, an annual simulation was run with the corresponding simulation tool in order to identify and eliminate any errors, before the model was passed onto JEPlus+EA.

4 The reason for using JEPlus+EA was its underlying EnergyPlus simulation engine that uses IDF files, and its capability to define bespoke objective functions using formulae, and to use these formulae in multi-objective optimisation. In this particular case, the IDF file generated in the preceding steps was used for simulation. The objective functions were defined as Electricity consumption: abs (Measured − Simulated)/Measured × 100 [%], and Gas consumption: abs (Measured − Simulated)/Measured × 100 [%] (Figure 306). Thus the optimisation results that occur near the origin of the Electricity consumption/Gas consumption coordinate system are close to the measured values of energy consumption, and the corresponding parameter set is a 'recipe' for a calibrated model.

Why were the objective functions used in this form? Expressing the error in absolute values does not provide us with a clear understanding of how close we are to the origin of the coordinate system. For instance, an absolute error in Joules can be a number of the order of 10^6 or greater, and an absolute error of say 100 kWh per annum does not give us much understanding of the quality of the solution. Relative errors expressed in percent give us that instant understanding, and hence that is how the objective functions were formulated.

Why were absolute values of relative errors used, instead of positive and negative values of relative errors? Multi-objective optimisation favours the minimum solutions, and the density of solution points in the scatter plot tends to decrease with distance from the minimum. Hence the presence of negative numbers in the scatter plot causes low density of solutions around the origin of the coordinate system, thus giving poorer choices for a calibration solution. This issue is eliminated when absolute values are used, concentrating the scatter plot points near the origin.

Figure 307 shows the results of JEPlus+EA multi-objective optimisation. These are typically expressed in the form of a Pareto front, a series of results that are the closest to the origin of the coordinate system. However, in this case, where optimisation is used for calibration purposes, we are not interested in the Pareto front, but in the values that are the closest to the origin of the coordinate system.

In the example in this figure, 691 candidate solutions were tested by the optimisation algorithm, filtered out from the initial solution space of 3,920 determined by the initial combination of design parameters. After the optimisation is completed, placing the cursor above the appropriate point in the JEPlus+EA scatter plot output gives the calibration parameter set for the simulation model (Figure 307) and the errors corresponding to this parameter set.

The final solution contains 0.17 % error in respect of electricity consumption and 0.33 % error in respect of gas consumption (Figure 307). This solution is therefore 99.83% accurate with respect to electricity consumption and 99.67% accurate with respect to gas consumption, and the model with the corresponding parameter set is considered to be sufficiently good for our purpose. See also Jankovic and Basurra (2017).

In cases where relative errors are much higher than those obtained in this analysis, improvements can be made by narrowing down the ranges of specific parameters, and by reducing the steps within parameter ranges, thus increasing the resolution of the simulation grid. Additional improvements can be made by using the K Nearest Neighbour (KNN) algorithm with high-density avoidance, targeting non-overlapping results of different parametric simulations (see Basurra and Jankovic, 2016).

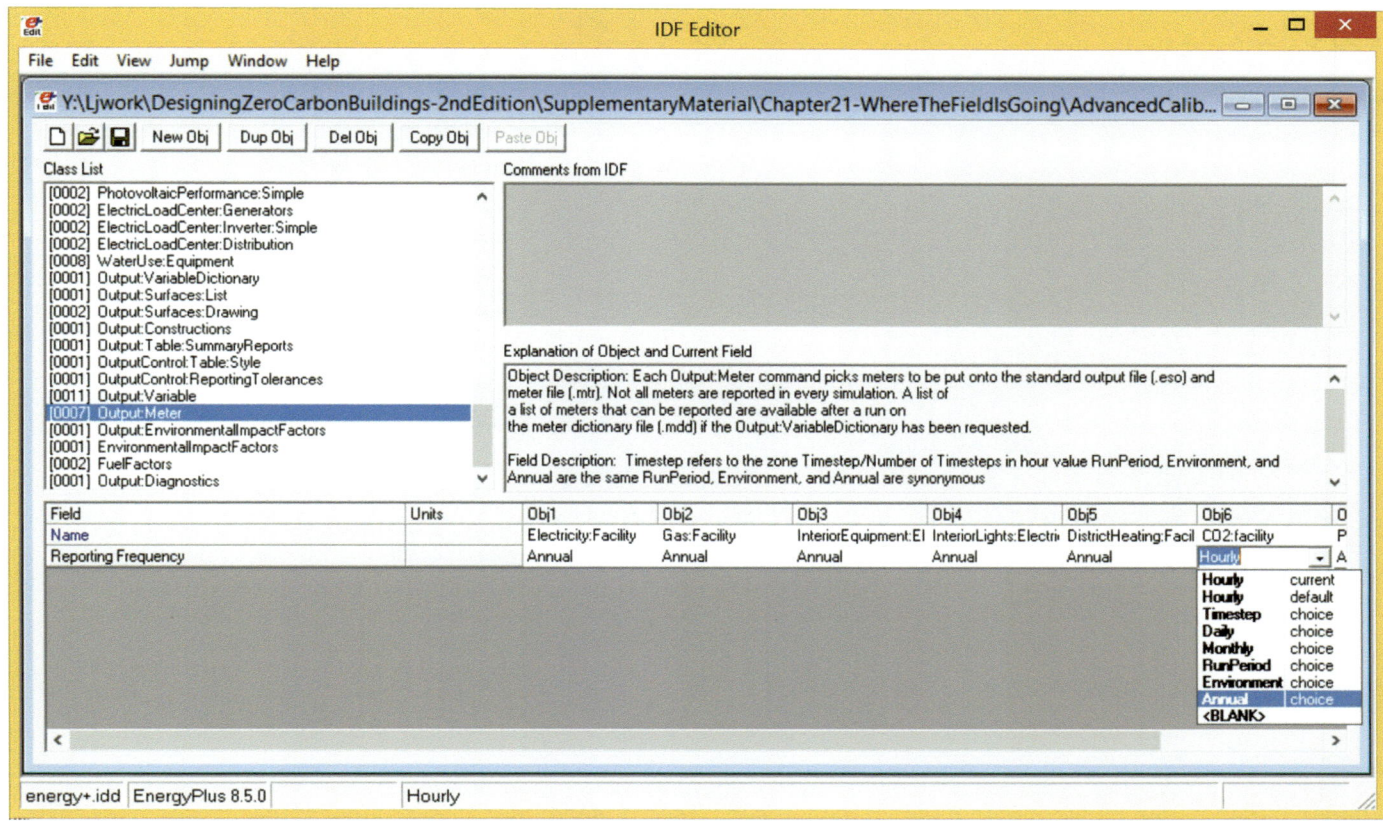

Figure 305   EnergyPlus IDF editor: modifications of EnergyPlus model imported from DesignBuilder export (from Jankovic and Basurra (2017))

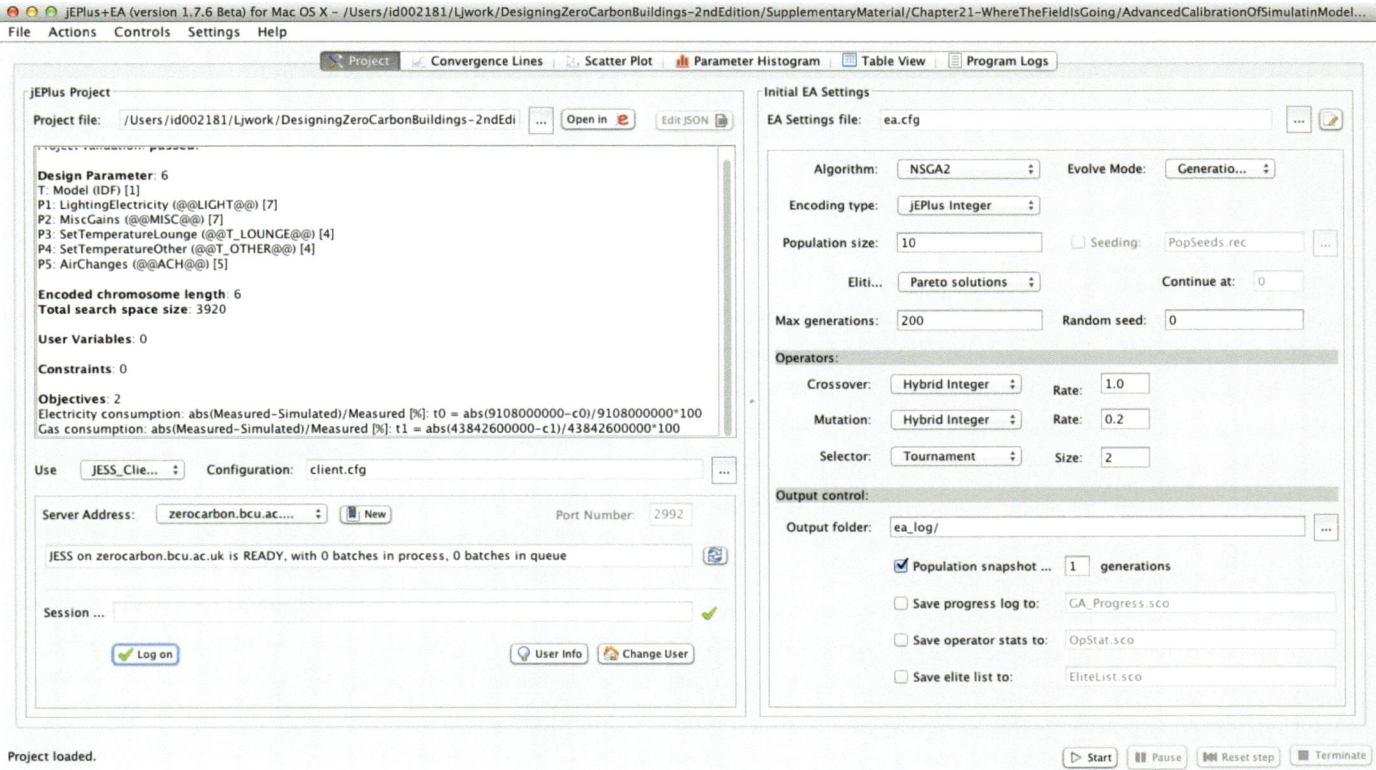

Figure 306   jEPlus+EA multi-objective optimisation settings for calibration (from Jankovic and Basurra (2017))

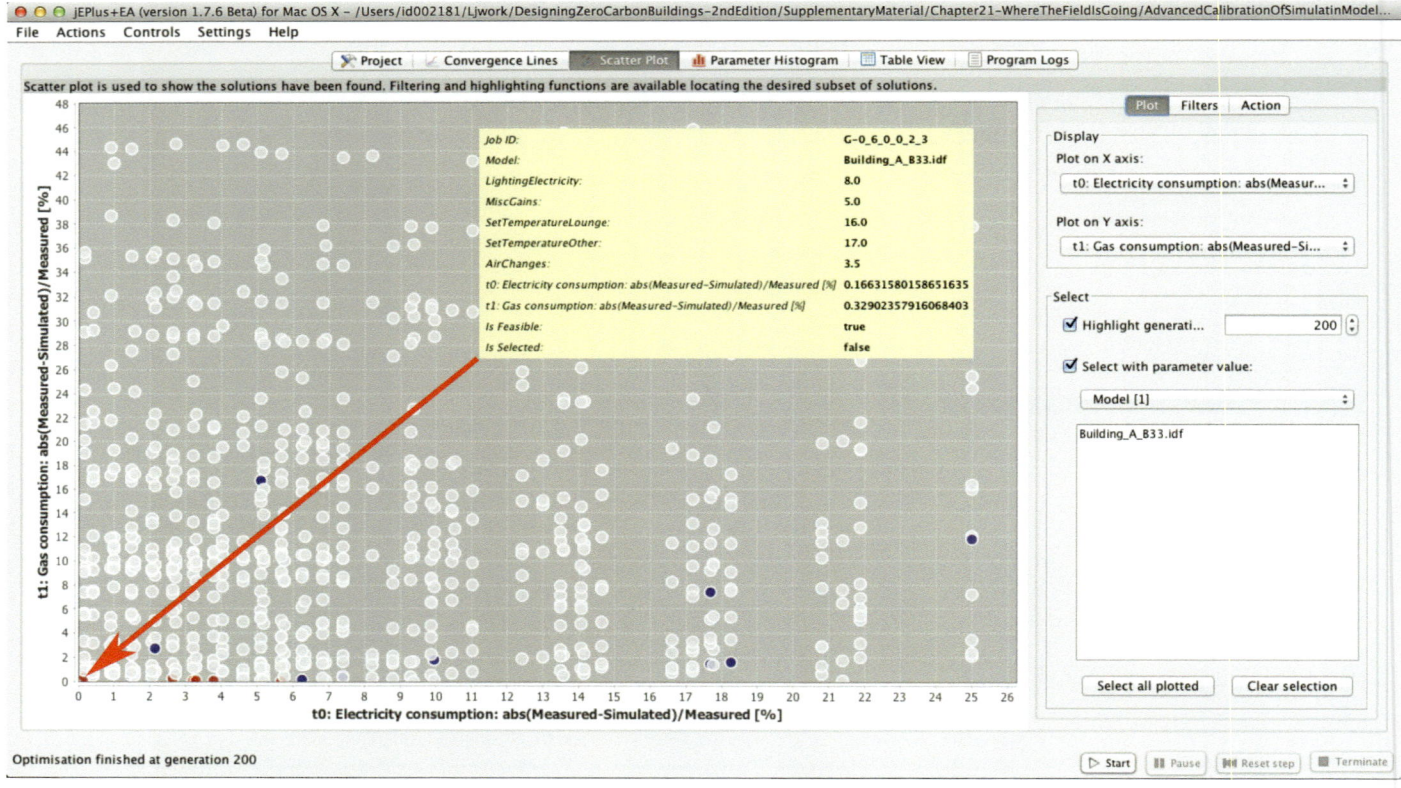

Figure 307   JEPlus+EA results scatter plot with the point nearest to the origin containing the calibration parameter set (from Jankovic and Basurra (2017))

22.1.3 Putting a retrofit design onto a zero carbon trajectory

In the example in the previous section, funding for retrofitting the envelope was available, but funding for other improvements, such as the heating system, the lighting and renewable energy was not. The approach taken was therefore to carry out multi-objective optimisation of the retrofit, to create a prerequisite for zero carbon design when funds for renewable energy become available.

22.1.3.1 Multi-objective optimisation

The objective functions chosen were discomfort hours and carbon emissions, and the independent variables included a range of technical and behavioural parameters. The technical parameters were: three different thicknesses of 'TCosy'[1] wall insulation: 150, 200, and 225 mm, combined in pairs with the identical TCosy roof insulation thicknesses; infiltration air changes per hour; fuel type (gas or biomass); lighting power density; and two different PV arrays (East side of the roof only, and East and West sides combined). The parameters that were left to the occupants to adjust are deemed to be behavioural parameters as follows: room set temperature and clothing level.

The results of multi-objective optimisation are shown in Figure 308 together with the point on the Pareto front that satisfies the maximum discomfort hours chosen.

This is a holistic approach that not only deals with technical but also with behavioural parameters. The reason for this is that dwellings with poor thermal insulation are expensive to heat, and occupants live under less comfortable conditions. The approach taken here shows that the occupants can adjust their behaviour after the retrofit, by increasing the set temperature in their home and reducing the amount of clothing, whilst spending less energy.

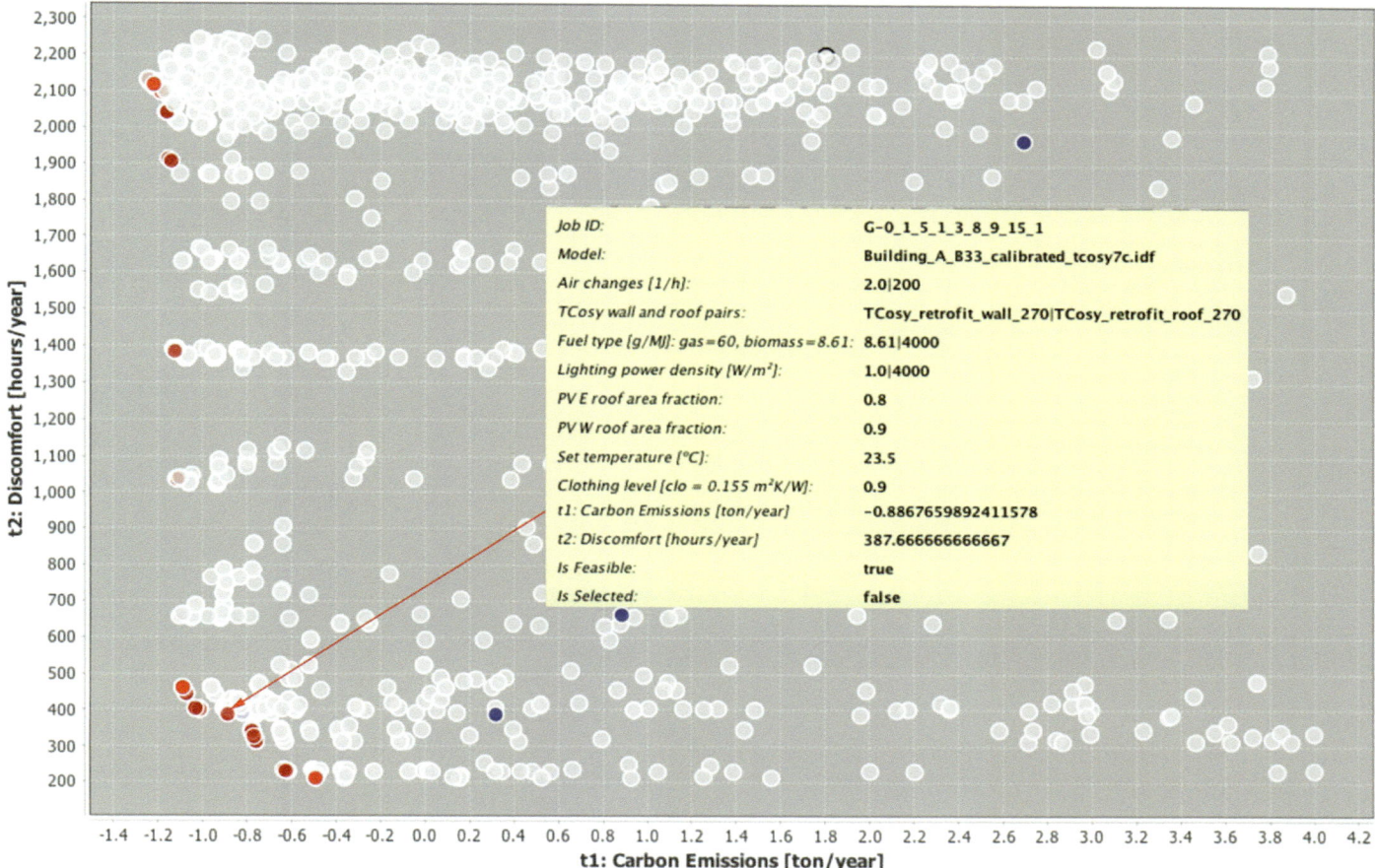

The figure shows a scatter plot with axes labeled "t2: Discomfort [hours/year]" (y-axis, ranging from 200 to 2,300) and "t1: Carbon Emissions [ton/year]" (x-axis, ranging from -1.4 to 4.2). An information box contains:

Job ID:	G-0_1_5_1_3_8_9_15_1
Model:	Building_A_B33_calibrated_tcosy7c.idf
Air changes [1/h]:	2.0|200
TCosy wall and roof pairs:	TCosy_retrofit_wall_270|TCosy_retrofit_roof_270
Fuel type [g/MJ]: gas=60, biomass=8.61:	8.61|4000
Lighting power density [W/m²]:	1.0|4000
PV E roof area fraction:	0.8
PV W roof area fraction:	0.9
Set temperature [°C]:	23.5
Clothing level [clo = 0.155 m²K/W]:	0.9
t1: Carbon Emissions [ton/year]	-0.8867659892411578
t2: Discomfort [hours/year]	387.666666666667
Is Feasible:	true
Is Selected:	false

Figure 308   Results of design optimisation

Source: (Jankovic, 2019b)

Although discomfort hours in Figure 308 appear to be high, this is due to constant clothing levels kept throughout each annual simulation. Further significant improvements in the assessment of thermal comfort can be achieved through the application of the adaptive clothing algorithm introduced in Section 17.4.

22.1.3.2 Summary

The analysis in this section shows how multi-objective optimisation can be used for the calibration of a simulation model, and how the calibration parameters can be carried forward into the simulation design analysis of a retrofit project. The multitude of results obtained from the design optimisation can be used to plot a trajectory of staged design interventions that will ultimately result in carbon negative performance. This work was used as one of the steps towards the preparation of retrofit in a 'Retro-fitPlus' project (Jankovic, 2019b).

22.1.4 Lessons from a completed retrofit

The retrofit discussed in section was carried out and subsequently monitored for two years. In this section, I introduce details of the retrofit process, and of the main performance findings after the retrofit. Further details can be found in my journal article 'Lessons learnt from design, off-site construction and performance analysis of deep energy retrofit of residential buildings' (Jankovic, 2019b).

22.1.4.1 Retrofit execution

Based on the design optimisation introduced in the previous section, a completely new building envelope was manufactured off-site (Beattie, 2017) and delivered to site to carry out installation. Figure 309a–c shows stages of installation of external pre-fabricated thermal envelope manels. Figure 309d shows injection of thermal insulation into one of the installed panels.

a)

b)

c)

d)

Figure 309   Installation of the prefabricated retrofit thermal envelope

Source: (Jankovic, 2018b)

This process led to a completed retrofit, as shown in Figure 310.

Figure 310   Photo of the completed retrofit

Source: (Jankovic, 2018b)

22.1.4.2 Retrofit performance evaluation

The project was monitored before and after the retrofit, using several types of instruments. A weather station provided information on external solar radiation, air temperature, relative humidity, wind velocity and direction, solar radiation, and rainfall, all in 15-minute intervals. A sample of weather data is shown in Figure 311.

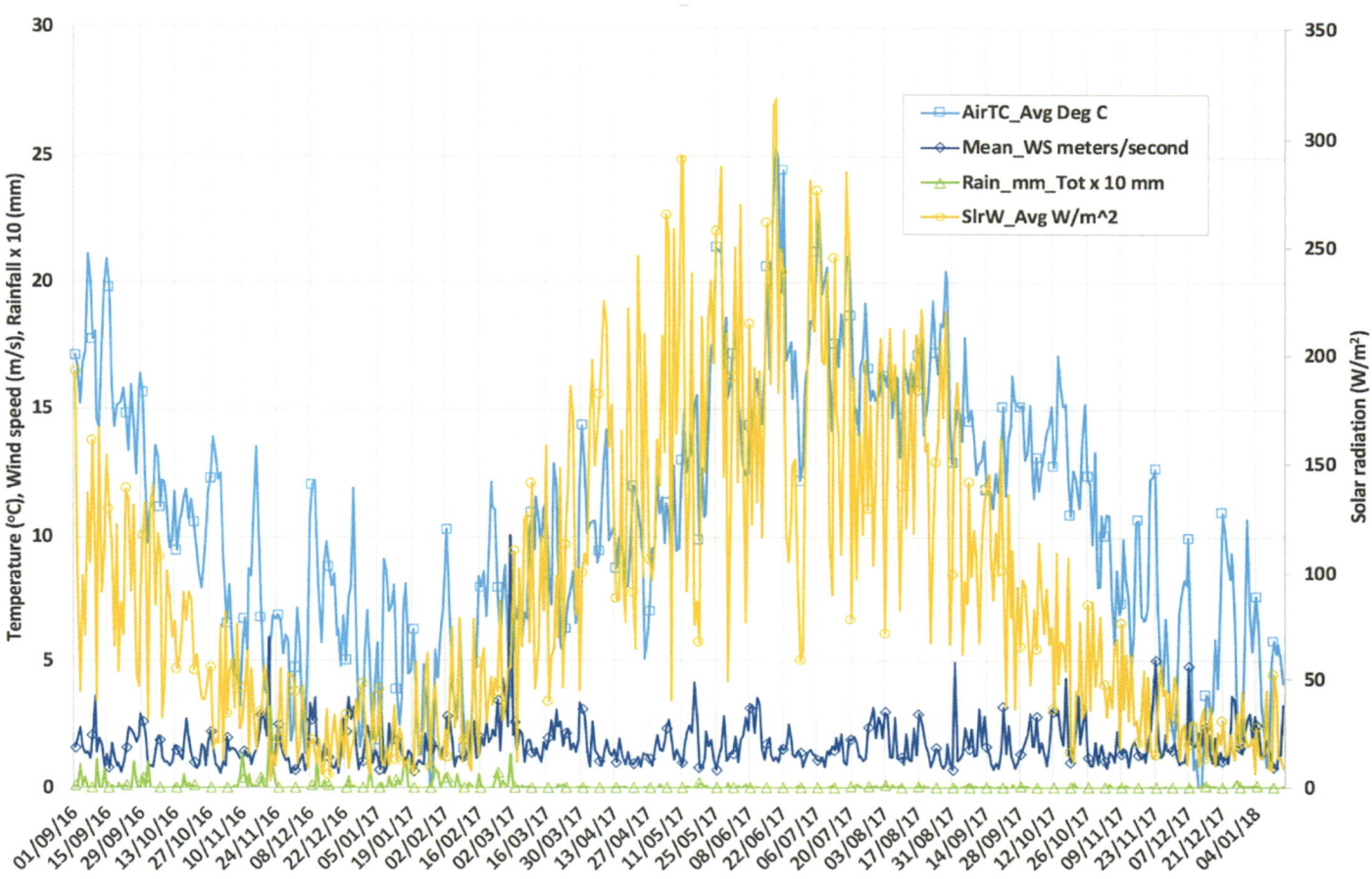

Figure 311   Monitored weather data

Source: (Jankovic, 2019b)

Energy consumption (Figure 312) and internal air temperatures (Figure 313) were monitored using a locally installed and network-connected data logger. Recordings were made in intervals of one minute. The energy consumption figures were cross-checked with smart meters installed in the houses.

Quantitative results of energy and carbon emissions performance are shown in Table 111. As it can be seen from this table, energy savings directly calculated from the monitoring results are 56%. However, when these savings are normalised to longer-term degree days, the savings are 45% and 42%, corresponding to either weather data degree days or CIBSE degree days respectively. It is worth noting that electricity consumption actually increased by 4% after the retrofit. This is attributed to the introduction of mechanical ventilation with heat recovery (MVHR) system, which was constantly in operation after the retrofit. Carbon emission savings follow a similar pattern as energy savings, so that 55% of savings are directly calculated from the monitoring results, with 43% and 40% of longer-term emissions savings normalised to weather data file degree days and CIBSE degree days respectively. Although achieving Passivhaus or EnerPHit standards were not the aims of this retrofit project, Table 111 shows where the project stands in relation to these standards. Despite the significant improvement in energy performance, the long-term normalised performance benchmarks of between 84 and 89 kWh/m²/yr show that the Passivhaus and EnerPHit standards have not been achieved.

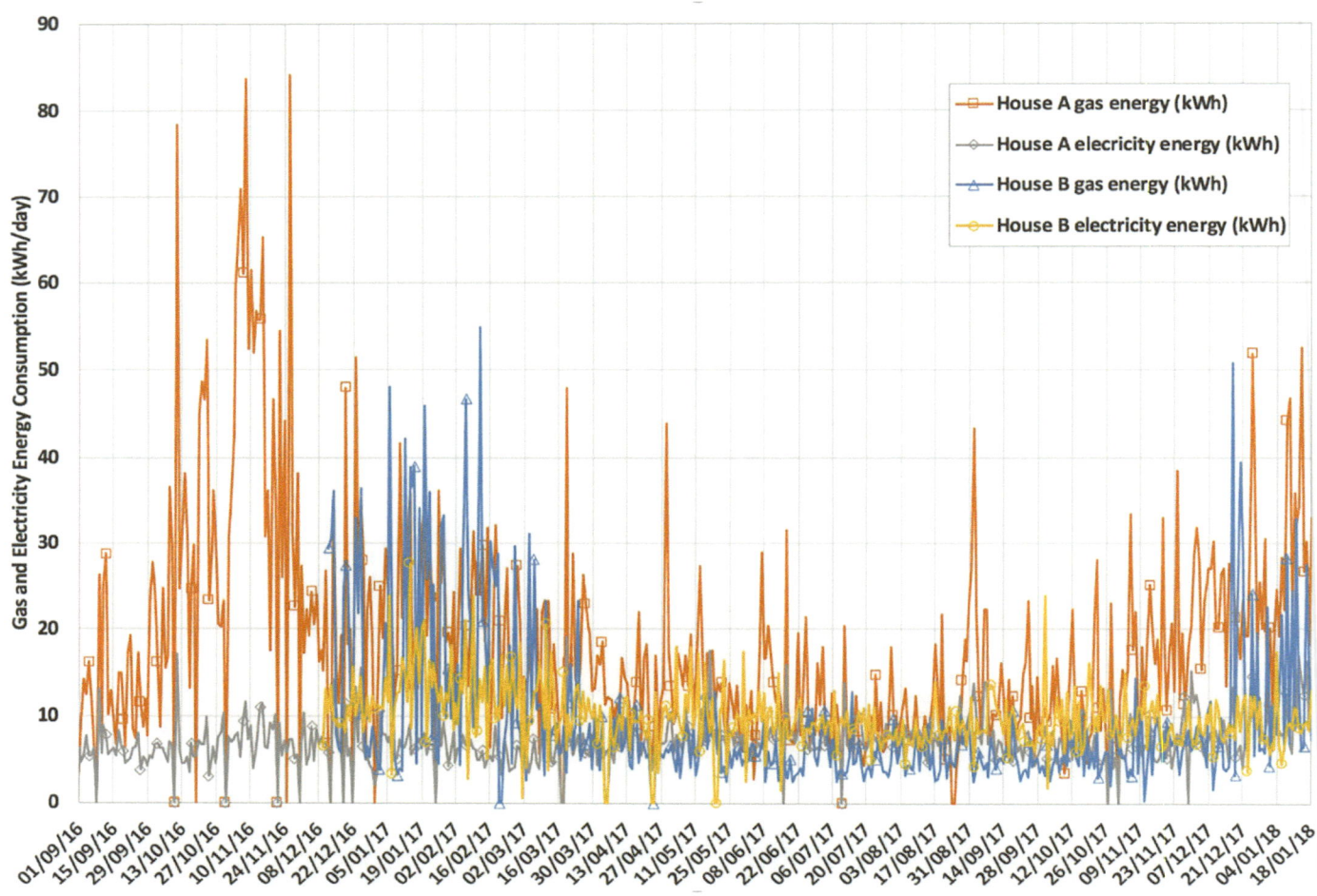

Figure 312   Monitored energy consumption

Source: (Jankovic, 2019b)

Internal conditions before and after retrofit were subjected to frequency of occurrence analysis of internal temperatures. The result of this analysis shows approximately a 3°C internal temperature increase resulting from the retrofit. We can visualise this by selecting a horizontal line in Figure 314 and observing on the horizontal axis where that line crosses the blue and red curve.

22.1.4.3 Evaluation of wellbeing of occupants

Periodic questionnaires were issued to the occupants to assess the changes to their wellbeing as a result of the retrofit. This was also complemented by interviews to cross-check and elaborate on some of the responses. In summary, all occupants reported high energy consumption before the retrofit and low energy consumption after the retrofit. The perception of thermal comfort was inverse to that, with low thermal comfort reported before the retrofit and high thermal comfort reported after the retrofit. The perception of air quality was similar to that of thermal comfort: it was considered to be low before the retrofit, caused by poor air tightness, and high after the retrofit, caused by increased air tightness and the effect of MVHR operation. The following comments in the questionnaires were particularly revealing of the improvements to the occupants' wellbeing:

> 'I did not need heating when outside temperature dropped below freezing yesterday';
> 'The house feels like home now – no damp, no dust, no noise';
> 'I have stopped using my asthma puffer'.

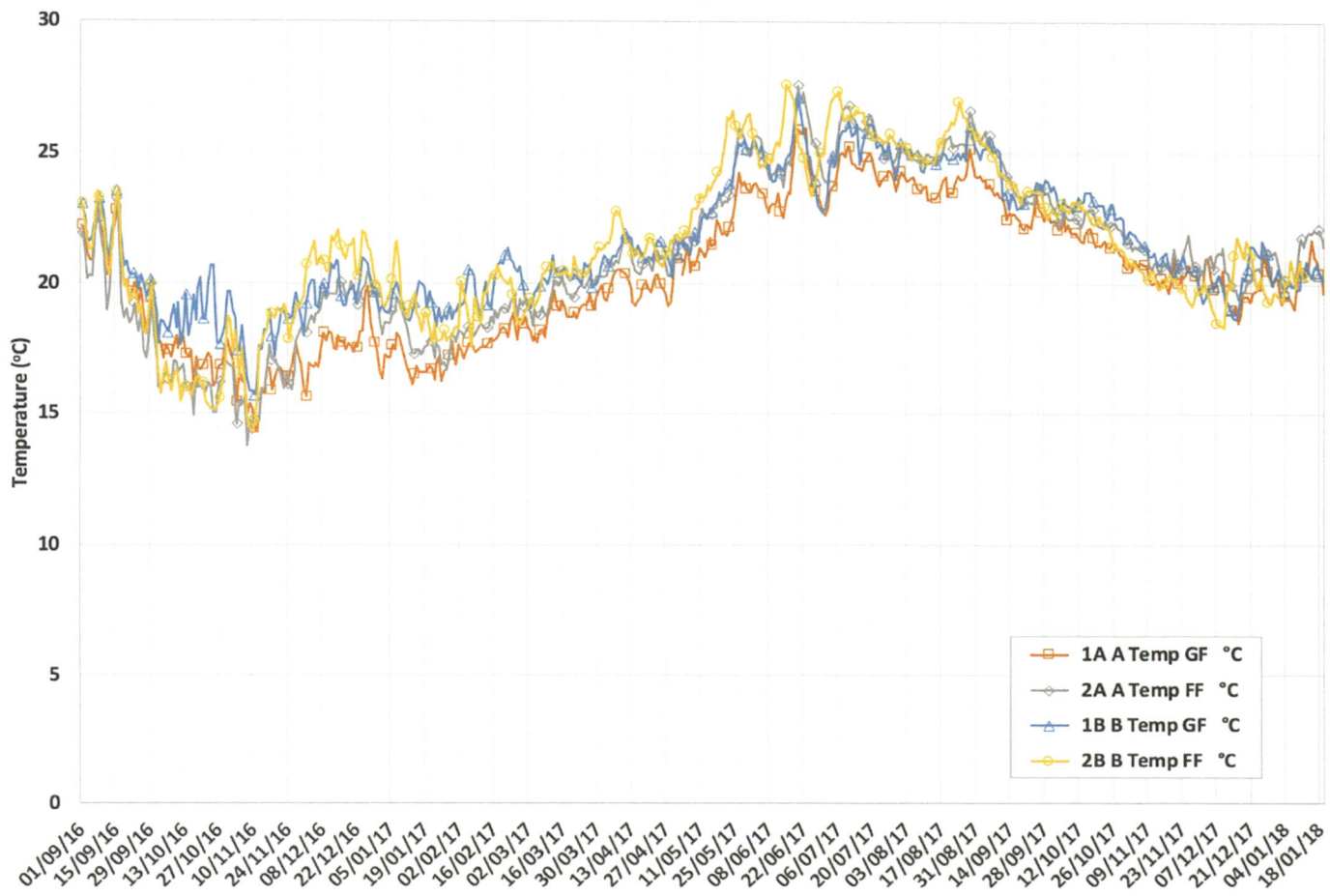

Figure 313   Monitored internal air temperatures

Source: (Jankovic, 2019b)

The last statement represents a significant health co-benefit, which should be taken into account when retrofit payback is evaluated (Jankovic, 2019b).

TABLE 111 SUMMARY OF ENERGY AND CARBON EMISSIONS PERFORMANCE					
Energy consumption and savings	**Pre-retrofit consumption (kWh)**	**Post-retrofit consumption (kWh)**	**Unadjusted savings (%)**	**Degree day normalised saving (Weather file DD) (%)**	**Degree day normalised saving (CIBSE DD) (%)**
Energy consumption and savings					
Gas	11,511	5,032	56	45	42
Electricity	2,071	2,155	−4		
Carbon emissions and savings					
Total space heating emissions (kgCO2/yr)	2,486	1,130	55	43	40
Energy performance benchmarks					
Space heating (kWh/m²/yr)	153	67	56	84	89
Passivhaus space heating demand (kWh/m²/yr)					≤15
EnerPHit space heating demand (kWh/m²/yr)					≤25

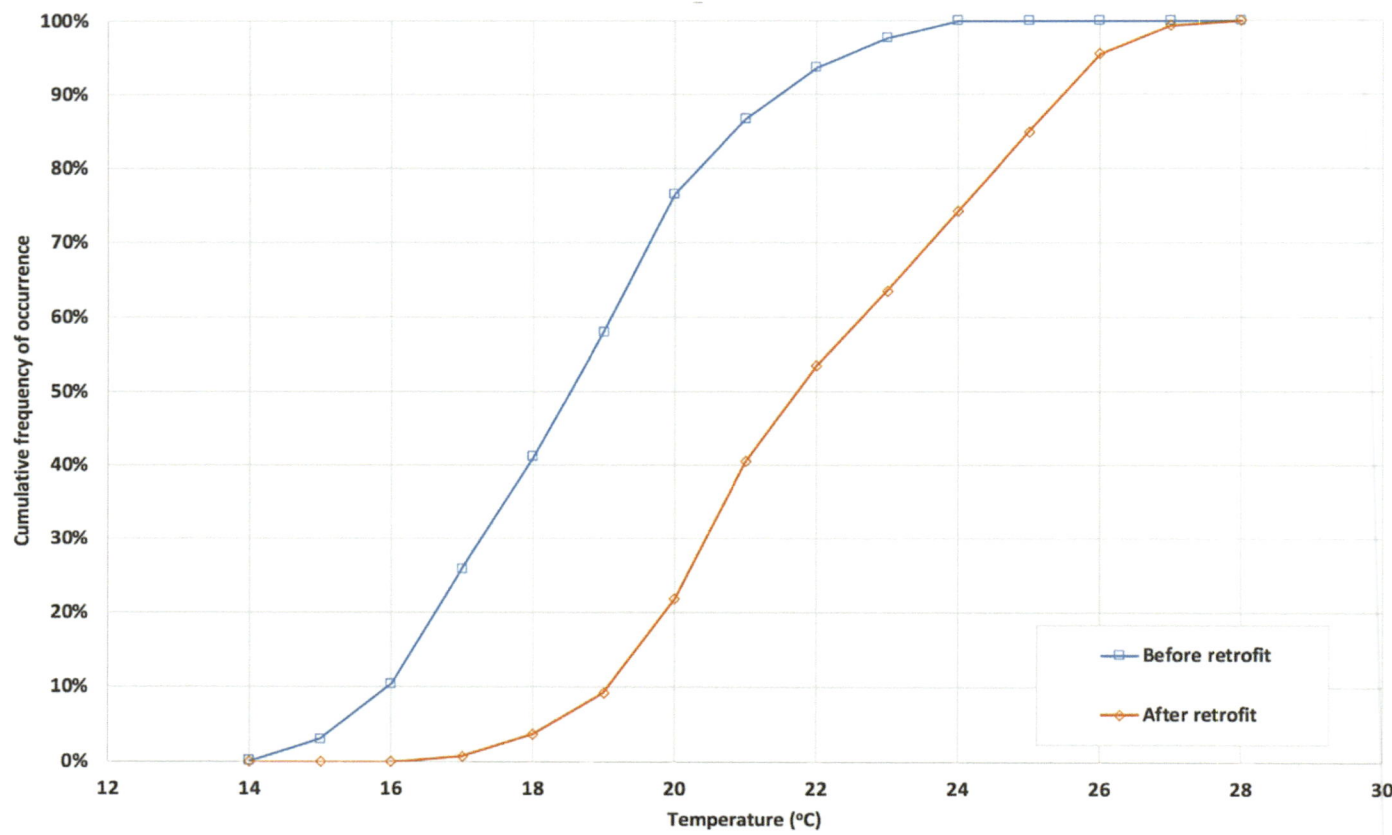

Figure 314   Internal temperature change as result of retrofit

Source: (Jankovic, 2019b)

22.1.4.4 Quality assurance measures

The retrofit project reported here was part-funded by Innovate UK. In the closing stages of the project, the funding body project officer asked a question: 'How can we be sure that the specified characteristics of the thermal envelope have been actually achieved?' In other words, how can we be sure that 'what is written on the tin is actually in the tin?'

There are different ways of determining the quality of the building thermal envelope: co-heating tests and dynamic heating tests, as explained in Chapter 20, Sections 20.1.1 and 20.3.1. However, these tests can only be carried out in unoccupied buildings, and therefore could not be applied to these houses, which were actually occupied throughout the retrofit process. That led to the development of a non-invasive way of determining the building physics parameters using calibrated simulation models. Thus, a calibration of simulation models before and after the retrofit was carried out, similarly to the calibration explained in Figure 307. The accuracy of these calibrated models was in excess of 99.5%, and these models were subsequently used to conduct simulated dynamic heating tests of the buildings before and after the retrofit. I explain the development of this method in a separate journal article entitled 'Improving Building Energy Efficiency through Measurement of Building Physics Properties Using Dynamic Heating Tests' (Jankovic, 2019a). The application of this method on this retrofit project is illustrated in Figure 315.

As can be seen from this figure, a step change of heat input of 8 kW occurred at time $t = 24$ hours, and the simulation model was left to run until internal temperatures stabilised. Results of building physics parameters obtained from these tests are

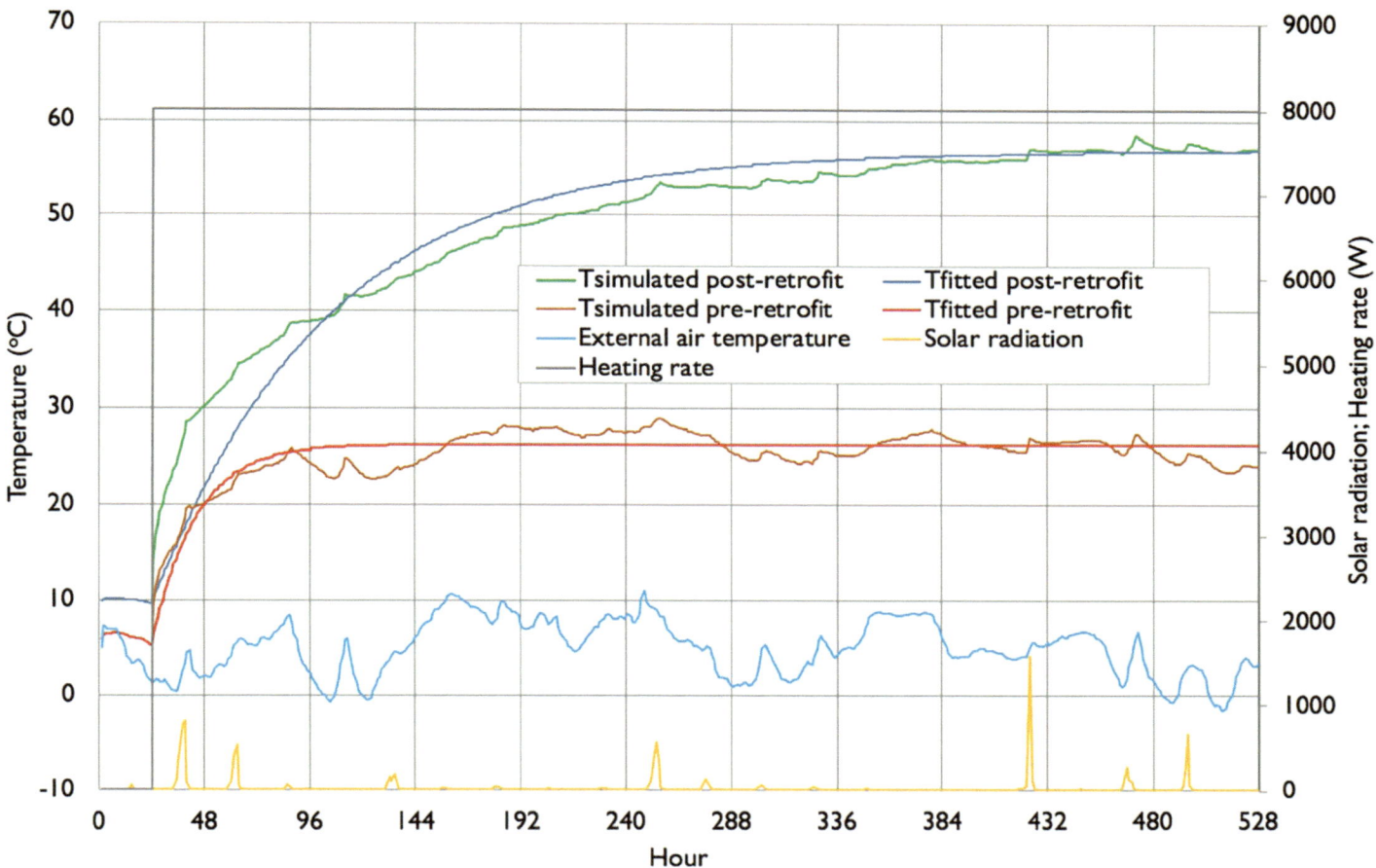

Figure 315   Simulation of a dynamic heating test before and after the retrofit

Source: (Jankovic, 2019b)

summarised in Table 112. The first three rows of this table show direct results from the tests. The time constant, which corresponds to the time it takes to go through 63% of temperature change from the time heat input is applied, has increased 3.58 times. This means that the building has become less responsive to heat perturbances and it therefore achieves more stable conditions. The results also show that the overall transmittance-area product (the UA value) has reduced by a factor of 0.45, demonstrating a significant reduction of heat loss. Effective thermal capacitance has increased by a factor of 1.63, demonstrating that the building has become much better at storing heat. The three last rows of this table are derived and comparison parameters. Row (4) shows the average U-value after the retrofit, which is 0.45 times better than the U-value from before the retrofit. Row (5) of this table shows theoretical transmittance-area product before and after the retrofit. Row (6) compares measured and theoretical values of the transmittance-area product. It shows that in both cases, theoretical values are significantly lower than the corresponding measured values. This suggests that UA values obtained from technical reference tables and manufacturers' specifications could make the energy performance of buildings appear a lot better at design stage than in reality. This is a cause of a significant performance gap between design and operation, and a call for action to update technical reference tables and the way manufacturers' specifications are reported.

22.1.4.5 Summary

The analysis of a completed retrofit in this section provides an overview of the retrofit execution using an off-site manufactured building envelope. The instrumental monitoring of the buildings before and after the retrofit enabled the analysis of improvements to building energy and carbon emission performance, as well as improvements of wellbeing to the building occupants.

TABLE 112 RESULTS OF DYNAMIC HEATING TEST SIMULATIONS

Row	Building physics parameter	Post-retrofit	Pre-retrofit	Ratio post-retrofit/pre-retrofit
(1)	Time constant C/UA (h)	78.4	21.9	3.58
(2)	Overall transmittance-area product UA (W/K)	147.3	328.7	0.45
(3)	Effective thermal capacitance C (MJ/K)	41.6	25.6	1.63
(4)	Overall thermal transmittance U (W/m^2·K)	0.69	1.54	0.45
(5)	Theoretical transmittance-area product UA (W/K)	86.0	288.5	0.30
(6)	Measured vs. theoretical [1 − (5)/(2)] × 100 (%)	42	12	

Despite significant improvements, the retrofit did not achieve Passivhaus performance. The measurement of building physics properties using a specially developed method of simulated dynamic heating tests showed significant discrepancies between theoretical and measured building envelope characteristics. This calls for re-evaluation of building technical reference tables used by designers.

22.1.5 Bottom-up modelling approaches

Over the past decade and more I have been making an argument for better design tools. The reason is that nature does not do design calculations in the way that people do. Instead of creating an overall model of a system using equations, which is computationally intensive and therefore slow, nature's systems emerge from the bottom up on the basis of interactions between objects at different scales, from molecules to galaxies, with timescales much closer to those observed in nature.

For instance, traditional methods of modelling air flow are based on Navier-Stokes equations. Although they were developed in 1820s, no complete solutions to these equations exist even today. In order to obtain any kind of solution for particular cases, the equations need to be simplified. This means that, having made a description of a system very and perhaps unnecessarily complicated, we need to simplify it in order to make the problem fit the solution method, rather than making the solution method fit the problem. For the reasons described, Navier-Stokes equations contain one of the seven most important open problems in mathematics that are still awaiting solutions (Clay Mathematics Institute, 2023).

As particles of air do not solve Navier-Stokes equations, here are a modelling examples based on principles analogous to those occurring in nature. In 'Changing the Culture of Building Simulation with Emergent Modelling' (Jankovic, 2017a), I developed an interactive sketch of a building enclosure with a heat source, which could be progressively modified as the heat transfer and consequent air movement take place. A linear horizontal heat source was first drawn in an enclosure representing a room and the simulation was initiated (Figure 316a), showing upward heat flow (Figure 316b and c). Subsequently, an obstacle was inserted into the rising heat (Figure 316d), and a flow pattern developed around it (Figure 316e). The beginning of a chimney was subsequently drawn in Figure 316f.

The interaction sequence then continued by completing the chimney in Figure 317a, and air flow out of the room through the chimney developed in Figure 317b. West wind (from the left) was switched on in Figure 317c and the air plume swayed to the right. Wind was then switched off, allowing the plume to become vertical in Figure 317d. Subsequently, an obstacle was inserted above the chimney in Figure 317e, and a flow pattern developed around it, with eddies emerging past the obstacle (Figure 317f).

In another piece of work entitled 'Modelling Computational Fluid Dynamics with Swarm Behaviour' (Jankovic, 2018a), I modified Reynolds' flocking algorithm (Reynolds, 1987) and implemented Newtonian physics on an individual agent level. That led to the creation of self-organised and emergent behaviour of particle flow in an enclosure, as shown in Figure 318.

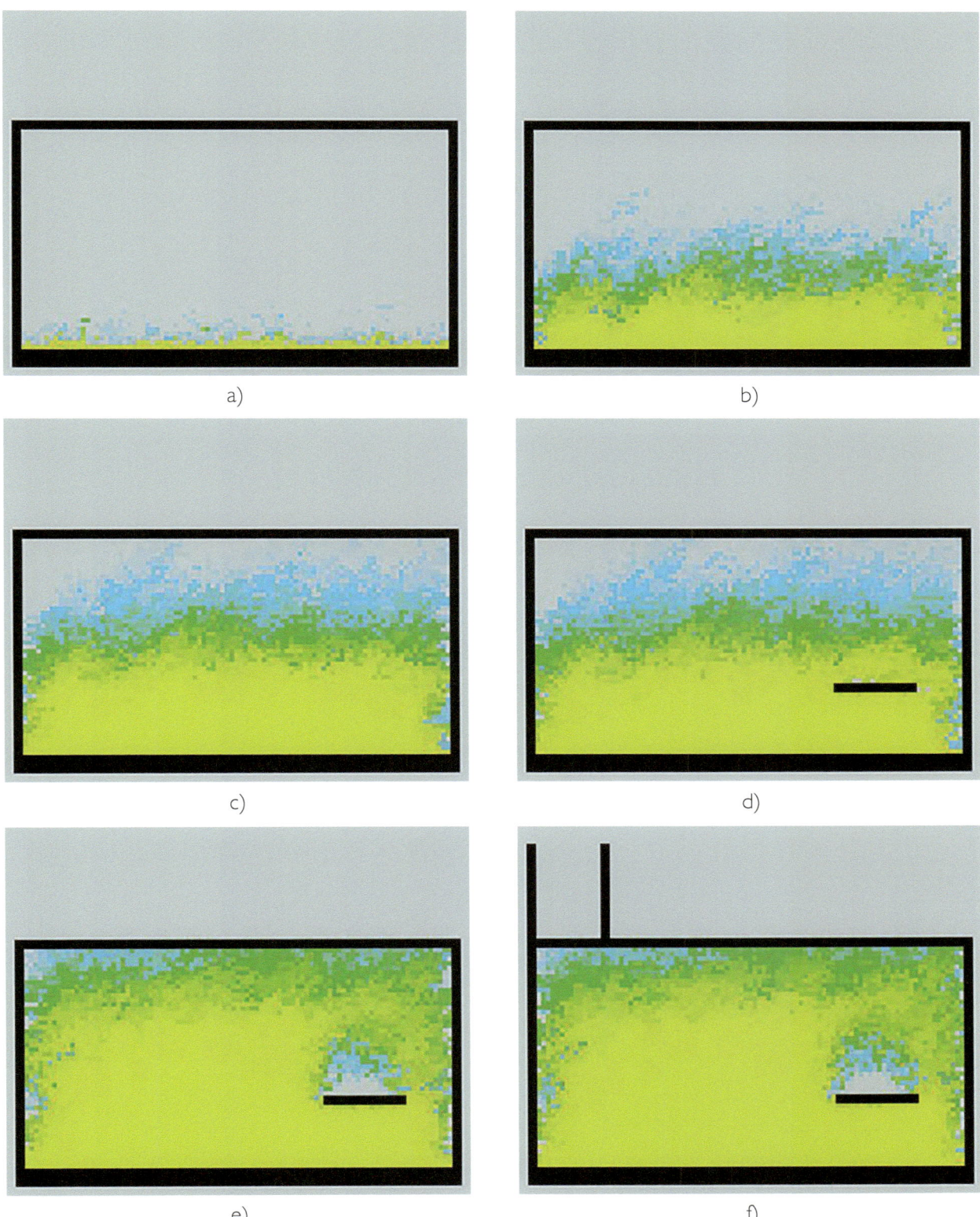

Figure 316   Emergence-based simulation of air flow in a room

Source: (Jankovic, 2017a)

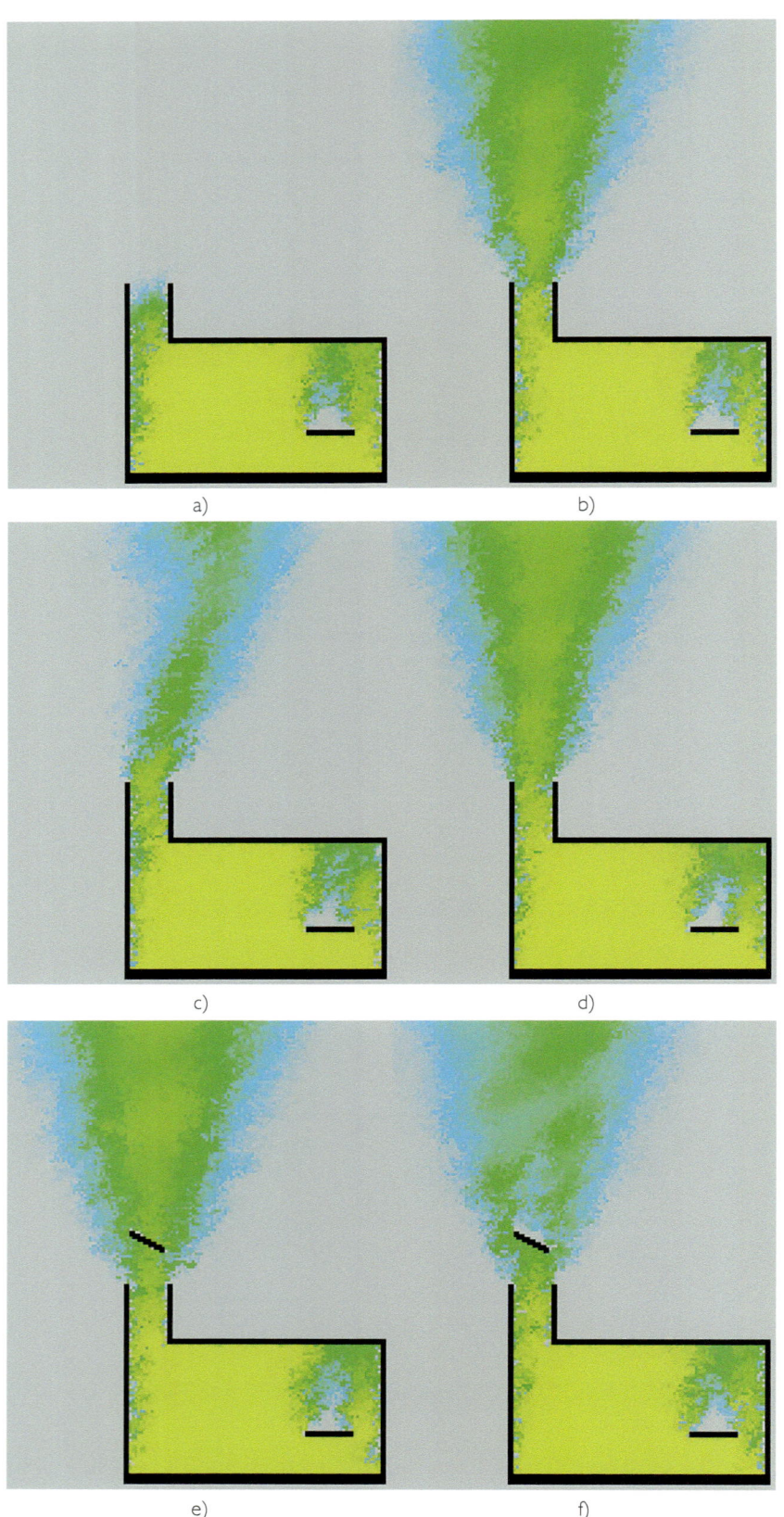

Figure 317   Emergence-based simulation of air flow out of a room

Source: (Jankovic, 2017a)

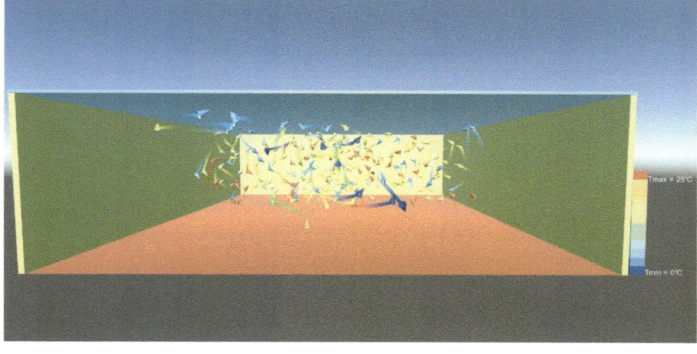
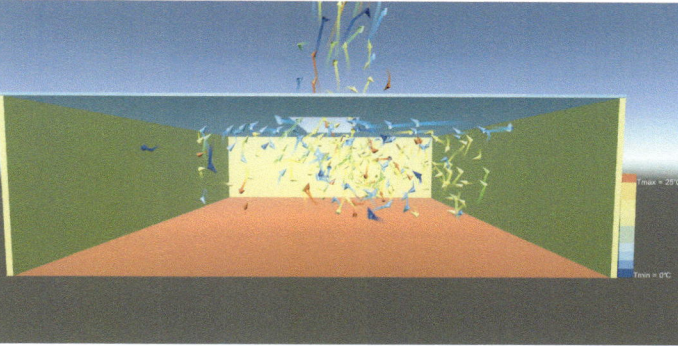

a) Flow pattern is formed

b) Particles find a way to escape through opening

Figure 318   Swarming based air flow simulation

Source: (Jankovic, 2018a)

In both of these examples, the particle movement was based on the sum of forces acting on each individual particle. Unlike in conventional modelling where the system model is based on a top-down system of equations, the system model in these examples emerged spontaneously from interaction between the particles. In this way, the solution method was brought much closer to the system in comparison with the conventional approaches, and consequently, the compelling behaviour that emerged was not explicitly programmed. These models produced profoundly more complex behaviour than that was explicitly programmed. They captured what 'traditional mathematics failed to capture' (Wolfram, 2002), and were guided by the idea that 'when things are what they seem, we can safely let them do what they must' (Toffoli & Margolus, 1987).

22.1.6 New business and economic models

Numerous efforts into improving building energy efficiency have mainly focused on technological solutions, whilst the concept of money as it is has been taken for granted and has not been questioned. This affects all sectors and therefore the world sustainability.

A pertinent question is therefore: Can a different kind of money, i.e. alternative currency, help to make a more sustainable world?

I investigated this in more detail in a technical paper entitled 'Opportunities for financing sustainable development using complementary local currencies' (Jankovic, 2019c). The work explores a system consisting of the following actors: a housing association, a retrofit provider, a PV manufacturer, and a housing association tenant (Figure 319). The system is driven by solar radiation falling on the roofs of buildings of the housing association (Figure 319a). On this basis, the housing association issues its own electronic 'eMoney' and uses it to pay the retrofit provider for PV installation on its roofs (Figure 319b). The retrofit provider uses the eMoney earnings from this work to purchase more PV from the PV manufacturer, whilst keeping an 'X' amount of eMoney as its salary (Figure 319c). Subsequently, the housing association has started electricity production from its newly installed PV system and the PV manufacturer uses its earnings of 'eMoney $- X$' to purchase electricity from the housing association, whilst keeping a 'Y' amount for its salaries. It therefore pays the housing association 'eMoney $- X - Y$' in exchange of 'Energy $- Z$' supplied by the housing association (Figure 319d). The housing association also supplies a 'Z' amount of energy to its tenant, for which it receives a direct payment of 'Money' from the tenant (Figure 319d). After the first iteration of this circular economy, every actor in the system has achieved some benefits: the housing association has energy and it has 'eMoney $- X - Y + $ Money'; the retrofit provider has 'Income $X + PV$'; the *PV* manufacturer has 'Income Y' and 'Energy $- Z$'; and the housing association tenant has 'Z Energy' (Figure 319e). Looking at Figure 319b–d, it is possible to notice that money and utility obtained in exchange for money always flow in opposite directions. In order to make this system work as part of a wider economy, it is necessary to make eMoney exchangeable with conventional money, as shown in Figure 319f.

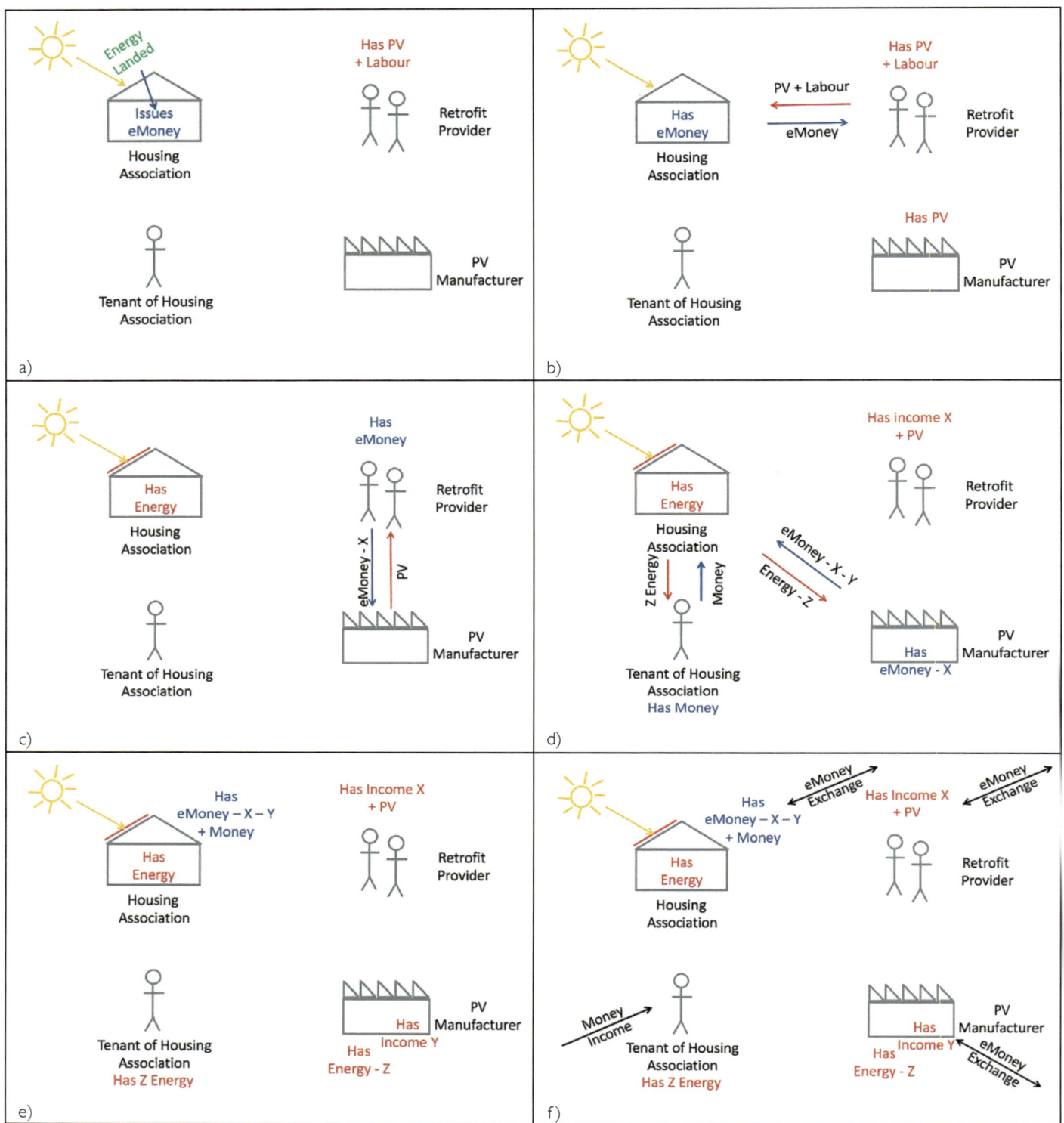

Figure 319   Circular economy example driven by solar radiation falling on the roofs of buildings of a housing association

Source: (Jankovic, 2019c)

The example of circular economy shown here is a qualitative illustration of how the new kind of money can be put to useful work, and achieve a change, driven entirely by the natural resource of solar radiation. The source publication also contains a quantitative illustration of the same system, showing how the quantities of money and utility add up in a spreadsheet (Jankovic, 2019c). Independently from this work, Thomas Greco worked on 'Solar Dollars: A Complementary Currency That Incentivizes Renewable Energy' (Greco, 2021). That work reenforces the idea of value creation from natural resources, and this idea therefore deserves serious consideration.

Despite the difficulties and resistance to change, I believe that it will be just a matter of time before we need to change our ways radically in order to survive as humanity. The way forward would therefore be not to take anything for granted, but to take a view outside of the box and consider alternatives as part of a holistic way of achieving the sustainability paradigm.

22.2 POLICY DEVELOPMENT SUPPORT

In 2023, I undertook a short project to provide policy development support to one of the local authorities/municipalities in the English county of Hertfordshire. The aim was to find out how the lessons learnt from the background research towards the preparation of this book can be used for policy development support. On this basis, property developers would be requested to provide information on upfront carbon and operational carbon to the local authority. A calculation of a cumulative total would then determine financial charges payable by the developer to the local authority towards a carbon offsetting fund, should the development not achieve cumulative zero by 2050. The payments would be made into an offset fund, to be used for interventions that reverse the outstanding emissions in relation to specific developments. I used the Zero Equation for analysis, and this enabled to demonstrate scenarios where a development would either overshoot the zero target, achieve it, or undershoot it, as shown in Figure 320.

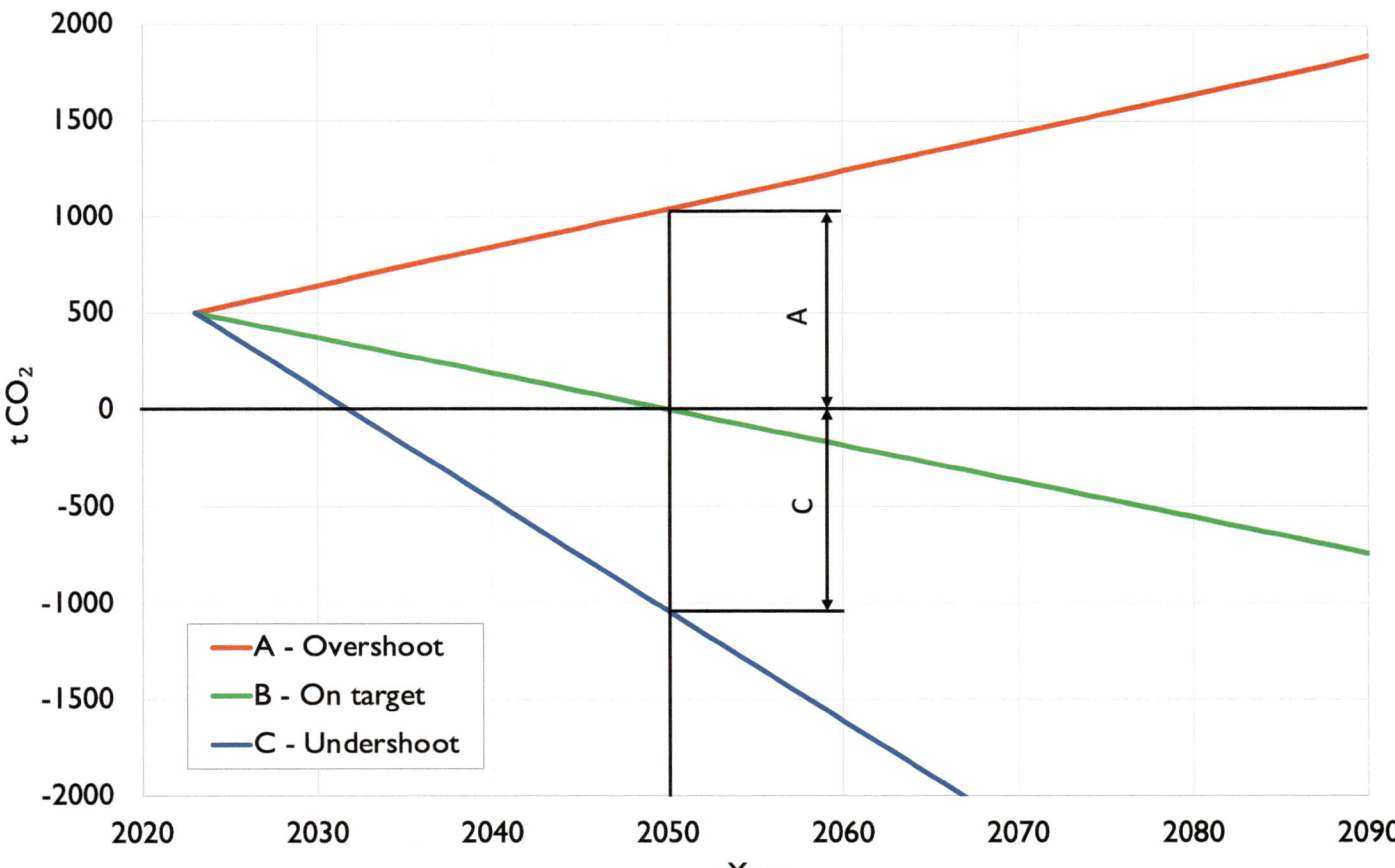

Figure 320   Charges and incentives with respect to cumulative emissions status

Thus, in Scenario A where there is an overshoot of the emissions target, the developer would be charged A × cost per tCO_2 towards the offset fund. In Scenario B, where the target is achieved, the developer would not be charged, and no money will be received by the offset fund. If these are the only two options, then this could be off-putting for developers who may be only interested in their profits, and the authorities in charge of the offset fund would need to find offsetting interventions that would generate a like-for-like emissions offset without costing more than received by the fund. Both options could be problematic on their own, without an additional option that offers incentives. Thus, option C could be introduced, where charging one developer for an overshoot would generate an incentive fund for a developer that that has achieved an undershoot and gone below zero cumulative emissions before 2050. Thus, developers would have an incentive to go over the minimum target of zero cumulative emissions by 2050. If $A = C$, then a like-for-like offset would be achieved and the offset intervention would break even financially. If $A > C$, then the offset would only be partially achieved and the amount of $(A - C)$ × cost per tCO_2 would be available for further offset interventions.

The existence of incentives for developers could lead to their interests becoming aligned with the interests of building users. When these interests are not aligned, developers construct buildings that fulfil the minimum performance standard in line with building regulations, and building users then have increased running costs throughout the building lifetime. With incentives for developers made available from an offset fund, buildings could be constructed to better standards to achieve better performance; developers could be paid for their effort that goes beyond the minimum performance; and building users would have lower running costs. These are just some ideas about how the material from this book could facilitate future policy development support that reduces both embodied and operational emissions.

NOTE

1 TCosy™ is an innovative approach to retrofit in which an existing building is entirely surrounded with a new Passivhaus envelope. See www.beattiepassiveretrofit.com for details.

CHAPTER 23

CONCLUSIONS AND FUTURE PROSPECTS

The main thesis of this book is that true zero can only be achieved if both embodied and operational emissions are taken into account. This is supported by evidence from the Zero Equation, developed in Chapter 1 as a jam jar analogy and as a mathematical equation. The Zero Equation is subsequently used in Chapter 19 to demonstrate the integration of design principles and in Chapter 21 to demonstrate how true zero is achieved in case studies of several existing buildings. As evidenced in this book, using bio-based materials reduces the upfront carbon as a result of carbon storage in these materials, and retrofitting buildings can lead to significant reduction of emissions because of the replacement of emissions prior to retrofit and reduced use of new materials in comparison with newbuild. Both bio-based materials and retrofit result in achieving true zero sooner and easier.

I believe that a realistic and perhaps a necessary goal for the way forward is for every design, newbuild and retrofit, will be a zero cumulative emissions building, to be achieved with certainty and by a specified year. This goal will significantly contribute to the overarching goal of keeping the planet within the global warming limit of 1.5°C and prevent catastrophic changes to the biosphere.

There are a few intermediate outcomes that will lead towards delivering this goal:

• Increased demand for zero cumulative emissions buildings
• Improved design methods and tools
• Improved embodied and operational emissions data
• Improved building performance monitoring
• Improved education
• Advanced regulation
• Advanced business and economic models
• Advanced research and technological development

These outcomes are linked in multiple ways. **Increased demand for zero cumulative emissions buildings** will become possible with **improved education** that increases the understanding of benefits of better building performance, and with **improved business and economic models** that facilitate the finance and incentives for better buildings. **Improved embodied and operational emissions data** and **improved building performance monitoring** will be facilitated by **advanced research and technological development**. **Improved building performance monitoring** will also become more affordable through **advanced business and economic models**. **Advanced research and technological development** will lead to **improved design methods and tools** and the latter, together with **improved business and economic models** will make **advanced regulation** easier to achieve.

Which activities will facilitate these outcomes?

INCREASED DEMAND FOR ZERO CUMULATIVE EMISSIONS BUILDINGS

Educating general population about building energy efficiency will create an incentive for construction industry to develop better buildings. There is a general appreciation of lower running costs of electric cars, and raising awareness of building energy efficiency and therefore lower running costs and lower emissions, would create higher demand for better buildings.

IMPROVED DESIGN METHODS AND TOOLS

Nature's processes, including heat transfer in buildings, emerge from the bottom up, and building simulation tools are developed from the top down. This leads to building simulation tools representing processes in buildings in a way that does not facilitate interactive design, such as interactive sketches that enable interactive heat and mass transfer simulation in the real time. In Section 22.1.5 Bottom-up design approaches, I demonstrated a way that can, with further development, enable a more intuitive building performance design whilst drawing a building. These improved methods require a paradigm shift in how building heat transfer is modelled and would facilitate wider access to building design tools and ultimately better buildings.

IMPROVED EMBODIED AND OPERATIONAL EMISSIONS DATA

There are uncertainties in embodied and operational emissions data. Embodied emissions in materials and equipment generally do not have environmental product declarations (EPDs), except in a small number of cases, and databases of embodied emissions do not reflect real-time upfront carbon emissions. Operational emissions do not reflect real-time grid carbon status in different countries. Despite these uncertainties, using available data for embodied and operational emissions brings us closer to calculating and achieving true zero. However, a better real-time tracking of grid carbon, such as in the example by Rogers and Parson (2022) would facilitate more accurate operational emissions information (stages B1–B7). This would also facilitate information on energy and emissions used for upfront embodied carbon (stages A1–A5), as well as for end of life carbon (stages C1–C4). In an increasingly connected world, where manufacturing of materials might occur far away from where these materials are used, a distributed ledgers could be used to track energy and emissions from future real-time grid carbon tracking systems. These distributed ledgers could be based on blockchain technology, which is increasingly finding dependable applications in industry (Welfare, 2019). In that way, each material or a component would have an ID, a passport of embodied emissions generated during various material stages, stored in a tamper-proof and encrypted ledger that adds more certainty to embodied emissions.

IMPROVED BUILDING PERFORMANCE MONITORING

Increasing the understanding of real-time building performance through post occupancy evaluation and instrumental monitoring of existing buildings would provide the essential feedback loop needed for improving future designs and for the calibration of simulation models. In my view, performance of every building should be continuously monitored with instruments, and the results should be available to building users and building designers through dedicated online databases. Closing the feedback loop between design and performance would enable a greater use of the methods for reverse modelling, for the reduction of simulation performance gap, predictive control, and calibration of simulation models of buildings in the design stage. These issues are discussed in Chapters 20 and 22.

IMPROVED EDUCATION

On the supply side, more extensive education and the training of designers and planning and building regulations inspectors in the use of dynamic simulation models, needs to take place. A small proportion of designers use dynamic simulation models today, yet these are the only tools that can be used to improve the energy and carbon efficiency of buildings whilst providing comfortable conditions for building occupants. Planning and building regulation inspectors would benefit from a greater understanding of the simulation based design process, as this would enable them to assess merits of proposed buildings much more accurately and make decisions accordingly. Advanced training of construction operatives would ultimately result in increased quality of delivery, where different trades have a better understanding of the impact of their work on building performance. For instance, my experience of monitoring some recent housing developments tells me that designing more airtight buildings will not necessarily make them more airtight if the operatives on the ground are not briefed about where they can and where they should not drill holes.

ADVANCED REGULATION

Building regulations are minimum mandatory requirements that designers and developers must comply with in order to get design and construction of buildings approved. Whilst these mandatory requirements are becoming stricter, as they have done in the UK for instance, buildings are not yet required to achieve zero cumulative emissions, and embodied emissions are not yet one of the criteria for compliance with building regulations. It is therefore imperative to include embodied emissions in future building regulations, and to target zero emissions year in the way introduced by the Zero Equation.

ADVANCED BUSINESS AND ECONOMIC MODELS

Advanced business and economic models will facilitate increased demand for zero cumulative emissions buildings. In Section 22.1.6 New business and economic models, we saw that natural resources, such as solar energy falling on roofs of buildings, could lead to value creation and complementary currencies to work in parallel with national currencies. This can make it easier for property owners to aspire to zero emissions buildings and overcome the general impression that zero emissions buildings are more expensive and less affordable. One of the problems that hinder building performance is a misalignment between the interests of building developers and building owners. The developers do not benefit financially from better building performance and therefore are not generally motivated to deliver best performing buildings. In Section 22.2 Policy development support, we saw an example of how a framework for developer incentives could be established, and this could be one of the ways of aligning the stakeholders' interests that could lead to better performing buildings.

ADVANCED RESEARCH AND TECHNOLOGICAL DEVELOPMENT

As showed in Section 4.6, the total world energy consumption is several orders of magnitude smaller than the available solar radiation intercepted by the earth. Earth receives in 42 minutes enough energy from the sun to meet the world annual energy demand. As the energy demand increases by the year 2050, the duration of solar radiation to meet the world annual energy demand will increase to 53 minutes. Another energy source to consider is earth's magma, which at temperatures well in in excess of 1,000°C deep in the ground can provide worldwide energy indefinitely. We just need to develop technology that makes its use more widely available.

Therefore, there is abundant energy above our heads and below our feet. But we need cutting edge research and development to bring the civilisation level to a stage where it can begin to use this energy and completely abandon fossil fuels, leading to drastically reduced carbon emissions. This requires a response on a scale of the planet, and a considerable re-focusing of global efforts. The vast amount of energy coming from the sun cannot be captured by us continuing to do things in the way we are used to. We need new approaches to old things and a new thinking. We need a significant percentage of the GDP targeted in that direction. If we manage to develop this technology soon enough, climate crisis will be a thing of the past. But we need to make a momentous effort to get there.

A WIDER CONTEXT

In a wider context, we need to overcome the economic growth imperative. We are currently using more than one planet worth of resources, and such overuse cannot last indefinitely. A step change is required to a transition to a non-growth imperative that balances the consumption with regeneration of resources and a steady state economy. We need to understand that infinite economic growth on a finite planet can only have one outcome – a total collapse.

Our greatest risk is not an unknown future but something that is much closer to us and that we know very well: the business as usual and the resistance to change. It is becoming increasingly apparent that we can no longer afford to resist change, but we need to embrace it.

Eliminating conflicts around the world would free up the resources and help to enable humanity to achieve the current challenges sooner, and for that we all need to learn how to pull in the same direction. We only need to look at the environmental conditions on Mars or the Moon to realise that our planet has the conditions of a true paradise. I hope that we can look after it well to keep it like that indefinitely.

Architype. (2023). *Entopia—No ordinary building*. https://architype.co.uk/project/entopia/

ASHRAE. (2020). *ASHRAE position document on infectious aerosols*. https://www.ashrae.org/file%20library/about/position%20 documents/pd_infectiousaerosols_2020.pdf

ASHRAE. (2021a). *2021 ASHRAE® handbook: Fundamentals* (SI edition). ASHRAE.

ASHRAE. (2021b). *Standard 55 – Thermal environmental conditions for human occupancy*. https://www.ashrae.org/technical-resources/bookstore/standard-55-thermal-environmental-conditions-for-human-occupancy

ASHRAE. (2022). Standard 90.1-2022—*Energy standard for sites and buildings except low-rise residential buildings*. https://www.ashrae.org/technical-resources/bookstore/standard-90-1

ASHRAE.org. (2021). *ASHRAE global headquarters reaches 'fully' net-zero-energy milestone*. https://www.ashrae.org/about/news/2021/ashrae-global-headquarters-reaches-fully-net-zero-energy-milestone

Awbi, H. B., Parker, J., Butcher, K., & Chartered Institution of Building Services Engineers (Eds.). (2015). *Building performance modelling* (Second edition). The Chartered Institution of Building Services Engineers.

Baker, N., & Steemers, K. (2002). *Daylight design of buildings*. James & James.

Basurra, S., & Jankovic, L. (2016). Performance comparison between KNN and NSGA-II algorithms as calibration approaches for building simulation models. *Proceedings of the 2016 Building Simulation and Optimization Conference*. IBPSA England. Newcastle Upon Tyne.

Beattie, R. (2017). *TCosy™ – Transforming homes, improving lifestyles*. http://www.beattiepassiveretrofit.com/

BEIS. (2022). *Domestic energy price indices*. GOV.UK. https://www.gov.uk/government/statistical-data-sets/monthly-domestic-energy-price-stastics

Bevan, R., & Woolley, T. (2008). *Hemp lime construction: A guide to building with hemp lime composites*. IHS BRE Press.

Blaylock, J., Courtice, P., Forman, T., French, J., Reynolds, J., & Cole, J. (2022). *Building Entopia—The story behind the ultra-sustainable retrofit of CISL's new home in Cambridge*. University of Cambridge Institute for Sustainability Leadership. https://www.cisl.cam.ac.uk/files/entopia_case_study_12_12_22.pdf

BIPM. (2023). *SI base unit: Candela (cd)*. https://www.bipm.org/en/si-base-units/candela

BRE Group. (2018, July 10). *SAP - Standard assessment procedure*. https://bregroup.com/sap/

BRE Group. (2020, January 21). *IMPACT*. https://bregroup.com/products/impact/

BRE Group. (2022, April 11). *How BREEAM works*. https://bregroup.com/products/breeam/how-breeam-works/

British Standards Institution. (2011). *The guide to PAS 2050:2011: How to carbon footprint your products, identify hotspots and reduce emissions in your supply chain*. BSI.

Buckley, L. (2023). *ASHRAE headquarters exemplifies net-zero energy building design*. https://www.iesve.com/discoveries/casestudy/27424/ashrae-headquarters

Butcher, K. (2011). *Natural ventilation in non-domestic buildings*. CIBSE.

Butcher, K., Craig, B., & Chartered Institution of Building Services Engineers (Eds.). (2015). *Environmental design: CIBSE guide A* (Eighth edition). Chartered Institution of Building Services Engineers.

Carbon Footprint. (2022). *Country specific electricity grid greenhouse gas emission factors*. https://www.carbonfootprint.com/docs/2022_03_emissions_factors_sources_for_2021_electricity_v11.pdf

CEN European Committee for Standardization. (2014). *EN 16449:2014 wood and wood-based products – Calculation of the biogenic carbon content of wood and conversion to carbon dioxide*. CEN, Brussels.

Christophers, J., & Jankovic, L. (2017). Birmingham Zero Carbon House – Retrofit Beyond Passivhaus Standard. In *Zero carbon buildings today and in the future 2016: Proceedings of a conference held at Birmingham City University, 8-9 September 2016*. Birmingham City University, Birmingham.

CIE. (2003). *CIE standard overcast sky and clear sky*. https://cie.co.at/publications/cie-standard-overcast-sky-and-clear-sky

Circular Ecology. (2021). Solar PV embodied carbon. In *Circular ecology*. https://circularecology.com/solar-pv-embodied-carbon.html

Circular Ecology. (2022). *Embodied carbon footprint calculators for construction*. https://circularecology.com/carbon-footprint-calculators-for-construction.html

Clay Mathematics Institute. (2023). *Millennium problems*. https://www.claymath.org/millennium-problems

Copernicus. (2022). *How close are we to reaching a global warming of 1.5°C?* https://climate.copernicus.eu/how-close-are-we-reaching-global-warming-15degc

Crawley, D. (1998). Which weather data should you use for energy simulations of commercial buildings? *ASHRAE Transactions, 104*, 498–515.

Crawley, D., Hand, J., & Lawrie, L. (1999). Improving the weather information available to simulation programs. *Proceedings of Building Simulation'99, 2*. https://doi.org/10.26868/25222708.1999.P-03

Cronshaw, A. (2022, September 30). *Building Entopia* [Text]. https://www.cisl.cam.ac.uk/building-entopia

Department for Business, Energy & Industrial Strategy. (2021). Green book supplementary guidance: Valuation of energy use and greenhouse gas emissions for appraisal. In *GOV.UK*. https://www.gov.uk/government/publications/valuation-of-energy-use-and-greenhouse-gas-emissions-for-appraisal

DesignBuilder Software Ltd. (2022). *DesignBuilder*. https://designbuilder.co.uk/

DOE. (2022). *EnergyPlus*. https://energyplus.net/

DOE. (2023). *EnergyPlus weather*. https://energyplus.net/weather

Duffie, J. A., & Beckman, W. A. (2013). *Solar engineering of thermal processes*. John Wiley & Sons, Inc. https://doi.org/10.1002/9781118671603

Dunn, N. (2014). *Architectural modelmaking* (Second edition). Laurence King.

Eames, M., Kershaw, T., & Coley, D. (2011). On the creation of future probabilistic design weather years from UKCP09. *Building Services Engineering Research and Technology, 32*(2), 127–142. https://doi.org/10.1177/0143624410379934

Eley, C. (2017). *Design professional's guide to zero net energy buildings*. Island Press. (ISBN 978-1-61091-765-0)

eTool. (2016, March 15). EToolLCD life cycle assessment app. *ETool*. https://etoolglobal.com/eblog/media-release/etoollcd-is-impact-compliant/

Fanger, P. O. (1972). *Thermal comfort: Analysis and applications in environmental engineering*. McGraw-Hill.

Fanger, P. O. (1988). Fundamentals of thermal comfort. In W. H. Bloss & F. Pfisterer (Eds.), *Advances in solar energy technology: Proceedings of the Biennial congress of the International Solar Energy Society*, Hamburg, Federal Republic of Germany, 13–18 September 1987 (First edition). Pergamon Press.

Finnegan, S., Jones, C., & Sharples, S. (2018). The embodied CO2e of sustainable energy technologies used in buildings: A review article. *Energy and Buildings, 181*, 50–61. https://doi.org/10.1016/j.enbuild.2018.09.037

Fourier, J. B. J. (2009). *The analytical theory of heat* (A. Freeman, Trans.; First edition). Cambridge University Press. https://doi.org/10.1017/CBO9780511693205

Germano, M., Ghiaus, C., & Roulet, C. (2016). Natural ventilation potential. In C. Ghiaus & F. Allard (Eds.), *Natural ventilation in the urban environment: Assessment and design*. Routledge.

Ghiaus, C., & Allard, F. (Eds.). (2016). *Natural ventilation in the urban environment: Assessment and design*. Routledge.

Goldberg, D. E. (1989). *Genetic algorithms in search, optimization, and machine learning*. Addison-Wesley Pub. Co.

Greco, T. H. (2021). Solar dollars: A complementary currency that incentivizes renewable energy. *Frontiers in Built Environment*, 7, 785145. https://doi.org/10.3389/fbuil.2021.785145

Hammond, G., Jones, C., Lowrie, F., Tse, P., Building Services Research and Information Association, & University of Bath. (2011). *Embodied carbon: The Inventory of Carbon and Energy (ICE)*. BSRIA.

Harnot, L., & George, C. B. (2021). *Embodied carbon in building services: A calculation methodology*. Chartered Institution of Building Services Engineers.

Heidari, B., & Marr, L. C. (2015). Real-time emissions from construction equipment compared with model predictions. *Journal of the Air & Waste Management Association*, 65(2), 115–125. https://doi.org/10.1080/10962247.2014.978485

Holbrook, T. (2022, October 27). Building study: Entopia, Cambridge, by Architype. *The Architects' Journal*. https://www.architectsjournal.co.uk/buildings/building-study-entopia-by-architype

Holland, J. H. (2000). *Emergence: From chaos to order*. Oxford Univ. Press.

Hopfe, C. J., & McLeod, R. S. (2015). *The Passivhaus designer's manual: A technical guide to low and zero energy buildings*. Routledge.

Hu, M. (2019). *Net zero energy building: Predicted and unintended consequences*. Routledge.

Huws, H., & Jankovic, L. (2014, June). Optimisation of zero carbon retrofit in the context of current and future climate. *Proceedings of the 2014 Building Simulation and Optimization Conference*. IBPSA England.

IES. (2022). *VE 2022*. https://www.iesve.com/VE2022

IES Ltd. (2022). *IES virtual environment 2022*. https://www.iesve.com/ve2022

International Energy Agency. (2022). *World energy outlook 2022*. IEA. https://www.iea.org/reports/world-energy-outlook-2022

IWBI. (2022). *WELL standard*. WELL Standard. https://v2.wellcertified.com/en/wellv2/overview

Jacques, K. (2022, April 25). *Lowering embodied carbon in buildings*. U.S. Green Building Council. https://www.usgbc.org/articles/lowering-embodied-carbon-buildings

Jankovic, L. (1988). *Solar energy monitoring, control and analysis in buildings* [PhD Thesis]. University of Birmingham.

Jankovic, L. (2016). Reducing simulation performance gap in hemp-lime buildings using Fourier filtering. *Sustainability (Switzerland)*, 8(9). 864; https://doi.org/10.3390/su8090864

Jankovic, L. (2017a). Changing the culture of building simulation with emergent modelling. *Building Simulation 2017 - Proceedings of the 15th IBPSA Conference*, 8. https://doi.org/10.26868/25222708.2017.062

Jankovic, L. (2017b). *Designing zero carbon buildings using dynamic simulation methods* (Second edition). Routledge.

Jankovic, L. (2018a). Modelling computational fluid dynamics with swarm behaviour. *Proceedings Building Simulation and Optimization 2018*, 112–118. Session 2A: Ventilation. IBPSA England. https://publications.ibpsa.org/conference/paper/?id=bso2018_2A-1

Jankovic, L. (2018b). Designing resilience of the built environment to extreme weather events. *Sustainability*, 10(1), 141. https://doi.org/10.3390/su10010141

Jankovic, L. (2019a). Improving building energy efficiency through measurement of building physics properties using dynamic heating tests. *Energies*, 12(8). https://doi.org/10.3390/en12081450

Jankovic, L. (2019b). Lessons learnt from design, off-site construction and performance analysis of deep energy retrofit of residential buildings. *Energy and Buildings*, 186, 319–338. https://doi.org/10.1016/j.enbuild.2019.01.011

Jankovic, L. (2019c). Opportunities for financing sustainable development using complementary local currencies. *IOP Conference Series: Earth and Environmental Science*, 297, 012023. https://doi.org/10.1088/1755-1315/297/1/012023

Jankovic, L., & Basurra, S. (2017). Taking a Passivhaus certified retrofit system onto scaled-up zero carbon trajectory. In *Zero carbon buildings today and in the future 2016: Proceedings of a conference held at Birmingham City University, 8-9 September 2016.* Birmingham City University, Birmingham.

Jankovic, L., Bharadwaj, P., & Carta, S. (2021). How can UK housing projects be brought in line with net-zero carbon emission targets? *Frontiers in Built Environment, 7,* 754733. https://doi.org/10.3389/fbuil.2021.754733

Jankovic, L., & Christophers, J. (2022). Cumulative embodied and operational emissions of retrofit in Birmingham zero carbon house. *Frontiers in Built Environment, Section Urban Science.* https://doi.org/10.3389/fbuil.2022.826265

Johnson, C., Affolter, M. D., Inkenbrandt, P., & Mosher, C. (2021). *Magma generation.* Geosciences LibreTexts. https://geo.libretexts.org/Bookshelves/Geology/Book%3A_An_Introduction_to_Geology_(Johnson_Affolter_Inkenbrandt_and_Mosher)/04%3A_Igneous_Processes_and_Volcanoes/4.03%3A_Magma_Generation

Jones, C., & Hammond, G. P. (2019). Embodied carbon footprint database. In *Embodied Carbon Database.* https://circularecology.com/embodied-carbon-footprint-database.html

Jones, W. P. (2020). *Air conditioning engineering* (Fifth edition). Routledge.

Jones, W. P., Butcher, K., & Chartered Institution of Building Services Engineers (Eds.). (2007). *Reference data: CIBSE guide C* (Eighth edition). Chartered Institution of Building Services Engineers.

Kiamili, C., Hollberg, A., & Habert, G. (2020). Detailed assessment of embodied carbon of HVAC systems for a new office building based on BIM. *Sustainability, 12*(8), 3372. https://doi.org/10.3390/su12083372

Kildesø, J., Wyon, D., Skov, T., & Schneider, T. (1999). Visual analogue scales for detection of changes in symptoms of the sick building syndrome in an intervention study. *Scandinavian Journal of Work, Environment & Health, 25*(4), 361–367. https://doi.org/10.5271/sjweh.446

Lamlom, S. H., & Savidge, R. A. (2003). A reassessment of carbon content in wood: Variation within and between 41 North American species. *Biomass and Bioenergy, 25*(4), 381–388. https://doi.org/10.1016/S0961-9534(03)00033-3

Lechner, N., & Andrasik, P. (2022). *Heating, cooling, lighting: Sustainable design strategies towards net zero architecture* (5th edition). John Wiley & Sons.

Machnouk, Y. (2021). *Embodied energy: The whole picture – CIBSE Journal.* https://www.cibsejournal.com/technical/embodied-energy-the-whole-picture/

Masson-Delmotte, V., Zhai, P., Pirani, A., Connors, S. L., Péan, C., Berger, S., Caud, N., Chen, Y., Goldfarb, L., Gomis, M. I., Huang, M., Leitzell, K., Lonnoy, E., Matthews, J. B. R., Maycock, T. K., Waterfield, T., Yelekçi, Ö., Yu, R., & Zhou, B. (Eds.). (2021). *Climate change 2021: The physical science basis. Contribution of working group I to the sixth assessment report of the intergovernmental panel on climate change.* Cambridge University Press.

Masson-Delmotte, V., Zhai, P., Pörtner, H.-O., Roberts, D., Skea, J., Shukla, P. R., Pirani, A., Moufouma-Okia, W., Péan, C., Pidcock, R., Connors, S., Matthews, J. B. R., Chen, Y., Zhou, X., Gomis, M. I., Lonnoy, E., Maycock, T., Tignor, M., & Waterfield, T. (2018). *Global warming of 1.5°C. An IPCC special report on the impacts of global warming of 1.5°C above pre-industrial levels and related global greenhouse gas emission pathways, in the context of strengthening the global response to the threat of climate change, sustainable development, and efforts to eradicate poverty.* IPCC.

Menezes, A. C., Cripps, A., Bouchlaghem, D., & Buswell, R. (2012). Predicted vs. actual energy performance of non-domestic buildings: Using post-occupancy evaluation data to reduce the performance gap. *Applied Energy, 97,* 355–364. https://doi.org/10.1016/j.apenergy.2011.11.075

Menzies, G. F., & Roderick, Y. (2010). Energy and carbon impact analysis of a solar thermal collector system. *International Journal of Sustainable Engineering, 3*(1), 9–16. https://doi.org/10.1080/19397030903362869

Nicol, F. (2013). *The limits of thermal comfort: Avoiding overheating in European buildings.* London : The Chartered Institution of Building Services Engineers.

Nicol, F., Humphreys, M. A., & Roaf, S. (2012). *Adaptive thermal comfort: Principles and practice.* Routledge.

OneClickLCA. (2022a). *Supported integrations – One Click LCA Help Centre*. https://oneclicklca.zendesk.com/hc/en-us/sections/360004321480-Supported-Integrations

OneClickLCA. (2022b). *World's fastest building life cycle assessment software—One Click LCA*. One Click LCA® Software. https://www.oneclicklca.com/

One Click LCA Ltd. (2018). One Click LCA - A life-cycle platform for the built environment. *One Click LCA® Software*. https://www.oneclicklca.com/construction/

Palmer, E. (Ed.). (2016). *Heating: CIBSE guide B1 : 2016*. Chartered Institution of Building Services Engineers. London.

Parry, S. (Ed.). (2022). *The SLL code for lighting*. CIBSE. London.

Passive House Institute. (2023). *PHPP – Passive house planning package [Passipedia EN]*. https://passipedia.org/planning/calculating_energy_efficiency/phpp_-_the_passive_house_planning_package

Passivhaus Institut. (2022a). *EnerPHit – the passive house certificate for retrofits [Passipedia EN]*. https://passipedia.org/certification/enerphit

Passivhaus Institut. (2022b). *PHPP – the energy balance and passive house planning tool*. https://passivehouse.com/04_phpp/04_phpp.htm

Passivhaus Trust. (2023). *What is Passivhaus?* https://www.passivhaustrust.org.uk/what_is_passivhaus.php#2

Passivhaus Trust. (2022). *Entopia building*. https://www.passivhaustrust.org.uk/projects/detail/?cId=123

Pearson, A. (2021). Case study: ASHRAE aims for net zero at HQ retrofit. *CIBSE Journal*. https://www.cibsejournal.com/case-studies/case-study-ashrae-aims-for-net-zero-at-hq-retrofit/

PRé Sustainability B. V. (2022). *SimaPro | LCA software for informed-change makers*. SimaPro. https://simapro.com/

Press, W. H. (Ed.). (2007). *Numerical recipes: The art of scientific computing* (Third edition). Cambridge University Press.

Reeder, L. (2016). *Net zero energy buildings: Case studies and lessons learned*. Routledge, Taylor & Francis Group.

Reynolds, C. (1987). Flocks, herds, and schools: A distributed behavioral model. *Computer Graphics*, 21(4), 24–34.

RICS. (2017). *Whole life carbon assessment for the built environment*. Royal Institution of Chartered Surveyors.

Robinson, D. (2019). *Future weather files*. http://emps.exeter.ac.uk/engineering/research/cee/research/prometheus/termsandconditions/futureweatherfiles/

Rodrigues, J. F. D., Marques, A. P. S., & Domingos, T. M. D. (2011). *Carbon responsibility and embodied emissions theory and measurement*. https://www.vlebooks.com/vleweb/product/openreader?id=none&isbn=9781136999703

Rogers, A., & Parson, D. O. (2022). *GridCarbon: A smartphone app to calculate the carbon intensity of the GB electricity grid*. http://www.cs.ox.ac.uk/people/alex.rogers/gridcarbon/gridcarbon.pdf

Scholand, M., & Dillon, H. E. (2012). *Life-cycle assessment of energy and environmental impacts of LED lighting products—Part 2: LED manufacturing and performance* (PNNL-21443, 1044508; p. PNNL-21443, 1044508). US Department of Energy. https://doi.org/10.2172/1044508

Scoggins, G., Walker, G., & Stafford, S. (2021). *Seminar 19: ASHRAE HQ: From conception to reception WITH LIVE Q&A - details*. https://events.rdmobile.com/Sessions/Details/1098923

Siviour, J. B. (1981). *Experimental thermal calibration of houses*. Electricity Council Research Centre.

Sol Voltaics. (2022, August 25). *Calculate solar panel KWp (KWh Vs. KWp + Meanings)—2022*. https://solvoltaics.com/calculate-kwp-solar-panel/

Sphera. (2022). *GaBi Life Cycle Assessment LCA software*. https://gabi.sphera.com/uk-ireland/index/

Statista. (2022a). *Projected electricity prices 2020–2050 Statistic*. Statista. https://www.statista.com/statistics/720679/projected-electricity-price-united-kingdom/

Statista. (2022b). *UK: Projected wholesale gas price 2040*. Statista. https://www.statista.com/statistics/720581/uk-projected-gas-price/

Sturgis, S. (2017). *Targeting zero: Embodied and whole life carbon explained*. RIBA Publishing.

Suhr, M., Hunt, R., & McCloud, K. (2019). *Old house eco handbook: A practical guide to retrofitting for energy efficiency and sustainability* (Second edition). White Lion Publishing.

Szokolay, S. V. (2014). *Introduction to architectural science: The basis of sustainable design* (Third edition). Routledge.

Tartarini, F., Schiavon, S., Cheung, T., & Hoyt, T. (2020). CBE Thermal Comfort Tool: Online tool for thermal comfort calculations and visualizations. *SoftwareX, 12*, 100563. https://doi.org/10.1016/j.softx.2020.100563

The Construction Wiki. (2022). *Performance gap between building design and operation* [Definition; Guidance; Research, Development and Innovation]. https://www.designingbuildings.co.uk. https://www.designingbuildings.co.uk/wiki/Performance_gap_between_building_design_and_operation

The Passivhaus Trust. (2022). *What is Passivhaus?* http://passivhaustrust.org.uk/what_is_passivhaus.php#2

TheyWorkForYou. (2013). *Large goods vehicles: Exhaust emissions*. https://www.theyworkforyou.com/wrans/?id=2013-03-01a.144740.h

Todde, G., Murgia, L., Carrelo, I., Hogan, R., Pazzona, A., Ledda, L., & Narvarte, L. (2018). Embodied energy and environmental impact of large-power stand-alone photovoltaic irrigation systems. *Energies, 11*(8), 2110. https://doi.org/10.3390/en11082110

Toffoli, T., & Margolus, N. (1987). *Cellular automata machines: A new environment for modeling*. MIT Press.

UK DLUHC. (2021). *National Calculation Methodology (NCM) modelling guide (for buildings other than dwellings in England)*. https://www.uk-ncm.org.uk/

unitconverters.net. (2023). *Convert Btu (IT)/hour/square foot to Watt/square meter*. https://www.unitconverters.net/heat-flux-density/btu-it-hour-square-foot-to-watt-square-meter.htm

University of Exeter. (2012). *Future weather files*. Centre for Energy and the Environment. https://engineering.exeter.ac.uk/research/cee/research/prometheus/termsandconditions/futureweatherfiles/

US EPA. (2023). *GHG emission factors Hub* [Overviews and factsheets]. https://www.epa.gov/climateleadership/ghg-emission-factors-hub

USGBC. (2019). *LEED v4.1*. U.S. Green Building Council. https://www.usgbc.org/leed/v41

USGBC. (2022). *LEED credit library*. U.S. Green Building Council. https://www.usgbc.org/credits

Vaisala. (2022). *Humidity calculator*. Vaisala. https://www.vaisala.com/en/lp/humidity-calculator.

VCA. (2016). Https://vanfueldata.vehicle-certification-agency.gov.uk/vehicles.aspx.

Vindian Solar. (2023). *Viridian solar in world first for environmental product declaration*. https://www.viridiansolar.co.uk/news/2023/02-23-Viridian-Solar-International-EPD-Certification.html

Weisstein, E. W. (2023). *Danielson-Lanczos lemma* [Text]. Wolfram Research, Inc. https://mathworld.wolfram.com/Danielson-LanczosLemma.html

Welfare, A. (Ed.). (2019). *Commercializing blockchain: Strategic applications in the real world* (First edition). Wiley.

Wolfram, S. (2002). *A new kind of science*. Wolfram Media. https://www.wolframscience.com

Worboys, C. (2021, July 15). *The rapid fall of solar's embodied carbon*. https://www.linkedin.com/pulse/rapid-fall-solars-embodied-carbon-chris-worboys

Zhang, Y. (2020a). *JePlus+EA*. https://www.jeplus.org/wiki/doku.php?id=docs:jeplus_ea:start

Zhang, Y. (2020b). *JEPlus – A parametric tool for EnergyPlus and TRNSYS*. http://www.jeplus.org/wiki/doku.php

Zhang, Y. (2022). *JESS online*. http://cms.ensims.com/index.php/jess-online

OCCUPANT SURVEY QUESTIONNAIRE FROM CASE STUDY 1, CHAPTER 21

The questionnaire below will be used to complement instrumental and simulation assessment of the house performance and it will be treated in confidence. Please mark visual analogue scales below with vertical lines that correspond to your perception of comfort or performance. Thank you for your help in advance.

Heating energy consciousness

Windows/doors/internal openings are opened or closed to capture or conserve heat

| Regularly | Rarely |

Natural daylight consciousness

Electrical lights are switched on/off in response to available daylight

| Regularly | Rarely |

Electrical lights are dimmed in response to available daylight

| Regularly | Rarely |

Natural ventilation energy consciousness

Natural ventilation is used to replace the operation of MVHR during mild external temperatures

| Regularly | Rarely |

Electricity energy consciousness

When electrical lights and electrical devices are not used, they are switched off, including devices on standby

| Regularly | Rarely |

Thermal comfort

What is your perception of your personal thermal comfort in the house in summer?

| It is too hot   It is too cold |

What is your perception of your personal thermal comfort in the house in winter?

| It is too hot   It is too cold |

Performance of systems

How would you rate the space heating system performance?

| Excellent   Poor |

How would you rate the water heating system performance?

| Excellent   Poor |

How would you rate the mechanical ventilation system performance?

| Excellent   Poor |

How would you rate the solar thermal system performance?

| Excellent   Poor |

How would you rate the PV system performance?

| Excellent   Poor |

House overall performance

How would you rate the overall performance of the house, including comfort and energy?

| Excellent   Poor |

Use of electric lights and appliances

Kitchen lights and appliances are typically used

Monday to Friday	**Saturday/Sunday**
between _____ hrs and _____ hrs a.m. and between _____ hrs and _____ hrs p.m.	between _____ hrs and _____ hrs a.m. and between _____ hrs and _____ hrs p.m.

Bathroom lights and appliances are typically used

Monday to Friday	**Saturday/Sunday**
between _____ hrs and _____ hrs a.m. and between _____ hrs and _____ hrs p.m.	between _____ hrs and _____ hrs a.m. and between _____ hrs and _____ hrs p.m.

Bedroom lights and appliances are typically used

Monday to Friday	**Saturday/Sunday**
between _____ hrs and _____ hrs a.m. and between _____ hrs and _____ hrs p.m.	between _____ hrs and _____ hrs a.m. and between _____ hrs and _____ hrs p.m.

Living room/studio lights and appliances are typically used

Monday to Friday	**Saturday/Sunday**
between _____ hrs and _____ hrs a.m. and between _____ hrs and _____ hrs p.m.	between _____ hrs and _____ hrs a.m. and between _____ hrs and _____ hrs p.m.

Are there any other performance or comfort issues that you wish to highlight

SUPPLEMENTARY MATERIAL

Supplementary material web site www.ljankovic.com contains the following material:

-Dynamic Simulation Model examples from the book

-Excel calculation spreadsheet examples

Eley, C. 11, 388
embodied: carbon 14, 244, 245, 247, 388, 389, 390;
    Carbon Footprint Calculators 16; carbon footprint
    database 390
Emergence 377, 378, 389
emergent behaviour 376
emissions savings 207, 336, 371
emissivity 100, 112, 232, 238
emittance 110–122
EMS 30, 31
end of life 13, 320, 338
energy-efficient 141, 177, 184, 339, 341, 349
EnergyManagementSystem 30
EnergyPlus 22, 27–31, 36, 37, 40, 43, 51, 56–58, 98, 229,
    360, 364, 366, 367, 388, 392
energy-positive 341
EnerPHit 316, 319, 349, 352, 371, 373, 391
engine 22, 27, 28, 152, 195, 196, 211, 366
engineering 50, 100, 388, 390–392
England 268, 291, 387, 389, 392
Entopia 387, 388, 389, 391; Building 348, 349, 350, 351,
    352, 354
Environmental Design 104
EP-Launch 29
EPS 123, 124, 244
EPW 22, 31, 37, 49–51, 56, 322, 332, 341, 349, 362
equation 84, 96, 102, 131, 150, 170, 176, 213, 215, 221, 242,
    257, 262, 275, 277, 306, 359, 383
equilibrium 185, 211
equinox 80
Erl 30, 31
eTool 16, 388
Europe 16
European 390
evacuated tube 67, 198, 199, 208, 296
evaporation 97, 190, 192, 211, 214, 215
evaporative cooling 36, 192, 193, 194
evaporator 190–192, 193, 200
evolution 224, 227, 240
evolutionary 40, 43, 46, 48, 231
exceedance 216
Excel spreadsheet 13, 30
exhaust 154, 164, 335
expansion 28, 31, 190, 268, 286, 331, 352
external insulation 104, 107, 108, 293, 355
eyestrain 169

fabric-first approach 349, 354
Fanger, P. O. 211–215, 283, 312, 313, 388
Fast Fourier Transform 359, 360

fatigue 91, 169
feedback loop 60, 262, 384
feed-in tariff 302, 308, 309
FFT 359, 360, 361
filter(s) 49, 153, 269, 286, 360–363, 366
filtering 153, 269, 270, 275–277, 389
finance 321, 383
financial 305, 320, 321, 381
financing 379, 389
Finland 16
Finnegan, S. 245, 388
firebox 301
fitness 227, 228
fittest 227
flicker 91, 183
Floating Office Rotterdam 321–325, 327, 328, 330
fluctuation 148, 359
flue 295, 297
fluid 22, 34, 160
fluorescent 177, 183, 187, 194, 294
flux 52, 53, 167, 207, 392
forest residues 196
Forman, T. 387
FORTRAN 29
fossil fuels 6, 9, 59, 177, 221, 242, 255, 385
Fourier 358, 361, 363, 389; transform 357, 359, 360, 362
Fourier, J. B. J. 388
free-floating 130
free-running 138, 216, 217, 360–362
freezing 356, 372
French, J. 387
Frenchmen 116
frequencies 359, 361
frequency 70, 77, 167, 177, 183, 184, 255, 366
frequency of occurrence 57, 138, 159, 163, 164, 189, 194,
    313, 314, 372
fridge 190, 191, 199, 200
fruit-growing 84
funnel 223
future weather data 56, 57

GA 227, 228, 332
GaBi 16, 391
GAs 227, 228
Gas-filled 112
gbXML 24, 31, 184, 226, 364, 365
GDP 385
generator 192–195, 202, 204, 206
genes 227, 228, 241
genetic 41, 48, 227, 228, 240, 389

predicted percentage of dissatisfied 213, 220, 312, 313, 315; see also PPD

predictive control 384

preheating 113, 114

Press, W. H. 359, 387, 388, 389, 390, 391, 392

product stage 13

productivity 36, 91

PROMETHEUS project 56, 58

psychological effect 91

pull in the same direction 386

PV-generated 356

PVS 204

PWF 306

pyramid 92

pyranometers 271

questionnaire(s) 284, 285, 311, 312, 372, 393

radiance 34, 89, 181, 182, 184, 298

RadianceIES 31, 34, 86, 172, 179

rainwater 155, 223, 267, 296

rammed earth 66, 293–297, 355

random: mutations 227; positions 227

randomly: distributed 228; oriented 113

recycled newspaper 142, 293, 295

recycling 13, 323

reflectance 84, 89, 110, 170

refurbishment 13, 352

regression analysis 275, 277

renewables 204, 206

re-radiated 9, 10, 115

resistance of clothing 212

resultant temperature 158, 159, 215

retrofit 10, 12, 60, 97, 221, 291, 304, 310, 318, 320, 321, 331–334, 338–341, 348–351, 353–355, 364, 368–376, 379, 382, 383, 387–391

return on investment 306, 310, 311

reuse 13, 60, 72, 73, 295, 323, 336, 349

reverse modelling 12, 49, 357, 360, 361, 362, 384

Revit 31

Reynolds, C. 376, 387, 391

Reynolds, J. 387

RIBA 48, 392

RMSE 360, 363

Roaf, S. 390

Roberts, D. 390

Robinson, D. 349, 391

Roderick, Y. 14, 67, 390

Rodrigues, J. F. D. 11, 391

Rogers, A. 19, 384, 391

ROI 311

room thermostat 30, 140

Roulet, C. 388

Rowheath Solar Village 267, 270–272, 277

Royal Institute of British Architects 291

RVI 37, 40, 41

RVX 41

SAP 21, 387

Savidge, R. A. 15, 390

SBEM 21

scale model experiment(s) 105, 143, 144, 155

Scandinavian 198, 201, 390

Schiavon, S. 392

Schneider, T. 390

Scholand, M. 246, 391

Scoggins, G. 339, 391

SCOP 192, 193

seasonal 52, 158, 193, 204, 316

SEER 192, 193

seesaw 6–8

selective surface 111, 152, 198, 199

self-destruction trajectory 58

self-regulation of solar gain 113

self-sufficient 322

sensible heat 141, 188, 194

Seven point scale 211, 283

shading: devices 138, 356; geometry 135, 138; ring 271

shadow prints 85

shallow plan 58, 60, 90, 91, 157, 173, 174, 291, 294, 304, 355

Sharples, S. 388

shelter planting 78–80, 88, 355

Shukla, P. R. 390

SI system 167

sick building syndrome 183, 390

signal: conditioning 272; filtering 269

silica gel 113

silicon 113, 201, 268

SimaPro 16, 391

simulation experiments 77, 83, 96, 104, 105, 139, 144, 157, 172

single-objective 45, 48, 51; optimisation 46

Siviour, J. B. 274, 391

Skanska 341, 347, 348

Skea, J. 390

sketches 12, 384

Sketchup 31

Skov, T. 390

SLL 168, 169, 173, 175, 177, 186, 391